Fracture and Fatigue Control in Structures: *Applications of Fracture Mechanics*

Third Edition

John M. Barsom
Stanley T. Rolfe

Butterworth-Heinemann
225 Wildwood Avenue
Woburn, MA 01801-2041

Originally printed in the U.S.A. by

ASTM
100 Barr Harbor Drive
West Conshohocken, PA 19428-2959
ASTM Stock Number: MNL41

British Library Cataloguing-in-Publication Data
A catalogue record for this book is available from the British Library.

The publisher offers special discounts on bulk orders of this book.
For information, please contact:
Manager of Special Sales
Butterworth-Heinemann
225 Wildwood Avenue
Woburn, MA 01801-2041
Tel: 781-904-2500
Fax: 781-904-2620

For information on all Butterworth-Heinemann publications available, contact our World Wide Web home page at: http://www.bh.com

Originally published in the U.S.A. by ASTM

Library of Congress Cataloging-in-Publication Data

Barsom, John M., 1938—
Fracture and fatigue control in structures: applications of fracture mechanics / John M. Barsom, Stanley T. Rolfe.—3rd ed.
p. cm.—(ASTM manual series: MNL 41)
ASTM stock number: MNL41
Includes bibliographical references and index.
ISBN 0-8031-2082-6
1. Fracture mechanics. 2. Metals—Fatigue. 3. Fracture mechanics—Case Studies.
I. Title. II. Rolfe, S. T. (Stanley Theodore), 1934—

TA409.B37 1999
620.1'126 21—dc21 99-045439

NOTE: This publication does not purport to address all of the safety problems associated with its use. It is the responsibility of the user of this publication to establish appropriate safety and health practices and determine the applicability of regulatory limitations prior to use.

Printed in Philadelphia, PA
November 1999

To Valentina and Phyllis

Contents

PART II: FRACTURE BEHAVIOR

PART III: FATIGUE AND ENVIRONMENTAL BEHAVIOR

PART IV: FRACTURE AND FATIGUE CONTROL

Foreword

(George Irwin wrote the following foreword for the first and second editions of this book in 1977 andd 1987. Dr. Irwin, the father of fracture mechanics, passed away in 1998.)

IN HIS WELL-KNOWN TEST on "Mathematical Theory of Elasticity," Love inserted brief discussions of several topics of engineering importance for which linear elastic treatment appeared inadequate. One of these topics was rupture. Love noted that various safety factors, ranging from 6 to 12 and based upon ultimate tensile strength, were in common use. He commented that "the conditions of rupture are but vaguely understood." The first edition of Love's treatise was published in 1892. Fifty years later, structural materials had been improved with a corresponding decrease in the size of safety factors. Although Love's comment was still applicable in terms of engineering practice in 1946, it is possible to see in retrospect that most of the ideas needed to formulate the mechanics of fracturing on a sound basis were available. The basic content of modern fracture mechanics was developed in the 1946 to 1966 period. Serious fracture problems supplied adequate motivation and the development effort was natural to that time of intensive technological progress.

Mainly what was needed was a simplifying viewpoint, progressive crack extension, along with recogniition of the fact that real structures contain discontinuities. Some discontinuities are prior cracks and others develop into cracks with applications of stress. The general ideas is as follows. Suppose a structural component breaks after some general plastic yield. Clearly a failure of this kind could be traced to a design error which caused inadequate section strength or to the application of an overload. The fracture failures which were difficult to understand are those which occur in a rather brittle manner at stress levels no larger than were expected when the structure was designed. Fractures of this second kind, in a special way, are also due to overloads. If one considers the stress redistribution around a pre-existing crack subjected to tension, it is clear that the region adjacent to the perimeter of the crack is overloaded due to the severe stress concentration and that local plastic strains must occur. If the toughness is limited, the plastic strains at the crack border may be accompanied by crack extension. However, from similitude, the crack border overload increases with crack size. Thus progressive crack extension tends to be self stimulating.

Given a prior crack, and a material of limited toughness, the possibility for development of rapid fracturing prior to general yielding is therefore evident.

Analytical fracture mechanics provides methods for characterizing the "overload" at the leading edge of a crack. Experimental fracture mechanics collects information of practical importance relative to fracture toughness, fatigue cracking, and corrosion cracking. By centering attention on the active region involved in progressive fracturing, the collected laboratory data are in a form which can be transferred to the leading edge of a crack in a structural component. Use of fracture mechanics analysis and data has explained many service fracture failures with a satisfactory degree of quantitative accuracy. By studying the possibilities for such fractures in advance, effective fracture control plans have been developed.

Currently the most important task is educational. It must be granted that all aspects of fracture control are not yet understood. However, the information now available is basic, widely applicable, and should be integrated into courses of instruction in strength of materials. The special value of this book is the emphasis on practical use of available information. The basic concepts of fracture mechanics are presented in a direct and simple manner. The descriptions of test methods are clear with regard to the essential experimental details and are accompanied by pertinent illustrative data. The discussions of fracture control are well-balanced. Readers will learn that fracture control with real structures is not a simple task. This should be expected and pertains to other aspects of real structures in equal degree. The book provides helpful fracture control suggestions and a sound viewpoint. Beyond this the engineer must deal with actual problems with such resources as are needed. The adage "experience is the best teacher" does not seem to be altered by the publication of books. However, the present book by two highly respected experts in applications of fracture mechanics provides the required background training. Clearly the book serves its intended purpose and will be of lasting value.

George R. Irwin

University of Maryland
College Park, Maryland

Preface

THE FIELD OF FRACTURE MECHANICS has become the primary approach to controlling fracture and fatigue failures in structures of all types. This book introduces the field of fracture mechanics from an applications viewpoint. Then it focuses on fitness for service, or life extension, of existing structures. Finally, it provides case studies to allow the practicing professional engineer or student to see the applications of fracture mechanics directly to various types of structures.

Since the first publication of this book in 1977, and the second edition in 1987, the field of fracture mechanics has grown significantly. Several specifications for fracture and fatigue control now either use fracture mechanics directly or are based on concepts of fracture mechanics. In this book, we emphasize applications of fracture mechanics to prevent fracture and fatigue failures in structures, rather than the theoretical aspects of fracture mechanics.

The concepts of *driving force* and *resistance force,* widely used in structural engineering, are used to help the reader differentiate between the mathematical side of fracture mechanics and the materials side of fracture mechanics. The driving force, K_I, is a calculated value dependent only upon the structure (or specimen) geometry, the applied load, and the size and shape of a flaw. Material properties are *not* needed to calculate values of K_I. It is analogous to the calculation of the applied stress, σ, in an unflawed structure. In fatigue, the driving force is $\Delta K = K_{I_{max}} - K_{I_{min}}$, analogous to $\Delta\sigma = \sigma_{max} - \sigma_{min}$.

In contrast, the resistance force, K_c(or K_{Ic}, or δ_c, or J_{Ic}, etc.), is a material property that can be obtained only by testing. Furthermore, this property can vary widely within a given ASTM composition, depending upon thermomechanical processing as well as a function of temperature, loading rate, and constraint, depending on the material. It is analogous to the measurement of yield strength.

By focusing on whether fracture mechanics is being used to *calculate* the driving force or to *measure* the resistance force, much of the mystery of fracture mechanics is eliminated. In the same manner that the driving stress, σ, is kept below the resistance stress, σ_{ys}, to prevent yielding, K_I should be kept below K_c to prevent fracture.

We believe the book will serve as an introduction to the field of fracture mechanics to practicing engineers, as well as seniors or beginning graduate students. This field has become increasingly important to the engineering community. In recent years, structural failures and the desire for increased safety and reliability of structures have led to the development of various fracture and fatigue criteria for many types of structures, including bridges, planes, pipelines, ships, buildings, pressure vessels, and nuclear pressure vessels.

In addition, the development of fracture-control plans for new and unusual types of structures has become more widespread. More importantly, the growing age of all types of structures, coupled with the economic fact that they may not be able to be replaced, necessitates a close look at the current safety and reliability of existing structures, i.e, a fitness for service or life extension consideration.

In this book, each of the topics of fracture criteria and fracture control is developed from an engineering viewpoint, including some economic and practical considerations. The book should assist engineers to become aware of the fundamentals of fracture mechanics and, in particular, of controlling fracture and fatigue failures in structures. Finally, the use of fracture mechanics in determining fitness for service or life extension of existing structures whose *design* life may have expired but whose *actual* life can be continued is covered.

In Parts I and II, the fundamentals of fracture mechanics theory are developed. In describing fracture behavior, the concepts of driving force (K_I), Part I, and the resistance force (K_c), Part II, are introduced. Examples of the calculations or the measurement of these two basic parts of fracture mechanics are presented for both linear-elastic and elastic-plastic conditions.

The effects of temperature, loading rate, and constraint on the measurement of various resistance forces (K_c, K_{Ic}, or δ_c, or J_{Ic}, etc.) are presented in Part II. Correlations between various types of fracture tests also are described.

In Part III, fatigue behavior (i.e., repeated loading) in structures is introduced by separating fatigue into initiation and propagation lives. The total fatigue life of a test specimen, member or structure, N_t, is composed of the initiation life, N_i, and the propagation life, N_p. Analysis of both of these components is presented as separate topics. In calculating the driving force, ΔK_I, the same K_I expressions developed in Part I for fracture are used in fatigue analyses of members with cracks subjected to repeated loading. Fatigue of weldments is also treated as a separate topic. Environmental effects (K_{Iscc}) complete the topics covered in Part III.

Parts I, II, and III focus on an introduction to the complex field of fracture mechanics as applied to fracture and fatigue in a straightforward, logical manner. The authors believe that Parts I, II, and III will serve the very vital function of introducing the topic to students and practicing engineers from an applied viewpoint.

Part IV focuses on applying the principles described in Parts I, II, and III to fracture and fatigue control as well as fitness for service of existing structures. Also called life extension, fitness for service is becoming widely used in many fields.

Many of today's existing bridges, ships, pressure vessels, pipelines, etc. have reached their original design life. If, from an economic viewpoint, it is desirable to continue to keep these structures in service, fracture mechanics concepts can be used to evaluate the structural integrity and reliability of existing structures. This important engineering field has been referred to as *fitness for service* or *life extension* and is described in Part IV.

Part V, Applications of Fracture Mechanics—Case Studies, should be invaluable to practicing engineers responsible for assessing the safety and reliability of existing structures, as well as showing students real world applications. The importance of the factors affecting fracture and fatigue failures is illustrated by case studies of actual failures. Case studies are described in terms of the importance of factors such as fracture toughness, fabrication, constraint, loading rate, etc. in the particular case study. Thus, for example, a case study describing the importance of constraint in a failure can easily be used in other types of structures where constraint is important.

Finally, the authors wish to acknowledge the support of our many colleagues, some of whom are former students who have contributed to the development of this book as well as to the continued encouragement and support of our families.

John Barsom
Stan Rolfe

Part I: Introduction to Fracture Mechanics

1

Overview of the Problem of Fracture and Fatigue in Structures

1.1 Historical Background

ALTHOUGH THE TOTAL number of structures that have failed by brittle fracture is low, brittle fractures have occurred and do occur in structures. The following limited historical review illustrates the fact that brittle fractures can occur in all types of engineering structures such as tanks, pressure vessels, ships, bridges, airplanes, and buildings.

Brittle fracture is a type of failure in structural materials that usually occurs without prior plastic deformation and at extremely high speeds (as high as 7000 ft/s in steels). The fracture is usually characterized by a flat cleavage fracture surface with little or no shear lips, as shown in Figure 1.1, and at average stress levels below those of general yielding. Brittle fractures are not as common as fatigue, yielding, or buckling failures, but when they do occur, they may be more costly in terms of human life and/or property damage.

Shank [1] and Parker [2] have reviewed many structural failures, beginning in the late 1800s when members of the British Iron and Steel Institute reported the mysterious cracking of steel in a brittle manner. In 1886, a 250-ft-high standpipe in Gravesend, Long Island, failed by brittle fracture during its hydrostatic acceptance test. During this same period, other brittle failures of riveted structures such as gas holders, water tanks, and oil tanks were reported even though the materials used in these structures had met all existing tensile and ductility requirements.

One of the most famous tank failures was that of the Boston molasses tank, which failed in January 1919 while it contained 2,300,000 gal of molasses. Twelve persons drowned in molasses or died of injuries, 40 others were injured, and several horses drowned. Houses were damaged, and a portion of the Boston Elevated Railway structure was knocked over. An extensive lawsuit followed, and many well-known engineers and scientists were called to testify. After years

FIG. 1.1 Photograph of typical brittle-fracture surface.

of testimony, the court-appointed auditor handed down the decision that the tank failed by overstress. In commenting on the conflicting technical testimony, the auditor stated in his decision, "amid this swirl of polemical scientific waters, it is not strange that the auditor has at times felt that the only rock to which he could safely cling was the obvious fact that at least one half of the scientists must be wrong . . . ." His statement fairly well summarized the state of knowledge among engineers regarding the phenomenon of brittle fracture. At times, it seems that his statement is still true today.

Prior to World War II, several welded vierendeel truss bridges in Europe failed shortly after being put into service. All the bridges were lightly loaded, the temperatures were low, the failures were sudden, and the fractures were brittle. Results of a thorough investigation indicated that most failures were initiated in welds and that many welds were defective (discontinuities were present). The Charpy V-notch impact test results showed that most steels were brittle at the service temperature.

However, in spite of these and other brittle failures, it was not until the large number of World War II ship failures that the problem of brittle fracture was fully appreciated by the engineering profession. Of the approximately 5000 merchant ships built during World War II, more than 1000 had developed cracks of considerable size by 1946. Between 1942 and 1952, more than 200 ships had sustained fractures classified as serious, and at least 9 T-2 tankers and 7 Liberty ships had broken completely in two as a result of brittle fractures. The majority of fractures in the Liberty ships started at square hatch corners or square cutouts at the top of the sheer strake. Design changes involving rounding and strengthening of the hatch corners, removing square cutouts in the shear strake, and adding riveted crack arresters in various locations led to immediate reductions in the incidence of failures [3,4].

Most of the fractures in the T-2 tankers originated in defects located in the bottom-shell butt welds. The use of crack arresters and improved work quality reduced the incidence of failures in these vessels. Studies indicated that in addition to design faults, steel quality also was a primary factor that contributed to brittle fracture in welded ship hulls [5].

Therefore, in 1947, the American Bureau of Shipping introduced restrictions on the chemical composition of steels, and in 1949 Lloyds Register stated that "when the main structure of a ship is intended to be wholly or partially welded, the committee may require parts of primary structural importance to be steel, the properties and process of manufacture of which have been specially approved for this purpose" [6].

In spite of design improvements, the increased use of crack arresters, improvements in quality of work, and restrictions on the chemical composition of ship steels during the late 1940s, brittle fractures still occurred in ships in the early 1950s [2]. Between 1951 and 1953, two comparatively new all-welded cargo ships and a transversely framed welded tanker broke in two. In the winter of 1954, a longitudinally framed welded tanker constructed of improved steel quality using up-to-date concepts of good design and welding quality broke in two [7].

Since the late 1950s (although the actual number has been low) brittle fractures continued to occur in ships. This is shown by Boyd's description of ten such failures between 1960 and 1965 and a number of unpublished reports of brittle fractures in welded ships since 1965 [8].

The brittle fracture of the 584-ft-long Tank Barge I.O.S. 3301 in 1972 [9], in which the 1-year-old vessel suddenly broke almost completely in half while in port with calm seas (Figure 1.2), shows that this type of failure continues to be a problem. In this particular failure, the material had very good notch toughness as measured by one method of testing (Charpy V-notch) and marginal toughness as measured by another more severe method of testing (dynamic tear). However, the primary cause of failure was established to be an unusually high loading stress caused by improper ballasting at a highly constrained welded detail.

In the mid-1950s two De Havilland Comet planes failed catastrophically while at high altitudes [10]. An exhaustive investigation indicated that the failures originated from very small fatigue cracks near the window openings in the fuselage. Numerous other failures of aircraft landing gear and rocket motor cases have occurred from undetected defects or from subcritical crack growth either by fatigue or stress corrosion. The failures of F-111 aircraft were attributed to brittle fractures of members with preexisting flaws. Also in the 1950s, several failures of steam turbines and generator rotors occurred, leading to extensive brittle-fracture studies by manufacturers and users of this equipment.

In 1962, the Kings Bridge in Melbourne, Australia failed by brittle fracture at a temperature of 40°F [11]. Poor details and fabrication resulted in cracks which

FIG. 1.2 Photograph of I.O.S. 3301 barge failure.

were nearly through the flange *prior* to any service loading. Although this and other bridges that failed previously by brittle fracture were studied extensively, the bridge-building industry did not pay particular attention to the possibility of brittle fractures in bridges until the failure of the Point Pleasant Bridge at Point Pleasant, West Virginia. On December 15, 1967, this bridge collapsed without warning, resulting in the loss of 46 lives. Photographs of an identical eyebar suspension bridge before the collapse and of the Point Pleasant Bridge after the collapse are shown in Figures 1.3 and 1.4, respectively. An extensive investigation of the collapse was conducted by the National Transportation Safety Board (NTSB) [*12*], and its conclusion was "that the cause of the bridge collapse was the cleavage fracture in the lower limb of the eye of eyebar 330." Because the failure was unique in several ways, numerous investigations of the failure were conducted.

Extensive use of fracture mechanics was made by Bennett and Mindlin [*13*] in their metallurgical investigation of the Point Pleasant Bridge fracture. They concluded that:

1. "The fracture in the lower limb of the eye of eyebar 330 was caused by the growth of a flaw to a critical size for fracture under normal working stress.
2. The initial flaw was due to stress-corrosion cracking from the surface of the hole in the eye. There is some evidence that hydrogen sulfide was the reagent responsible for the stress-corrosion cracking. The final report indicates that the initial flaw was due to fatigue, stress-corrosion cracking, and/or corrosion fatigue [*12*].

FIG. 1.3 Photograph of St. Mary's Bridge similar to the Point Pleasant Bridge.

FIG. 1.4 Photograph of Point Pleasant Bridge after collapse.

3. The composition and heat treatment of the eyebar produced a steel with very low fracture toughness at the failure temperature.
4. The fracture resulted from a combination of factors; in the absence of any of these, it probably would not have occurred: (a) the high hardness of the steel which rendered it susceptible to stress-corrosion cracking; (b) the close spacing of the components in the joint which made it impossible to apply paint to the most highly stressed region of the eye, yet provided a crevice in this region where water could collect; (c) the high design load of the eyebar chain, which resulted in a local stress at the inside of the eye greater than the yield strength of the steel; and (d) the low fracture toughness of the steel which permitted the initiation of complete fracture from the slowly propagating stress-corrosion crack when it had reached a depth of only 0.12 in. (3.0 mm) [Figure 1.5]."

Since the time of the Point Pleasant Bridge failure, other brittle fractures have occurred in steel bridges and other types of structures as a result of unsatisfactory fabrication methods, design details, or material properties [*14,15*]. Fisher [*16*] has described numerous fractures in a text on case studies.

These and other brittle fractures led to an increasing concern about the possibility of brittle fractures in steel bridges and resulted in the AASHTO (American Association of State Highway and Transportation Officials) Material Toughness Requirements being adopted for bridge steels. Other industries have developed fracture-control plans for arctic construction, offshore drilling rigs, and more specific applications such as the space shuttle.

FIG. 1.5 Photograph showing origin of failure in Point Pleasant Bridge.

Fracture mechanics has shown that because of the *interrelation* among *materials, design, fabrication,* and *loading,* brittle fractures cannot be eliminated in structures merely by using materials with improved notch toughness. The designer still has the fundamental responsibility for the overall safety and reliability of his or her structure. It is the objective of this book to describe the fracture, fatigue, and stress-corrosion behavior of structural materials and to show how fracture mechanics can be used in design to *prevent* brittle fractures and fatigue failures of engineering structures.

Furthermore, as existing structures reach their design life, there is considerable pressure to extend the life of these structures. Fracture mechanics can be used to establish the fitness-for-service or life extension of these structures on a rational technical basis.

As will be described throughout this textbook, the science of *fracture mechanics* can be used to describe *quantitatively* the tradeoffs among stress, material fracture toughness, and flaw size so that the designer can determine the relative importance of each during the *design* process. However, fracture mechanics also can be used during fitness-for-service or life-extension evaluations, as described in Part IV.

1.2 Ductile vs. Brittle Behavior

Brittle fractures occur with little or no elongation or reduction in area and with very little energy absorption. Brittle fracture is a type of failure that usually occurs without prior plastic deformation and at extremely high speeds [as fast as 2000 m/s (7000 ft/s) in steels].

Schematic examples of the stress-strain behavior for ductile and brittle types of failure are presented in Figure 1.6. Most structural materials exhibit considerable strain (deformation) before reaching the tensile or ultimate strength, σ_{tens} (Figure 1.6*a*). In contrast, brittle materials exhibit almost no deformation before failure (Figure 1.6*b*). However, under conditions of low temperature, rapid loading and/or high constraint (e.g., when the principal stresses σ_1, σ_2, and σ_3 are essentially equal), even ductile materials may not exhibit any deformation before fracture. In these cases, the stress-strain curve of a normally ductile material resembles that shown in Figure 1.6*b*. Obviously, ductile behavior is much more desirable than brittle behavior because of the energy absorption and deformation that occurs before failure.

Ductile failures normally are characterized by large shear lips and considerable deformation (Figure 1.7*a*). In contrast, brittle fractures are usually characterized by a flat cleavage fracture surface with little or no ductility (Figure 1.7*b* and Figure 1.1) and often at average stress levels below those of general yielding. Brittle fractures are not so common as yielding, buckling, or fatigue failures, but when they do occur they may be more costly in terms of human life and property damage.

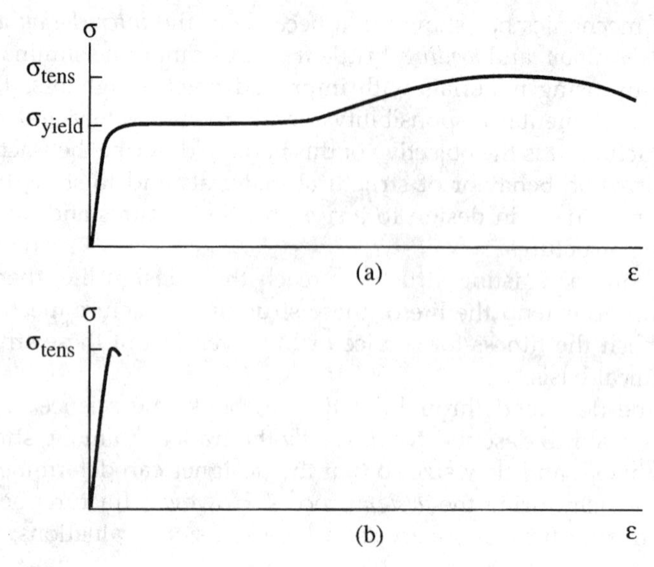

FIG. 1.6 Comparison of σ-ε curves for ductile and brittle materials: (*a*) ductile material; (*b*) brittle material.

Figure 1.2 shows a ship that failed because of a brittle fracture. This ship was subjected to above normal loads (yet the nominal stress was below yielding) in the presence of a severe stress concentration. The stress concentration increased the local constraint and restricted yielding, resulting in principal stresses that were essentially equal. Thus, the local stress reached the tensile strength of the steel with little or no yielding, and brittle fracture occurred. Once the brittle fracture was initiated, the loading condition was such that the fracture propagated completely around the ship in less than 1 s. The steel in this ship had very good ductility and notch toughness (e.g., CVN impact value of 55 ft-lb at the service temperature), indicating that brittle fractures can be caused by severe loading and high constraint, not just by materials with low notch toughness.

The 1994 experience with buildings in the Northridge earthquake [*17*], where fractures occurred in moment connections in highly constrained joints, emphasizes the importance of many factors, including loading, design, fabrication, inspection, and material properties.

1.3 Notch Toughness

Because it is very difficult to fabricate large welded structures without introducing some type of notch, flaw, discontinuity, or stress concentration, the design engineer must be aware of the effect of notches and constraint on material behavior. Thus, in addition to the material properties such as yield strength, modulus of elasticity, and tensile strength, there is another very important material

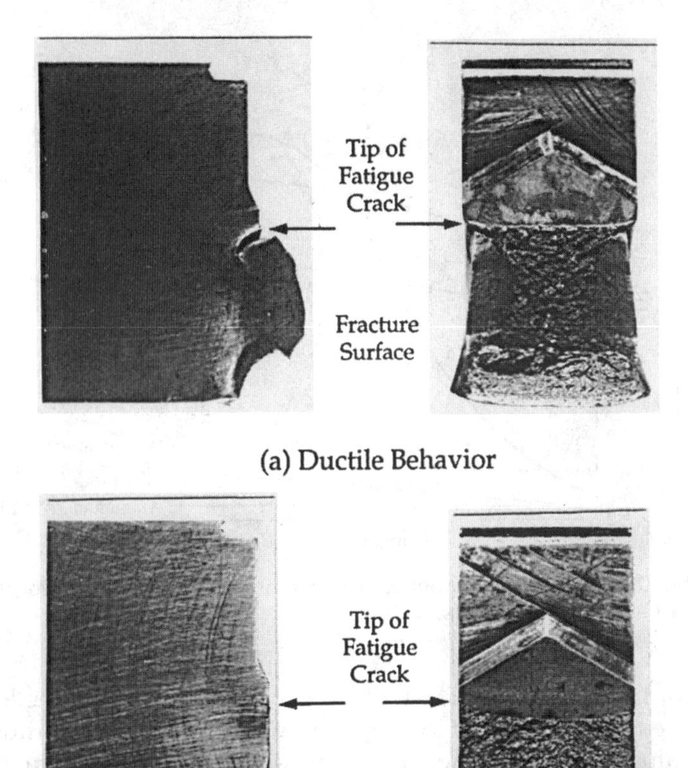

FIG. 1.7 Ductile and brittle fracture surfaces.

property, namely notch toughness, that may be related to the behavior of a structure. Notch toughness is defined as the ability of a material to absorb energy in the presence of a sharp notch, often when subjected to an impact load. Notch toughness is usually measured as the amount of energy (joules or foot-pounds) required to fracture a particular notch-toughness specimen at a particular temperature and loading rate.

Notch toughness is measured with a variety of test specimens. One of the most widely used is the Charpy V-notch (CVN) impact specimen. A test machine with a pendulum is used to impact the specimens at various temperatures. The absorbed energy required to fracture the specimen is plotted as a function of temperature. Typical CVN results for common structural materials are shown in Figure 1.8, which shows the transition from brittle to ductile behavior under

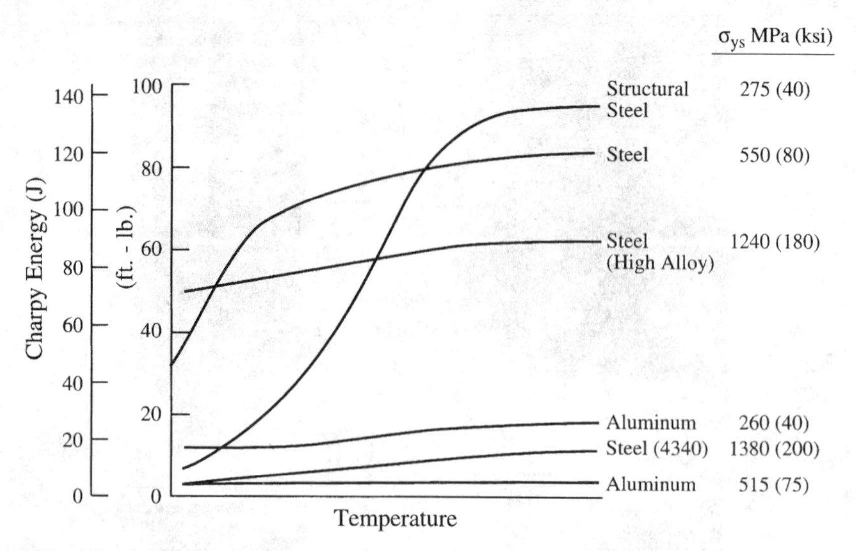

FIG. 1.8 Charpy V-notch impact energy versus temperature behavior for selected structural materials.

conditions of impact loading. The CVN impact values shown at the lower left of Figure 1.8 are representative of low levels of notch toughness or brittle behavior, while the values at higher temperatures (upper right) are representative of ductile-type behavior. It should be noted that some materials (such as aluminum and very high-strength steels) do not exhibit a distinct transition behavior. Also, some materials have low notch toughness at all temperatures (e.g., 75-ksi-yield aluminum), whereas some materials have a high level of notch toughness at all temperatures (e.g., 180-ksi yield strength alloy steel).

The change in absorbed energy, ductility (lateral expansion or contraction at the root of the notch), and fracture appearance (as measured by percent shear on the fracture surface) for a structural steel is shown in Figure 1.9. At +140°F, completely ductile behavior is observed. At −200°F, completely brittle behavior is observed. The region between these two extremes is called the transition region. Note that the transition region is different for the two different loading rates, slow and impact. This effect of loading rates has a very significant influence on the fracture behavior of structures as described later.

Various "transition temperatures" are often established as an indication of the notch toughness of a structural material. For example, the 15 ft-lb impact transition temperature for the steel shown in Figure 1.9*a* is about 30°F. The 20-mil lateral expansion transition temperature is about 30°F. Also, the 50% impact fracture-appearance transition temperature for this steel is about 30°F, as shown in Figure 1.9*c*. Obviously, these transition temperatures need not occur at the same temperature and will vary from material to material, depending on the particular notch-toughness characteristics of each material. One traditional

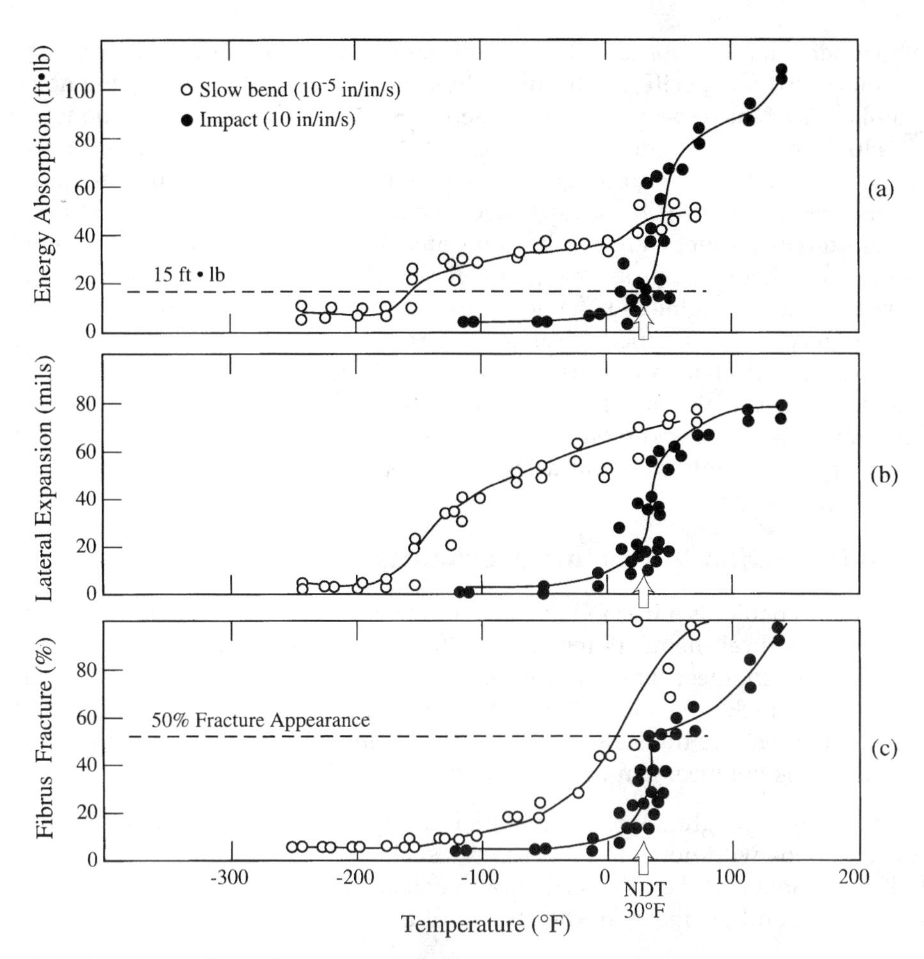

FIG. 1.9 Charpy V-notch energy absorption, lateral expansion, and fibrous fracture for impact and slow-bend test of standard CVN specimens for a low-strength structural steel.

method used to prevent brittle fracture in a member has been to specify that it can be used only above some particular transition temperature such as the 15 ft-lb impact transition temperature.

The NDT (nil-ductility temperature) test is another ASTM test method used to predict behavior of structural steels. Below the NDT temperature, the steel is considered to be brittle under conditions of impact loading. At slow or intermediate loading rates, the steel can still exhibit satisfactory notch toughness levels at lower temperatures as shown in Figure 1.9.

All these notch-toughness tests generally have one thing in common, however, and that is to produce fracture in steels under carefully controlled laboratory conditions. It is hoped that the results of the tests can be correlated with service performance to establish levels of performance for various materials being considered for specific applications. In fact, the results of the foregoing notch-toughness tests have been extremely useful in many structural applications.

However, even if correlations are developed for existing structures, they do not necessarily hold for all designs, new operating conditions, or new materials because the results, which are expressed in terms of energy, fracture appearance, or deformation, cannot always be translated into structural design and engineering parameters such as stress and flaw size. Thus, a much better way to analyze fracture toughness behavior is to use the science of fracture mechanics. Fracture mechanics is a method of characterizing the fracture behavior in structural parameters that can be used directly by the engineer, namely, stress and flaw size. Fracture mechanics is based on a stress analysis as described in Chapter 2 and thus does not depend on the use of extensive service experience to translate laboratory results into practical design information.

1.4 Introduction to Fracture Mechanics

Fracture mechanics is a method of characterizing the fracture behavior of sharply notched structural members (cracked or flawed) in terms that can be used directly by the engineer. Fracture mechanics is based on a stress analysis in the vicinity of a notch or crack. It does not, therefore, depend on the use of extensive service experience to translate laboratory results into practical design information as long as the engineer can obtain or determine:

1. The fracture toughness of the material, using fracture-mechanics tests or correlations with notch toughness tests such as the CVN impact test.
2. The nominal stress on the structural member being analyzed.
3. Flaw size and geometry of the structural member being analyzed.

Many large, complex structures such as bridges, ships, buildings, aircraft, and pressure vessels can have crack-like imperfections, sharp notches, or discontinuities of various kinds. Using fracture mechanics, an engineer can quantitatively establish allowable stress levels and inspection requirements to design against the occurrence of fractures in such structures. In addition, fracture mechanics may be used to analyze the growth of small cracks to critical size by fatigue loading or by stress corrosion cracking. Therefore, fracture-mechanics testing and analysis techniques have several advantages over traditional notch-toughness test methods and offer the designer a method of quantitative design to prevent fracture in structures. In addition, fracture mechanics can be used to evaluate the fitness-for-service, or life extension, of existing structures.

1.4.1 Driving Force, K_I

The fundamental concept of linear-elastic fracture mechanics is that the stress field ahead of a sharp crack can be characterized in terms of a single parameter, K_I, the stress intensity factor having units of ksi$\sqrt{in.}$ This single parameter, K_I, is related to both the stress level, σ, and the crack or flaw size, a. It is analogous to the driving force, σ, in structural design. When the particular combination of σ and a leads to a critical value of K_I, called K_c, unstable crack growth fracture occurs. Equations that describe the elastic-stress field in the vicinity of a crack tip in a body subjected to tensile stresses can be used to establish the relation between K_I, σ, and crack size, a, for different structural configurations, as shown in Figure 1.10. K_I values for these and other crack geometries, as well as different structural configurations, are described in Chapter 2.

In fatigue, the driving force is ΔK_I, where $\Delta K_I = K_{IMAX} - K_{IMIN}$, analogous to the case of $\Delta\sigma = \sigma_{MAX} - \sigma_{MIN}$.

1.4.2 Resistance Force, K$_c$

The critical value of a stress-intensity factor at failure, K_c, is a material property. It is analogous to the resistance force, σ_{ys}, to prevent yielding in structural design. From testing, the critical value of K_I at failure, K_c, can be determined for

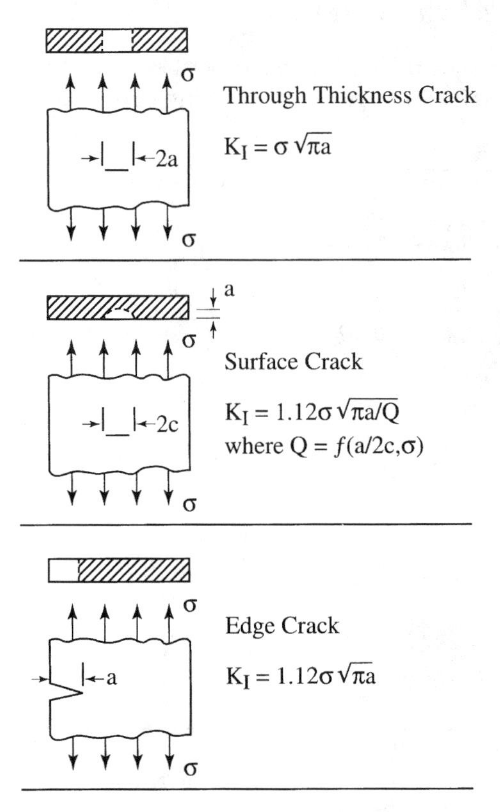

FIG. 1.10 *K*$_I$ values for different crack geometries.

a given material at a particular thickness and at a specific temperature and loading rate. Using this critical material property, the designer can determine theoretically the flaw size that can be tolerated in structural members for a given design stress level, temperature, and loading rate. Conversely, the engineer can determine the design stress level that can be safely used for a flaw size that may already be present in an existing structure.

The critical stress-intensity factor for structural materials is highly dependent on such service conditions as temperature, loading rate, and constraint. Thus, the critical value must be obtained by testing actual structural materials to failure at various temperatures and loading rates as described in Chapter 3.

Examples of various K_c values for a structural steel having a room-temperature yield strength of 50 ksi (345 MPa) are presented in Figure 1.11. These results, obtained at three different loading rates, show the large effect that temperature and loading rate can have on the critical stress-intensity factors for a particular structural material.

1.5 Fracture Mechanics Design

In addition to the major brittle failures described in Section 1.1, there have been *numerous* "minor" failures of structures during construction or service that have

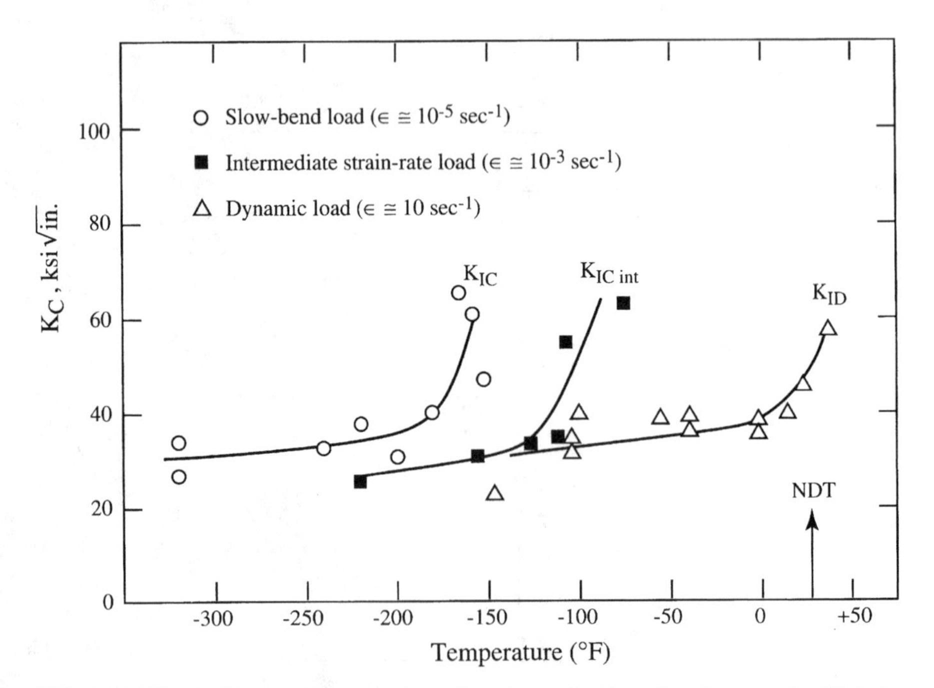

FIG. 1.11 Effect of temperature and strain rate on fracture toughness of a 50-ksi yield strength structural steel.

resulted in delays, repairs, and inconveniences, some of which are very expensive. Nonetheless, compared with the total number of engineering structures that have been built throughout the world, the number of brittle fractures has been very small. As a result, the designer seldom concerns himself or herself with the notch toughness of structural materials because the failure rate of structures due to brittle fracture is very low. Nonetheless,

1. As designs become more complex.
2. As the use of high-strength thick welded plates becomes more common compared with the use of lower-strength thin riveted or bolted plates.
3. As the choice of construction practices becomes more dependent on minimum cost.
4. As the magnitude of loadings increases.
5. As actual factors of safety decrease because of more precise computer designs.
6. As fatigue or stress corrosion cracks grow in existing structures.

the possibility of fractures in large complex structures must be considered. Thus the engineer must become more aware of available methods to prevent failures.

The state of the art is that fracture mechanics concepts that can be used in the design of structures to prevent fractures, as well as to extend the life of existing structures through fitness-for-service analyses, are available.

The fundamental design approach to preventing fracture in structural materials is to keep the calculated stress-intensity factor, K_I (the driving force), below the critical stress-intensity factor, K_c (the resistance force). This is analogous to keeping $\sigma < \sigma_{ys}$ to prevent yielding.

A general design procedure to prevent fracture in structural members is as follows:

1. Calculate the maximum nominal stress, σ, for the member being analyzed.
2. Estimate the most likely flaw geometry and initial crack size a_0. To design against fracture during the expected lifetime of a structure, estimate the maximum probable crack size during the expected lifetime.
3. Calculate K_I for the stress, σ, and flaw size, a, using the appropriate K_I relation. (K_I relations are presented in Chapter 2.)
4. Determine or estimate the critical stress-intensity factor, K_c, by testing the material from which the member is to be built, as described in Chapter 3. These critical stress intensity values are a function of the appropriate service temperature and loading rate as described in Chapter 4. Alternatively, approximate critical stress-intensity values can be estimated from CVN impact test results as described in Chapter 5.
5. Compare K_I with K_c. To design against fracture, insure that K_I will be less than the critical stress-intensity factor, K_c, throughout the entire life of the structure. This may require the selection of a different material or reduction of the maximum nominal service stress as described in Chapter 6. Also, it

may require better quality control during fabrication or periodic inspection for cracks throughout the life of the structure.

The general relationship among material fracture toughness, K_c, nominal stress, σ, and crack size, a, is shown schematically in Figure 1.12. If, for a particular combination of stress and crack size in a structure, K_I reaches the critical K_c level, fracture can occur. Thus, there are many combinations of stress and flaw size which may cause fracture in a structure that is fabricated from a material having a particular value of K_c at a particular service temperature and loading rate. Conversely, there are many combinations of stress and flaw size that will not cause fracture of a particular structural material.

As an example of the design application of fracture mechanics, consider the equation $K_I = \sigma\sqrt{\pi a}$ relating K_I to the applied stress and flaw size for a through-thickness crack in a wide plate (Figure 1.13). Assume that laboratory test results show that for a particular structural steel with a yield strength of 80 ksi the K_c is 60 ksi$\sqrt{\text{in.}}$ at the service temperature, loading rate, and plate thickness. Furthermore, assume that the design stress is 20 ksi. Substituting $K_I = K_c = 60$ ksi$\sqrt{\text{in.}}$ and $\sigma = 20$ ksi results in $2a_{CR} = 5.7$ in. Thus, for these conditions the maximum tolerable flaw size would be about 5.7 in. (145 mm). For a design stress of 45 ksi, the same material could tolerate a flaw size of only about 1.1 in. (28 mm). If residual stresses, such as those that might be caused by welding, are present so that the total stress in the vicinity of a crack is approximately 80 ksi, the tolerable flaw size is reduced considerably. Note that if a material with higher

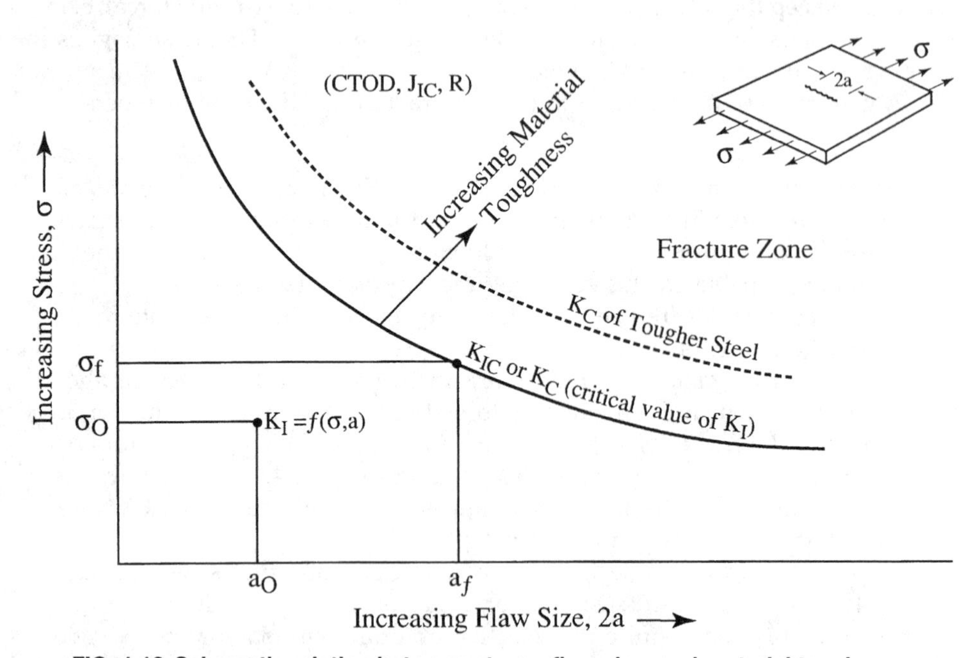

FIG. 1.12 Schematic relation between stress, flaw size, and material toughness.

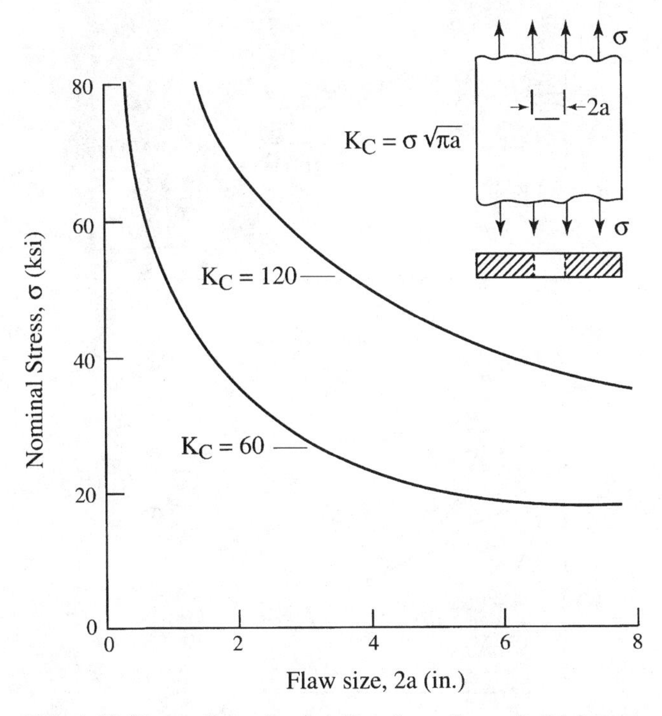

FIG. 1.13 Stress-flaw size relation for a through-thickness crack in materials with K_c = 60 and K_c = 120 ksi$\sqrt{\text{in.}}$

fracture toughness is used, one with a K_c of 120 ksi$\sqrt{\text{in.}}$, the tolerable flaw sizes at all stress levels are increased significantly.

An analogy that may be useful in understanding the fundamental aspects of fracture-mechanics design is the comparison with Euler column instability (Figure 1.14). The stress level required to cause instability in a column (buckling) decreases as the L/r ratio increases. Similarly, the stress level required to cause instability (fracture) in a flawed or cracked tension member decreases as the flaw size increases. As the stress level in either case approaches the yield strength, both the Euler analysis and the K_c analysis are invalidated because of general yielding. To prevent buckling, the actual stress and L/r value must be below the Euler curve. To prevent fracture, the actual stress and flaw size, a, must be below the K_c level shown in Figure 1.14. Other design considerations are described in Chapter 6.

1.6 Fatigue and Stress-Corrosion Crack Growth

Fatigue behavior is described in Chapters 7 through 10, which are in Part III. Conventional procedures that are used to design structural components subjected

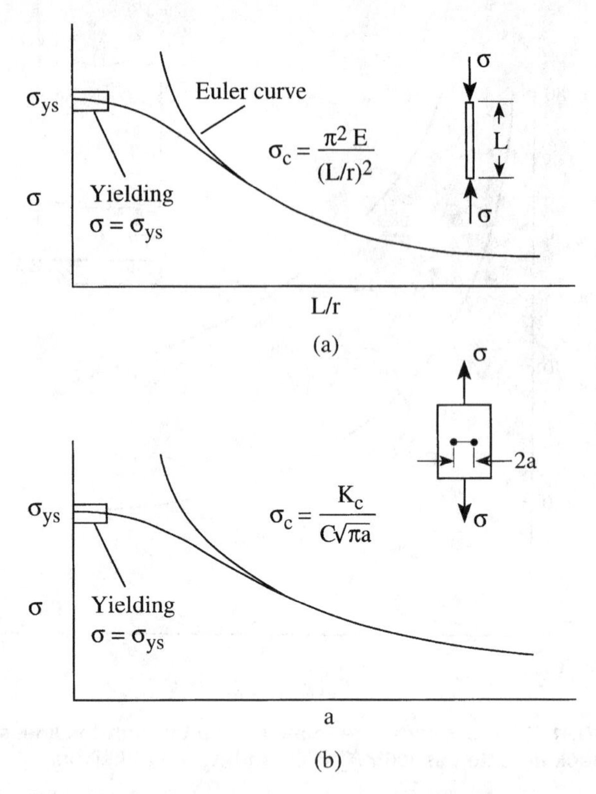

FIG. 1.14 Analogy between column instability and crack instability: (*a*) Column instability (*b*) Crack instability.

to fluctuating loads provide the engineer with design fatigue curves. These curves characterize the basic unnotched fatigue properties of the material. A fatigue reduction factor is used to account for the effects of all the different parameters characteristic of the specific structural component that make it more susceptible to fatigue failure than an unnotched specimen. The design fatigue curves are based on the prediction of cyclic life from data on nominal stress versus elapsed cycles to failure (*S-N* curves) as determined from laboratory test specimens. It should be emphasized that the primary factor affecting fatigue behavior is $\Delta\sigma$, which is equal to $\sigma_{MAX} - \sigma_{MIN}$. *S-N* data generally combine both the number of cycles required to initiate a crack in the specimen and the number of cycles required to propagate the crack from a subcritical size to a critical dimension. The dimension of the critical crack required to cause "failure" in the fatigue specimen depends on the magnitude of the applied stress and on the test specimen size. Fatigue specimens that incorporate the actual geometry, welding, as well as other characteristics can be tested to obtain an *S-N* curve specifically for that member.

Figure 1.15 is a schematic *S-N* curve for smooth specimens divided into an initiation component and a propagation component. The number of cycles corresponding to the fatigue limit represents initiation life primarily, whereas the number of cycles expended in crack initiation at a high value of applied stress is negligible. Consequently, *S-N* type data for smooth specimens do not necessarily provide information regarding safe-life predictions in structural components (particularly in components having surface irregularities different from those of the test specimens) and in components containing crack-like discontinuities because the existence of surface irregularities and crack-like discontinuities reduces and may eliminate the crack-initiation portion of the fatigue life of structural components.

Although *S-N* curves have been used widely to analyze the fatigue behavior of steels and weldments, closer inspection of the overall fatigue process in complex welded structures indicates that a more rational analysis of fatigue behavior may be possible by using concepts of fracture mechanics. Specifically, small and possibly large fabrication discontinuities may be present in welded structures, even though the structure has been "inspected" and "all injurious flaws removed" according to some specifications. Accordingly, a conservative approach to designing to prevent fatigue failure is to assume the presence of an initial flaw and analyze the fatigue-crack-growth behavior of the structural member. The size of the initial flaw is obviously highly dependent on the quality of fabrication and inspection. However, such an analysis would minimize the need for expensive fatigue testing for many different types of structural details. In this case, the primary driving force is $\Delta K_I = K_{IMAX} - K_{IMIN}$. Note the analogy to $\Delta\sigma = \sigma_{MAX} - \sigma_{MIN}$.

A schematic diagram showing the general relation between fatigue-crack initiation and propagation is given in Figure 1.16. The question of when does a

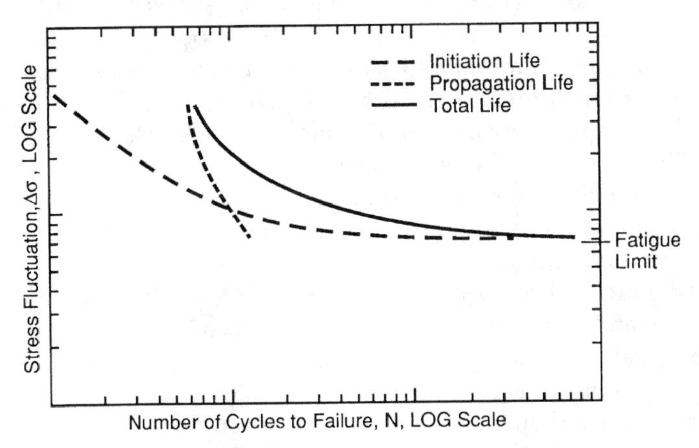

FIG. 1.15 Schematic *S-N* curve divided into "initiation" and "propagation" components.

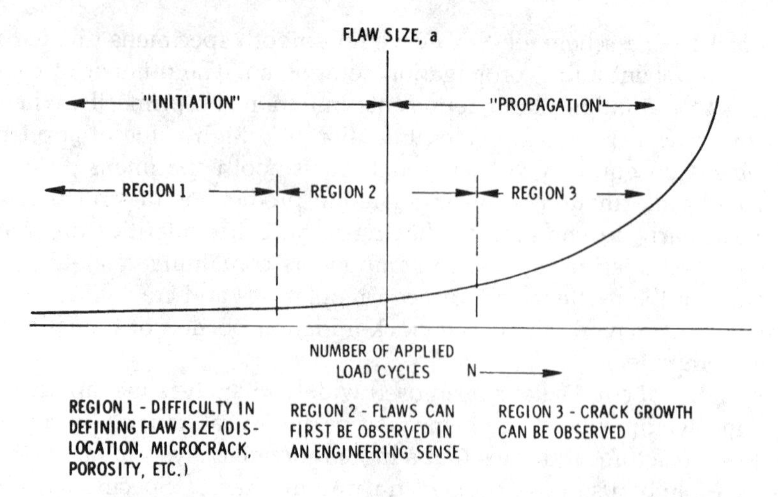

FIG. 1.16 Schematic showing relation between "initiation" life and "propagation" life.

crack "initiate" to become a "propagating" crack is somewhat philosophical and depends on the level of observation of a crack, that is, crystal imperfection, dislocation, micro crack, and macro crack. The fracture-mechanics approach to fatigue is to assume an initial imperfection on the basis of the quality of fabrication or inspection and then to calculate the number of cycles required to initiate a sharp crack from that imperfection (N_i) and then to grow that crack to the critical size for brittle fracture, N_p. Obviously if the initial imperfection is very sharp, the initiation life can be very short. Using this approach, inspection requirements can be established logically.

In addition to subcritical crack growth by fatigue, small cracks also can grow by stress corrosion during the life of structures, as described in Chapter 11 of Part III. Although crack growth by either fatigue or stress corrosion does not represent catastrophic failure for structures fabricated from materials having reasonable levels of notch toughness, in both mechanisms small cracks can become large enough to require repairs and, if neglected, can cause failure. Furthermore, the possibility of both mechanisms operating at once by corrosion fatigue also exists. Thus, a knowledge of the fatigue, corrosion-fatigue, and stress-corrosion behavior of materials is required to establish an overall fracture-control plan that includes inspection requirements.

By testing precracked specimens under static loads in specific environments (such as salt water) and analyzing the results according to fracture-mechanics concepts, a K_I value can be determined, below which subcritical crack propagation does not occur. This threshold value is called K_{Iscc}. The K_{Iscc} value for a particular material and environment is the plane-strain stress-intensity threshold that describes the value below which subcritical cracks (scc) will not propagate under static (dead) loads and also has units of ksi$\sqrt{\text{in.}}$, as described in Chapter 11.

1.7 Fracture and Fatigue Control

Parts I and II of this book, which encompass Chapters 1 through 6, present the basics of fracture mechanics as applied to fracture. Part III, which encompasses Chapters 7 through 11, presents the basics of fatigue, stress corrosion cracking, and corrosion fatigue as analyzed by fracture mechanics. Part IV (Chapters 12 through 14) focuses on fracture and fatigue control and fitness for service or life extension. Part V (Chapters 15 through 19) presents practical examples of the use of fracture mechanics in a variety of structural situations.

Fracture control in structures can be accomplished in several ways. Basically, fracture occurs when the driving force, K_I, exceeds the resistance force, K_c. Thus, to prevent fracture, the engineer needs to keep K_I less than K_c. In practice, this can be done either by reducing the applied stress, σ, and/or the crack size, a, factors that affect the driving force, or by increasing the resistance force, K_c, which is the material toughness at a given temperature, loading rate, and constraint level.

Most engineering structures are performing safely and reliably. The comparative few service failures in structures indicate that present-day practices governing material properties, design, and fabrication procedures are generally satisfactory. However, the occurrence of infrequent failures indicates that further understanding and possible modifications in present-day practices still are needed. The identification of the specific modifications needed requires a thorough study of material properties, design, fabrication, inspection, erection, and service conditions. In addition, the current conditions of a structure must be evaluated when performing a fitness-for-service evaluation.

Because of the wide variety of structures, it is very difficult to develop a set of rules that would ensure the safety and reliability of all structures when different structures are subjected to different operating conditions. The use of data obtained by testing laboratory specimens to predict the behavior of complex structural components may result in approximations or in excessively conservative estimates of the life of such components but does not guarantee a correct prediction of the fracture or fatigue behavior. The safety and reliability of structures and the correct prediction of their overall resistance to failure by fracture or fatigue can be approximated best by using a fracture-control plan. A fracture-control plan is a detailed procedure used:

1. To identify all the factors that may contribute to the fracture of a structural detail or to the failure of the entire structure.
2. To assess the contribution of each factor and the synergistic contribution of these factors to the fracture process.

3. To determine the relative efficiency and trade-off of various methods to minimize the probability of fracture.
4. To assign responsibility for each task that must be undertaken to ensure the safety and reliability of the structure.

The development of a fracture-control plan for complex structures is very difficult. Despite the difficulties, attempts to formulate a fracture-control plan for a given application, even if only partly successful, should result in a better understanding of the fracture characteristics of the structure under consideration.

A fracture-control *plan* is a procedure tailored for a given application and cannot be extended indiscriminately to other applications. However, general fracture-control *guidelines* that pertain to classes of structures (such as bridges, ships, pressure vessels, etc.) can be formulated for consideration in the development of a fracture-control plan for a particular structure within a class of structures.

The correspondence among fracture-control plans based on crack initiation, crack propagation, and fracture toughness of materials can be readily demonstrated by using fracture-mechanics concepts. The fact that crack initiation, crack propagation, and fracture toughness are functions of the stress-intensity fluctuation, ΔK_I, and of the critical-stress-intensity factor, K_c, which are in turn related to the applied nominal stress (or stress fluctuation), demonstrates that a fracture-control plan for various structural applications depends on:

1. The fracture toughness, K_c, of the material at the temperature and loading rate representative of the intended application. The fracture toughness can be modified by changing the material used in the structure.
2. The applied stress, loading rate, stress concentration, and stress fluctuation, which can be altered by design changes, loading changes, and by proper detailing.
3. The initial size of the discontinuity and the size and shape of the critical crack, which can be controlled by design changes, fabrication, and inspection.

Fatigue control is based on the fact that the total useful life of structural components is determined by the time necessary to initiate a crack and to propagate the crack from subcritical dimensions to the critical size. The life of the component can be prolonged by extending the crack-initiation life and the subcritical-crack-propagation life. Consequently, crack initiation, subcritical crack propagation, and fracture characteristics of structural materials are primary considerations in the formulation of fracture-control guidelines for structures.

Fracture-control guidelines and the principles of fracture-control plans are presented in Chapter 12.

1.8 Fracture Criteria

Fracture criterion selection is the determination of how much fracture toughness is necessary for a particular structural application. Some of the factors involved in the development of a fracture criterion are:

1. A knowledge of service conditions (temperature, loading, loading rate, etc.) to which the structure will be subjected.
2. The desired level of performance in the structure (plane-strain, elastic-plastic, or plastic).
3. The consequences of structural failure.

Several of the most common technical procedures available to develop a fracture criterion are presented in Chapter 13.

1.9 Fitness for Service

Although the concept of "fitness for service" was proposed around 1960, common-sense engineering has been around for a long time. Both concepts rely heavily on principles of fracture mechanics, although early engineers certainly did not use the term "fracture mechanics." However, when faced with either structural failures or situations that might lead to failures, early engineers did apply common-sense engineering and thus principles of fitness for service as well as fracture mechanics concepts.

Fitness for service (purpose) has been defined by Alan Wells [*18*] as:

> "Fitness for purpose is deemed to be that which is consciously chosen to be the right level of material [that is, having the appropriate fracture toughness, e.g., K_{Ic}, CTOD, *J*-integral, CVN, etc.] and fabrication quality [that is, the appropriate loading or stress level for the given application], having regard to the risks and consequences of failure; it may be contrasted with the best quality that can be achieved within a given set of circumstances, which may be inadequate for some exacting requirements, and needlessly uneconomic for others which are less demanding. A characteristic of the fitness-for-purpose approach is that it is required to be defined beforehand according to known facts, and by agreement with purchasers which will subsequently seek to be national and eventually international.
>
> The need for such an approach has already been seen with the development and application of fracture mechanics, but the Paper [by Wells] draws attention to a wider scope which also embraces the evolution of the design process, risk analysis and reliability.
>
> It is considered that the assessment of fitness for purpose should relate well to the quality assurance approach, since the latter aims to be comprehensive, and makes provision for updating its own procedures."

It is proposed that fitness-for-service is indeed common-sense engineering and that both concepts rely heavily on the field of fracture mechanics. The fitness-for-service approach has been used both during the design process as well as to analyze the remaining life of an existing structure in which a crack has developed. Although several standards are available to analyze the fitness for service of existing structures, there is no single all encompassing methodology. Fitness-for-service also encompasses risk analysis or probability, nondestructive examination, quality assurance, and quality control as well as other related technologies, most of which incorporate concepts of fracture mechanics. In short, one of the major new directions in fracture mechanics research deals with fitness for service and the application of fracture mechanics concepts to extending the life of existing structures. Fitness for service is described in Chapter 14.

1.10 Case Studies

Several case studies illustrating concepts and applications of fracture mechanics are presented in Chapters 15 through 19 in Part V of this book. Case studies are not aimed at describing examples of fracture and fatigue failures of specific types of structures. Rather, they are aimed at illustrating principles discussed in Parts I through IV of this book.

For example, Chapter 15 describes the brittle fracture in the Bryte Bend Bridge in California [*14*]. It illustrates the importance of details and how these details can lead to premature failure of a structure. This case study also illustrates one type of retrofit or repair in which the principle of replacement of function rather than replacement of actual members is used.

Chapter 16 describes the difference in two particular structures, namely the Ingram Barge in New York [*9*], which failed, and the surge tanks in North Dakota [*19*], which have not failed. The Ingram Barge failed after nine months of service even though its notch toughness was considered to be quite good, e.g., 55 ft-lb CVN impact toughness at service temperature. The surge tanks in North Dakota, which have not failed after more than 40 years of service, have only 2 to 3 ft-lb CVN impact toughness at the service temperature. These two contrasting situations illustrate the importance of constraint and factors other than notch toughness.

Chapter 17 describes the Trans-Atlantic Taps Vessels which developed cracks during service going between the west coast of the United States and Valdez [*20*]. The importance of loading and inspection is illustrated in this case study.

Chapter 18 describes the importance of using the proper failure analysis, and Chapter 19 describes the importance of loading rate in failure analysis.

Once again it should be emphasized that these case studies are meant to illustrate the principles of successful fracture and fatigue control of structures rather than to merely serve as examples of failure in a particular type of structure. The principles are emphasized in each of the particular chapters so that the practicing engineer, as well as the student, can appreciate the importance of the proper application of the fundamentals of fracture mechanics described in Parts I, II, and III of this book.

1.11 References

[*1*] Shank, M. E., "A Critical Review of Brittle Failure in Carbon Plate Steel Structures Other than Ships," *Ship Structure Committee Report, Serial No. SSC-65,* National Academy of Sciences—

National Research Council, Washington, DC, Dec. 1, 1953 (also reprinted as *Welding Research Council Bulletin,* No. 17).
[2] Parker, E. R., *Brittle Behavior of Engineering Structures,* prepared for the Ship Structure Committee under the general direction of the Committee on Ship Steel—National Academy of Sciences—National Research Council, John Wiley, New York, 1957.
[3] Bannerman, D. B. and Young, R. T., "Some Improvements Resulting from Studies of Welded Ship Failures," *Welding Journal,* Vol. 25, No. 3, March 1946.
[4] Acker, H. G., "Review of Welded Ship Failures," *Ship Structure Committee Report, Serial No. SSC-63,* National Academy of Sciences—National Research Council, Washington, DC, Dec. 15, 1953.
[5] *Final Report of a Board of Investigation—The Design and Methods of Construction of Welded Steel Merchant Vessels,* 15 July 1946, GPO, Washington, DC, 1947.
[6] Boyd, G. M. and Bushell, T. W., "Hull Structural Steel—The Unification of the Requirements of Seven Classification Societies," *Quarterly Transactions: The Royal Institution of Naval Architects (London),* Vol. 103, No. 3, March 1961.
[7] Turnbull, J., "Hull Structures," *The Institution of Engineers and Shipbuilders of Scotland, Transactions,* Vol. 100, Pt. 4, Dec. 1956–1957, pp. 301–316.
[8] Boyd, G. M., "Fracture Design Practices for Ship Structures," in *Fracture,* Vol. V: *Fracture Design of Structures,* edited by H. Libowitz, Academic Press, New York, 1969, pp. 383–470.
[9] Marine Casualty Report, "Structural Failure of the Tank Barge I.O.S. 3301 Involving the Motor Vessel Martha R. Ingram on 10 January 1972 Without Loss of Life," *Report No. SDCG/NTSB,* March 1974.
[10] Bishop, T., "Fatigue and the Comet Disasters," *Metal Progress,* May 1955, p. 79.
[11] Madison, R. B., "Application of Fracture Mechanics to Bridges," Ph.D. thesis, Lehigh University, 1969, and *Fritz Engineering Laboratory Report No. 335.2,* Bethlehem, PA, June 1969.
[12] "Collapse of U.S. 35 Highway Bridge, Point Pleasant, West Virginia," *NTSB Report No. NTSB-HAR-71-1,* Oct. 4, 1968.
[13] Bennet, J. A. and Mindlin, H., "Metallurgical Aspects of the Failure of the Pt. Pleasant Bridge," *Journal of Testing and Evaluation,* March 1973, pp. 152–161.
[14] "State Cites Defective Steel in Bryte Bend Failure," *Engineering News Record,* Vol. 185, No. 8, Aug. 20, 1970.
[15] "Joint Redesign on Cracked Box Girder Cuts into Record Tied Arch's Beauty," *Engineering News Record,* Vol. 188, No. 13, March 30, 1972.
[16] Fisher, J. W., *Fatigue and Fracture in Steel Bridges—Case Studies,* John Wiley, New York, 1984.
[17] Background Reports: Metallurgy, Fracture Mechanics, Welding, Moment Connections and Frame Systems Behavior, *SAC Joint Venture Report No. SAC 95-09,* FEMA-288, March 1997.
[18] Wells, A. A., "The Meaning of Fitness-for-Purpose and Concept of Defect Tolerance: International Conference," *Fitness-for-Purpose Validation of Welded Constructions,* The Welding Institute, London, 1981.
[19] Rolfe, S. T., "Fitness for Service—Common Sense Engineering, The Art and Science of Structural Engineering," Symposium Honoring William J. Hall, University of Illinois, Urbana Campaign, April 1993.
[20] Rolfe, S. T., Henn, A. E., and Hays, K. T., "Fracture Mechanics Methodology for Fracture Control in Oil Tankers," *Ship Structures Symposium 1993,* Nov. 16–17, 1993, Arlington, VA.

2

Stress Analysis for Members with Cracks—K_I

2.1 Introduction

FRACTURE MECHANICS is a method of characterizing the fracture and fatigue behavior of sharply notched structural members (cracked or flawed) in terms that can be used by the engineer, namely, stress (σ) and flaw size (a).

This chapter describes the stress analysis procedures for structural members with cracks. The difference between stress-*concentration* factors (used to analyze stress at a point in the vicinity of well-defined notches) and stress-*intensity* factors (used to analyze the stress field ahead of a sharp crack) is first described. Then, the stress analysis of members with sharp cracks is introduced. The complete stress analysis involves complex variables and other forms of higher mathematics and may be found in various references [*1–7*]. Emphasis in this book is on the applications and use of fracture mechanics.

Stress-intensity factors (K_I) are presented for most of the cracks commonly found in structures. Stress-intensity factors for other crack geometries can be found in various handbooks [*8–10*]. In fatigue, where the driving force is $\Delta K = K_{MAX} - K_{MIN}$, the same K_I relations described in this chapter also are used.

The stress-intensity factor is a mathematical calculation relating the applied load and crack size for a particular geometry. The calculation of K_I is analogous to the calculation of applied stress, σ, in an unflawed member. To prevent yielding, the engineer keeps the applied stress, σ, below the material yield strength, σ_{ys}. In an unflawed member, σ is the "driving force," and σ_{ys} (the yield strength) is the "resistance force." The driving force is a calculated quantity while the resistance force is a measured value. In the same sense, K_I is a calculated "driving force" and K_c (described in Chapter 3) is a measured fracture toughness value and represents the "resistance force" to crack extension. To prevent brittle fracture, the engineer keeps the calculated applied stress intensity factor, K_I, below the measured fracture toughness value, K_c, in the same manner that σ is kept below σ_{ys} to prevent yielding.

This chapter describes the calculation of K_I for various crack geometries. Chapter 3 describes the measurement of fracture toughness values, K_c, under various conditions of loading and constraint.

2.2 Stress-Concentration Factor—k_t

Most structural members have discontinuities of some type, for example, holes, fillets, notches, etc. If these discontinuities have well-defined geometries, it is usually possible to determine a stress-concentration factor, k_t for these geometries [11]. Then, the engineer can account for the local elevation of stress using the well-known relation between the local maximum stress and the applied nominal stress:

$$\sigma_{max} = k_t \cdot \sigma_{nom} \tag{2.1}$$

In fact, for most structures, the engineer usually relies on the ductility of the material to redistribute the load around a mild stress concentration. The effects of mild stress concentrations (holes, smooth fillets, etc.) in ductile materials are addressed in various codes and standards. However, if the stress concentration is severe, for example, approaching a sharp crack in which the radius of the crack tip approaches zero, the use of fracture mechanics becomes necessary to analyze structural performance.

To illustrate this point, let us analyze the stress concentration at the edge of an ellipse as shown in Figure 2.1. This is given as:

$$k_t = \frac{\sigma_{max}}{\sigma_{nom}} = \left(1 + 2\frac{a}{b}\right) \tag{2.2}$$

Rearranging Equation 2.2 gives σ_{max} for an ellipse,

$$\sigma_{max} = \sigma_{nom}\left(1 + 2\frac{a}{b}\right) \tag{2.3}$$

For very sharp cracks,

$$a >> b \therefore \sigma_{max} \simeq \sigma_{nom}\left(\frac{2a}{b}\right) \tag{2.4}$$

As shown in Figure 2.1, k_t becomes very large as a/b becomes large. The radius at the end of the major axis can be approximated by $\rho = b^2/a$. For sharp cracks,

$$a >> \rho, \sigma_{max} \simeq \sigma_{nom} \cdot 2\sqrt{\frac{a}{\rho}} \tag{2.5}$$

and

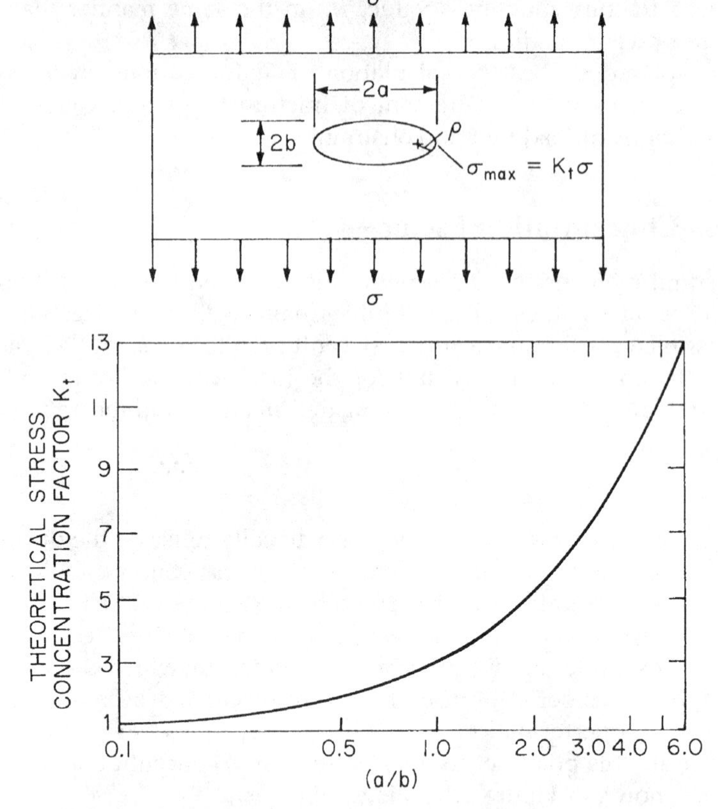

FIG. 2.1 Stress-concentration factor for an elliptical hole.

$$k_t = \frac{\sigma_{\max}}{\sigma_{\text{nom}}} = 2\sqrt{\frac{a}{\rho}} \tag{2.6}$$

For sharp cracks, $\rho \to 0$ and $k_t \to \infty$. Thus, the use of the stress-concentration approach becomes meaningless. Consequently, an analytical method different from the stress-concentration approach is needed to analyze the behavior of structural or machine components that contain cracks or sharp imperfections. The first analysis of fracture behavior for components containing cracks was developed by Griffith [*1*] as described in the appendix to this chapter. Presently, the fracture behavior for such components can be analyzed best by using fracture-mechanics technology.

2.3 Stress-Intensity Factor, K_I

Linear-elastic fracture-mechanics technology is based on an analytical procedure. This procedure relates the stress-field magnitude and distribution in the vicinity of a crack tip, the nominal stress applied to a test specimen or a structural member, and the size, shape, and orientation of a crack or crack-like discontinuity.

The fundamental principle of fracture mechanics is that the stress field ahead of a sharp crack in a test specimen or a structural member can be characterized as a single parameter, K, which is the stress-intensity factor. This parameter, K, is related to both the nominal stress level (σ) in the member and the size of the crack (a), and has units of ksi$\sqrt{\text{in.}}$ (MPa$\sqrt{\text{m}}$). Thus, all structural members or test specimens that have flaws can be loaded to various levels of K. This is analogous to the situation where unflawed structural or mechanical members can be loaded to various levels of stress, σ.

To establish methods of stress analysis for cracks in elastic solids, it is convenient to define three types of relative movements of two crack surfaces [5]. These displacement modes (Figure 2.2) represent the local deformation ahead of a crack. The opening mode, Mode I, is characterized by local displacements that

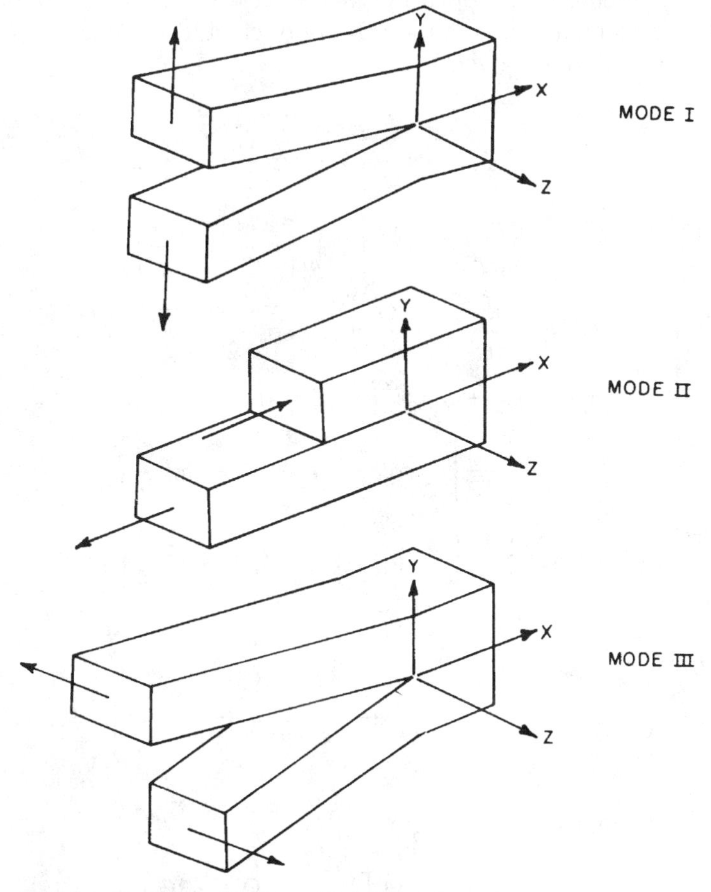

FIG. 2.2 The three basic modes of crack surface displacements (Ref. 5).

are symmetric with respect to the x-y and x-z planes. The two fracture surfaces are displaced perpendicular to each other in opposite directions. Local displacements in the sliding or shear mode, Mode II, are symmetric with respect to the x-y plane, and skew symmetric with respect to the x-z plane. The two fracture surfaces slide over each other in a direction perpendicular to the line of the crack tip. Mode III, the tearing mode, is associated with local displacements that are skew symmetric with respect to both x-y and x-z planes. The two fracture surfaces slide over each other in a direction parallel to the line of the crack front. Each of these modes of deformation corresponds to a basic type of stress field in the vicinity of crack tips. In any analysis, the deformations at the crack tip can be treated as one or a combination of these local displacement modes. Moreover, the stress field at the crack tip can be treated as one or a combination of the three basic types of stress fields.

Most practical design situations and failures correspond to Mode I displacements. Accordingly, Mode I is emphasized in this book.

By using a method developed by Westergaard [6], Irwin [7] found that the stress and displacement fields in the vicinity of crack tips subjected to the three modes of deformation are given by:

Mode I

$$
\begin{aligned}
\sigma_x &= \frac{K_I}{(2\pi r)^{1/2}} \cos\frac{\Theta}{2}\left[1 - \sin\frac{\Theta}{2}\sin\frac{3\Theta}{2}\right] \\
\sigma_y &= \frac{K_I}{(2\pi r)^{1/2}} \cos\frac{\Theta}{2}\left[1 + \sin\frac{\Theta}{2}\sin\frac{3\Theta}{2}\right] \\
\tau_{xy} &= \frac{K_I}{(2\pi r)^{1/2}} \sin\frac{\Theta}{2}\cos\frac{\Theta}{2}\cos\frac{3\Theta}{2} \\
\sigma_z &= \upsilon(\sigma_x + \sigma_y),\ \tau_{xz} = \tau_{yz} = 0 \qquad (2.7) \\
u &= \frac{K_I}{G}\left[\frac{r}{2\pi}\right]^{1/2} \cos\frac{\Theta}{2}\left[1 - 2\upsilon + \sin^2\frac{\Theta}{2}\right] \\
v &= \frac{K_I}{G}\left[\frac{r}{2\pi}\right]^{1/2} \sin\frac{\Theta}{2}\left[2 - 2\upsilon - \cos^2\frac{\Theta}{2}\right] \\
w &= 0
\end{aligned}
$$

Mode II

$$
\begin{aligned}
\sigma_x &= -\frac{K_{II}}{(2\pi r)^{1/2}} \sin\frac{\Theta}{2}\left[2 + \cos\frac{\Theta}{2}\cos\frac{3\Theta}{2}\right] \\
\sigma_y &= \frac{K_{II}}{2\pi r^{1/2}} \sin\frac{\Theta}{2}\cos\frac{\Theta}{2}\cos\frac{3\Theta}{2} \qquad (2.8) \\
\tau_{xy} &= \frac{K_{II}}{(2\pi r)^{1/2}} \cos\frac{\Theta}{2}\left[1 - \sin\frac{\Theta}{2}\sin\frac{3\Theta}{2}\right]
\end{aligned}
$$

$$\sigma_z = \upsilon(\sigma_x + \sigma_y),\ \tau_{xz} = \tau_{yz} = 0$$

$$u = \frac{K_{II}}{G}\left[\frac{r}{2\pi}\right]^{1/2} \sin\frac{\Theta}{2}\left[2 - 2\upsilon + \cos^2\frac{\Theta}{2}\right]$$

$$v = \frac{K_{II}}{G}\left[\frac{r}{2\pi}\right]^{1/2} \cos\frac{\Theta}{2}\left[-1 + 2\upsilon + \sin^2\frac{\Theta}{2}\right]$$

$$w = 0$$

Mode III

$$\tau_{xz} = -\frac{K_{III}}{(2\pi r)^{1/2}} \sin\frac{\Theta}{2}$$

$$\tau_{yz} = \frac{K_{III}}{(2\pi r)^{1/2}} \cos\frac{\Theta}{2}$$

$$\sigma_x = \sigma_y = \sigma_z = \tau_{xy} = 0 \tag{2.9}$$

$$w = \frac{K_{III}}{G}\left[2\frac{r}{\pi}\right]^{1/2} \sin\frac{\Theta}{2}$$

$$u = v = 0$$

where the stress components and the coordinates r and θ are shown in Figure 2.3; u, v, and w are the displacements in the x, y, and z directions, respectively; υ is Poisson's ratio, and G is the shear modulus of elasticity.

Equations (2.7) and (2.8) represent the case of plane strain ($w = 0$) and neglect higher-order terms in r. Because higher-order terms in r are neglected, these equations are exact in the limit as r approaches zero and are a good approximation in the region where r is small compared with other x-y planar dimensions. These field equations show that the distribution of the elastic-stress fields, and the deformation fields in the vicinity of the crack tip, are invariant in all components subjected to a given mode of deformation.

The magnitude of the elastic-stress field can be described by single-term parameters, K_I, K_{II}, and K_{III}, that correspond to Modes I, II, and III, respectively. Consequently, the applied stress, the crack shape, size, and the structural configuration associated with structural components subjected to a given mode of deformation affect the value of the stress-intensity factor but do not alter the stress-field distribution.

Dimensional analysis of Equations (2.7), (2.8), and (2.9) indicates that the stress-intensity factor must be linearly related to stress and must be related to the square root of a characteristic length. Based on Griffith's original analysis of glass components with cracks and the subsequent extension of that work to more ductile materials, the characteristic length in a structural member is the crack length. Consequently, the magnitude of the stress-intensity factor must be related directly to the magnitude of the applied nominal stress, σ_{nom}, and the square root

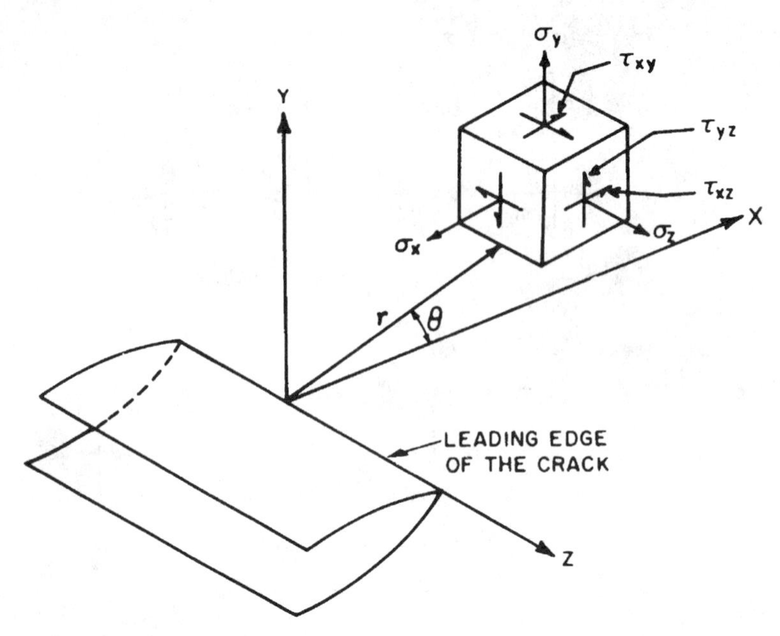

FIG. 2.3 Coordinate system and stress components ahead of a crack tip.

of the crack length, *a*. In all cases, the general form of the stress-intensity factor is given by:

$$K = \sigma_{nom}\sqrt{a}\cdot f(g) \tag{2.10}$$

where *f*(*g*) is a parameter that depends on the geometry of the particular member and the crack geometry. Fortunately, a large number of relationships between the stress-intensity factor and various body configurations, crack sizes, orientations, shapes, and loading conditions have been published [5,8–10]. The more common ones are presented in this chapter. One key aspect of the stress-intensity factor, K_I, is that it relates the *local* stress field ahead of a sharp crack in a structural member to the *global* (or nominal) stress applied to that structural member away from the crack. Specifically, Figure 2.3 shows the stresses just ahead of a sharp crack. Most fractures occur under conditions of Mode I loading (Figure 2.2). Accordingly, the stress of primary interest in Figure 2.3 and in most practical applications is σ_y. For σ_y to be a maximum in Equation (2.7), let $\theta = 0$,

$$\sigma_y = \frac{K_I}{\sqrt{2\pi r}} \tag{2.11}$$

Rearranging this expression shows that:

$$K_I = \sigma_y\sqrt{2\pi r} \tag{2.12}$$

At locations farther and farther from the crack tip (increasing *r*), the stress,

σ_y, decreases. However, K_I remains constant and describes the intensity of the stress *field* ahead of a sharp crack. This same stress-intensity factor is also related to the global stress by Equation (2.10) for various crack geometries as described in this chapter. Hence, K_I describes the stress field intensity ahead of a sharp crack in any structural member (plates, beams, airplane wings, pressure vessels, etc.) as long as the correct geometrical parameter, $f(g)$, can be determined. Expressions for different crack geometries in various structural members are presented in the next section.

2.4 Stress-Intensity-Factor Equations

Stress-intensity-factor equations for the more common basic specimen geometries that are subjected to Mode I deformations are presented in this section. Extensive stress-intensity-factor equations for various geometries and loading conditions are available in the literature in tabular form [5,8,9].

2.4.1 *Through-Thickness Crack*

The stress-intensity factor for an infinite plate subjected to uniform tensile stress, σ, that contains a through-thickness crack of length $2a$ is:

$$K_I = \sigma\sqrt{\pi a} \tag{2.13}$$

Note that from the Griffith analysis given in the appendix to this chapter there is a direct relationship between Griffith's analysis and Equation (2.13).

A tangent-correlation factor having the form:

$$\left(\frac{2b}{\pi a}\tan\frac{\pi a}{2b}\right)^{1/2} \tag{2.14}$$

is used to approximate the K_I values for a through-thickness crack in plates of finite width, $2b$ (Figure 2.4). Thus, the stress-intensity factor for a plate of finite width $2b$ subjected to uniform tensile stress, σ, and that contains a through-thickness crack of length $2a$ (Figure 2.4) is:

$$K_I = \sigma\sqrt{\pi a}\left(\frac{2b}{\pi a}\tan\frac{\pi a}{2b}\right)^{1/2} \tag{2.15}$$

The values for the tangent correction factor for various ratios of crack length to plate widths are given in Table 2.1. Equation 2.14 is accurate within 7% for $a/b \leq 0.5$.

2.4.2 *Single-Edge Notch*

The stress-intensity-factor equation for a single-edge-notch in an infinite width specimen (Figure 2.5) is

$$K_I = 1.12\sigma\sqrt{\pi a} \tag{2.16}$$

The 1.12 factor is referred to as the free surface correction factor. Thus, when

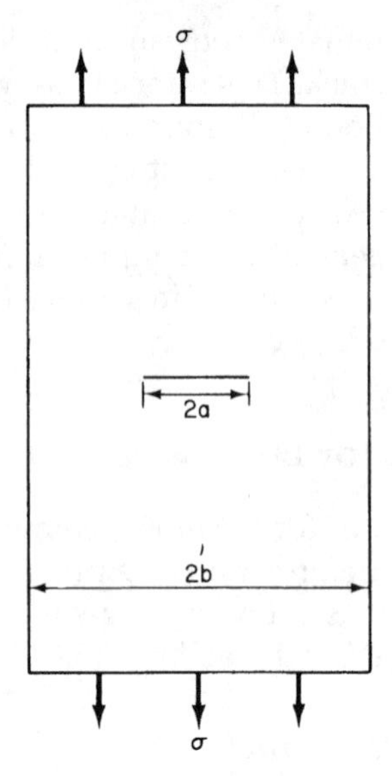

FIG. 2.4 Finite-width plate containing a through-thickness crack.

the through-thickness crack of Figure 2.4 is "cut" in half to form an edge crack in a plate, the K_I value is increased by about 12%.

For single-edge-notch specimens having finite width, an additional correction factor is necessary to account for bending stresses caused by lack of symmetry in the single-edge-notched specimen. Equation (2.17) incorporates the various correction factors for a single-edge-notch in a finite width plate, Figure 2.5.

$$K_I = 1.12\sigma\sqrt{\pi a} \cdot k\left(\frac{a}{b}\right) \tag{2.17}$$

The values of the function $k(a/b)$ are tabulated in Table 2.2 for various values

TABLE 2.1. Correction Factors for a Finite-width Plate Containing a Through-thickness Crack (Ref. 5).

a/b	$[2b/\pi a \cdot \tan \pi a/2b]^{1/2}$
0.074	1.00
0.207	1.02
0.275	1.03
0.337	1.05
0.410	1.08
0.466	1.11
0.535	1.15
0.592	1.20

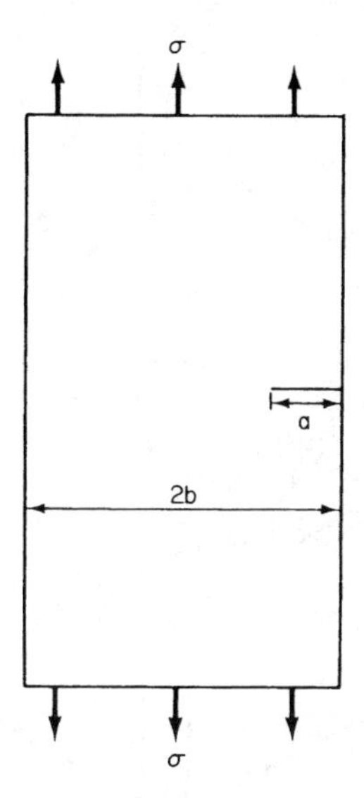

FIG. 2.5 Single-edge-notched plate of finite width.

of the ratio of crack length to specimen width. Note that for $a/b = 1.0$, the notch depth is one-half the width of the plate, and the correction factor is quite large.

TABLE 2.2. Correction Factors for a Single-edge Notched Plate (Ref. *5*).

a/b	$k(a/b)$
0.00	1.00
0.10	1.03
0.20	1.07
0.30	1.15
0.40	1.22
0.50	1.35
0.60	1.50
0.70	1.69
0.80	1.91
0.90	2.20
1.00	2.55

2.4.3 Embedded Elliptical or Circular Crack in Infinite Plate

The stress-intensity factor at any point along the perimeter of elliptical or circular cracks embedded in an infinite body subjected to uniform tensile stress (Figure 2.6) is given by [*12*]:

$$K_I = \frac{\sigma\sqrt{\pi a}}{Q}\left(\sin^2\beta + \frac{a^2}{c^2}\cos^2\beta\right)^{1/4} \tag{2.18}$$

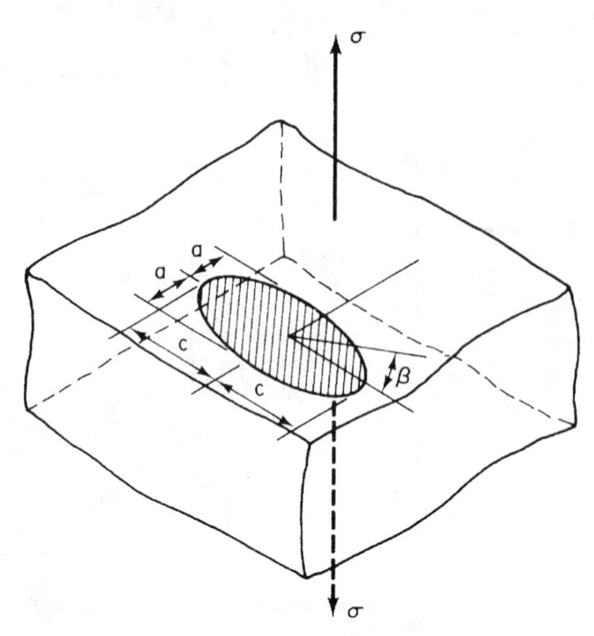

FIG. 2.6 Elliptical crack in an infinite body subjected to uniform tension (Ref. 5).

In Equation (2.18), K_I corresponds to the value of the stress-intensity factor for a point on the perimeter of the crack whose location is defined by the angle β. Q is a flaw-shape parameter that depends on σ/σ_{ys} and $a/2c$, as shown in Figure 2.7, where σ_{ys} is the yield strength of the material. This figure is the "Boeing Design Curve," [5], which is based on the solution of an elliptical integral, ϕ, that is equal to $\sqrt{Q}$ as described in various handbooks [8–10]. This curve is widely used to solve for K_I factors for cracks with various $a/2c$ ratios.

The stress-intensity factor for an embedded elliptical crack reaches a maximum at $\beta = \pi/2$ and is given by the equation:

$$K_I = \sigma\sqrt{\pi \frac{a}{Q}} \tag{2.19}$$

Values of the ratio of nominal applied stress, σ, to yield stress, σ_{ys}, are intended to account for the effects of plastic deformation in the vicinity of the crack tip on the stress-intensity-factor value and are incorporated into the value of Q which can be obtained from Figure 2.7.

For a circular crack where $a = c$ and $Q \simeq 2.4$, Equation (2.19) becomes:

$$K_I = 0.65\,\sigma\sqrt{\pi a} = 1.15\,\sigma\sqrt{a} \tag{2.20}$$

The exact expression for an embedded circular crack [5] is:

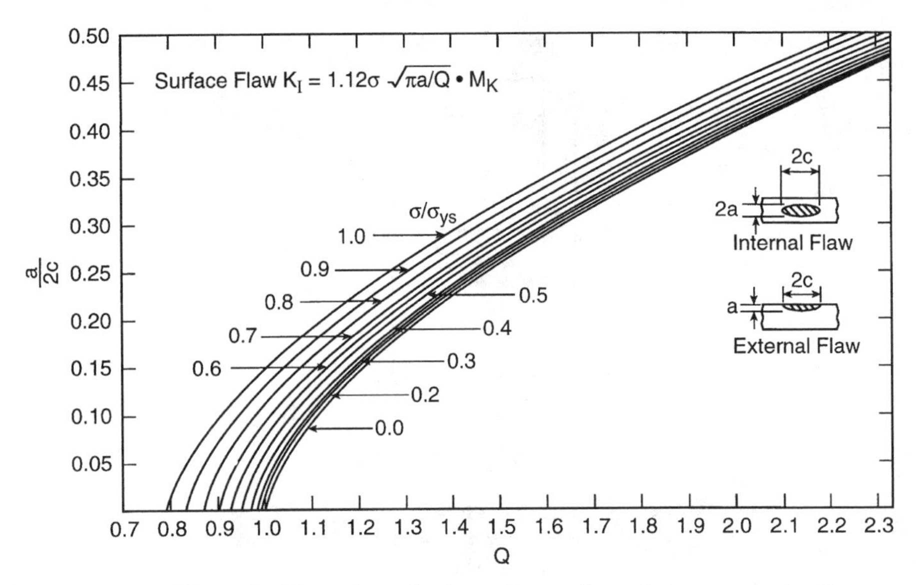

FIG. 2.7 Effect of $a/2c$ ratio and σ/σ_{ys} ratio on flaw-shape parameter Q.

$$K_I = \frac{2}{\sqrt{\pi}} \sigma\sqrt{a} = 1.13\ \sigma\sqrt{a} \tag{2.21}$$

Note that the two values agree within 2% of each other.

2.4.4 Surface Crack

The stress-intensity factor for a part-through "thumbnail" crack in a plate subjected to uniform tensile stress (Figure 2.8) can be calculated by using Equation (2.19) and a free-surface-correction factor equal to 1.12. Remember that this 1.12 factor occurs any time a crack originates at a free surface. The stress-intensity factor for $\beta = \pi/2$ (which is the location of maximum stress intensity) is given by:

$$K_I = 1.12\ \sigma\sqrt{\pi\frac{a}{Q}} \cdot M_k \tag{2.22}$$

This equation is identical to Equation (2.19) except for the 1.12 multiplier that corresponds to the front free-surface correction factor and M_K, which corresponds to a back free-surface correction factor. M_K is approximately 1.0 as long as the crack depth, a, is less than one-half the wall thickness, t. As a approaches t, M_K approaches approximately 1.6, and a useful approximation is:

$$M_K = 1.0 + 1.2\left(\frac{a}{t} - 0.5\right) \tag{2.23}$$

for values of $a/t \geq 0.5$.

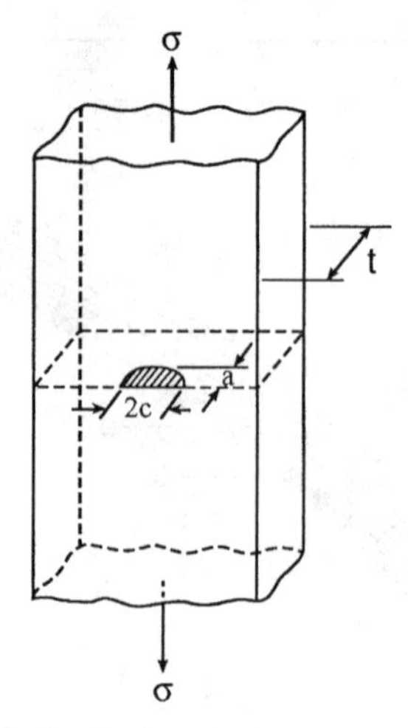

FIG. 2.8 Plate containing a semi-elliptical surface crack.

2.4.5 *Cracks Growing from Round Holes*

The stress-intensity factor for cracks growing from a circular hole (e.g., a bolt hole) in an infinite plate (Figure 2.9) is given by:

$$K_I = \sigma\sqrt{\pi a} \cdot f\left(\frac{a}{r}\right) \tag{2.24}$$

where r = radius of hole, and
a = crack length from the side of the hole.

Note that as a/r approaches zero, $f(a/r)$ for either one or two cracks approach a value of about 3. This is the value of the stress-concentration factor, k_t, at a round hole. Thus, the K_I factor can be thought of as being equal to $K_I \simeq k_t\sigma\sqrt{\pi a}$ for very short cracks near a stress concentration where the local stress is elevated by the stress concentration.

As the crack grows and (a/r) becomes large, the effect of the stress concentration (in this case the round hole) decreases, and the expression approaches that of a through-thickness crack in an infinite plate.

2.4.6 *Single Crack in Beam in Bending*

The stress-intensity factor for a beam in bending that contains an edge crack (Figure 2.10) is:

$$K_I = \frac{6M}{B(W-a)^{3/2}} \cdot g\left(\frac{a}{W}\right) \tag{2.25}$$

where M is the moment, and the values of $g(a/W)$ are presented in Table 2.3 for various ratios of crack length, a, to beam depth, W.

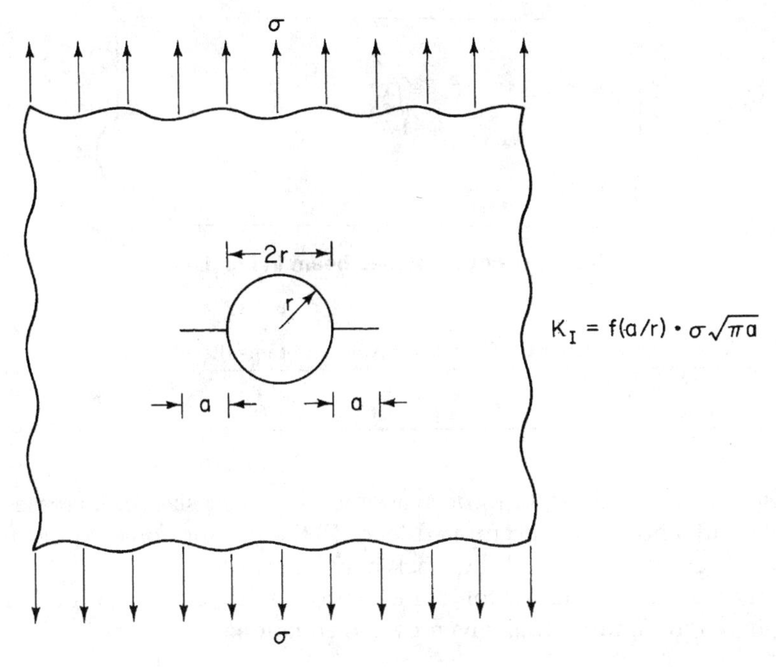

a/r	f(a/r), ONE CRACK	f(a/r), TWO CRACKS
0.00	3.39	3.39
0.10	2.73	2.73
0.20	2.30	2.41
0.30	2.04	2.15
0.40	1.86	1.96
0.50	1.73	1.83
0.60	1.64	1.71
0.80	1.47	1.58
1.0	1.37	1.45
1.5	1.18	1.29
2.0	1.06	1.21
3.0	0.94	1.14
5.0	0.81	1.07
10.0	0.75	1.03
∞	0.707	1.00

FIG. 2.9 Cracks growing from a round hole (Ref. 5).

2.4.7 Holes or Cracks Subjected to Point or Pressure Loading

The relation for K_I for a crack growing from a hole subjected to centrally applied point forces, P, such as a bolt, Figure 2.11 is:

$$K_I = \frac{P}{\sqrt{\pi a}} \tag{2.26}$$

where P = force/thickness of plate.

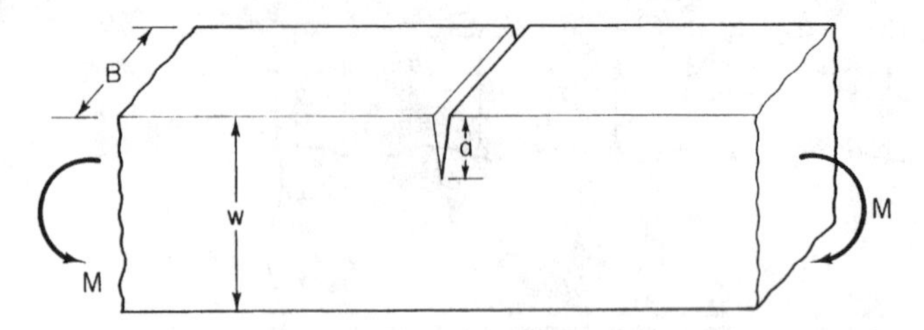

FIG. 2.10 Edge-notched beam in bending.

TABLE 2.3. Stress-intensity-factor Coefficients for Notched Beams (Ref. 5).

a/W	0.05	0.1	0.2	0.3	0.4	0.5	0.6 (and larger)
$g(a/W)$	0.36	0.49	0.60	0.66	0.69	0.72	0.73

Note that for a given P, K_I decreases as the crack size, a, increases. Thus, under the influence of a concentrated load, P, the driving force, K_I decreases as the crack length increases and the crack may be arrested.

The stress-intensity factor for a crack subjected to an internal pressure p, in psi, as shown in Figure 2.12 is given by the equation:

$$K_I = 1.12\, p\sqrt{\pi a} \tag{2.27}$$

This expression can be used to calculate the additional effect of a very high pressure in a thick-walled pressure vessel with an internal surface crack as:

$$K_I \simeq 1.12\, p\sqrt{\frac{\pi a}{Q}} \tag{2.28}$$

2.4.8 *Estimation of Other* K_I *Factors*

The preceding stress intensity factors have been sufficient to analyze most of the design cases and structural failures familiar to the authors. For very unusual cases, the reader is referred to the various handbooks of stress-intensity factors available in the literature [*8–10*]. However, there are several common-sense guidelines that can be applied to estimate stress-intensity factors for crack geometries not described above.

These guidelines are as follows:

1. *Free-Surface Correction*

   For every free-surface at which a crack might originate, increase K_I by 1.12. For example, the K_I for an embedded circular crack (Figure 2.6) was given by Equation 2.21 as:

$$K_I = \frac{2}{\sqrt{\pi}}\sigma\sqrt{a} \tag{2.21}$$

   If this were to occur at the corner of a plate, Figure 2.13, the K_I expression

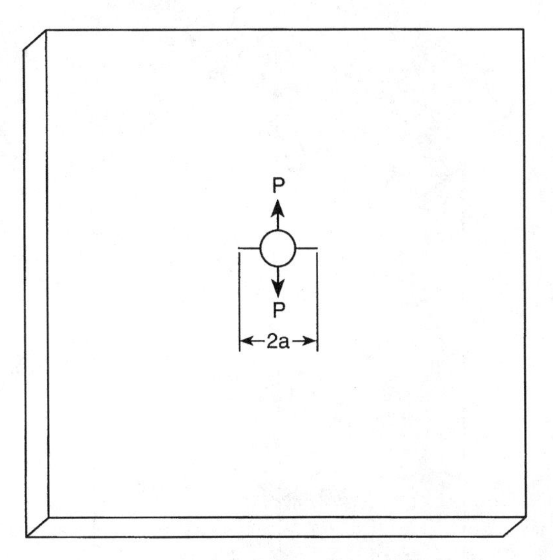

FIG. 2.11 Crack subjected to point loads.

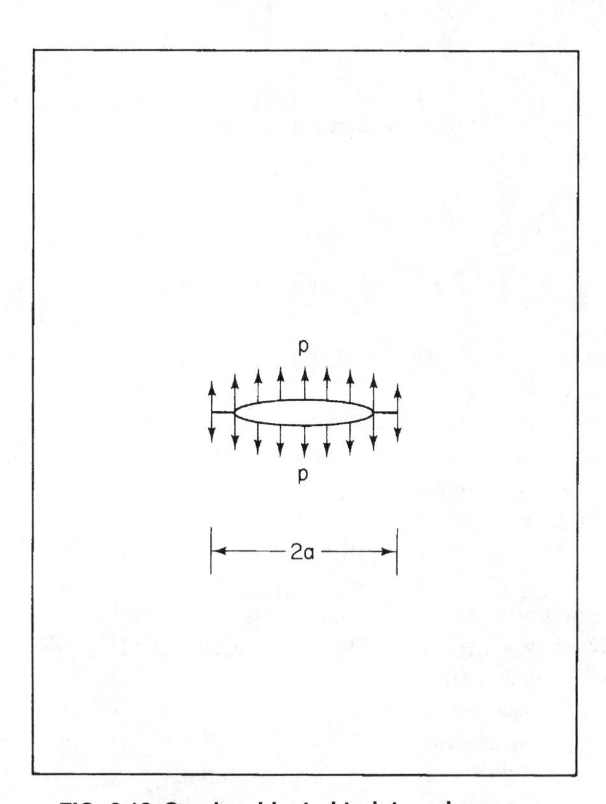

FIG. 2.12 Crack subjected to internal pressure.

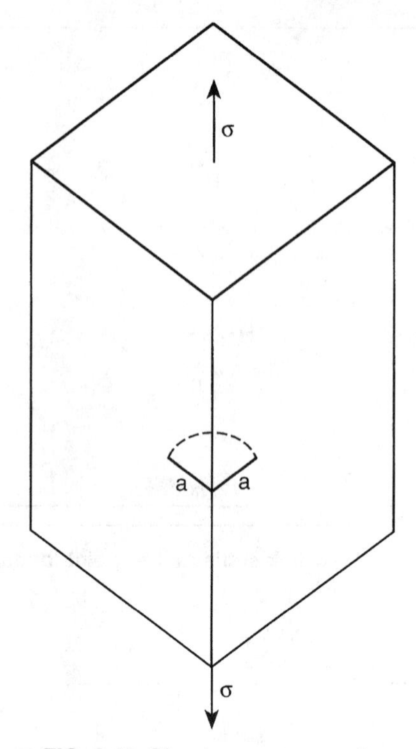

FIG. 2.13 Circular corner crack.

would be:

$$K_I = (1.12)(1.12)\frac{2}{\sqrt{\pi}}\sigma\sqrt{a} \tag{2.29}$$

because there are two (1.12) free surface flaw corrections.

2. *Inclined Cracks*
   Experience has shown that cracks tend to grow perpendicular to the direction of applied stress, particularly in fatigue or repeated loading. Accordingly, inclined cracks of length L, such as is shown in Figure 2.14, can be treated as a projected crack length of $2a$. If θ approaches 90°, the crack is parallel to the direction of applied stress and generally is harmless, i.e., $K_I \to 0$. This fact will be particularly important in Chapter 10, Fatigue and Fracture Behavior of Welded Components, where some discontinuities in welded structures are parallel to the direction of the applied stress and thus are essentially harmless.
3. *Bounding Using Existing Solutions*
   In many circumstances, the particular crack of interest has an irregular shape. For most of these cases, the stress-intensity factor can be estimated

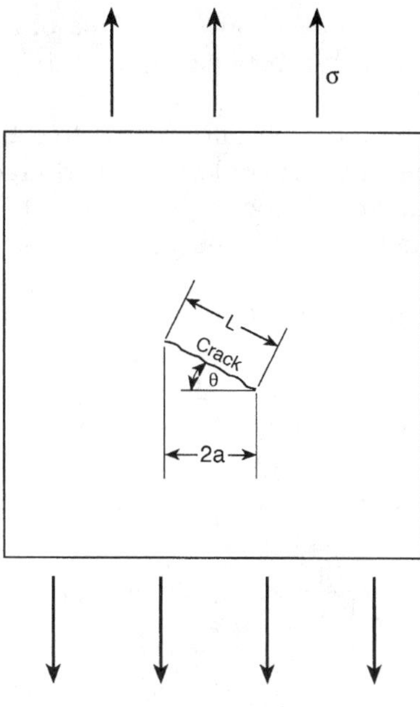

FIG. 2.14 Inclined crack.

from equations for simple and familiar shapes such as tunnel, circular, or elliptical cracks.

Estimating stress-intensity factors for irregularly shaped cracks can be made easier if one remembers some basic relationships and trends. A tunnel crack is a special case of the equation for an embedded elliptical crack (Equation 2.18) with the crack-shape parameter, Q, equal to 1.0. Furthermore, the equation for an embedded circular crack is also a special case of Equation (2.18) for $a = c$ ($Q \approx 2.4$). Thus, as the crack shape changes from a circular crack front to a straight crack front, the value of Q changes from 2.4 to 1.0 and the magnitude of the stress-intensity factor, K_I, increases by a factor of about 1.5. A corollary to these observations is that, for a tunnel crack that is curved, K_I is higher for the convex front of the crack than for the concave front.

A study of the crack-shape parameter, Q (Figure 2.7), shows that when $a/2c$ is less than about 0.15 (i.e., when the crack length on the surface, $2c$, is larger than about six times its depth, a), the crack-shape parameter approaches 1.0 and the crack behavior approaches that for a long tunnel crack where $a/2c \rightarrow 0$. An approximate linear relationship between the change in crack geometry, $a/2c$, and the change in the crack-shape parameter, Q, between the extreme of a circular crack and a long crack is as follows:

$$2.4(\text{circular crack, } a/2c = 0.5) > Q > 1.1(\text{long crack, } a/2c \approx 0.15) \quad (2.30)$$

Thus, for every 0.1 decrease in the magnitude of $a/2c$, the value of the crack-shape parameter, Q, decreases by about 0.35. This observation is derived from Figure 2.7, where the relationship between $a/2c$ and Q is shown to be approximately linear in this range.

The application of the preceding information to estimate the stress-intensity factor along the perimeter of an irregularly shaped crack can be demonstrated by analyzing the embedded crack shown in Figure 2.15. The crack has an irregular shape and is subjected to a uniform tensile stress, σ, that is perpendicular to the crack plane.

The K_I for the crack in region 1 (where the crack is denoted as a_1 in Figure 2.15) can be represented by a circular crack having a radius a_1 because the crack is essentially circular in shape. The K_{I1} value for this perimeter, which opens into a long tunnel crack, should be slightly higher than the K_I for a circular crack of radius a_1 and should be less than the K_I value for a long tunnel crack having a width equal to $2a_1$. Therefore,

$$\sigma\sqrt{\frac{\pi a_1}{Q}} < K_{I1} << \sigma\sqrt{\pi a_1} \tag{2.31}$$

where $Q = 2.4$ or:

$$0.65\ \sigma\sqrt{\pi a_1} < K_{I1} << \sigma\sqrt{\pi a_1} \tag{2.32}$$

Thus, a reasonable estimate for K_I is closer to a circular crack, or:

$$K_{I1} \approx 0.75\ \sigma\sqrt{\pi a_1} \tag{2.33}$$

is a reasonable estimate for that portion of the crack perimeter.

The crack front in region 2 corresponds closely to a long tunnel crack and, therefore,

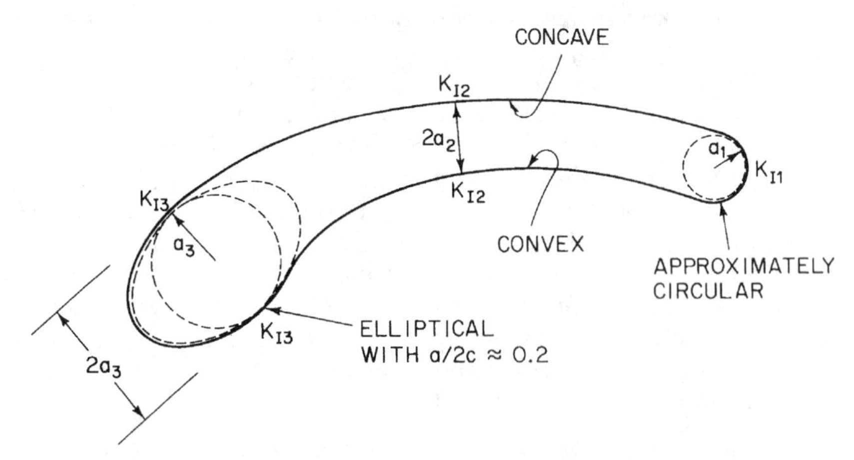

FIG. 2.15 Estimating stress-intensity factors for an embedded crack having an irregular shape.

$$K_{I2} \approx \sigma\sqrt{\pi a_2} \tag{2.34}$$

The elliptical crack front in region 3 can be represented by a circular crack of radius a_3, or a tunnel crack of width $2a_3$. Because of the general elliptical shape, the K_I for this region is better described for a tunnel crack than for a circular crack. Therefore,

$$0.65\ \sigma\sqrt{\pi a_3} << K_{I3} < \sigma\sqrt{\pi a_3} \tag{2.35}$$

A reasonable estimate for K_{I3} is closer to a tunnel crack, or:

$$K_{I3} \approx 0.9\ \sigma\sqrt{\pi a_3} \tag{2.36}$$

The crack front in region 3 can be approximated also by an elliptical crack with a minor axis equal to a_3, and for example a crack geometry, $a/2c$, of about 0.2. The decrease in the value of $a/2c$ from 0.5 for a circular crack to a value of 0.2 for the ellipse is equal to 0.3. Thus, the decrease in the value of Q from 2.4 for a circular crack is 1.05 (i.e., $0.3 \times 0.35/0.1 = 1.05$). Consequently, the value of Q for the ellipse is approximately 1.35 (i.e., $2.4 - 1.05 = 1.35$), and

$$K_{I3} \approx \sigma\sqrt{\frac{\pi a_3}{1.35}} = 0.86\ \sigma\sqrt{\pi a_3} \tag{2.37}$$

This value is a reasonable estimate of K_{I3}, especially for the lower portion of the crack front, which fits the ellipse more closely than the upper portion. Because the upper portion of the perimeter has less curvature than the ellipse, the value for K_{I3} in that region should be increased slightly. Consequently,

$$K_{I3} \approx 0.9\ \sigma\sqrt{\pi a_3} \tag{2.38}$$

should be a good estimate for this region.

2.4.9 Superposition of Stress-Intensity Factors

Components that contain cracks may be subjected to one or more different types of Mode I loads such as uniform tensile loads, concentrated tensile loads, or bending loads. The stress-field distributions in the vicinity of the crack tip subjected to these loads are identical and are represented by Equation (2.7). Consequently, the total stress-intensity factor can be obtained by algebraically adding the stress-intensity factors that correspond to each load.

For example, Figure 2.16 shows a plate with an edge crack subjected to tension plus bending loads. The combined stress intensity factor for that case is

$$K_{I\ TOTAL} = K_I\ (\text{EDGE}) + K_I\ (\text{BENDING}) \tag{2.39}$$

$$K_{I\ TOTAL} = 1.12\ \sigma\sqrt{\pi a} \cdot k\left(\frac{a}{b}\right) + \frac{6M}{B(W-a)^{3/2}} \cdot g\left(\frac{a}{W}\right)$$

A fairly common situation is a plate loaded in tension at one end and this force being withstood by a bolt force P. The solution, K_I, is a superposition of K_I

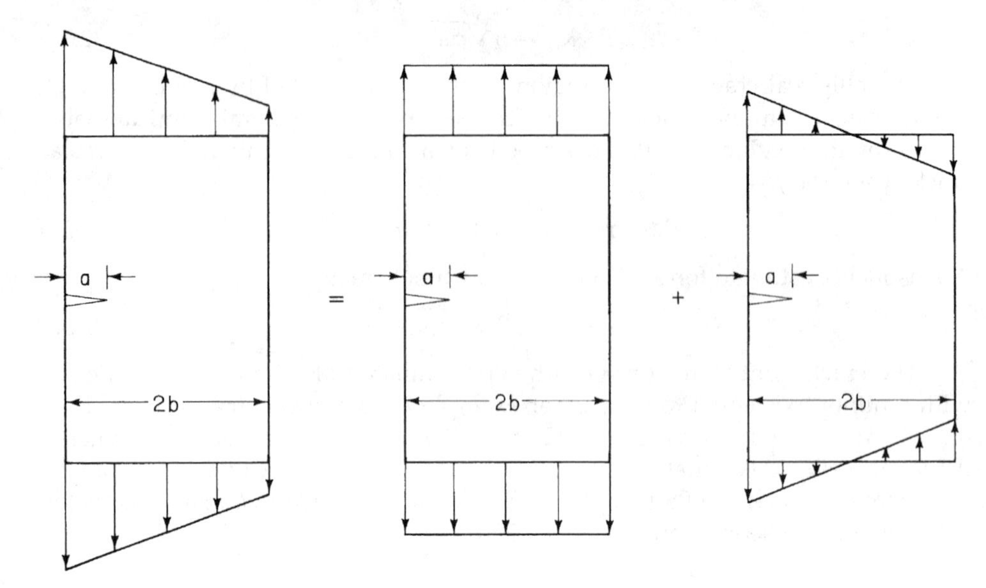

FIG. 2.16 Superposition solution for tensile and bending stresses applied to a single-edge notched plate.

solutions for a through-thickness crack (Equation 2.13), and a crack subjected to point loads (Equation 2.26), as shown in Figure 2.17.

Figure 2.18 shows a portion of a thin-walled pressure vessel subjected to a hoop stress, σ_{HOOP}, and the crack surface subjected to the internal pressure, p.

$$K_I = K_I \text{ (HOOP)} + K_I \text{ (PRESSURE)}$$

$$K_I = 1.12 \left[\sigma_{HOOP} \sqrt{\frac{\pi a}{Q}} + p \sqrt{\frac{\pi a}{Q}} \right] \tag{2.40}$$

$$K_I = 1.12 \left[\left(\frac{pD}{2t} \right) \sqrt{\frac{\pi a}{Q}} + p \sqrt{\frac{\pi a}{Q}} \right]$$

Note that as D/t becomes large, the effect on K_I of the pressure stress on the crack becomes 10% or less and thus generally is neglected, particularly for shallow flaws.

Some components may be subjected to loads that correspond to various modes of deformation. Because the stress-field distributions in the vicinity of a crack, Equations (2.7), (2.8), and (2.9) are different for different modes of deformation, the stress-intensity factors for different modes of deformation cannot be added. Under these loading conditions, the total energy-release rate, G, described in the appendix, rather than stress-intensity factors, K, can be calculated by algebraically adding the energy-release rate for the various modes of deformation.

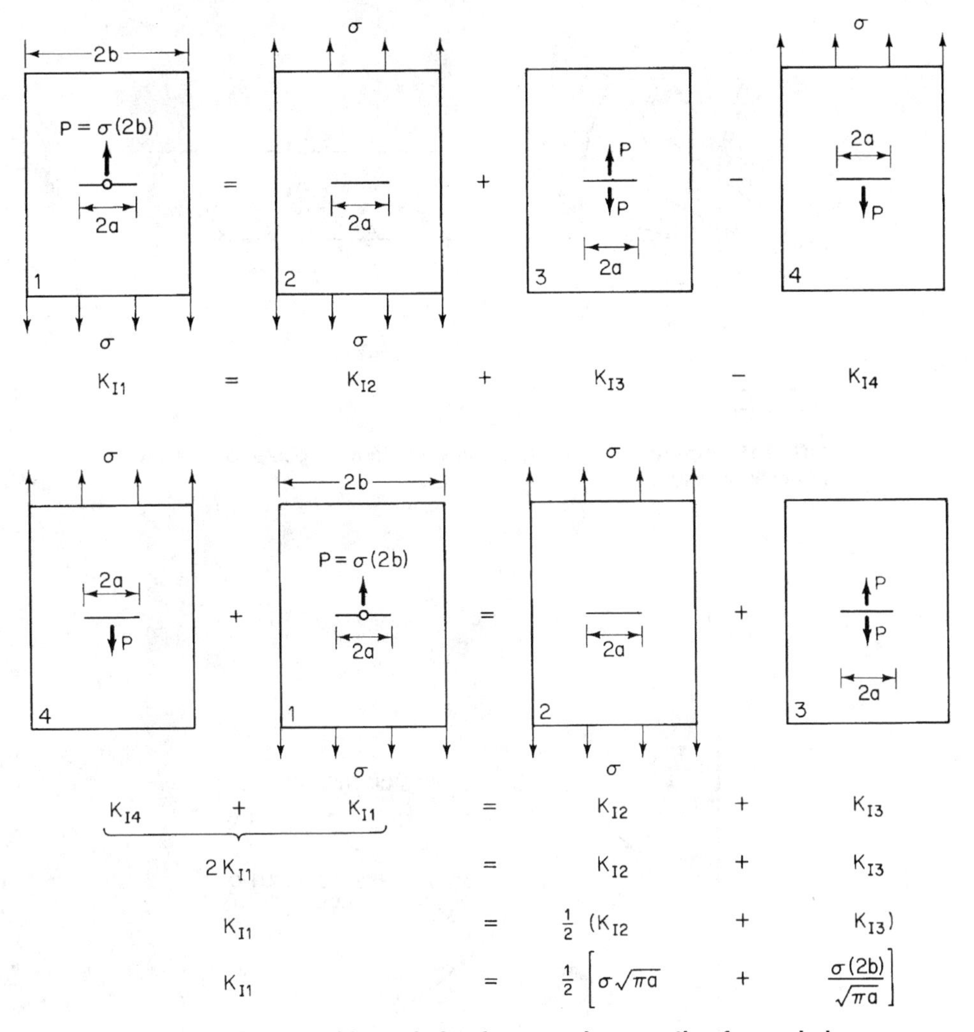

FIG. 2.17 Superposition solution for a crack emanating from a hole.

2.5 Crack-Tip Deformation and Plastic Zone Size

The stress-field equations, Equations (2.7), (2.8), and (2.9), show that the elastic stresses in the vicinity of a crack tip where $r << a$ can be very large. In reality, such high stress magnitudes do not occur because the material in this region undergoes plastic deformation, thus creating a plastic zone that surrounds the crack tip. Figure 2.19 is a schematic presentation of the change in the distribution of the y component of the stress caused by the localized plastic deformation in the vicinity of the crack tip.

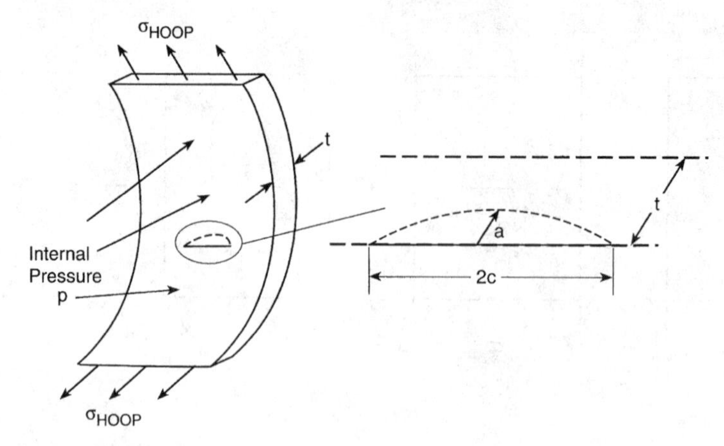

FIG. 2.18 Superposition of K_I factors for hoop stress and internal pressure.

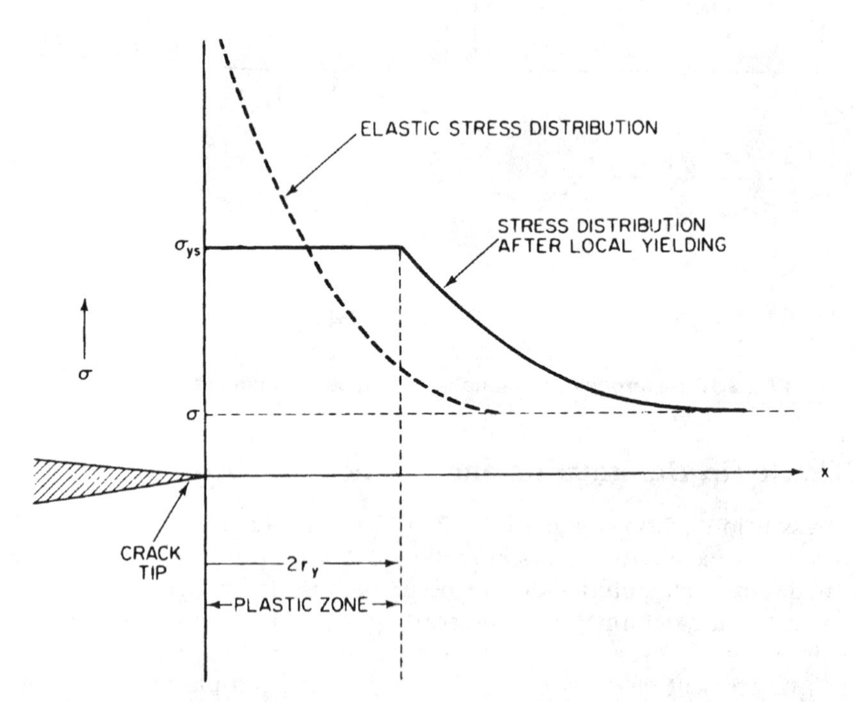

FIG. 2.19 Distribution of the σ_y stress component in the crack-tip region.

The size of the plastic zone, r_y, can be estimated from the stress-field equations by setting the y component of stress, σ_y, equal to the yield strength, σ_{ys}, which results in:

$$r_y = \frac{1}{2\pi}\left(\frac{K}{\sigma_{ys}}\right)^2 \tag{2.41}$$

Irwin [13] suggested that the plastic-zone size under plane-strain* conditions can be obtained by considering the increase in the tensile stress for plastic yielding caused by plane-strain elastic constraint. Under these conditions, the yield strength is estimated to increase by a factor of $\sqrt{3}$. Consequently, the plane-strain plastic-zone size becomes:

$$r_y = \frac{1}{6\pi}\left(\frac{K}{\sigma_{ys}}\right)^2 \tag{2.42}$$

The plastic zone along the crack front in a thick specimen is subjected to plane-strain conditions in the center portion of the crack front where $w = 0$ and to plane-stress conditions near the surface of the specimen where $\sigma_z = 0$. Consequently, Equations (2.41) and (2.42) indicate that the plastic zone in the center of a thick specimen is smaller than at the surface of the specimen. A schematic representation of the variation of the plastic-zone size along the front of a crack in a thick specimen is shown in Figure 2.20. Irwin [13] suggested that the effect of small plastic zones corresponds to an apparent increase of the elastic crack length by an increment equal to r_y. This plastic-zone correction factor is valid for small plastic-zone sizes and as described in the next section can be used to estimate $K_{I\,eff}$ (effective) factors for loadings that result in moderate plastic zone sizes.

2.6 Effective K_I Factor for Large Plastic Zone Size

Figure 2.21 shows the effective crack length, $2a + 2r_y$, for a wide plate loaded in tension. As $\sigma_{applied}$ is increased and the plane stress plastic zone size increases, use $a_{eff} = a_0 + r_y$ to establish K_I:

$$K_{I_{eff}} \simeq \sigma_{applied}\sqrt{\pi a_{eff}} \tag{2.43}$$

$$K_{I_{eff}} \simeq \sigma_{applied}\sqrt{\pi\left(a + \frac{1}{2\pi}\left(\frac{K_{I_{eff}}}{\sigma_{ys}}\right)^2\right)} \tag{2.44}$$

*Plane strain and plane stress are conditions of maximum and variable constraint, respectively, as discussed in Chapters 3 and 4.

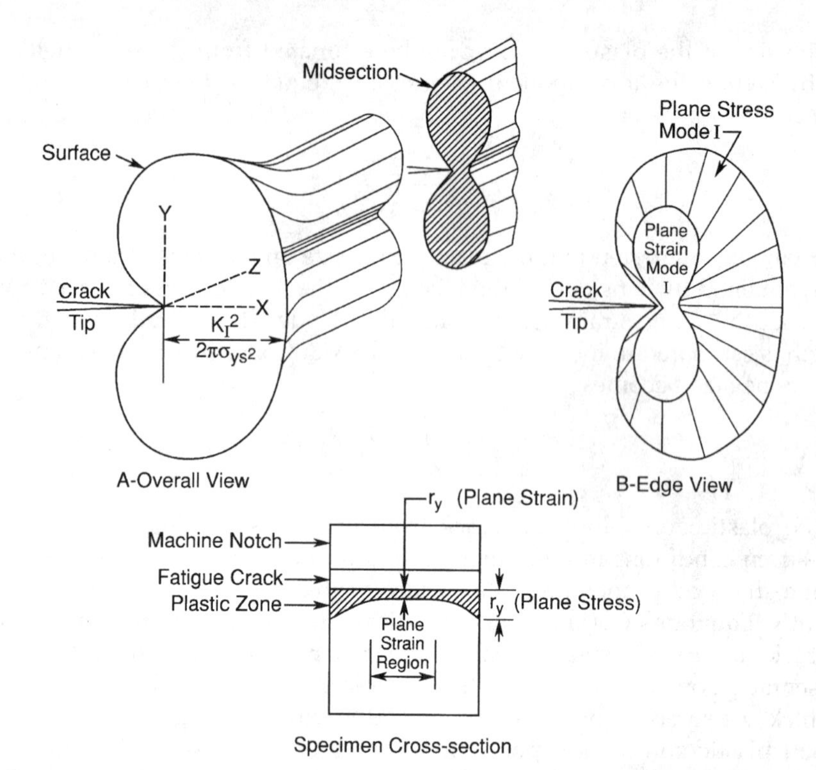

FIG. 2.20 Schematic representation of plastic zone ahead of crack tip.

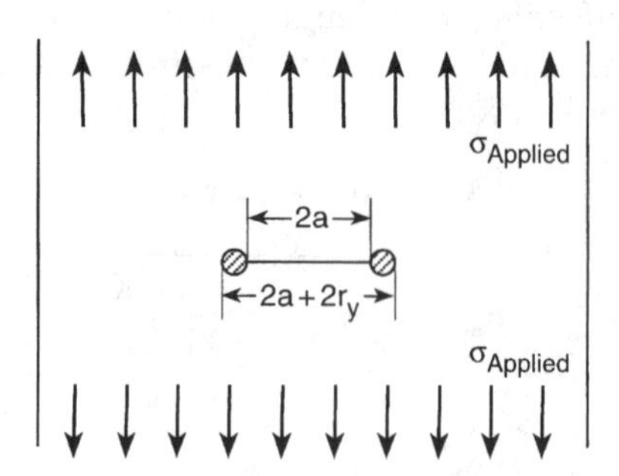

FIG. 2.21 Effective crack length, $2a_{eff} = 2a + 2r_y$.

Solving Equation (2.44) for K_I as $K_{I_{eff}}$ because of the larger plastic zone size, we obtain:

$$\left(\frac{K_{I_{eff}}}{\sigma_{app}}\right)^2 = \pi a + \frac{1}{2} \cdot \left(\frac{K_{I_{eff}}}{\sigma_{ys}}\right)^2$$

$$(K_{I_{eff}})^2 \left[1 - \frac{1 \cdot \sigma_{app}^2}{2\sigma_{ys}^2}\right] = \pi a \cdot \sigma_{app}^2 \tag{2.45}$$

$$K_{I_{eff}} = \frac{\sigma_{app}\sqrt{\pi a}}{\left[1 - \frac{1}{2}\left(\frac{\sigma_{app}}{\sigma_{ys}}\right)^2\right]^{1/2}}$$

For $\sigma_{app} \cong 3/4\ \sigma_{ys}$ (a reasonable upper limit to the applied nominal stress in a structure):

$$K_{I_{eff}} = \frac{\sigma_{app}\sqrt{\pi a}}{\left[1 - \frac{1}{2}\left(\frac{3}{4}\frac{\sigma_{ys}}{\sigma_{ys}}\right)^2\right]^{1/2}} \simeq \frac{\sigma_{app}\sqrt{\pi a}}{(1 - .28)^{1/2}} \tag{2.46}$$

$$K_{I_{eff}} = \frac{\sigma\sqrt{\pi a}}{.85} \sim 1.18\ \sigma_{app}\sqrt{\pi a}$$

Note that this is only an 18% increase in K_I compared with K_I for $r_y \simeq 0$, and thus should give a reasonable approximation to the driving force for fairly large stress levels.

For this same case, and a crack length $2a = 2''$, the absolute value of r_p is still fairly small:

$$r_p \simeq \frac{1}{2\pi}\left(\frac{K_I}{\sigma_{ys}}\right)^2 \simeq \frac{1}{2\pi}\left(\frac{.75\sigma_{ys}\sqrt{\pi(1)}}{\sigma_{ys}}\right)^2 \tag{2.47}$$

$$r_p \simeq .28 \text{ in.}$$

Thus, calculating K_I as a driving force for $\sigma \simeq 3/4\ \sigma_{ys}$ is not an unrealistic extension of linear-elastic fracture mechanics into the beginning of the elastic-plastic regime. In fact, for $\sigma_{app} = \sigma_{ys}$, as might be the case for weldments where the residual stress may be equal to σ_{ys}, the approximation of $K_{I_{eff}}$ is:

$$K_I \simeq \frac{\sigma_{ys}\sqrt{\pi a}}{\left[1 - \frac{1}{2}((1)^2)\right]^{1/2}} \tag{2.48}$$

$$K_I \simeq 1.4\ \sigma_{ys}\sqrt{\pi a}$$

If the structural component is heavily constrained, the linear portion of the

σ-ε curve would be elevated such that the yield stress, σ_{ys}, may well be elevated 40%, and the behavior would still be close to linear-elastic.

In summary, as a first approximation, use of $K_I = K_{I_{eff}} = f(\sigma_{app}, a + r_y)$ appears to be a reasonable engineering approach, even for stresses approaching the yield strength.

2.7 J_I and δ_I Driving Forces

Two other fracture mechanics parameters that can be used to calculate driving forces are the path-independent integral, J_I, and the crack-tip-opening displacement (CTOD), δ_I. These two parameters are used primarily in the elastic-plastic regime, whereas the stress intensity parameter, K_I, is used primarily in the linear-elastic regime.

2.7.1 J *Integral*

The path-independent *J*-integral proposed by Rice [*14*] is a method of characterizing the stress-strain field at the tip of a crack by an integral path taken sufficiently far from the crack tip to be analyzed elastically, and then substituted for the inelastic region close to the crack-tip region. Thus, even though considerable yielding (elastic-plastic behavior) may occur in the vicinity of the crack tip, if the region away from the crack tip can be analyzed elastically, behavior of the crack-tip region can be inferred. Elastic stress strain behavior is assumed, even though the stress-strain curve may be nonlinear. This technique can be used to estimate the fracture characteristics of materials exhibiting elastic-plastic behavior and is a means of extending fracture-mechanics concepts from linear-elastic (K_{Ic}) behavior to elastic-plastic behavior.

For linear-elastic behavior, the *J*-integral is identical to *G*, the energy release rate per unit crack extension (see Appendix). Therefore,

$$J_I = G_I = \frac{(1 - \nu^2)K_I^2}{E} \tag{2.49}$$

There are procedures available to calculate J_I driving forces and these are introduced in the appendix to this chapter. The most widely used procedure is to use 2D or 3D finite element analysis programs to evaluate the *J*-integral ahead of a crack in the geometry of interest. It requires that a stress-strain curve be available and was first used in the pressure vessel industry, primarily for nuclear pressure vessels. However, in most cases, the K_I driving force is used, recognizing that:

$$K_I = \sqrt{\frac{J_I E}{1 - \nu^2}}$$

for plane strain conditions and,

$$K_I = \sqrt{J_I E} \tag{2.50}$$

for plane stress conditions.

As will be discussed in Chapter 3, it is common in most structural situations to measure the critical resistance fracture toughness using the *J*-integral procedure (J_c or J_{Ic}), convert these values to K_c or K_{Ic}, and then compare the driving and resistance forces in terms of K_I and K_c. For complex structures, such as nuclear pressure vessels, where elastic-plastic analyses can be justified, J_I is determined using finite element analyses. J_c is measured as described in Chapter 3. Fracture control then consists of keeping $J_I < J_c$.

2.7.2 *CTOD* (δ_I)

In 1961, Wells [15] proposed that the fracture behavior in the vicinity of a sharp crack could be characterized by the opening of the notch faces, namely, the crack-tip opening displacement, CTOD, as shown in Figure 2.22. Furthermore, he showed that the concept of crack-opening displacement was analogous to the concept of critical crack extension force (G_c as described in the appendix to this chapter), and thus the CTOD values could be related to the plane-strain fracture toughness, K_{Ic}. Because CTOD measurements can be made even when there is considerable plastic flow ahead of a crack, such as would be expected for elastic-plastic or fully plastic behavior, this technique can be used to establish critical design stresses or crack sizes in a quantitative manner similar to that of linear-elastic fracture mechanics.

The CTOD relationship for a center crack in a wide plate is developed in the appendix and is as follows:

$$\delta_I = \frac{\pi \sigma_{app}^2 a}{E \sigma_{ys}} \tag{2.51}$$

Because,

$$K_I = \sigma_{app} \sqrt{\pi a} \tag{2.52}$$

$$\delta_I = \frac{K_I^2}{E \sigma_{ys}}$$

Thus, in the same manner as the design use of the *J*-integral, it is common to *measure* δ_c (see Chapter 3) and then convert δ_c values to K_c values. These K_c resistance values obtained from CTOD test results are then compared to the driving force in terms of K_I.

2.8 Summary

One of the underlying principles of fracture mechanics is that unstable fracture occurs when the stress intensity factor at the crack tip, K_I, reaches a critical value, K_c. K_I represents the stress intensity ahead of a sharp crack in any material and

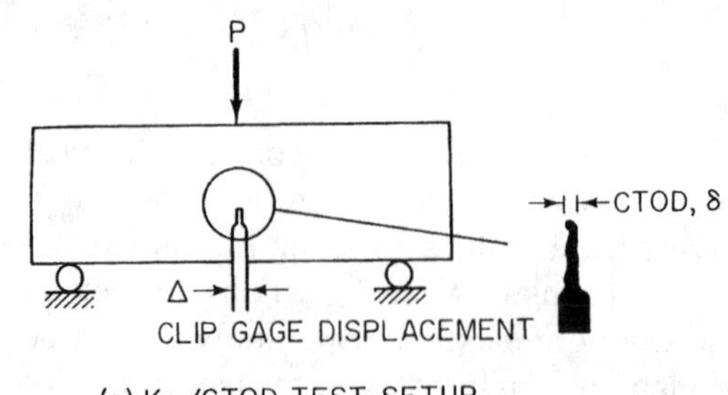

(a) K_{Ic}/CTOD TEST SETUP

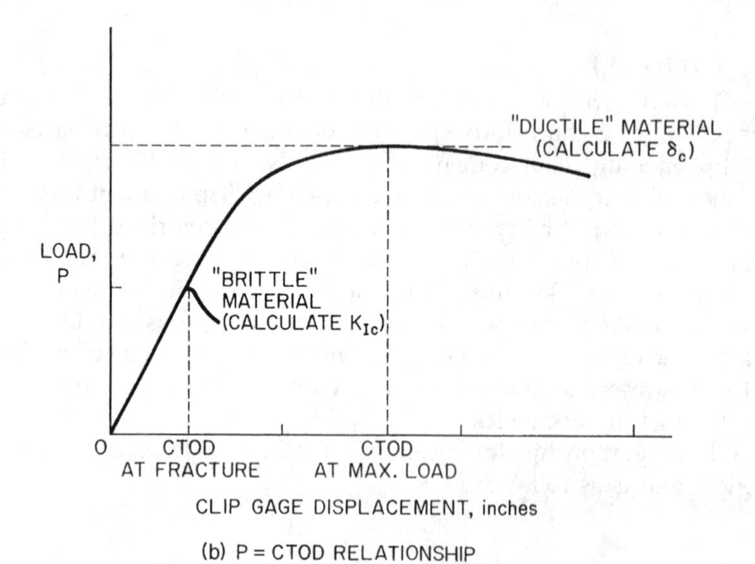

(b) P = CTOD RELATIONSHIP

FIG. 2.22 Relation between K_{Ic} and CTOD test behavior, K_{Ic}-CTOD test setup, and load displacement records from K_{Ic}-CTOD tests, as described in Chapter 3.

is a mathematical calculation based on the crack size and the geometry of the member, as well as the applied load. K_I relations for various crack geometries have been presented in this chapter. K_c represents the fracture toughness of a particular material at a given temperature, loading rate, and constraint level, and must be measured. Procedures used to measure K_c are presented in Chapter 3.

The stress intensity factor, K_I, although based on linear-elastic concepts, is quite useful for calculating driving forces in structures even if the plastic zone size ahead of the crack becomes moderately large. However, most structures are loaded in the elastic range at stresses less than yield. Therefore, as described in this chapter, the use of either K_I or K_{Ieff} as a driving force appears to be a realistic engineering approach.

Elastic-plastic methods, such as the J-integral and the CTOD (crack tip opening displacement, δ) parameter, also can be used to calculate driving forces. However, most engineers will *measure* the fracture toughness in terms of J_c or δ_c and convert those values to a K_c. Then the comparison of driving force and resistance force can be made by using the stress intensity factors K_I and K_c. The advantage of this approach is that K_I, the driving force, is expressed as a function of stress and flaw size, terms familiar to engineers. Also, there are many K_I relations available for various crack geometries and structural configurations.

As mentioned previously, Chapter 3 describes the procedures used to measure K_c under various conditions of loading rate and constraint. Chapter 4 then will describe the effect of temperature, loading rate, and constraint on the resistance force, K_c.

As will be described in Part IV, K_I should be kept below K_c at all times to prevent fracture of members with flaws in the same manner that the applied stress, σ, is kept below the yield strength, σ_{ys}, to prevent yielding in the design of members that do not have flaws.

2.9 References

[1] Griffith, A. A., "The Phenomena of Rupture and Flaw in Solids," *Transactions, Royal Society of London,* Vol. A-221, 1920.

[2] Inglis, C. E., "Stresses in a Plate due to the Presence of Cracks and Sharp Corners," *Proceedings, Institute of Naval Architects,* Vol. 60, 1913.

[3] Irwin, G. R., "Fracture Dynamics," in *Fracturing of Metals,* American Society of Metals, Cleveland, 1948.

[4] Orowan, E., "Fracture Strength of Solids," in *Report of Progress in Physics,* Vol. 12, Physical Society of London, 1949.

[5] Paris, C. P. and Sih, G. C., "Stress Analysis of Cracks," in *Fracture Toughness Testing and Its Applications, ASTM STP 381,* American Society for Testing and Materials, Philadelphia, 1965.

[6] Westergaard, H. M., "Bearing Pressures and Cracks," *Transactions, ASME, Journal of Applied Mechanics,* 1939.

[7] Irwin, G. R., "Analysis of Stresses and Strains Near the End of a Crack Transversing a Plate," *Transactions, ASME, Journal of Applied Mechanics,* Vol. 24, 1957.

[8] Tada, H., Paris, P. C., and Irwin, G. R., Eds., *Stress Analysis of Cracks Handbook,* Del Research Corporation, St. Louis, MO, 1985.

[9] Sih, G. C., *Handbook of Stress-Intensity Factors for Researchers and Engineers,* Institute of Fracture and Solid Mechanics, Lehigh University, Bethlehem, PA, 1973.

[10] Rooke, D. P. and Cartwright, D. J., "Compendium of Stress Intensity Factors," Her Majesty's Stationary Office, London, Hillingdon Press, 1976.

[11] Pilkey, W. D., "Petersons Stress Concentration Factors," Second Edition, John Wiley & Sons Inc., 1997.

[12] Irwin, G. R., "The Crack Extension Force for a Part Through Crack in a Plate," *Transactions, ASME, Journal of Applied Mechanics,* Vol. 29, No. 4, 1962.

[13] Irwin, G. R., "Plastic Zone Near a Crack and Fracture Toughness," *1960 Sagamore Ordnance Materials Conference,* Syracuse University, 1961.

[14] Rice, J. R., "A Path Independent Integral and the Approximate Analysis of Strain Concentration by Notches and Cracks," *Journal of Applied Mechanics, Transactions ASME,* Vol. 35, June 1968.

[15] Wells, A. A., "Unstable Crack Propagation in Metals—Cleavage and Fast Fracture: *Cranfield Crack Propagation Symposium,* Vol. 1, September 1961, p. 210.

[16] Dugdale, D. S., "Yielding of Steel Containing Slits," *Journal of Mechanics and Physics of Solids,* Wiley Interscience, New York, 1963, p. 103.

[17] Sorem, W. A., Dodds, R. H., and Rolfe, S. T., "An Analytical Comparison of Short Crack and Deep Crack CTOD Fracture Specimens of an A36 Steel," *Fracture Mechanics: 21st Symposium, ASTM STP 1074,* J. P. Gudas, J. A. Joyce, and E. Hackett, Eds., American Society for Testing and Materials, Philadelphia, 1990, pp. 3–23.
[18] Wilson, A. D. and Donald, K., "Evaluating Steel Toughness Using Various Elastic-Plastic Fracture Toughness Parameters," *Nonlinear Fracture Mechanics: Volume II: Elastic-Plastic Fracture, ASTM STP 995,* J. D. Landes, A. Saxena, and J. G. Merkle, Eds., American Society for Testing and Materials, Philadelphia, 1989, pp. 144–168.
[19] Dodds, R. H. and Read, D. T., "Elastic-Plastic Response of Highly Deformed Tensile Panels Containing Short Cracks," *ASME Special Publication, Computational Fracture Mechanics—Nonlinear and 3-D Problems,* Vol. 85, PVPD, June 1984, pp. 25–34.
[20] Dodds, R. H., Read, D. T., and Wellman, G. W., "Finite-Element and Experimental Evaluation of the J-Integral for Short Cracks," *Fracture Mechanics, Vol. I: Theory and Analysis, ASTM STP 791,* American Society for Testing and Materials, Philadelphia, 1983, pp. 520–542.
[21] API 579—Recommended Practice for Fitness—For Service, Appendix B: Stress Analysis Overview for a FFS Assessment, Draft in Preparation, 1999.

Appendix

2.10 GRIFFITH, CTOD AND *J*-INTEGRAL THEORIES

2.10.1 *The Griffith Theory*

The first analysis of fracture behavior of components that contain sharp discontinuities was developed by Griffith [*1*]. The analysis was based on the assumption that incipient fractures in ideally brittle materials occur when the magnitude of the elastic energy supplied at the crack tip during an incremental increase in crack length is equal to or greater than the magnitude of the elastic energy at the crack tip during an incremental increase in crack length. This energy approach can be presented best by considering the following example.

Consider an infinite plate of unit thickness that contains a through-thickness crack of length $2a$ (Figure 2.4) and that is subjected to uniform tensile stress, σ, applied at infinity. The total potential energy of the system, U, may be written as:

$$U = U_o - U_a + U_\gamma \tag{A-1}$$

where U_o = initial elastic energy of the uncracked plate,
U_a = decrease in the elastic energy caused by introducing the crack in the plate, and

U_γ = increase in the elastic-surface energy caused by the formation of the crack surfaces.

Griffith used a stress analysis that was developed by Inglis [2] to show that:

$$U_a = \frac{\pi\sigma^2 a^2}{E} \tag{A-2}$$

the elastic-surface energy, U_γ, is equal to the product of the elastic-surface energy of the material, γ_e, and the new surface area of the crack:

$$U_\gamma = 2(2a\gamma_e) \tag{A-3}$$

Consequently, the total elastic energy of the system, U, is:

$$U = U_o - \frac{\pi\sigma^2 a^2}{E} + 4a\gamma_e \tag{A-4}$$

The equilibrium condition for crack extension is obtained by setting the first derivative of U with respect to crack length, a, equal to zero. The resulting equation can be written as:

$$\frac{dU}{da} = 0 - \frac{2a\pi\sigma^2}{E} + 4\gamma_e = 0 \tag{A-5}$$

$$2a\pi\sigma^2 = 4\gamma_e E \tag{A-6}$$

$$\sigma\sqrt{a} = \left(\frac{2\gamma_e E}{\pi}\right)^{1/2} \tag{A-7}$$

which indicates that crack extension in ideally brittle materials is governed by the product of the applied nominal stress and the square root of the crack length as well as by material properties. Because E and γ_e are material properties, the right-hand side of Equation (A-7) is equal to a constant value characteristic of a given ideally brittle material. Consequently, Equation (A-7) indicates that crack extension in such materials occurs when the product $\sigma\sqrt{a}$ attains a critical value. This critical value can be determined experimentally by measuring the fracture stress for a large plate that contains a through-thickness crack of known length and that is subjected to a remotely applied uniform tensile stress. This value can also be measured by using other specimen geometries, which is what makes this approach to fracture analysis so powerful.

Equation (A-7) can be rearranged in the form:

$$\frac{\pi\sigma^2 a}{E} = 2\gamma_e \tag{A-8}$$

The left-hand side has been designated the energy-release rate, G, and represents the elastic energy per unit crack surface area available for infinitesimal crack extension. The right-hand side of Equation (A-8) represents the material's resistance to crack extension, R.

In 1948 Irwin suggested that the Griffith fracture criterion for ideally brittle materials could be modified and applied to brittle materials and to metals that exhibit plastic deformation. A similar modification was proposed by Orowan at about the same time. The modification recognized that a material's resistance to crack extension is equal to the sum of the elastic-surface energy and the plastic-strain work, γ_p, accompanying crack extension. Consequently, Equation (A-8) was modified to:

$$G = \frac{\pi\sigma^2 a}{E} = 2(\gamma_e + \gamma_p) \tag{A-9}$$

Because the left-hand side is the energy-release rate, G, and because $\sigma\sqrt{\pi a}$ represents the intensity, K_I, of the stress field at the tip of a through-thickness crack of length $2a$, the following relation exists between G and K_I for plane-strain conditions,

$$\frac{\pi\sigma^2 a}{E} = G = \frac{K_I^2}{E}(1 - \nu^2) \tag{A-10}$$

The energy-balance approach to crack extension defines the conditions required for instability of an ideally sharp crack. This approach is not applicable to analysis of stable crack extension such as occurs under cyclic-load fluctuation or under stress-corrosion-cracking conditions. However, the stress-intensity parameter, K, is applicable to stable crack extension, and therefore development of linear-elastic fracture-mechanics theory has assisted greatly in improving our understanding of subcritical crack extension and crack instability.

2.10.2 *Crack-Tip Opening Displacement (CTOD) and the Dugdale Model*

The tensile stresses applied to a body that contains a crack tend to open the crack and to displace its surfaces in a direction normal to its plane. For small crack-tip displacements and small plastic deformation at the crack tip, the stress and strain fields in the vicinity of the crack tip can be described by linear-elastic analyses. Under these loading conditions, the fracture instability can be predicted by using the critical plane-strain stress-intensity factor, K_{Ic}.

As the size of the plastic zone and of the displacements at the crack tip increase, the stress and strain distributions in that neighborhood can be characterized better by using elastic-plastic analyses rather than linear-elastic analyses. Wells [15] argued that the opening displacement at the crack tip reflects the strain distribution in that region. He also proposed that fracture would initiate when the strains in the crack-tip region reach a critical value, which can be characterized by a critical crack-tip opening displacement.

Using a crack-tip plasticity model proposed by Dugdale [*16*], referred to as the strip-yield-model analysis, it is possible to relate the CTOD to the applied stress and crack length. The strip yield model consists of a through-thickness crack in an infinite plate that is subjected to a tensile stress normal to the plane of the crack (Figure A.1). The crack is considered to have a length equal to $2a + 2r_y$. At each end of the crack there is a length, r_y, that is subjected to yield-point stresses that tend to close the crack or, in reality, to prevent it from opening. Another way of looking at the behavior of this model is to assume that yield zones of length r_y spread out from the tip of the real crack, a, as the loading is increased. Thus, the displacement at the original crack tip, δ, which is the CTOD, increases as the real crack length increases or as the applied loading increases. The basic relationship developed by Dugdale [*16*] is:

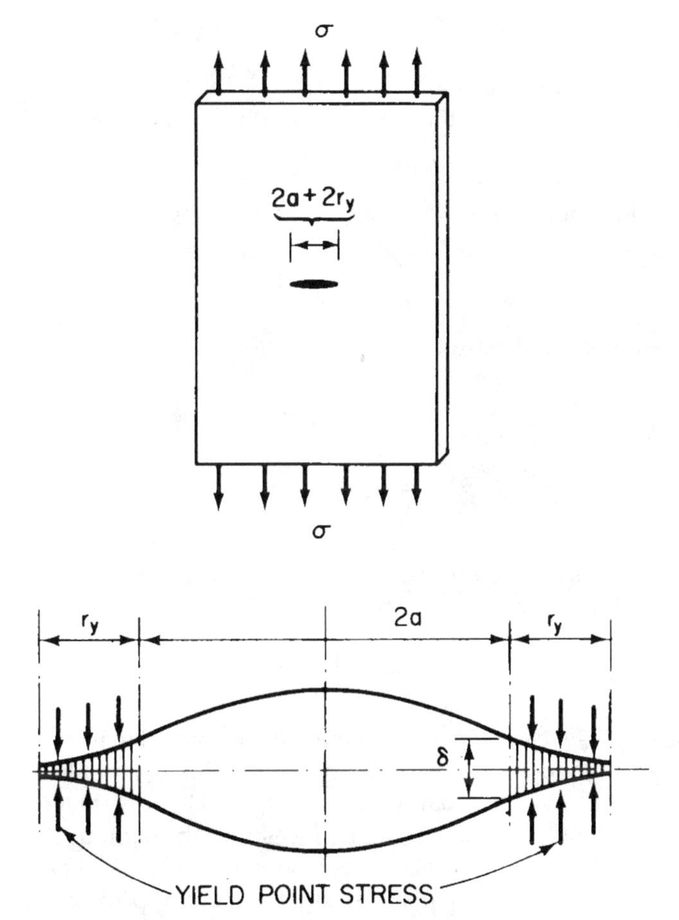

FIG. A.1 Dugdale strip-yield model.

$$\delta = 8\,\frac{\sigma_{ys}a}{\pi E}\ln\sec\left(\frac{\pi}{2}\frac{\sigma}{\sigma_{ys}}\right) \tag{A-11}$$

where σ_{ys} = yield strength of the material, ksi,
a = ½ real crack length, in.,
σ = nominal stress, ksi, and
E = modulus of elasticity of the material, ksi.

Using a series expansion for ln sec $[(\pi/2)(\sigma/\sigma_{ys})]$, this expression becomes:

$$\delta = \frac{8\sigma_{ys}a}{\pi E}\left[\frac{1}{2}\left(\frac{\pi}{2}\frac{\sigma}{\sigma_{ys}}\right)^2 + \frac{1}{12}\left(\frac{\pi}{2}\frac{\sigma}{\sigma_{ys}}\right)^4 + \frac{1}{45}\left(\frac{\pi}{2}\frac{\sigma}{\sigma_{ys}}\right)^6 \cdots\right] \tag{A-12}$$

For nominal stress values less than σ_{ys}, a reasonable approximation for δ, using only the first term of this series, is:

$$\delta = \frac{\pi\sigma^2 a}{E\sigma_{ys}} \tag{A-13}$$

In this chapter, it was shown that, for a through-thickness crack of length $2a$,

$$K_I = \sigma\sqrt{\pi a} \tag{A-14}$$

Thus $\delta E\sigma_{ys} = K_I^2$, and since $E = \sigma_{ys}/\varepsilon_{ys}$, the following relation exists:

$$\frac{\delta}{\varepsilon_{ys}} = \left(\frac{K_I}{\sigma_{ys}}\right)^2 \tag{A-15}$$

Also, the strain energy release rate, G, is:

$$G = \frac{\pi\sigma^2 a}{E} \tag{A-16}$$

Therefore,

$$G = \delta \cdot \sigma_{ys} \tag{A-17}$$

At the onset of crack instability under plane-strain conditions, where K_I reaches K_{Ic} and CTOD reaches a critical value, δ_c,

$$\frac{\delta_c}{\varepsilon_{ys}} = \left(\frac{K_{Ic}}{\sigma_{ys}}\right)^2 \tag{A-18}$$

Because $(K_{Ic}/\delta_{ys})^2$ can be related to the critical crack size in a particular structure, it is reasonable to assume that the parameter $\delta_c/\varepsilon_{ys}$ can likewise be related to the critical crack size in a particular structure. The advantage of the CTOD approach is that the CTOD values can be measured throughout the entire plane-strain, elastic-plastic, and fully plastic behavior regions, whereas K_{Ic} values can be measured only in the elastic plane-strain region or approximated in the early portions of the elastic-plastic region. These regions are described in Chapter 3.

As with the K_I analysis, the application of the CTOD approach to engineering structures requires the measurement of a fracture-toughness parameter, δ_c, which is a material property that is a function of temperature, loading rate, specimen thickness, and specimen geometry. The CTOD test methodology is described in Chapter 3.

2.10.3 J-Integral

For linear-elastic behavior, the J-integral is identical to G, the energy release rate per unit crack extension, described previously in this appendix. Therefore a J-failure criterion for the linear-elastic case is equivalent to the K_{Ic} failure criterion. For linear-elastic plane-strain conditions,

$$J_{Ic} = G_{Ic} = \frac{(1-\nu^2)K_{Ic}^2}{E} \tag{A-19}$$

The CTOD parameter, δ, also is related to J (and K) as follows:

$$\delta = \frac{G}{m\sigma_{ys}} = \frac{K^2}{m\sigma_{ys}E} \tag{A-20}$$

because $J = G$,

$$J \approx m\sigma_{ys}\delta \tag{A-21}$$

where $1 \leq m \leq 2$.

Recent studies [*17,18*] of the relations between J and δ indicate that $\sigma_{flow} = (\sigma_{ys} + \sigma_{ult})/2$ should be used in Equation (A-21). Also, $m = 1.7$ fits most experimental data. Thus,

$$J = 1.7\sigma_{flow}\delta \tag{A-22}$$

is a preferred relation between J and δ.

The J-integral, a mathematical expression, is a line or surface integral that encloses the crack front from one crack surface to the other as shown in Figure A.2. It is used to characterize the local stress-strain field around the crack front for either elastic or elastic-plastic behavior.

The line integral is defined as follows:

$$J = \int_\Gamma W\,dy - T\left(\frac{\partial \bar{u}}{\partial X}\right) ds \tag{A-23}$$

where Γ = any contour surrounding the crack tip as shown in Figure A.2 (note that the integral is evaluated in a counterclockwise manner starting from the lower flat notch surface and continuing along an arbitrary path Γ to the upper flat surface),

W = loading work per unit volume or, for elastic bodies, the strain energy density $= \int_o^\varepsilon \sigma d\varepsilon$,

T = the traction vector at ds defined according to the outward normal n along Γ, $T_i = \sigma_{ij}n_j$,

$\bar{u}$ = displacement vector at ds,

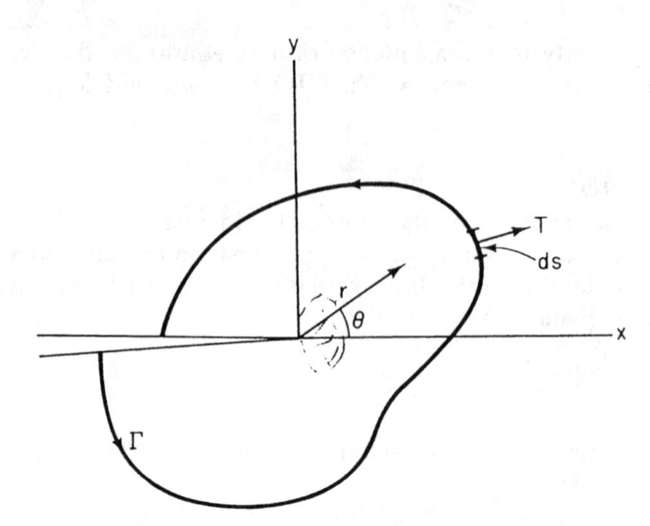

FIG. A.2 Crack-tip coordinate system and typical line integral contour.

ds = arc length along contour Γ,

$T\left(\frac{\partial \bar{u}}{\partial x}\right) ds$ = the rate of work input from the stress field into the area enclosed by Γ.

For any linear-elastic or elastic-plastic material treated by deformation theory of plasticity, Rice [*14*] has proven path independence of the *J*-integral. Dodds et al. [*19,20*] have described experimental and analytical methods for direct evaluation of the *J*-integral.

The most widely used analyses methods to determine the driving force, J_I, are either 2D or 3D finite element analyses, depending on the structure under consideration. API 579—Recommended Practice for Fitness-For-Service [*21*] describes this process in detail. Briefly, the steps involved include:

Step 1 Develop a finite element model of the structural component including all relevant geometry and flaw characteristics.

Step 2 Define all relevant loading conditions and apply them to the model of the structural component.

Step 3 An accurate stress-strain curve of the material used in the structural component should be included in the model.

Step 4 Perform the analysis using an evaluation of the *J*-integral defined in Equation A-23.

Step 5 Measure J_c as described in Chapter 3.

Step 6 As stated previously for fracture control, keep $J_I < J_c$.

Part II: Fracture Behavior

3

Resistance Forces—K_c-J_c-δ_c

3.1 General Overview

IN CHAPTER 2 we described various analytical relationships for determining stress-intensity factors in elastic bodies with different-shaped cracks. These stress-intensity factors are a function of load, crack size, and geometry, but *not* material properties. Ideally, a K_I stress-intensity factor associated with a specific crack geometry can be used to model an actual crack in a real structure. Three of the most common geometries are the edge crack, surface crack, and through-thickness crack, as shown in Figure 3.1.

As stated before, these particular stress-intensity values, K_I, are calculated for different load levels in the same general manner as stresses, σ, are calculated for different load levels in uncracked members subjected to tension. That is, for an uncracked member loaded in tension, the stress is calculated as $\sigma = P/A$ for various loads, P. In an analogous fashion, the stress-intensity factor is calculated in a cracked member as $K_I = C\sigma\sqrt{a}$ for various nominal stress levels, σ, and crack sizes, a. Note that the calculations for a stress-intensity factor, K_I, (analogous to the calculation of stress, σ) are the same for any structural material as long as the general boundary conditions described in Chapter 2 are satisfied.

However, because actual structural materials have certain limiting characteristics (e.g., yielding or fracture, there are limiting values of both σ and K_I, namely, σ_{ys} (yield strength) and K_c (critical stress-intensity factor).

The critical stress-intensity factor, K_c, at which unstable crack growth occurs in a particular material depends on the service conditions of temperature, loading rate, and constraint to a much greater degree than does the yield strength, σ_{ys}. Furthermore, as these service conditions change, e.g., increasing temperature, the fracture behavior of a material (and thus its level of fracture toughness or resistance force) can vary between linear-elastic with no ductility, elastic-plastic with variable ductility, and general yielding with considerable ductility. Examples of

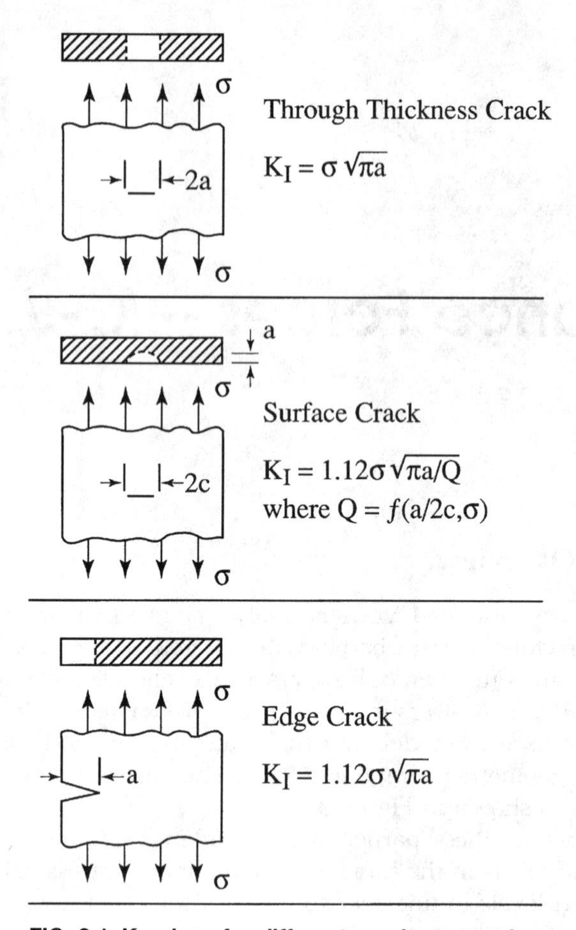

FIG. 3.1 K_I values for different crack geometries.

these three general types of behavior are shown in Figure 3.2, along with their corresponding load-displacement records as shown in Figure 3.3. Obviously there are structural materials that can be produced to high levels of toughness. The cost of these materials generally is higher than for typical structural materials.

The test method that should be used to obtain the critical K_c value for a specific material in a specific application depends on the type of expected (or desired) fracture behavior shown in Figures 3.2 and 3.3

Fracture toughness, K_c, is defined as the resistance to the propagation of a crack in a given test specimen or structural member, i.e., the resistance force. If the conditions in a structural member are modeled in a laboratory specimen, this laboratory specimen can be tested to obtain the appropriate K_c value to be used in the structural condition being analyzed.

The lowest value of fracture toughness occurs under linear-elastic conditions, resulting in sudden brittle fractures, Figure 3.2*a* and Figure 3.3*a*. As the

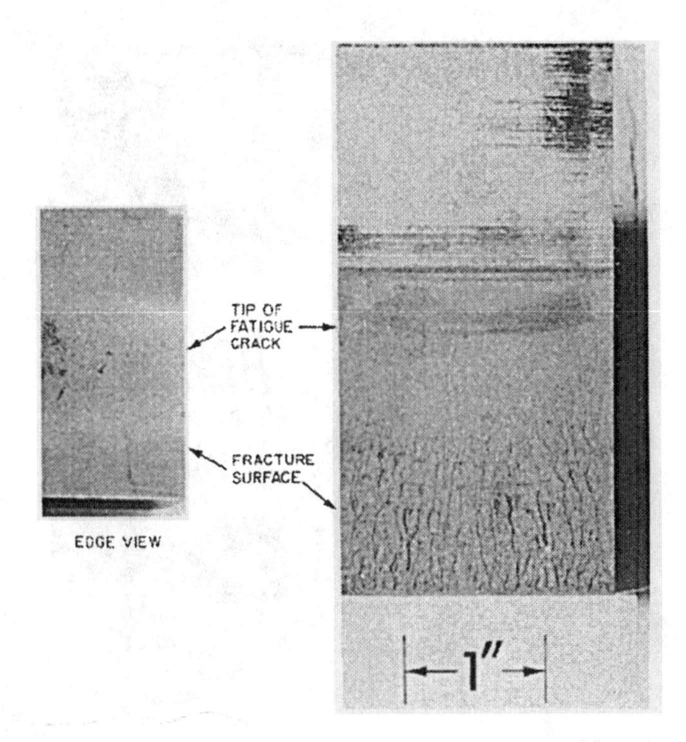

FIG. 3.2a Photograph of linear-elastic fracture surface and edge view.

toughness of a given structural material increases, either by increasing temperature, decreasing loading rate, or decreasing constraint, elastic-plastic behavior occurs, Figures 3.2*b* and 3.3*b*. This behavior includes the development of a large plastic zone ahead of the crack tip and may or may not include stable crack growth. Final fracture usually is by sudden brittle fracture. Further increase in temperature or decrease in loading rate generally result in general yielding or fully plastic behavior, Figure 3.2*c* and Figure 3.3*c*.

These various types of fracture behavior require different fracture tests. This requirement is *different* from determining the yield strength of a structural material, where a standard test specimen is used for all but very unusual service conditions. It should be emphasized that the inherent fracture toughness of a structural material is affected significantly by composition and thermomechanical processing. Thus, as stated earlier, some structural materials can have high levels of fracture toughness for all possible service conditions.

3.2 Service Conditions Affecting Fracture Toughness

Before describing the various ASTM standard fracture tests available to measure resistance forces, we need to discuss the various service conditions.

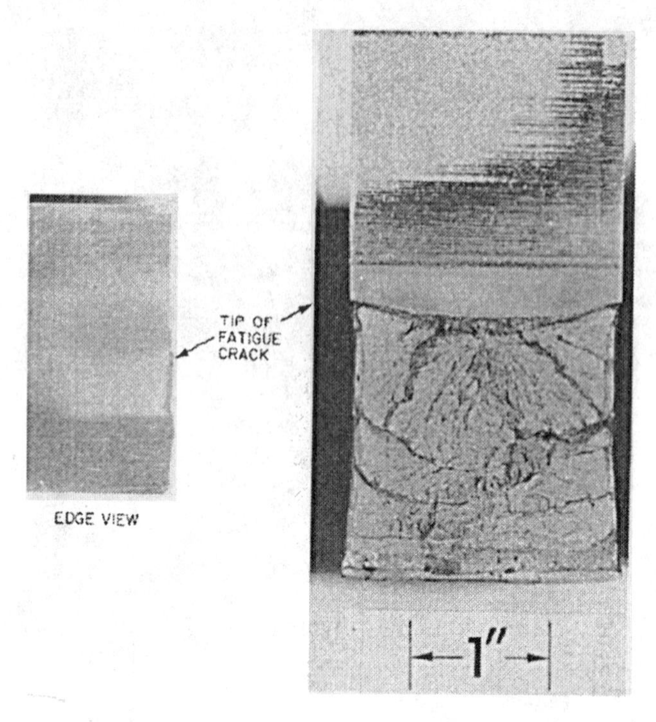

FIG. 3.2b Photograph of elastic-plastic (mixed model) fracture surface and edge view.

3.2.1 Temperature

This is perhaps the easiest to describe as it is merely either the minimum service temperature of a structure or the actual test temperature of a laboratory specimen.

If only one laboratory test temperature is used, it generally is the same as the minimum service temperature. One exception to this general rule occurs when a "loading-rate shift" is made to account for differences in the service loading rate and the laboratory test loading rate. This "loading rate shift" is described in the following section as well as in Chapter 4. However, it is preferable to test specimens at several temperatures to get an idea of how rapidly the fracture toughness changes with temperature.

3.2.2 Loading Rate

Loading rates in most structures can vary from "slow" (maximum load is reached in 10 or more seconds) to dynamic (or impact) where the maximum load is reached in about 0.001 s or less. Intermediate loading rates vary, but a typical one is about 1.0 s to maximum load, such as is found in bridge structures subjected to truck loadings.

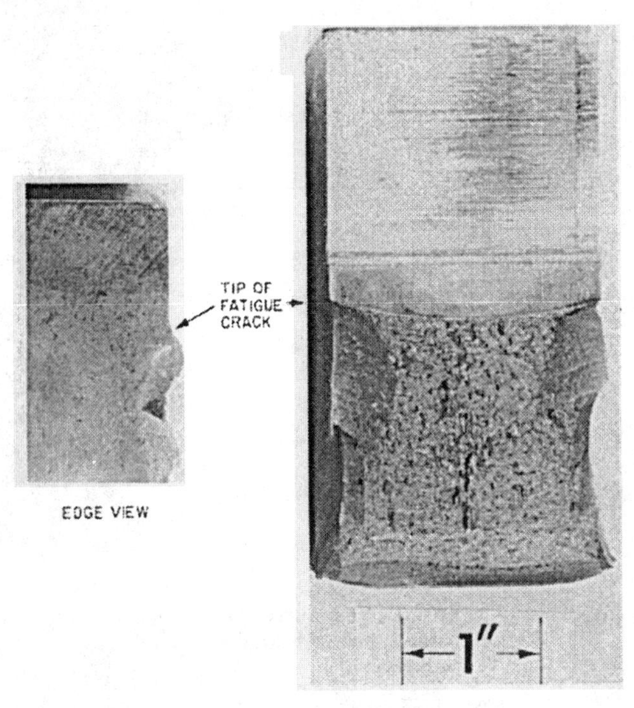

FIG. 3.2c Photograph of general yielding fracture surface and edge view.

In general, these loading rates are defined as follows:

Slow = $\dot{\varepsilon} \approx 10^{-5}$ in./in./s.
Intermediate = $\dot{\varepsilon} \approx 10^{-3}$ in./in./s.
Dynamic = $\dot{\varepsilon} \approx 10$ in./in./s.

where $\dot{\varepsilon}$ is the strain rate (in./in. per second) just ahead of the crack tip as described in Chapter 4. It should be noted that little change in fracture toughness occurs unless the loading rate is varied by at least an order of magnitude.

Figure 3.4 compares these three loading rates. Figure 3.5 shows representative fracture test results for a structural steel tested at three loading rates over a range of temperatures. Note that the temperatures at which the fracture toughness levels begin to increase significantly depend upon the loading rate.

3.2.3 *Constraint*

Of the three primary factors that affect the fracture toughness of a given structural material (i.e., temperature, loading rate, and constraint), constraint is the most difficult to establish quantitatively.

The primary definition of constraint deals with the plane strain to plane stress transition as defined by specimen thickness. Plane strain refers to maxi-

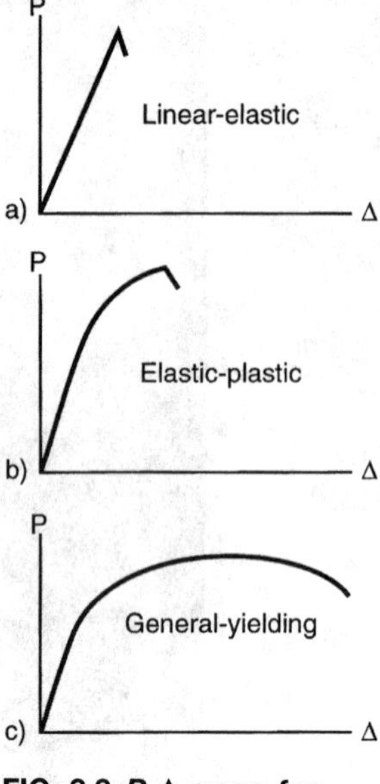

FIG. 3.3 *P*-Δ curve for three types of fracture behavior.

mum constraint and occurs in very thick test specimens that have deep cracks. In contrast, plane stress refers to minimum constraint and occurs in thin test specimens. The distinction between these two extremes has been established by an ASTM standard to define the conditions for K_{Ic}, which is the plane strain fracture toughness value under conditions of slow loading. [1].

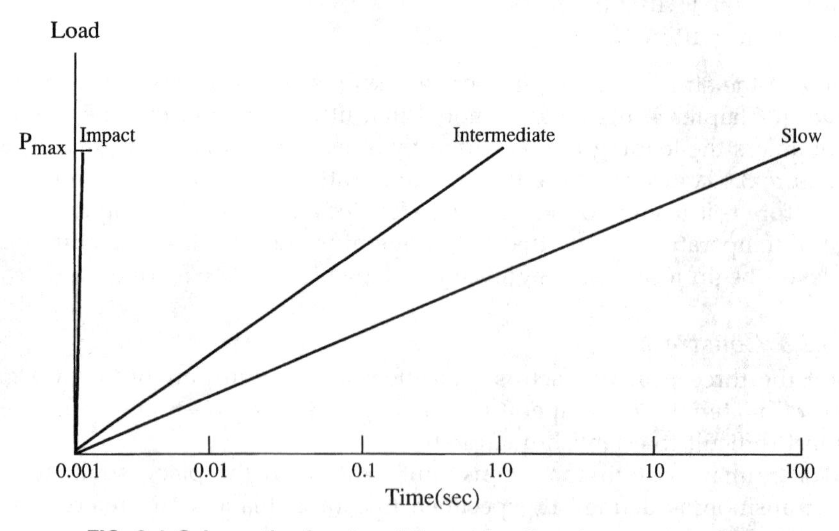

FIG. 3.4 Schematic showing different loading times to fracture.

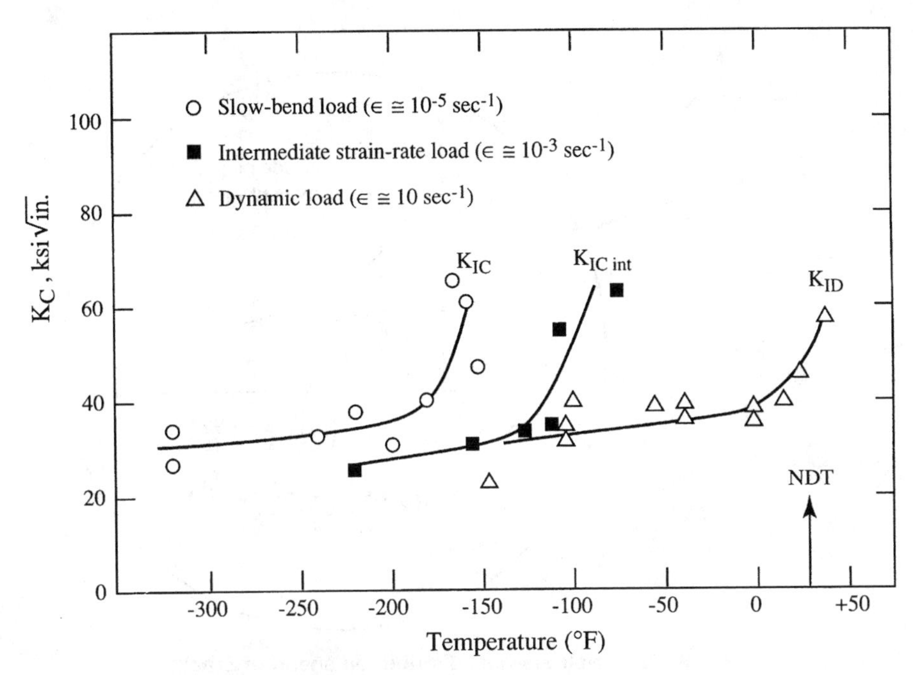

FIG. 3.5 Effect of temperature and strain rate on fracture toughness of a 50-ksi yield strength structural steel.

Figure 3.6 shows the elastic-stress-field distribution ahead of a crack as described in Chapter 2. The extent of the plastic zone ahead of the crack front can be estimated by using the following expression for stress in the y direction:

$$\sigma_y = \frac{K_I}{\sqrt{2\pi r}} \cos\frac{\theta}{2}\left(1 + \sin\frac{\theta}{2}\sin\frac{3\theta}{2}\right) \tag{3.1}$$

for $\theta = 0$ (along the x axis):

$$\sigma_y = \frac{K_I}{\sqrt{2\pi r}} \tag{3.2}$$

Letting σ_y equal to the yield stress, which is the 0.2% offset yield strength of the material at a particular temperature and loading rate, the extent of yielding ahead of the crack is:

$$r_y = \frac{1}{2\pi}\left(\frac{K_I}{\sigma_{ys}}\right)^2 \tag{3.3}$$

At instability, $K_I = K_c$, and the limiting value of r_y, or the plastic zone, is:

$$r_y = \frac{1}{2\pi}\left(\frac{K_c}{\sigma_{ys}}\right)^2 \tag{3.4}$$

This value of r_y is estimated to be the plastic-zone radius at instability under

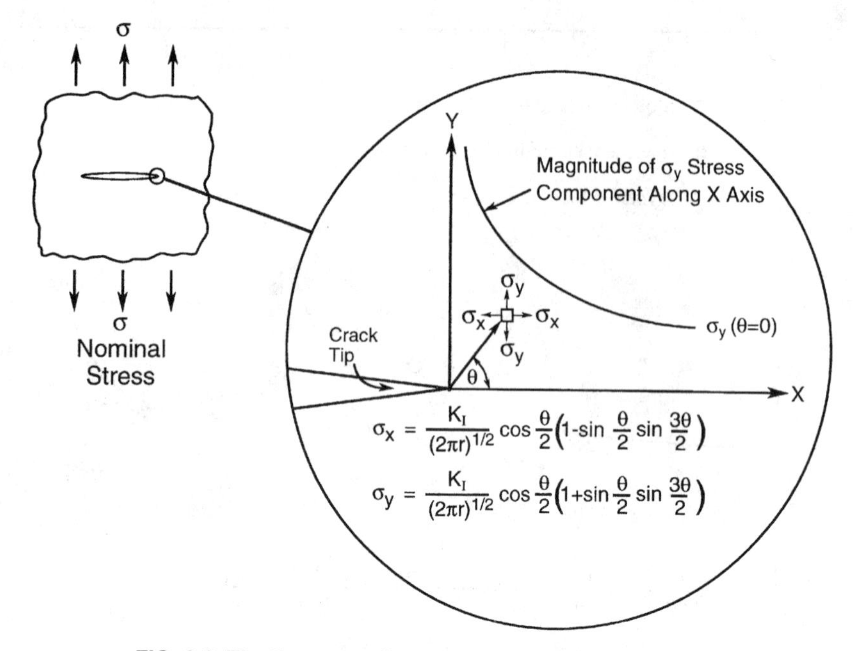

FIG. 3.6 Elastic-stress-field distribution ahead of crack.

plane-stress conditions. As shown in Figure 3.7, this value occurs at the surface of a plate where the lateral constraint is zero and plane-stress conditions exist. The plane-strain yield strength is assumed to be equal to $\sqrt{3}$ times greater than the uniaxial (plane-stress) yield strength. Because of this increase in the tensile stress for plastic yielding under plane-strain conditions, the plastic-zone radius at the center, where the constraint is greater and plane-strain conditions exist, is equal to one-third of this value, or:

$$r_y \text{ (plane strain)} \simeq \frac{1}{6\pi}\left(\frac{K_{Ic}}{\sigma_{ys}}\right)^2 \tag{3.5}$$

where K_{Ic} is the critical stress intensity factor under conditions of maximum constraint, e.g., plane strain.

Thus, the relative plastic-zone size ahead of a sharp crack is proportional to the $(K_{Ic}/\sigma_{ys})^2$ value of a particular structural material subjected to plane-strain constraint.

In establishing the specimen size requirements for plane-strain K_{Ic} tests, the specimen dimensions should be large enough compared with the plastic zone, r_y, so that any effects of the plastic zone on the K_I analysis can be neglected. The pertinent dimensions of plate specimens for K_{Ic} testing are crack length (a), thickness (denoted as B in the ASTM standard [*1*]), and the remaining uncracked ligament length ($W - a$, where W is the overall specimen depth). After consid-

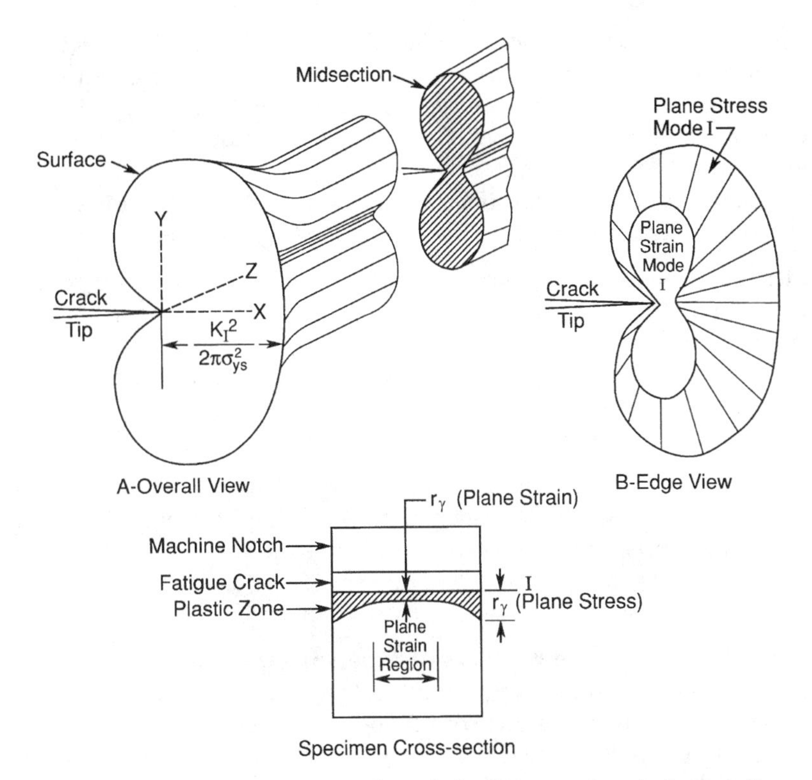

FIG. 3.7 Schematic representation of plastic zone ahead of crack tip.

erable experimental work, the following minimum specimen size requirements to ensure elastic plane-strain behavior were established by an ASTM Committee

$$a \geq 2.5 \left(\frac{K_{Ic}}{\sigma_{ys}}\right)^2 \tag{3.6}$$

$$B \geq 2.5 \left(\frac{K_{Ic}}{\sigma_{ys}}\right)^2 \tag{3.7}$$

$$W \geq 5.0 \left(\frac{K_{Ic}}{\sigma_{ys}}\right)^2 \tag{3.8}$$

The following calculation shows that for specimens meeting this requirement, the specimen thickness is approximately 50 times the radius of the plane-strain plastic-zone size.

$$\frac{\text{specimen thickness}}{\text{plastic-zone size}} = \frac{B}{r_y} \simeq \frac{2.5(K_{Ic}/\sigma_{ys})^2}{(1/6\pi)(K_{Ic}/\sigma_{ys})^2} \simeq 2.5\ (6\pi) \simeq 47 \tag{3.9}$$

Thus, the restriction that the plastic zone be "contained" within an elastic-stress field certainly appears to be satisfied. In fact, during the development of the recommended test method, there was considerable debate about whether or not this requirement was too conservative. Subsequent test results indicate that the requirements are conservative. Currently, these requirements are being re-analyzed by the ASTM Committee.

By adhering to the test specimen dimensions (a, B, W) described above, the following two essential conditions are satisfied for plane strain K_{Ic} behavior.

1. The test specimen is large enough so that linear-elastic behavior of the material being tested occurs over a large enough stress field so that any effect of the plastic zone ahead of the crack can be neglected.
2. There is a triaxial tensile stress field present such that the shear stress is very low compared to the maximum normal stress, and a plane-strain opening mode behavior would be expected (Mode I, Chapter 2).

Note that for a K_{Ic} test specimen to be sized, the K_{Ic} value *to be obtained* should already be known or at least estimated. This is not only a very unusual test specimen requirement, it is very difficult to satisfy.

Two general means of sizing test specimens before the K_{Ic} value is even known are as follows:

1. Estimate the K_{Ic} value on the basis of experience with similar materials and judgement based on other types of notch-toughness tests. In Chapter 5 we will describe various empirical correlations with other types of notch-toughness tests, such as the CVN impact test specimen, that can be used to estimate K_{Ic} values.
2. Use specimens that have as large a thickness as possible, namely, a thickness equal to that of the plates to be used in service.

Fortunately for the structural designer (because such an individual would like his or her structure to be built from materials that do *not* exhibit elastic, plane-strain behavior) but *unfortunately* for the materials engineer responsible for determining K_{Ic} values, many low- to medium-strength structural materials in the section sizes of interest for most large structures (such as ships, bridges, and pressure vessels) are of *insufficient* thickness to maintain plane-strain conditions under slow loading and at normal service temperatures. In those cases, the linear-elastic analysis used to calculate the K_{Ic} test values is invalidated by general yielding and the formation of large plastic zones. Under these conditions, elastic-plastic test methods must be used to measure the fracture toughness, as described later in this chapter. Nonetheless, the basis for most subsequent fracture criteria and fracture control rests on knowing or estimating how much a material exceeds K_{Ic}-type behavior at the service temperature and loading rate.

3.3 ASTM Standard Fracture Tests

There are several standard ASTM test methods developed to measure the various critical stress intensity factors for materials that exhibit different types of fracture behavior. The different types of fracture behavior and the corresponding service conditions are described in this section as well as in Section 3.4, Fracture Behavior Regions. Figure 3.8 is a schematic showing the general regions where each of these methods is used. The specific ASTM test methods referred to in Section 3.4 and Figure 3.8 are as follows:

1. K_{Ic} Plane-strain critical fracture toughness value obtained at slow loading rates. Plane-strain refers to conditions of maximum constraint, e.g., generally thick plates and deep cracks. Fracture is sudden, resulting in unstable brittle fracture with little or no deformation. ASTM Test Method E-399: Standard Test Method for Plane-Strain Fracture Toughness of Metallic Materials [*1*].
2. $K_{Ic(t)}$ Plane-strain critical fracture toughness value obtained at intermediate loading rates, where t = time to maximum load in seconds. This is generally about 1.0 s for intermediate loading rate tests for steel bridges and other structures. Constraint is maximum and failure is sudden, resulting in unstable brittle fracture with little or no deformation. ASTM Test Method E-399—Annex A7. Special Requirements for Rapid Load Plane-Strain Fracture Toughness $K_{Ic(t)}$ Testing [2].

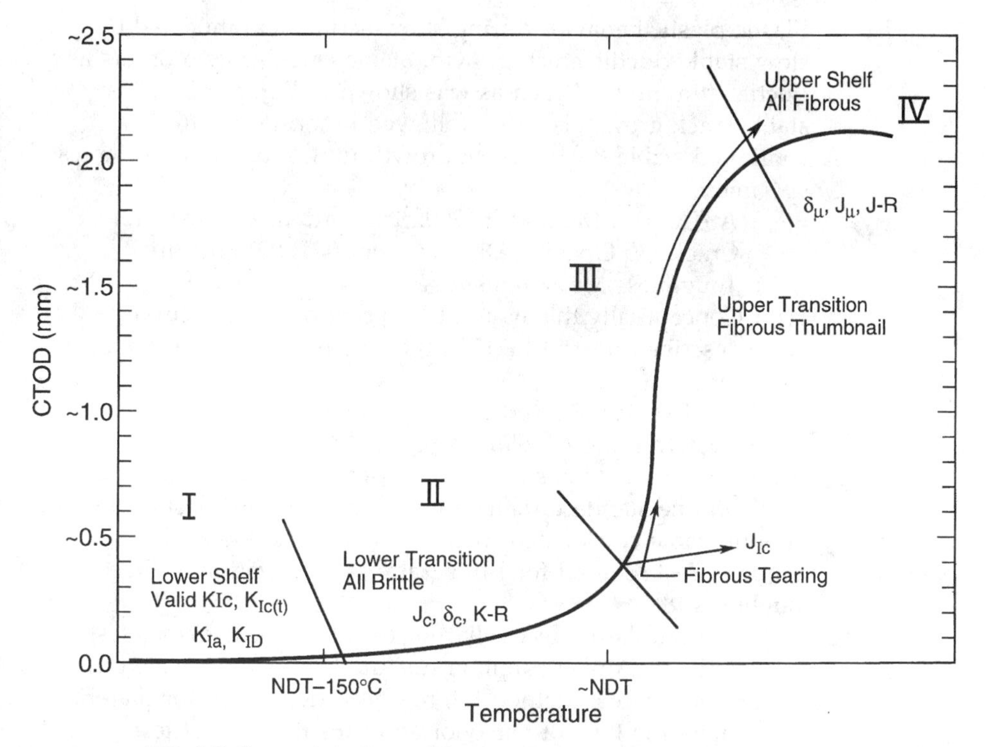

FIG. 3.8 General regions of fracture behavior for structural steels.

3. K_{Ia} (K_{ID}) Linear-elastic behavior during dynamic or impact loading results in rapid unstable brittle fracture. Referred to as K_{Ia}, the plane-strain crack-arrest toughness. $K_{I\ Dynamic}$ (K_{Id}) is assumed to be equivalent to $K_{I\ arrest}$ although there is no standard K_{Id} test method. ASTM Test Method E-1221: Standard Test Method for Determining Plane-Strain Crack-Arrest Fracture Toughness, K_{Ia}, of Ferritic Steels [3].
4. δ_c, J_c, *K-R* Elastic-plastic plane stress behavior during slow-loading accompanied by plastic zone development, but not stable crack growth. Failure is by rapid unstable brittle fracture.
   - (δ_c) ASTM Test Method E-1290: ASTM Standard Test Method for Crack-tip Opening Displacement (CTOD) Fracture Toughness Measurement [4].
   - (*K-R*) ASTM Test Method E 561: Standard Practice for R-Curve Determination [5].
   - (J_c) ASTM Test Method E-1737: Standard Test Method for *J*-Integral Characterization of Fracture Toughness [8].
5. J_{Ic} Critical value of the *J*-integral that describes the stress-strain field ahead of a crack. J_{Ic} is a measure of the fracture toughness at the onset of slow stable crack extension. Behavior is non-linear elastic plastic. ASTM Test Method E-813: Standard Test Method for J_{IC}, A Measure of Fracture Toughness [6].
6. δ_U, J_U, *J-R* Elastic-plastic behavior during slow-loading accompanied by slow stable ductile crack growth. Stable crack growth occurs as a ductile "thumbnail" such as was shown in Figure 3.2*b*. This stable crack growth is either followed by brittle fracture or continued stable ductile crack growth until separation of the test specimen.
   - (δ_U) ASTM Test Method E-1290: Standard Test Method for Crack-Tip Opening Displacement (CTOD) Fracture Toughness Measurement [4].
   - (J_U) Conceptually this would be a point on the *J-R* curve described next in E-1152 [7] as there is no single standard for J_U.
   - (*J-R*) ASTM Test Method E-1152: Standard Test Method for Determining *J-R* Curves [7].
7. J_c, J_{Ic}, *J-R* A new test method has been developed to cover all *J*-integral test results in one standard. Behavior would be elastic-plastic with or without stable crack extension. ASTM Test Method E-1737: Standard Test Method for *J*-Integral Characterization of Fracture Toughness [8].
8. *K*, *J*, *CTOD* (δ) This standard effectively replaces all of the previous test methods. A new common fracture test method, called the Standard Test Method [9], has been developed for materials where the type of behavior and thus the type of test needed

also is not known before testing. A bend or compact specimen is tested and the P-Δ_{CMOD} and P-Δ_{LLD} records, where CMOD is the crack mouth opening displacement and LLD is the load line displacement, are analyzed as either K, J, or δ values, depending on the test records. The standard is: ASTM Test Method E-1820-96: Standard Test Method for Measurement of Fracture Toughness [*9*].

The equations used to analyze K, J, or δ are described in Appendix A of this chapter.

9. K_{Jc} This test method covers the determination of a reference temperature, T_o, that characterizes the fracture toughness of ferritic steels that experience onset of cleavage cracking at elastic, or elastic-plastic K_{Jc} instabilities, or both [*10*]. This method treats the statistical effects of specimen size on K_{Jc} in the transition range using the weakest link theory applied to a three-parameter Weibull distribution of fracture toughness values [*11*]. Accordingly, it has advantages in dealing with the variability of test results. The test procedure is described briefly in Appendix B of this chapter. The test standard is: ASTM Test Method E-1921-97: Standard Test Method for the Determination of Reference Temperature T_o, for Ferritic Steels in the Transition Range [*10*].

3.4 Fracture Behavior Regions

The general regions over which each of these types of tests are used is shown schematically in Figure 3.8. This schematic shows the general increase in fracture toughness with increasing temperature for typical structural, ship, bridge, pressure vessel, etc. steels.

All behavior in Region I is essentially brittle, such as was shown in Figures 3.2*a* and 3.3*a*, regardless of loading rate. Materials that exhibit this behavior should be avoided if possible, as the toughness levels generally do not provide adequate margins of safety except under closely controlled service conditions.

Behavior in Region II is elastic-plastic with increasing plastic zone sizes but not stable crack extension. Final fracture is rapid brittle fracture. Depending on service conditions, this behavior generally is satisfactory for many structural applications. This behavior would be similar to that shown in Figures 3.2*b* and 3.3*b* but with no stable crack growth (ductile thumbnail) prior to the final brittle or mixed mode fast fracture shown in Figure 3.2*b*.

Behavior in Region III is elastic-plastic with large plastic zones and increasing amounts of ductile tearing (Figures 3.2*b* and 3.3*b*) followed by unstable mixed mode fast fracture, but only after considerable deformation. Stable crack growth, i.e., thumbnail behavior, would occur.

Behavior in Region IV is general yielding, and specimens exhibit stable tearing at fracture as shown in Figures 3.2*c* and 3.3*c*.

K_{Ic}, $K_{Ic(t)}$, and K_{Ia} values can be measured only in Region I. The size requirements established in ASTM E 399 are fairly conservative, and thus this region is generally fairly small. J_c and δ_U, as well as *K-R* values, are measured in Region II because there is no stable crack growth. Region III is defined as the start of various amounts of stable crack growth. Because J_{Ic} is defined as the fracture toughness level at the onset of stable crack growth, it defines the start of Region III. The distinction between Region III and IV is not well defined as J_U, δ_U, and *J-R* tests are conducted well into Region IV also.

Some materials, such as structural aluminums and titaniums, do not exhibit temperature or loading rate effects. Thus, rather than exhibit a change in behavior with changes in temperature or loading rate, these materials would exhibit the different types of behavior shown in Figures 3.2 and 3.3 at all temperatures and loading rates on the basis of their inherent metallurgical properties. Different test methods would still need to be used, depending on the particular levels of inherent fracture toughness of the metal. For example, if a particular aluminum had very low fracture toughness, K_{Ic} tests would be valid, where as if it had a high level of toughness, perhaps only J_U or δ_U tests could be conducted at all test temperatures.

3.5 General ASTM Fracture Test Methodology

All fracture mechanics test specimens have much in common. Accordingly, the general test methodology will be described in this section. Specific test procedures are found in the various ASTM Test Standards described in Section 3.3.

The generic steps in fracture testing are as follows:

3.5.1 Test Specimen Size

As described in the various ASTM standards, there are many types of specimens, e.g., bend, compact tension, arc, disk, etc. However, the two most common specimens used are the three-point bend specimen shown in Figure 3.9 and the compact-tension specimen shown in Figure 3.10.

The first step in testing a given structural material is to determine the size of the primary specimen dimensions of thickness (*B*), depth or width (*W*), and

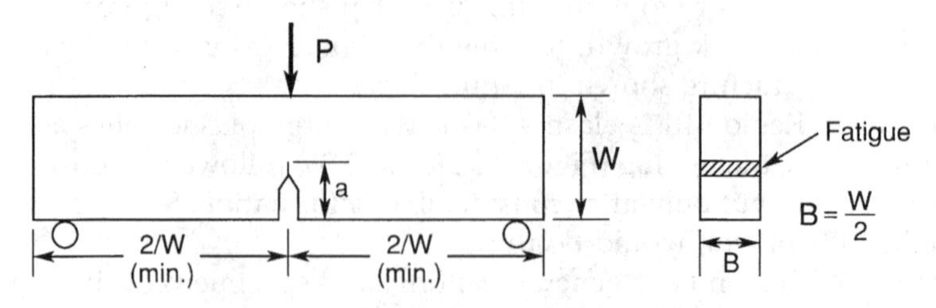

FIG. 3.9 Three-point bend specimen.

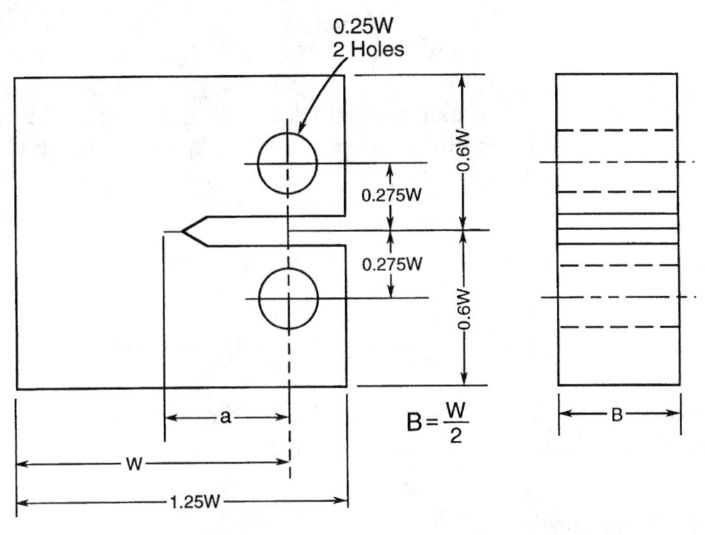

FIG. 3.10 Compact tension specimen.

crack length (a), as shown in Figures 3.9 and 3.10. The remaining specimen dimensions of length L and height H are of less importance, but still are prescribed in each ASTM test standard.

Ideally, the specimen thickness, B, should meet a given requirement such as the ones given below for K_{Ic} testing as described in E 399:

$$a = \text{crack depth} \geq 2.5\left(\frac{K_{Ic}}{\sigma_{ys}}\right)^2 \tag{3.10}$$

$$B = \text{specimen thickness} \geq 2.5\left(\frac{K_{Ic}}{\sigma_{ys}}\right)^2 \tag{3.11}$$

$$W = \text{specimen width} \geq 5.0\left(\frac{K_{Ic}}{\sigma_{ys}}\right)^2 \tag{3.12}$$

However, generally it is difficult to meet these specific size requirements. Thus, the engineer should size the test specimen for the plate thickness of interest, B. After this is done, W generally is set equal to $2B$, and the final crack length, a, generally is set equal to B.

As noted by rearranging Equation (3.11), the maximum K_{Ic} capacity that can be measured for a specimen of thickness B is:

$$K_{Ic} = \sqrt{\frac{B}{2.5}}\,\sigma_{ys} \tag{3.11}$$

From ASTM E 1921, the maximum K_{Jc} capacity of a J-integral specimen is:

$$K_{Jc} = \left(\frac{Eb_o\sigma_{ys}}{30}\right)^{1/2} \tag{3.13}$$

where b_o is $W - a$. For a specimen with thickness B and width of $2B$, $b_o = B$.

Comparing the measurement capacities of K_{Ic} to K_{Jc} for a 1-in.-thick specimen of 50-ksi yield strength steel,

$$K_{Ic} = \sqrt{\frac{1}{2.5}} \cdot 50 = 32 \text{ ksi } \sqrt{\text{in.}}$$

$$K_{Jc} = \left(\frac{30 \times 10^6 \cdot 1 \cdot 50}{30}\right)^{1/2} = 220 \text{ ksi } \sqrt{\text{in.}}$$

Thus, the $K_{CRITICAL}$ measurement capacity (in terms of K) for a J-integral test specimen is about seven times that of a K_{Ic} test specimen.

3.5.2 Test Specimen Notch

The purpose of notching the test specimen is to simulate an ideal plane crack with essentially zero root radius to agree with the assumptions made in the K_I analysis in Chapter 2. Because a fatigue crack is considered to be the sharpest crack that can be reproduced in the laboratory, the machined notch is extended by fatigue. The fatigue crack should extend at least 0.05 W ahead of the machined notch to eliminate any effects of the geometry of the machined notch.

The fatigue cracking procedure is a very important part of the preparation of the test specimens. The specifics of each test method should be followed closely so that there is as little effect as possible from the fatigue cracking process on the test result.

3.5.3 Test Fixtures and Instrumentation

The test fixtures used in a particular test should be designed to minimize friction during the test, e.g., use rollers and/or pins, as shown in Figures 3.11 and 3.12. Specifics of rollers and/or pin geometry and tolerances are given in each ASTM test method.

Preferably, the test machine used should be one in which either load or displacement can be controlled. Generally, fatigue cracking is done in load control, whereas most fracture tests are conducted under displacement control.

Instrumentation must be available that will continuously record load as well as one or more displacement measurements during the test. Load is usually obtained from the load cell on the testing machine.

The two types of displacement measurements that are required, depending on the type of test are crack mouth opening displacement (CMOD) or load-line displacement (LLD).

The CMOD is measured at the mouth of the crack (on the bottom surface) as shown in Figure 3.13*a*. An example of a Load-CMOD (P-Δ) curve is shown in

FIG. 3.11 Three-point bend test setup.

FIG. 3.12 Compact-specimen K_{Ic} test setup.

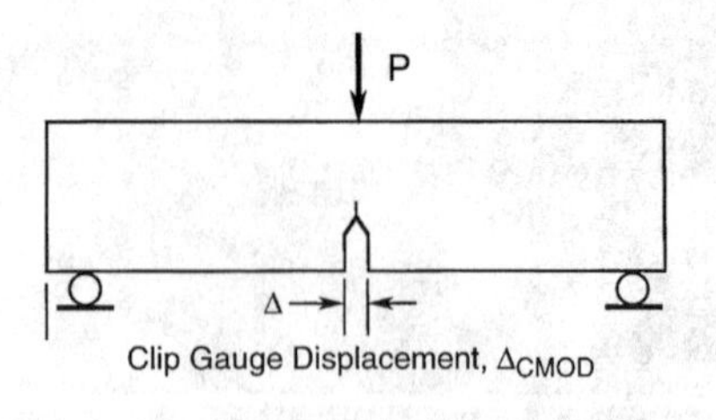

(a) K_{Ic} Test Setup

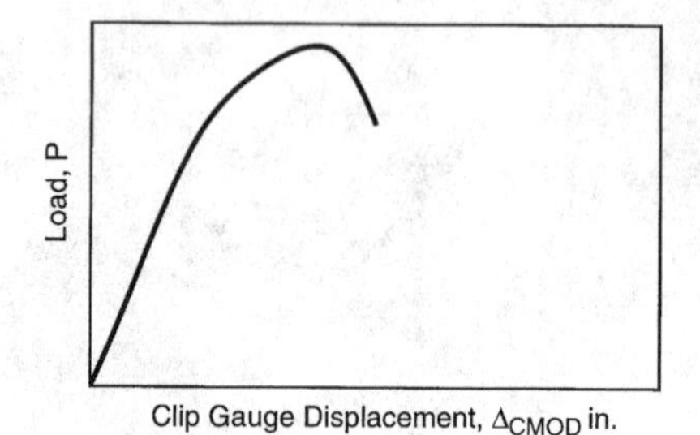

(b) Load Displacement Record For Fracture Test

FIG. 3.13 Schematic showing displacement measurement for K_{Ic} test: (*a*) K_{Ic} test setup; (*b*) load displacement record.

Figure 3.13*b*. Details of a representative CMOD gage are shown in Figure 3.14. This type gage, details of which are described in the various ASTM standards, will give a very accurate measurement of the movement of the two crack faces at the edge of a specimen. A test record consisting of an autographic plot of the output of the load-sensing transducer versus the output of the displacement gage should be obtained. The initial slope of the linear portion should be between 0.7 and 1.5. It is conventional to plot the load along the vertical axis, as in an ordinary tension test record. Select a combination of load-sensing transducer and autographic recorder so that the maximum load can be determined from the test record with an accuracy of ± 1 percent. With any given equipment, the accuracy of readout will be greater the larger the scale of the test record. Continue the test until the specimen can sustain no further increase in load. The use of the P-Δ record to analyze the test results is described later.

The load-line displacement is measured in the direction of the applied load so that a P-LLD (load-line displacement) record can be obtained and an energy analysis can be made to measure a J-integral test value. Normally a sensitive gage such as is shown in Figure 3.14 is used to measure the displacement of the crack tip in the exact direction of the load, as shown in Figure 3.15.

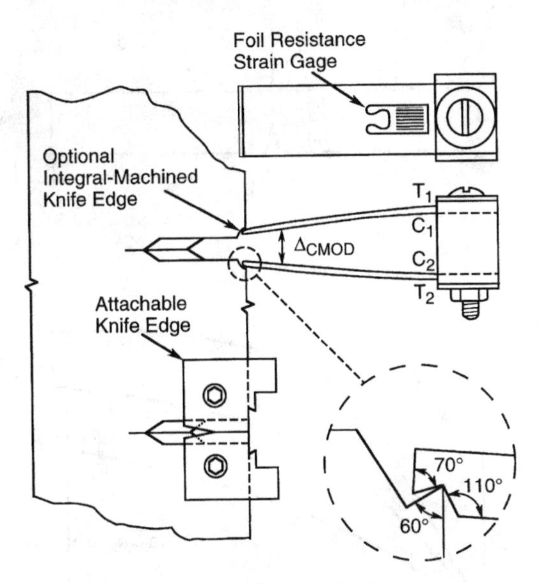

FIG. 3.14 Double cantilever clip-in displacement gage.

The loading rate is specified so that the test is either slow or intermediate, as established by the test method being used.

3.5.4 Analysis of Results

Although the appearance of the fracture surface of a completed fracture test is very useful in qualitatively analyzing the result of any fracture test (Figures 3.2*a,b,c*), the real analysis is of the particular load-displacement record (Figures 3.3*a,b,c*). The two types of records that are obtained are a load–crack-mouth-opening displacement (P-Δ_{CMOD}) or a load–load-line displacement (P-Δ_{LLD}). Recording both measurements in any test increases the analysis options.

The specific analysis of each record is outlined briefly below. For a complete description of each record analysis, a detailed review of the particular ASTM test standard is required.

1. K_{Ic}, $K_{Ic(t)}$, K_{Ia} If the P-Δ_{CMOD} is linear-elastic (Figure 3.16*a*), and the particular K_{Ic} size and other requirements of the standard are satisfied, a K_{Ic}, $K_{Ic(t)}$, or K_{Ia} value can be obtained from this type test record.
2. δ_c, δ_U, $K_{I\text{-}R}$ these P-Δ_{CMOD} records are non-linear (Figure 3.16*b*) and indicate either moderate to large plastic zone development or stable crack growth prior to fast fracture. δ_c values occur before stable crack growth and δ_U values occur after stable crack growth, Figure 3.16*b*. Note that J_C values, which are based on an energy

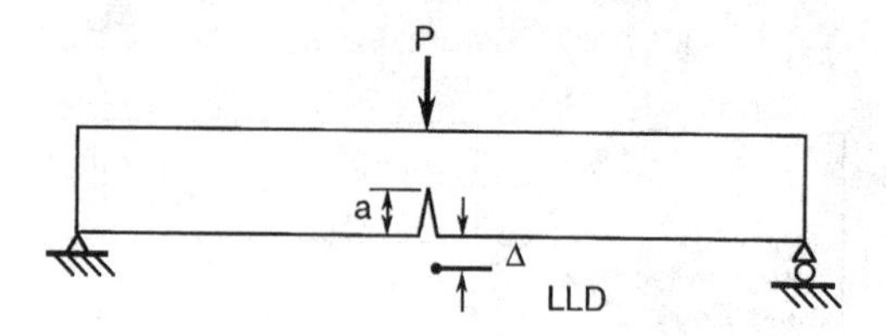

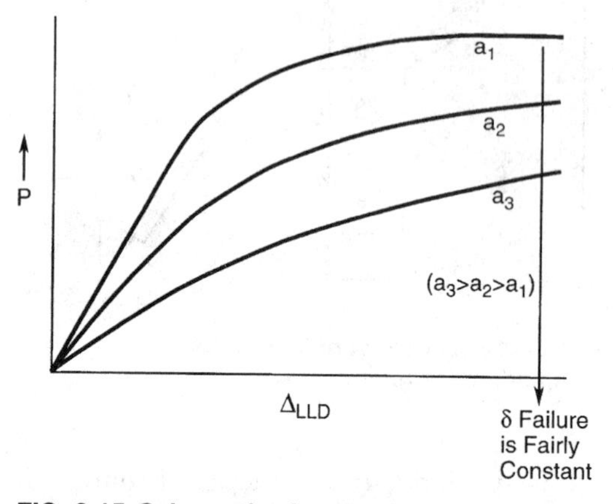

FIG. 3.15 Schematic showing load-displacement curves for various crack depths.

analysis, require a P-Δ_{LLD} measurement.

Both the δ_c and δ_U analyses (as well as other δ parameters described in the test standards) use the CMOD measurement to estimate the crack-tip opening displacement (CTOD). A schematic showing the relation between CMOD and CTOD is shown in Figure 3.17. It is the CTOD value that is used in analyses and specifications regarding the fracture resistance of a particular structural member.

3. J_c, J_{Ic} This test requires a P-Δ_{LLD} record and either multiple specimens, or, more typically, a contained partial unloading of the test specimen so that the onset of stable crack growth can be determined. This is done by a careful analysis of the changes in slope of the unloading portion of the P-Δ_{LLD} record as shown in Figure 3.16*c*.

4. J_C, J_U, J-R As stated earlier, all J fracture tests require P-Δ_{LLD} records because the J-Integral analysis is basically an energy analysis. All J test records are analyzed to obtain values of J (related to the area under the P-Δ_{LLD} curve, Figure 3.16*c*, vs. crack extension, Δa). The crack extension, Δa, is obtained from the slope of the partial unloading measurements, e.g., the slope of the curve is related to

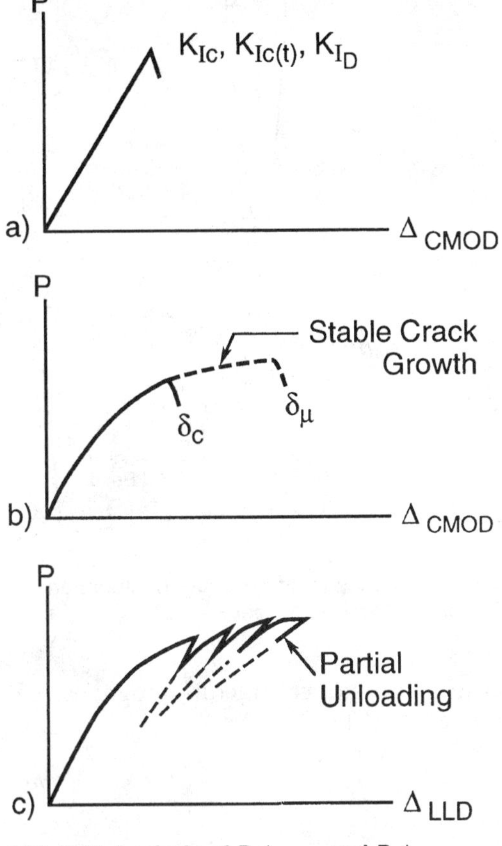

FIG. 3.16 Analysis of P-Δ_{CMOD} and P-Δ_{LLD} records.

the length of the crack. Other techniques sometimes are used to measure crack extension, such as an electrical potential crack growth procedure. These are described in detail in the ASTM Test Standards.

Because it is not always possible to anticipate the fracture behavior of a structural material before the test is completed, it is preferable to obtain *both* P-Δ_{CMOD} and P-Δ_{LLD} records for all tests. Then, the appropriate record can be used to analyze the appropriate fracture parameter.

3.6 Relations Between *K*-*J*-δ

Several theoretical relationships between the various fracture-mechanics parameters exist. In the linear-elastic regime, the critical-strain energy-release rate, G_{Ic}

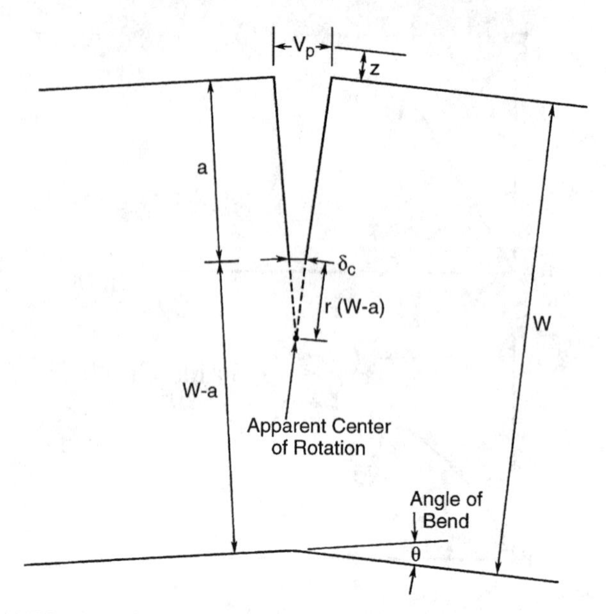

FIG. 3.17 Assumed rotation of bend specimen during plastic portion of test $r = 0.4$.

(Chapter 2), is equal to J_{Ic}, the critical fracture toughness value at the initiation of crack growth for metallic materials.

Furthermore, because

$$K_{Ic} = \sqrt{\frac{EG_{Ic}}{1 - \nu^2}} \tag{3.14}$$

for plane strain, it follows that:

$$K_{Ic} = \sqrt{\frac{E\,J_{Ic}}{1 - \nu^2}} \tag{3.15}$$

for plane strain. Also,

$$K_c \cong \sqrt{EJ_c} \tag{3.16}$$

for plane stress.

In Chapter 2 it was shown that

$$G = \frac{\pi\sigma^2 a}{E} \tag{3.17}$$

for a wide plate with a center crack of length $2a$.

In the Appendix of Chapter 2, it was shown that the crack-tip opening displacement, δ, is equal to:

$$\delta = \frac{\pi\sigma^2 a}{E\sigma_{ys}} \tag{3.18}$$

for a wide plate with center crack of length $2a$. Thus,

$$G = \delta\sigma_{ys} \tag{3.19}$$

Because

$$K_I = \sigma\sqrt{\pi a} \tag{3.20}$$

for this case,

$$\delta = \frac{K_I^2}{E\sigma_{ys}} \tag{3.21}$$

Also, because $E = \sigma_{ys}/\varepsilon_{ys}$,

$$\frac{\delta}{\varepsilon_{ys}} = \left(\frac{K_I}{\sigma_{ys}}\right)^2 \tag{3.22}$$

At the onset of crack instability, where K_I equals K_c and δ is equal to the critical value, δ_c,

$$\frac{\delta_c}{\varepsilon_{ys}} = \left(\frac{K_c}{\sigma_{ys}}\right)^2 \tag{3.23}$$

Equation (3.21) can be rewritten as follows:

$$K_I = \sqrt{E\sigma_{ys}\delta} \tag{3.24}$$

Because actual specimens are neither completely plane strain nor plane stress, a constraint factor, m, is introduced into Equation (3.24) to account for the actual state of stress in the specimen at failure.

$$K_c = \sqrt{mE\sigma_{ys}\delta_c} \tag{3.25}$$

Finite-element analyses [*11*] as well as experimental test results have shown that for moderate constraint levels between the limits of plane strain and plane stress, a constraint factor m of about 1.7 is a good value to use. Furthermore, Wellman [*12*] and Wilson and Donald [*13*] have shown that the use of the flow stress

$$\sigma_{flow} = \frac{\sigma_{ys} + \sigma_{ult}}{2} \tag{3.26}$$

gives a better correlation between K_c and CTOD values than the use of σ_{ys}. Hence the suggested relation between CTOD and K_c is

$$K_c = \sqrt{1.7\, E\sigma_{flow}\delta_c} \tag{3.27}$$

Because $K_c = \sqrt{EJ_c}$, it follows that

$$J_c = 1.7\, \sigma_{flow}\, \delta_c \tag{3.28}$$

Regardless of the particular fracture test conducted, the relations can be used to estimate a critical resistance force in terms of the stress-intensity factor, K_c. Thus, fracture control can be approached in the following manner:

1. Calculate the driving force K_I (Chapter 2).
2. Measure the resistance force in the appropriate K, J, δ test (Chapter 3).
3. Use the above relations to obtain the resistance force in terms of an equivalent K_c value.
4. Keep $K_I < K_c$ to prevent fracture in the same manner as the stress, σ, is kept less than σ_{ys} to prevent yielding.

This is obviously a simplification of the overall fracture control process in actual structures; however, it is the basic starting point. Part IV discusses fracture control in more detail.

As has been stated previously, critical K, J, δ resistance forces are affected by temperature, loading rate, and constraint. The effects of these service conditions on the resistance force of structural materials will be discussed in Chapter 4.

Finally, common sense tells us that if a material has a high (or low) level of fracture toughness in one of the numerous fracture mechanics type tests, then the material probably will exhibit a high (or low) level of notch toughness in a less-expensive test such as the widely used Charpy V-notch impact test. Thus, empirical relations ought to exist between fracture mechanics test results and less-expensive conventional fracture toughness tests. These relations do exist and are discussed in Chapter 5.

3.7 References

[1] E 399-90 (Reapproved 1997). Standard Test Method for Plane-Strain Fracture Toughness of Metallic Materials, ASTM, Vol. 03.01.

[2] E 399-A-7. Special Requirements for Rapid-Load Plane-Strain Fracture Toughness $K_{IC}(t)$ Testing, ASTM, Vol. 03.01.

[3] E 1221-96. Standard Test Method for Determining Plane-Strain Crack-Arrest Fracture Toughness, K_{Ia}, of Ferritic Steels, ASTM, Vol. 03.01.

[4] E 1290-93. Standard Test Method for Crack-Tip Opening Displacement (CTOD) Fracture Toughness Measurement, ASTM, Vol. 03.01.

[5] E 561-94. Standard Practice for R-Curve Determination, ASTM, Vol. 03.01.

[6] E 813-89. Standard Test Method for J_{IC}, A Measure of Fracture Toughness, ASTM, Vol. 03.01.

[7] E 1152-95. Standard Test Method for Determining J-R Curves, ASTM, Vol. 03.01.

[8] E 1737-96. Standard Test Method for J-Integral Characterization of Fracture Toughness, ASTM, Vol. 03.01.

[9] E 1820-96. Standard Test Method for Measurement of Fracture Toughness, ASTM, Vol. 03.01.

[10] E 1921-97. Standard Test Method for Determination of Reference Temperature, T_o, for Ferritic Steels in the Transition Range.

[11] Anderson, T. L., Steinstra, D., and Dodds, R. H., "A Theoretical Framework for Addressing Fracture in the Ductile-Brittle Transition Region," *Fracture Mechanics, 24th Volume, ASTM STP 1207*, ASTM, 1994, pp. 185–214.

[12] Wellman, G. W. and Rolfe, S. T., "Three Dimensional Elastic-Plastic Finite Element Analysis of Three-Point Bend Specimen," *Fracture Mechanics: Sixteenth Symposium, ASTM STP 868*, ASTM, 1985, pp. 214–237.

[13] Wilson, A. D. and Donald, K., "Evaluating Steel Toughness Using Various Elastic-Plastic Fracture Toughness Parameters," *Nonlinear Fracture Mechanics: Volume II: Elastic-Plastic Fracture, ASTM STP 995*, J. D. Landes, A. Saxena, and J. G. Merkle, Eds., American Society for Testing and Materials, Philadelphia, 1989, pp. 144–168.
[14] McCabe, D. E., Merkle, J. G., and Nanstad, R. K., "A Perspective on Transition Temperature and K_{Jc} Data Characterization," *Fracture Mechanics: Twenty-Fourth Volume, ASTM STP 1207*, 1994, pp. 215–232.

3.8 APPENDIX A

K, J, CTOD (δ) Standard Test Method E 1820

Often, the type of fracture behavior (Figure 3.3) is not known before testing, and thus the type of test needed also is not known. Fracture specimens can still be tested, *P*-CMOD and *P*-LLD records obtained, and the results analyzed in one of three ways, depending on the type of record obtained. That is, for linear elastic behavior, analyze for K_{Ic}. For non-linear elastic-plastic behavior, analyze for J or CTOD (δ).

Calculation of *K*

If the *P*-CMOD record is linear elastic (Figure 3.3*a*), calculate a conditional K_Q value that involves the construction of a 5% secant offset line on the *P*-CMOD record. For the bend specimen at a load, P_Q, calculate K_Q as follows:

$$K_Q = \left[\frac{P_Q S}{(BB_N)^{1/2} W^{3/2}}\right] f(a_i/W) \tag{A-1}$$

where:

$$f(a_i/W) = \frac{3(a_i/W)^{1/2}\,[1.99 - (a_i/W)(1 - a_i/W) \; x(2.15 - 3.93(a_i/W) + 2.7\,(a_i/W)^2)]}{2(1 + 2a_i/W)(1 - a_i/W)^{3/2}} \tag{A-2}$$

If K_Q meets all of the requirements of E 1820-A5, Method for K_{Ic} Determination, then $K_Q = K_{Ic}$.

For most structural materials, however, this will generally not be the case, and J and/or CTOD (δ) analyses must be used.

Calculation of *J*

The J test method uses the *P*-LLD test record (Figure 3.15). For the single edge-notch bend specimen:

$$J = J_{elastic} + J_{plastic} \tag{A-3}$$

where $J_{elastic}$ = elastic component of J, and
$J_{plastic}$ = plastic component of J.

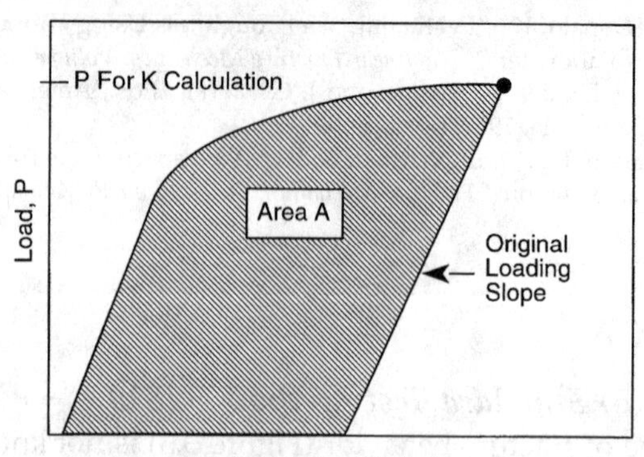

FIG. 3.A1 Definition of area for *J* calculation using the basic method.

At a point corresponding to v and P on the specimen load versus load-line displacement, Figure 3.A1, calculate J as follows:

$$J = \frac{K^2(1 - \upsilon^2)}{E} + J_{p\ell} \tag{A-4}$$

where K is calculated as discussed in the previous section with $a = a_o$, and

$$J_{p\ell} = \frac{2A_{p\ell}}{B_N b_o} \tag{A-5}$$

where $A_{p\ell}$ = area as shown in Figure 3A.1,
B_N = net specimen thickness ($B_N = B$ if no side grooves are present), and
$b_o = W - a_o$.

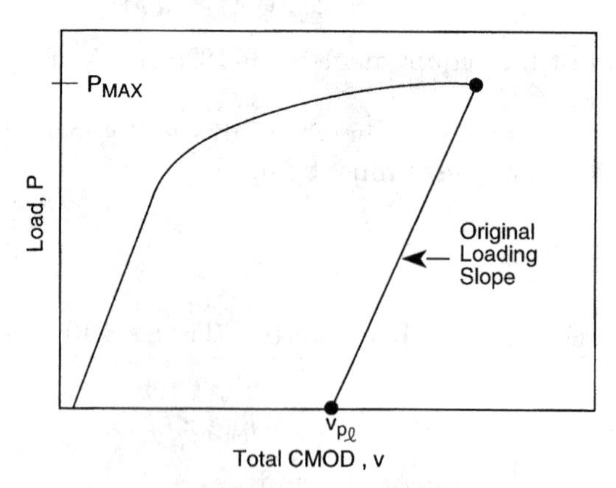

FIG. 3.A2 Definition of $V_{p\ell}$ and P_{MAX} for *K* calculation.

Calculation of CTOD (δ)

Calculation of CTOD for a point on the load-CMOD curve (Figure 3.A2) are made as follows:

$$\delta = \frac{K^2 (1 - \upsilon^2)}{2\sigma_{ys} E} + \frac{r_p (W - a_o) V_{p\ell}}{[r_p (W - a_o) + a_o + z]} \tag{A-6}$$

where a_0 = original crack length,
K = stress intensity factor as defined in Equation (A-1) with $a = a_0$ and $P = P_{max}$,
ν = Poisson's ration,
σ_{ys} = yield stress or 0.2% offset yield strength at the temperature of interest,
E = elastic modulus at the test temperature,
$V_{p\ell}$ = plastic component of crack mouth opening displacement at the point of evaluation on the load-displacement curve,
Z = distance of knife edge measurement point from the notched edge on the single edge bend specimen, and
r_p = plastic rotation factor = 0.44.

If stable crack growth has occurred, the current crack length, a_i, is used plus the value of Δa as described in the test standard [9].

3.9 APPENDIX B

Reference Temperature T_o, to Establish a Master Curve Using K_{Jc} Values In Standard Test Method E 1921

This test method uses statistical analysis to establish a K_{Jc} value from J_c test results at a reference temperature, T_o. This reference temperature is the temperature at which the K_{Jc} value is 100 MPa $\sqrt{m}$ (90.9 ksi$\sqrt{in.}$). Knowing the reference temperature, T_o, a master curve that describes the shape and location of the K_{Jc} transition temperature fracture toughness for 1-in.-thick specimens of ferritic steels can be established using the following equation:

$$K_{Jc\ (median)} = 30 + 70 \exp[0.019(T - T_o)] \tag{B-1}$$

where K_{Jc} = MPa $\sqrt{in.}$,
T = test temperature, °C, and
T_o = reference temperature, °C.

This master curve is similar to the ASME Section XI K_{Ic} and K_{IR} lower-bound design curves described in Chapter 14. Figure 3.B1 shows a master curve developed using this standard compared with the ASME Section XI K_{Ic} and K_{IR} lower bound design curves [14].

The test temperature, T_o, at which K_{Jc} is about 100 MPa $\sqrt{in.}$ (90.9 ksi $\sqrt{in.}$) can be estimated by determining the CVN impact temperature at

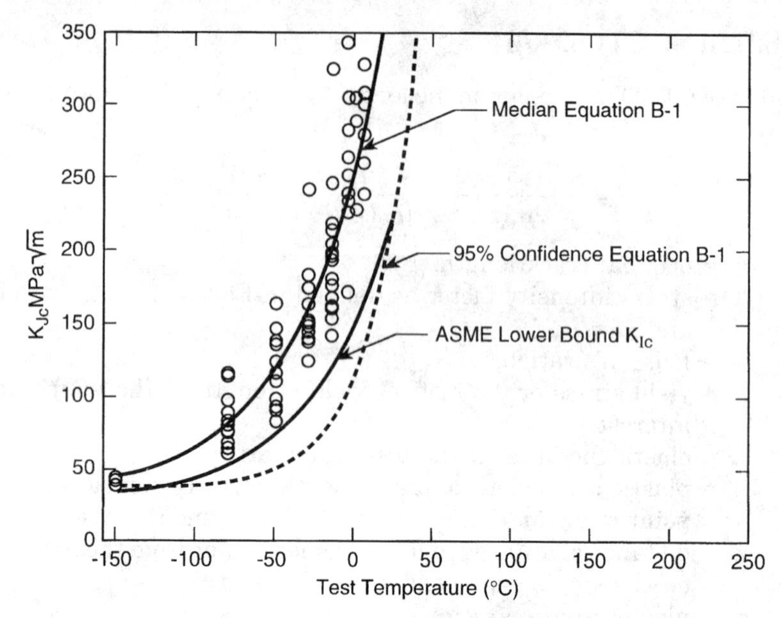

FIG. 3.B1 Data from HSST Fifth Irradiation Series normalized to ITC 9T equivalent data. Median curve fit, 95% lower-bond curve, and ASME lower-bound curve for RT_{NDT} = −34°C. (Ref. *14*).

28 J. Knowing that temperature, T_{28J}, the estimate of the T_o temperature can be made as follows:

T_o (estimate) $= T_{28J} + C$ where C depends on the J test specimen size as described in E 1921.

Knowing an estimate of T_o from CVN test results, six J_c specimens are tested as described in Appendix A. A Weibull analysis then is conducted to establish K_{Jc} at T_o.

4

Effects of Temperature, Loading Rate, and Constraint

4.1 Introduction

IN CHAPTER 3, VARIOUS fracture test methods for determining the fracture toughness under conditions of slow, intermediate, and dynamic loading rates were described. Generally, fracture toughness values are referred to as K_c, the *critical* stress intensity factor. For specific situations, there are specific K_c values, such as K_{Ic}, for slow-loading under plane strain conditions as defined by ASTM E 399. The fracture toughness under intermediate-load rate conditions is referred to as K_{Ic} (t), where the time to maximum load is given in the parentheses. The dynamic crack arrest fracture toughness is K_{Ia}, and the critical plane-strain stress-intensity factor under conditions of impact loading is referred to as K_{Id}. Furthermore, the K_{Ic}, K_{Ic} (t), and K_{Ia} tests are frequently conducted at various temperatures to determine the "static," "intermediate," and "dynamic" fracture toughness of various structural materials as a function of temperature.

The fact that the inherent fracture toughness of most structural steels increases with increasing test temperature is well known. This increase has been measured using various notch-toughness specimens such as the Charpy V-notch impact specimen, and it is certainly reasonable to expect a similar increase using fracture-mechanics-type test specimens. What is not so widely known is the fact that the same inherent fracture toughness can decrease significantly with increasing loading rate, that is, K_{Id} can be smaller than K_{Ic} at the same test temperature. Also, testing plates thinner than those required for plane-strain values may result in plane-stress K_c fracture toughness values that are considerably higher than the K_{Ic} values. Thus, before the engineer can use fracture-toughness values in design, fracture control, failure analysis, or fitness for service, the critical fracture-toughness value for the particular service temperature, loading rate, and constraint level must be known. In this chapter we shall describe the general effects of these three variables on the fracture toughness of various structural materials.

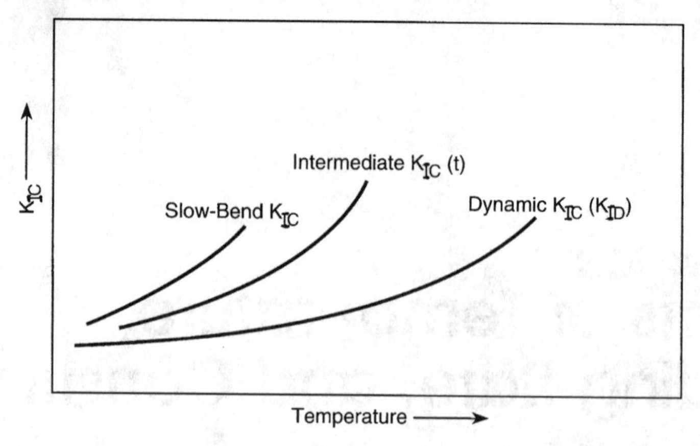

FIG. 4.1 Schematic showing effect of temperature and loading rate of K_{Ic}.

4.2 Effects of Temperature and Loading Rate on K_{Ic}, $K_{Ic}(t)$, and K_{Id}

In general, the fracture toughness of structural materials, particularly steels, increases with increasing temperature and decreasing loading rate. These two general types of behavior are shown schematically in Figures 4.1 and 4.2. Figure 4.1 shows that both K_{Ic} and K_{Id} increase with increasing test temperature. However, for any given temperature, the fracture toughness measured in an impact test, $K_{Id,}$ generally is lower than the fracture toughness measured in either of the other two types of tests. Figure 4.2 shows that, at a constant temperature, fracture-toughness tests conducted at higher loading rates generally result in lower fracture toughness values. Actual test results for three structural steels are shown in Figures 4.3, 4.4, and 4.5.

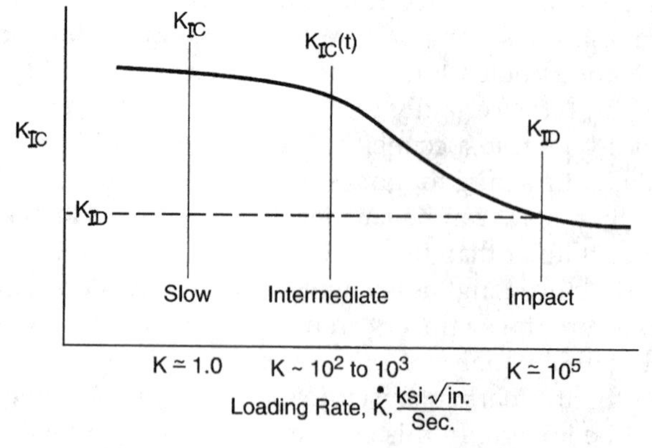

FIG. 4.2 Effect of loading rate on K_{Ic}.

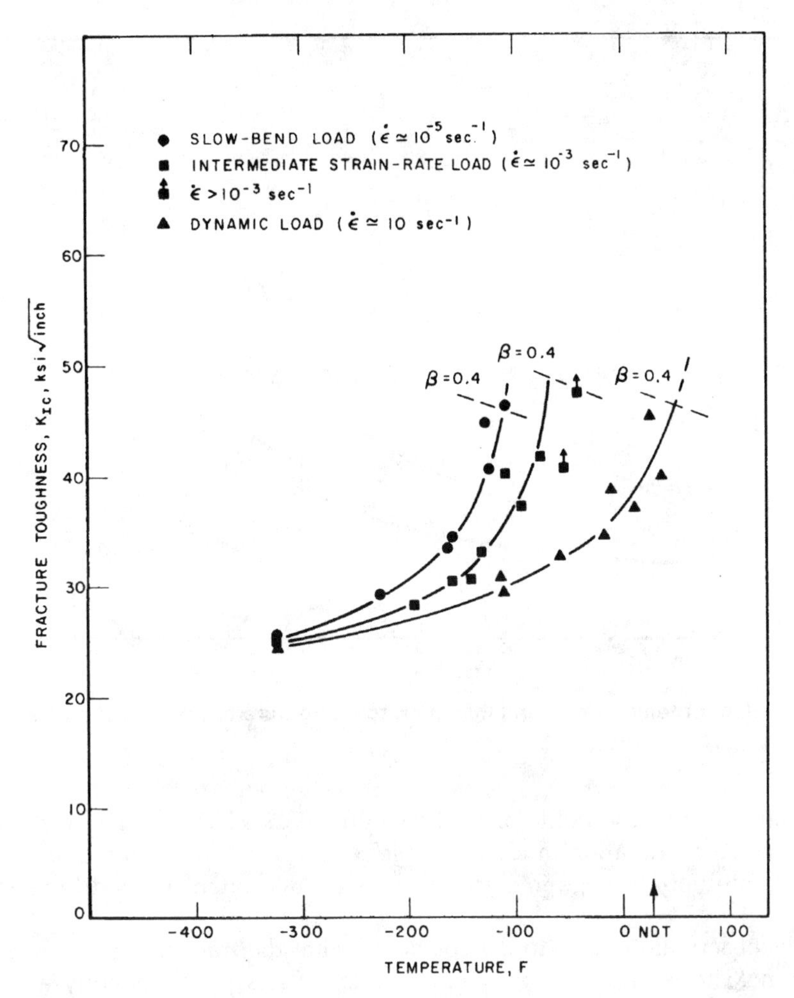

FIG. 4.3 Effect of temperature and loading rate on fracture toughness of an A36 steel.

This effect of temperature and loading rate also is observed with Charpy V-notch test results, as shown schematically in Figure 4.6. Figure 4.7 shows Charpy V-notch test results for the same A36 structural steel described in Figure 4.3. Thus, the transition from brittle to ductile behavior begins at lower temperatures for specimens tested at slow loading rates compared with specimens tested at impact loading rates, for both fracture-mechanics-type tests, as well as the more conventional CVN tests.

Fractographic analyses of numerous fracture surfaces show that the fracture-toughness transition behavior is associated with the onset of change in the microscopic-fracture mode at the crack tip. At the low end of the transition-

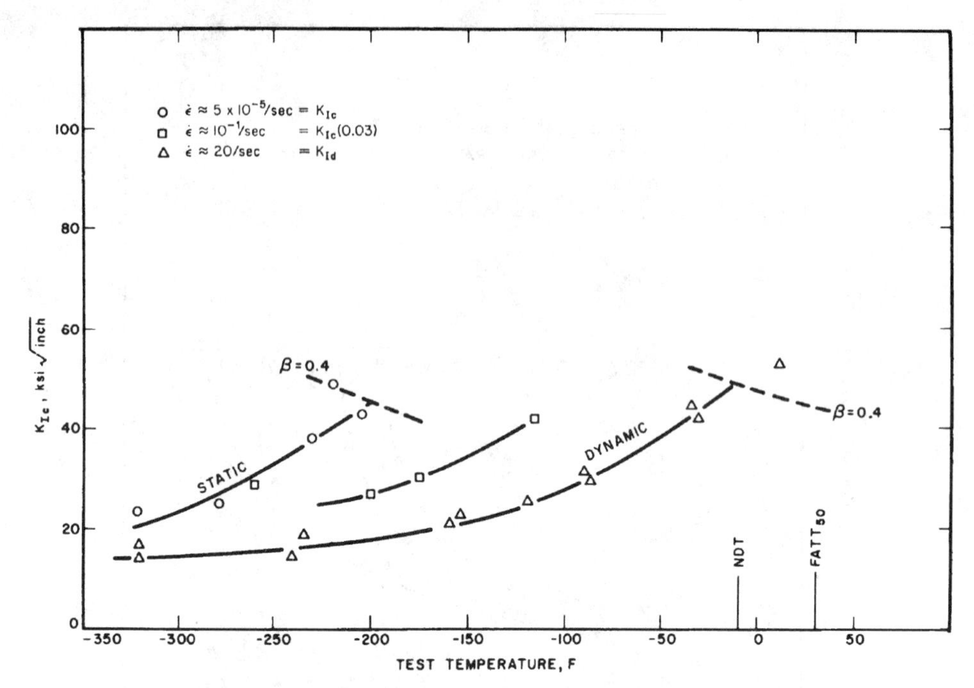

FIG. 4.4 Effect of temperature and loading rate on fracture toughness of an ABS-C steel.

temperature range, the mode of fracture initiation is cleavage, and at the upper end the fracture initiation mode is ductile tear. In the transition-temperature region, a continuous change in fracture mode occurs, often referred to as mixed-mode.

This observation leads to the conclusion that the transition-temperature behavior in K_c tests and Charpy V-notch tests reflects predominantly a gradual change in the microscopic mode of fracture from cleavage at very low temperatures to tear dimples at the upper-shelf region of the CVN test results.

4.3 Effect of Loading Rate on Fracture Toughness

K_{Ic} tests are conducted at "slow" loading rates such that the time to maximum load is in the range of about 10–60 s. Specifically, the loading rate is specified by ASTM to be within the range 30–150 ksi$\sqrt{\text{in.}}$/min. Because some structural materials are strain-rate sensitive, their fracture toughness at faster loading rates can be quite different from that measured in a "slow" K_{Ic} test. Low-strength structural steels exhibit a large change in fracture toughness for different loading rates, as was shown in Figures 4.3, 4.4, and 4.5. These figures show critical stress-intensity factor test results conducted according to ASTM E 399 (K_{Ic} tests), K_{Ic} tests conducted at intermediate strain rates (K_{Ic} (t)), and K_{Ic} tests conducted at

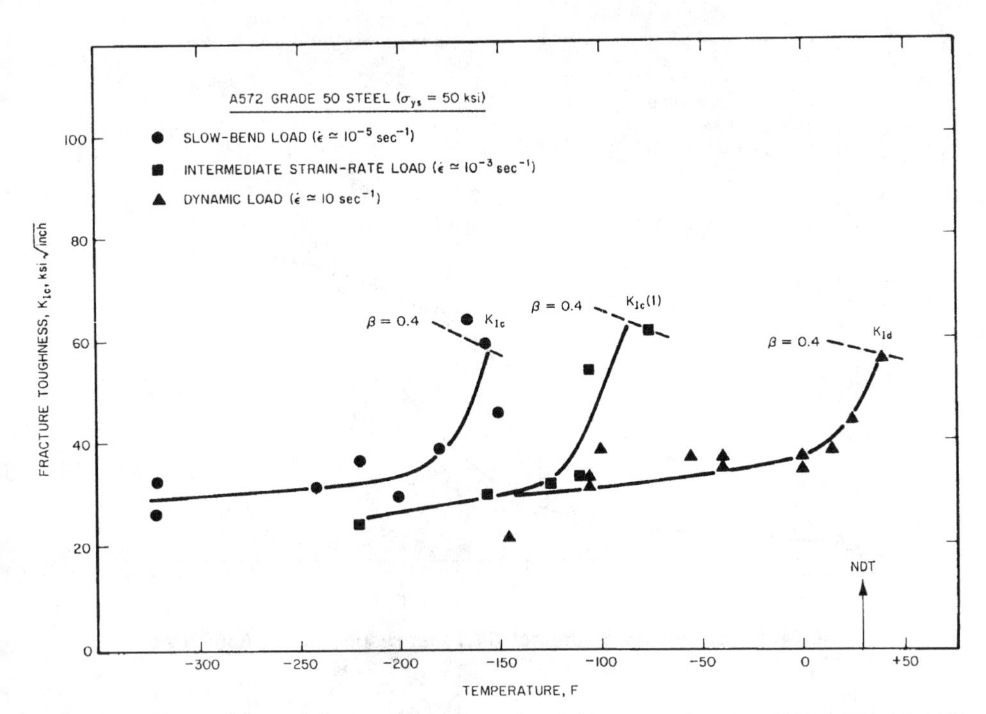

FIG. 4.5 Effect of temperature and strain rate of fracture toughness of an A572 Grade 50 steel (σ_{ys} = 50 ksi).

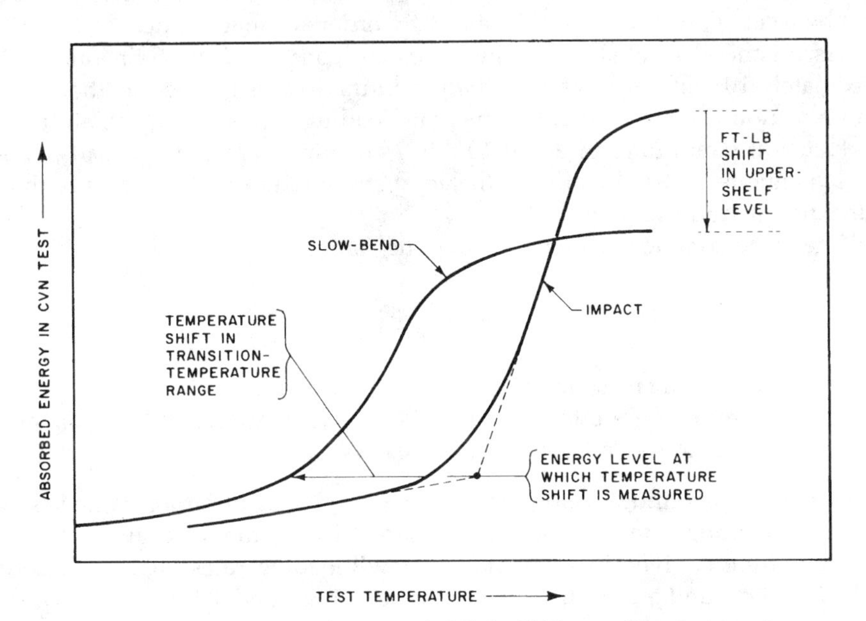

FIG. 4.6 Schematic representation of shift in CVN transition temperature and upper-shelf level due to strain rate.

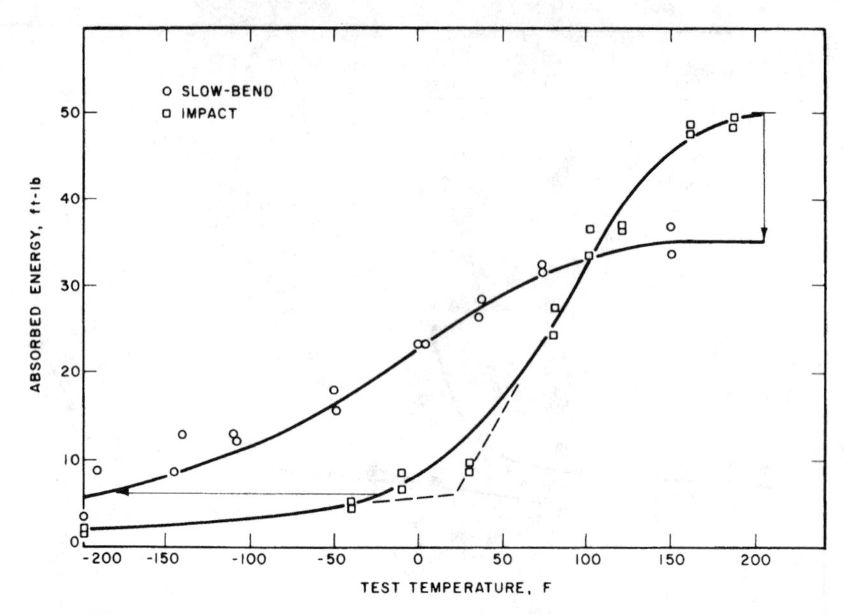

FIG. 4.7 Slow-bend and impact CVN test results for an A36 steel.

impact loading rates, referred to as K_{Id} (dynamic), similar to K_{Ia} results. Note that the shift between the slow-bend test results and the dynamic-load results for low-strength steels ($\sigma_{ys} < 50$ ksi) is over 150°F and that the difference in loading rates between slow and impact is about six orders of magnitude.

As a general rule, slow loading rates are conducted at strain rates of approximately 10^{-5} in./in./s, that is, the maximum load is reached in about 1 min, as in a standard tension test. Intermediate-loading-rate tests, $K_{Ic}(t)$, are usually conducted at strain rates of about 10^{-3} in./s or time to maximum load of about 1 s. Dynamic tests usually are conducted at strain rates of 10 in./in./s with time to maximum load of about 0.001 s.

The more common reference to loading rate is:

$$\dot{K} = \frac{K_{critical}}{t} \tag{4.1}$$

where $\dot{K}$ = rate of increase in the stress-intensity factor, ksi $\sqrt{\text{in.}}$/s,
$K_{critical}$ = critical stress-intensity factor (K_{Ic}, $K_{Ic}(t)$, K_{Id}, K_{Ia}, K_c), ksi $\sqrt{\text{in.}}$, and
t = time, in seconds, required to reach $K_{critical}$.

There is a continual change in fracture toughness of structural steels with increasing loading rates, as shown in Figure 4.2. The rate of change of K_I with respect to time is given in ksi $\sqrt{\text{in.}}$/s. "Slow" loading rates, that is, those prescribed in the standard method of K_{Ic} testing, are around 1 ksi $\sqrt{\text{in.}}$/s, whereas those loading rates generally obtained in K_{Id} testing are around 10^5 ksi $\sqrt{\text{in.}}$/s.

Other structural materials, for example, aluminums, titaniums, and very-high-strength steels (yield strengths of 140 ksi and higher), generally do not exhibit loading-rate effects. Thus, for these materials, there generally would be no difference between K_{Ic} and K_{Id} values tested at the same temperature.

4.4 Effect of Constraint on Fracture Toughness

Ahead of a sharp crack, the lateral constraint (which increases with increasing plate thickness) is such that through-thickness stresses are present. Because these through-thickness stresses must be zero at each surface of a plate, they are less for thin plates compared with thick plates. For very thick plates, the through-thickness stresses at the centerline are large, and a triaxial tensile state of stress occurs ahead of the crack. This triaxial state of stress reduces the apparent ductility of the steel by decreasing the shear stresses. Because yielding is restricted, the constraint ahead of the crack is increased and thus the fracture toughness is reduced. This decrease in fracture toughness is controlled by the thickness of the plate, even though the inherent metallurgical properties of the material may be unchanged. Thus, the fracture toughness is smaller for thick plates compared with thinner plates of the same material. This behavior is shown schematically in Figure 4.8, which indicates that the minimum fracture toughness of a particular material, K_{Ic}, is reached when the thickness of the specimen is large enough so that the state of stress is plane strain.

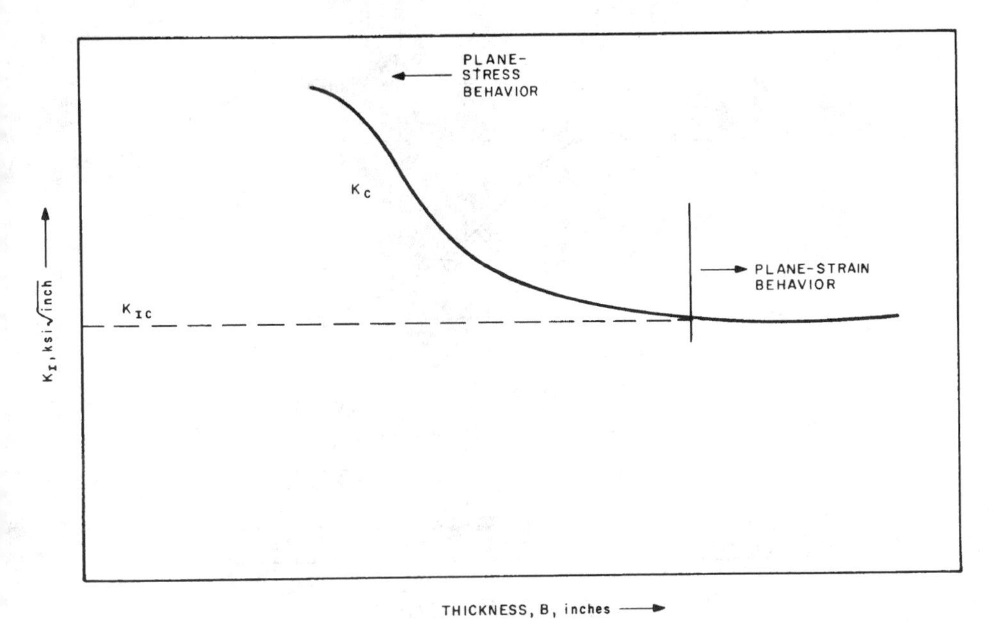

FIG. 4.8 Effect of thickness on K_c.

In Figure 4.9, actual test results are presented for a high-strength maraging steel that illustrate this behavior. For thicknesses greater than some value related to the fracture toughness and yield strength of individual materials, maximum constraint occurs and plane-strain, K_{Ic}, behavior results. In Chapter 3 it was shown that this limiting thickness has been defined to be $B \geq 2.5\ (K_{Ic}/\sigma_{ys})^2$, as given in the ASTM E 399 Standard Method of Test for K_{Ic}. Conversely, as the thickness of the plate is decreased, *even though the inherent metallurgical characteristics of the steel are not changed,* the fracture toughness increases. Thus plane-stress, K_c, behavior exists for thin plates, as shown in Figure 4.9, compared with thick plates.

Figure 4.10 shows the shear lips at the surface of fracture test specimens with different thicknesses machined from the center of a single plate. Thus the material at the centerline of each specimen where fracture initiated was the same. The percentage of shear lips as compared with the total fracture surface is a qualitative indication of fracture toughness. A small percentage of shear-lip area indicates a relatively brittle behavior. A comparison of the fracture surfaces in Figure 4.10 shows that thinner plates are more resistant to brittle fracture than thick plates in that the percentage of shear lips is larger for the thinner specimens compared with the thicker ones. This qualitative behavior is shown quantitatively in Figure 4.9.

Pellini [1] described the physical significance of constraint and plate thickness on fracture toughness in terms of plastic flow, as shown in Figure 4.11. This figure shows that the introduction of a circular notch in a bar loaded in tension.

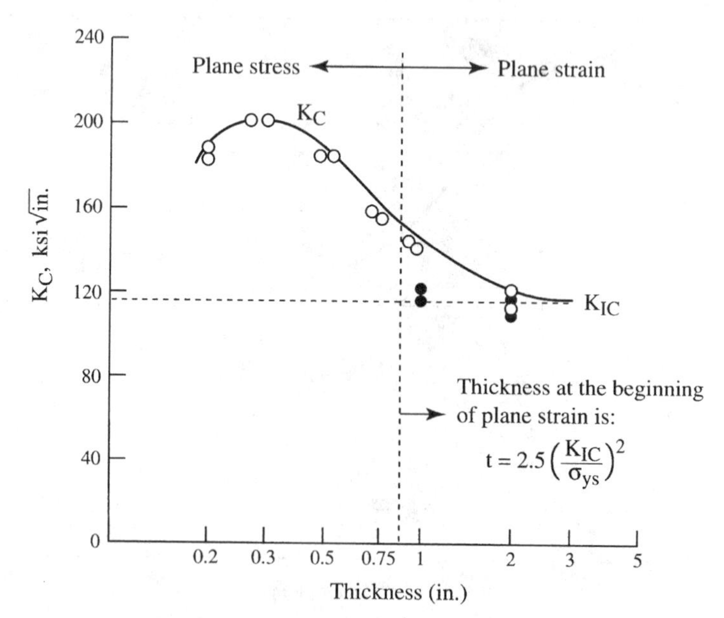

FIG. 4.9 Effect of thickness on K_c behavior.

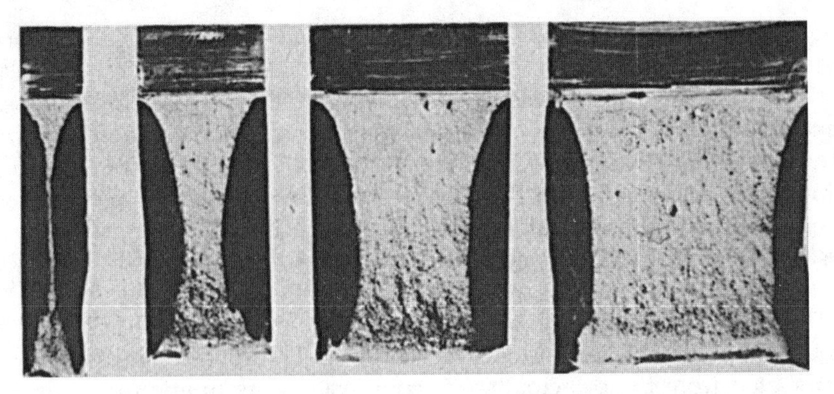

FIG. 4.10 Effect of specimen thickness (½, 1, 1½, 2 in.) on toughness as determined by size of shear lips.

causes an elevation of the stress-strain, or flow, curve. The plastic flow of the smooth tensile bar, which is usually used to develop conventional stress-strain curves, is "free" flow because lateral contraction is not constrained during the initial loading.

In the notched bar, however, the reduced section deforms inelastically while the ends of the specimen are still loaded elastically. Since the amount of elastic contraction (Poisson's ratio) is small compared to the inelastic contraction of the reduced section, a *restriction* to plastic flow is developed. This restriction is in the nature of a reaction-stress system such that the σ_x and σ_z stresses restrict or constrain the flow in the σ_y (load) direction. Thus, the uniaxial stress state of the smooth bar is changed to a triaxial tensile stress system in the notched bar com-

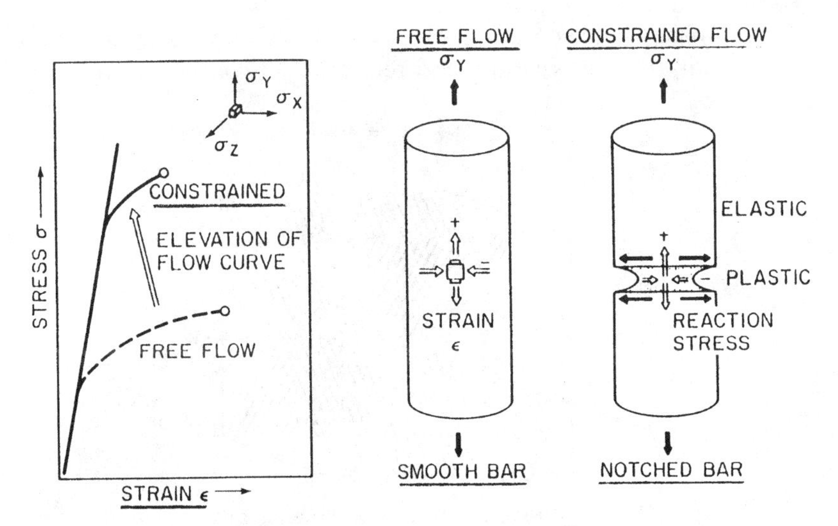

FIG. 4.11 Origins of constraint effects.

pared with the unnotched bar. As the notch becomes more sharp, the severity of the stress state increases.

For a uniform triaxial state of stress, where the three principle stresses, σ_x, σ_y, and σ_z are equal, there are no shear stresses. This results in almost complete constraint against plastic flow. Thus the elastic stresses at the tip of a sharp crack are increased compared with the lower "free"-flow stresses in an unrestrained tension specimen. In the case of most notched specimens, $\sigma_y > \sigma_x$ or σ_z, but the stresses are not equal. Thus, some shear stresses do occur, and there is some nonlinear behavior.

The stress field for an element within a structure can be described by three principal stresses that are normal to each other, Figure 4.12. Shear stresses can be calculated from the principal stress components. Assuming that σ_1 in Figure 4.12 is the largest principal stress and σ_3 is the smallest principal stress, the maximum shear stress component along the two shaded planes is:

$$\tau_{max} = \frac{1}{2}(\sigma_1 - \sigma_3) \tag{4.2}$$

In a uniaxial tension test used to specify material properties, $\sigma_1 = \sigma_{max}$ and $\sigma_2 = \sigma_3 = 0$. Therefore

$$\tau_{max} = \frac{\sigma_1}{2} = \frac{\sigma_{max}}{2} \tag{4.3}$$

Since the plastic deformation, i.e., yielding, begins when τ_{max} reaches a critical value, a change in the relationship between τ_{max} and σ_{max} represents a change in the plastic deformation behavior of the material. Note that yielding occurs when the shear stress, τ_{max}, reaches a critical value, not when σ_{max} reaches a critical value.

The relationship between the shear stress and the normal stresses, σ_1, σ_2, and σ_3, can result in either yielding and relaxation of constraint or not yielding

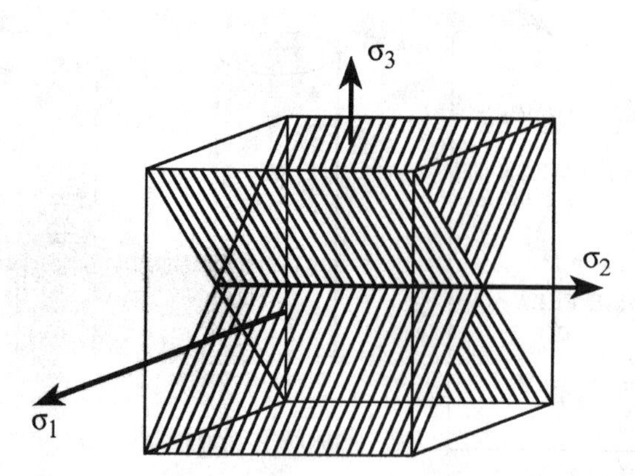

FIG. 4.12 Principal stresses and planes of maximum shear stress.

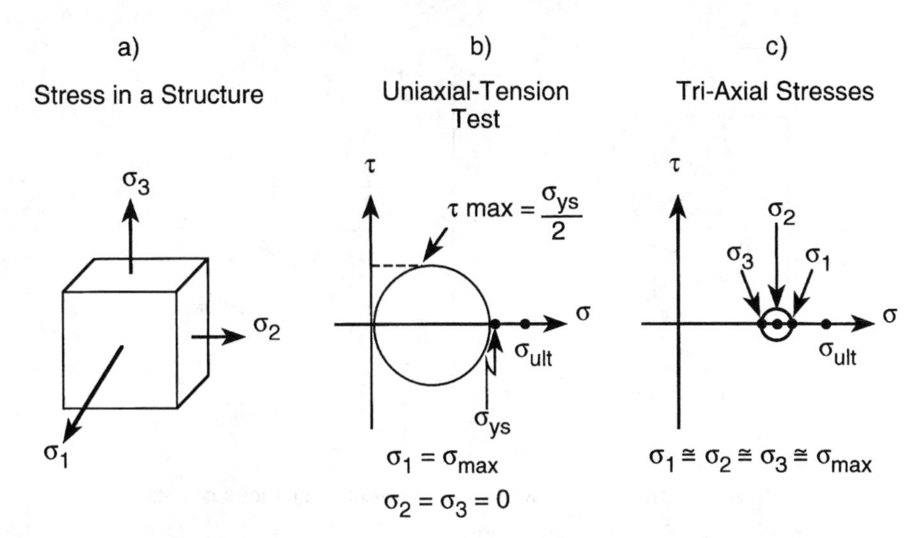

FIG. 4.13 Mohr's circle of stress analysis for stresses in a structure.

and increased constraint. This behavior is illustrated in Figure 4.13 using Mohr's circle of stress. Figure 4.13*a* shows the principle stress directions, with the largest being σ_1. For uniaxial loading, such as the case of a standard tension test, σ_1 = applied stress and $\sigma_2 = \sigma_3 = 0$. At $\sigma_1 = \sigma_{max}$, $\tau_{max} = \sigma_{max}/2$, as shown in Figure 4.13*b*, and yielding occurs when $\tau_{max} = \sigma_{ys}/2$.

In contrast to the simple tension test, Figure 4.13*c* represents a triaxial tensile state of stress such as would be expected in highly constrained connections (see Chapter 16). Because of the triaxial stress loading, the stresses approach the ultimate stress and fracture occurs. Yielding, which is prevented because τ_{max} is low, may never occur.

Figure 4.14 is a schematic description of the state of stress at the tip of a through-thickness crack in a sharply notched specimen loaded in tension. To satisfy compatibility conditions, the plastic "cylinder" (plastic-zone region defined in Chapter 2) that develops at the crack tip must increase in diameter with an increase in stress in the y direction due to load. However, this can happen only if through-thickness lateral contraction occurs in the z direction. This lateral contraction is constrained by elastically stressed materials surrounding the "cylinder" and leads to the triaxial state of stress that raises the flow stress. Furthermore, the material behind the notch is unstressed because of the free surface of the notch and adds to the lateral constraint ahead of the notch. Therefore, *as the thickness is increased, the constraint increases, and the flow stress curve* is raised, as shown schematically in Figure 4.14.

In summary, the constraint ahead of a sharp crack is increased by increasing the plate thickness. Thus the critical stress intensity, K_c, for a particular structural material tested at a particular temperature and loading rate decreases with increasing specimen thickness (Figures 4.8 and 4.9). Beyond some limiting thick-

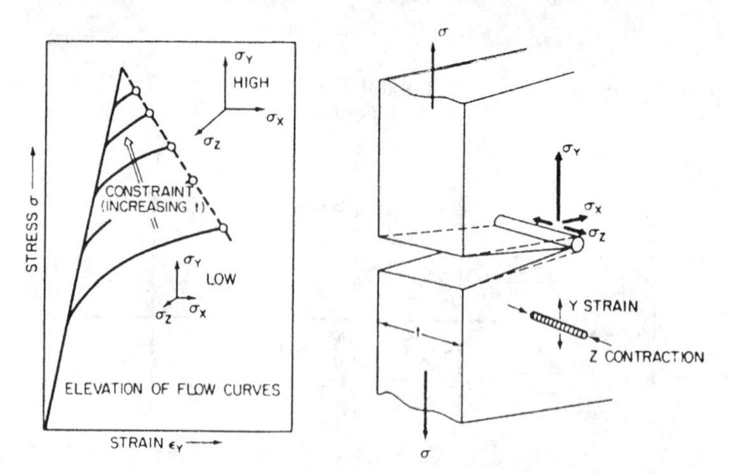

FIG. 4.14 Constraint conditions for through-thickness cracks.

ness, maximum constraint is attained, and the critical stress-intensity factor reaches the minimum plane-strain value, K_{Ic}. This maximum constraint occurs when the plate thickness is sufficiently large in a notched specimen of the particular material being tested at a particular test temperature and loading rate. As described previously, the limiting thickness for plane-strain behavior under slow loading conditions has been established by the ASTM Standard Test Method E-399 [2] as:

$$B \geq 2.5 \left(\frac{K_{Ic}}{\sigma_{ys}}\right)^2 \tag{4.4}$$

For dynamic loading, the limiting thickness would be:

$$B \geq 2.5 \left(\frac{K_{Id}}{\sigma_{yd}}\right)^2 \tag{4.5}$$

where B = thickness of test specimen.

K_{Ic} = critical plane-strain stress-intensity factor under conditions of static loading described in ASTM Method E 399, "Standard Method of Test for Plane Strain Fracture Toughness of Metallic Materials" [2] as described in Chapter 3.

σ_{ys} = static tensile yield strength obtained in "slow" tension test as described in ASTM Test Method E, Standard Methods of Tension Testing of Metallic Materials [3].

K_{Id} = critical plane-strain stress-intensity factor as measured by "dynamic" or "impact" test; the test specimen is similar to a K_{Ic} test specimen but is loaded rapidly. As described in Chapter 3, K_{Id} values are presumed to be equivalent to K_{Ia} values [4].

σ_{yd} = dynamic tensile yield strength obtained in "rapid" tension test at loading rates comparable to those obtained in K_{Id} tests; although extremely difficult to measure, a good engineering approximation based on experimental results of structural steels is:

$$\sigma_{yd} = \sigma_{ys} + (20 \text{ to } 30 \text{ ksi}) \quad (4.6)$$

This limiting constraint condition for K_{Ic} or K_{Id} is established for a crack tip of "infinite" sharpness, namely, $\rho = 0$. This "infinite" sharpness is obtained by fatigue cracking the test specimens at low-stress levels, as described in Chapter 3. As a test specimen is loaded during a fracture test, some local plastic flow will occur at the crack tip, and the crack tip will be blunted slightly. For a brittle material, that is, any structural material tested at a temperature and loading rate where it has very low fracture toughness, the degree of crack blunting is very small. Consequently, unstable crack *extension* will occur from a sharp crack. In essence, the material fractures under elastic loading and exhibits plane-strain behavior under conditions of maximum constraint.

However, if the inherent fracture toughness of the structural material is such that it is *not* brittle at the particular test temperature and loading rate (e.g., the start of elastic-plastic behavior), an increase in plastic deformation at the crack tip occurs and the crack tip is "blunted." As a result, the limit of plane-strain constraint is exceeded. At temperatures above this test temperature, the inherent fracture toughness begins to increase rapidly with increasing test temperature because the effects of the crack blunting and relaxation of plane-strain constraint are synergistic. That is, the crack blunting leads to a relaxation in constraint that causes increased plastic flow, which leads to additional crack blunting. Thus, elastic-plastic behavior begins to occur rapidly at increasing test temperatures once this plane-strain constraint (thickness plus notch acuity) is exceeded.

This behavior can be illustrated in terms of a constraint relaxation that changes the flow curve as shown schematically in Figure 4.15*a*. The degree of crack-tip blunting establishes the particular flow-stress curve, leading to plane-strain, elastic-plastic, or plastic behavior. For example, the dashed curves *A* and *B* in Fig. 4.15*a* represent flow curves of unconstrained material (as in standard tension tests) leading to plastic or elastic-plastic behavior, depending on the inherent ductility of the structural material. Curve *C* represents the flow curve of a notched fully constrained material leading to failure under conditions of plane strain. If the material is tested at a temperature above the limit of plane strain such that partial crack-tip blunting occurs, a partial relaxation of constraint occurs, leading to elastic-plastic behavior. If the material is tested at still higher temperatures, considerable crack-tip blunting occurs, and considerable relaxation of constraint occurs, leading to plastic behavior.

Figure 4.15*b* is a schematic of the metal-grain structure ahead of the crack tip and indicates the microscopic behavior of the particular structural material. The dark line tracings within grains indicate slip on crystal planes, which is

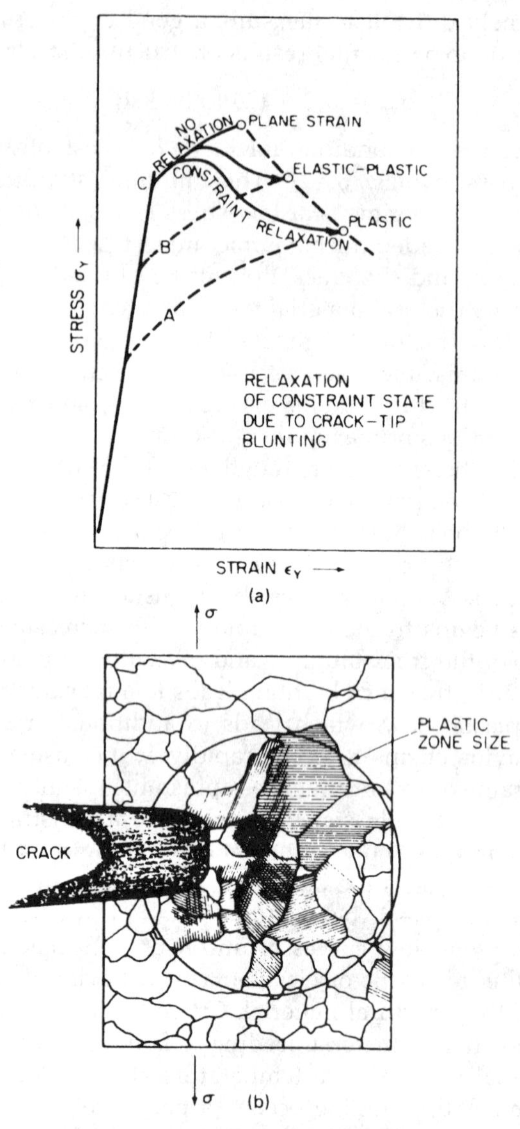

FIG. 4.15 Relaxation of constraint state due to crack-tip blunding: (*a*) relaxation of plane-strain constraint due to metal-grain flow, which causes crack-tip blunting; (*b*) effect of crack-tip blunting ahead of crack.

necessary to produce deformation. This deformation of individual grains is necessary to provide for growth of the plastic zone, r_y, (approximated by the circle) ahead of the crack tip. Depending on the inherent metallurgical structure of the metal, continued loading either increases slip and deformation (elastic-plastic or plastic behavior) or leads to the development of cracks and/or voids (plane-strain behavior). A material that is brittle at a particular temperature will develop microcracks or voids before the plastic-zone size is very large, resulting in rapid or unstable crack growth, that is, brittle fracture. As the test temperature is increased for successive test specimens, the metal becomes more ductile and slip occurs before microcracks occur, resulting in larger plastic-zone sizes. This is the beginning of the plane-strain transition where the individual grains begin to undergo large amounts of plasticity (*microscopic plasticity*), but the overall specimen is still elastic (*macroscopic plane strain*). This behavior was shown in Figures 4.3, 4.4, and 4.5. For each of the loading conditions, there is a short region where the fracture toughness begins to increase rapidly, but the overall behavior is still one of plane strain.

The transition in behavior from elastic (plane-strain) to plastic behavior occurs over the region known as the transition-temperature region. Note that K_{Ic} or K_{Id} specimens cannot be used to measure the entire range of behavior since they require essentially elastic plane-strain behavior to satisfy the restrictions placed on the analysis described in Chapter 2. Other fracture-mechanics-type specimens (J_{Ic}, CTOD, and R-curve) *can* be used to obtain quantitative estimates of the fracture toughness in the elastic-plastic region, as was described in Chapter 3.

4.5 Loading-Rate Shift for Structural Steels

4.5.1 CVN Temperature Shift

The K_{Ic}, $K_{Ic}(t)$, and K_{Id} test results presented in Section 4.4 demonstrated the general effect of temperature and loading rate on K_{Ic}. A similar effect of loading rate exists for CVN specimens tested in three-point slow-bend and standard-impact loading. The general effect of a slow loading rate (compared with standard-impact loading rates for CVN specimens) is to shift the CVN curve to the left and to lower the upper-shelf values. This behavior was shown schematically in Figure 4.6 and for actual CVN test results in Figure 4.7.

The shifts in the transition temperature for slow-bend and impact K_{Ic} and CVN test results for sixteen steels having yield strengths in the range 40–250 ksi are related to yield strength, as shown in Figure 4.16. Note the linear relation between yield strength and the temperature shift.

The magnitude of the temperature shift caused by high-strain-rate testing should be measured, at the same energy level, from the onset of the dynamic temperature transition to the onset of the transition on the static curve as was shown in Figure 4.6. This onset of the dynamic temperature transition for the

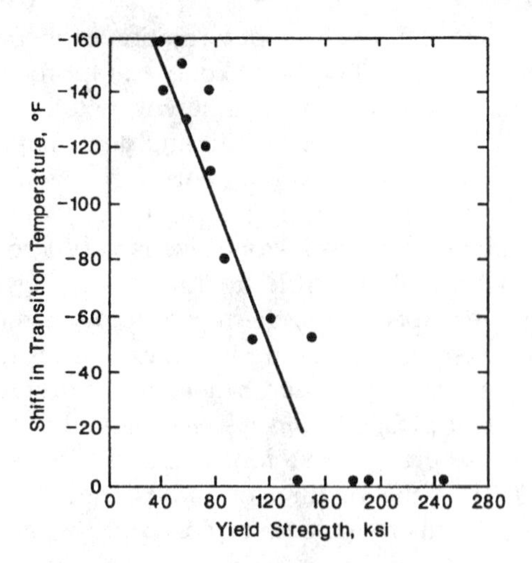

FIG. 4.16 Effect of yield strength on shift in transition temperature between impact and slow-bend K_{Ic} and CVN tests.

CVN tests is defined arbitrarily by the intersection of tangent lines drawn from the lower shelf level and the transition region (Figure 4.6). The loading-rate shift has been verified by Roberts et al. [5] using the 15-ft-lb level. Thus, the fact that the shift exists does not depend on the particular means used to measure it. However, the onset of dynamic transition seems to be the best reference point from which to measure strain-rate effects because this point is located in the energy-absorption region where a change in the microscopic mode of fracture starts to occur at the initial crack front for both static and dynamic testing. Also, because the onset of the static temperature transition occurs at a lower temperature than that marking the onset of the dynamic temperature transition and because the static upper energy-absorption shelf is usually of lower magnitude than that measured in the dynamic test, the static and dynamic energy-absorption curves usually intersect. Thus, measurements of the temperature shift at temperatures above that defined by the onset of the impact CVN temperature transition may underestimate the magnitude of the shift. Below this reference temperature, the slopes with respect to the temperature axis of both the static and the dynamic CVN energy become very small; consequently, it is difficult to measure the magnitude of the shift between the two curves in the lower-shelf region.

4.5.2 K_{Ic}-K_{Id} *Impact-Loading-Rate Shift*

The maximum difference in K_{Ic} and K_{Id} fracture behavior for a given steel occurs between static loading and full-impact loading that correspond to strain rates on the order 10^{-5} s^{-1} and 10 s^{-1}, respectively, as was shown in Figures 4.3, 4.4, and 4.5. Moreover, various studies [6,7] have shown that the temperature shift between static loading and impact loading decreases as the room temperature yield strength of the steel increases, as shown in Figure 4.16. The magnitude of the temperature shift between slow loading and impact loading (in both CVN and K_{Ic}-K_{Id} tests) in steels of various yield strengths can be approximated by:

$$T_{shift} = 215 - 1.5\sigma_{ys} \tag{4.7}$$

for 36 ksi $< \sigma_{ys} <$ 140 ksi (250 MPa $< \sigma_{ys} <$ 965 MPa) and

$$T_{shift} = 0 \tag{4.8}$$

for $\sigma_{ys} >$ 140 ksi

where T_{shift} = absolute magnitude of the shift in the transition temperature curves between slow loading and impact loading, °F, and
σ_{ys} = room temperature yield strength, ksi.

4.5.3 K_{Ic}(t) *Intermediate-Loading-Rate Shift*

For the various K_{Ic}, K_{Ic} (t), and K_{Id} tests, strain rates are calculated for a point on the elastic-plastic boundary as determined by Irwin using the following equation:

$$\dot{\varepsilon} = \frac{2\sigma_{ys}}{tE} \tag{4.9}$$

where σ_{ys} = yield strength for the test temperature and loading rate,
t = loading time for the test, e.g., the time stated in the brackets for the rapid-load $K_{Ic}(t)$ test results, and
E = elastic modulus of the material tested.

The rate of application of K is:

$$\dot{K} = \frac{K_{critical}}{t} \tag{4.10}$$

where $K_{critical}$ = the K_{Ic}, K_{Ic} (t), or K_{Id} value at instability, and
t = loading time for the test.

K_{Ic} transition curves obtained at intermediate loading rates (10^{-5} $s^{-1} < \dot{\varepsilon} <$ 10 s^{-1}) are always between those obtained under static ($\dot{\varepsilon} \simeq 10^{-5}$ s^{-1}) and under impact ($\dot{\varepsilon} \simeq$ 10 s^{-1}) loading. Test results at an intermediate loading rate that corresponded to a strain rate of about 10^{-3} s^{-1} at the elastic-plastic boundary were presented in Figures 4.3, 4.4, and 4.5 for ASTM A36, ABS-C, and A572 Grade 50 steels, respectively. The data suggested that the shift between a K_{Ic} curve obtained under static load ($\dot{\varepsilon} \simeq 10^{-5}$ s^{-1}) and under an intermediate loading rate ($\dot{\varepsilon} \simeq 10^{-3}$ s^{-1}) was approximately equal to 25% of the total shift, Equation

(4.5), between the curves obtained under static loading and under impact loading.

4.5.4 Predictive Relationship for Temperature Shift

Based on the preceding observations for K_{Ic} transition behavior under static, intermediate, and impact rates of loading, a generalized characterization for the dependence of temperature shift, T_{shift}, strain rate, $\dot{\varepsilon}$, and yield strength, σ_{ys}, for a material can be formulated, Figure 4.17. For intermediate strain rates in the range $10^{-3}\ s^{-1} \leq \dot{\varepsilon} \leq 10\ s^{-1}$ and for steels having yield strengths of less than 140 ksi (965 MPa), the relationship among T_{shift}, $\dot{\varepsilon}$, and σ_{ys}, Figure 4.18, can be approximated by:

$$T_{shift} = (150 - \sigma_{ys})(\dot{\varepsilon})^{0.17} \tag{4.11}$$

where T_{shift} is in °F, σ_{ys} is in ksi, and $\dot{\varepsilon}$ is in s^{-1}.

The plane-strain fracture-toughness value for a given steel tested at a given temperature and for strain rate equal to or less than $10^{-5}\ s^{-1}$ is constant and is equal to the static K_{Ic} value. Although the plane-strain fracture toughness behavior for a given steel tested at a given temperature and at a strain rate greater than $10\ s^{-1}$ is yet to be established, it is generally assumed that the value is almost constant and is independent of the rate of loading [6–11].

4.5.5 Significance of Temperature Shift

Because of the temperature shift, increasing the loading rate can decrease the fracture-toughness value at a particular temperature for steels having yield

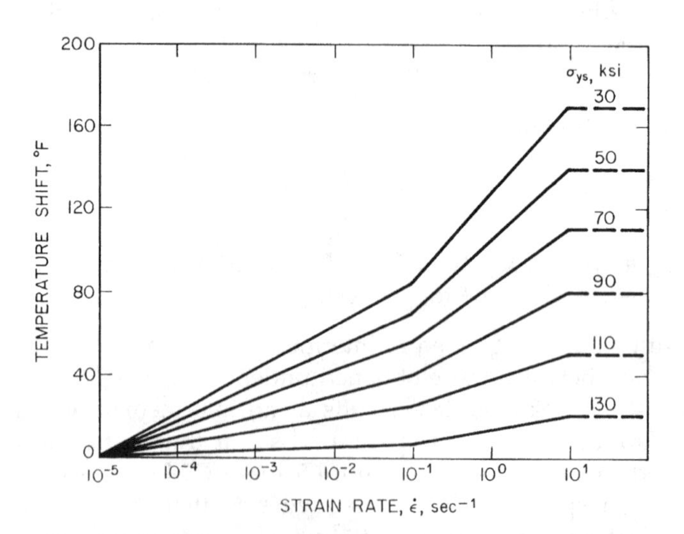

FIG. 4.17 Shift in transition temperature for various steels and for strain rates greater than $10^{-5}\ s^{-1}$.

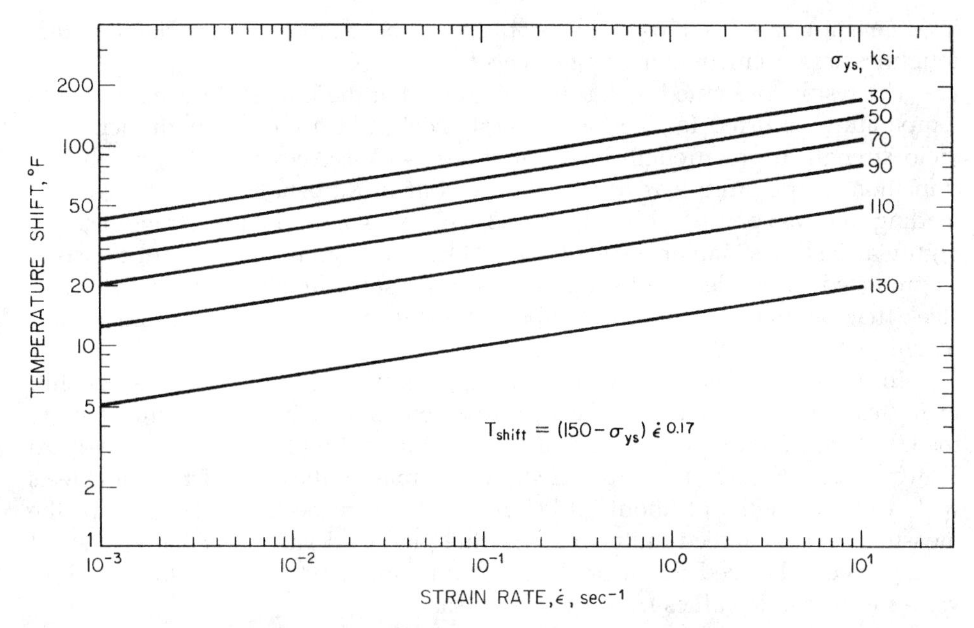

FIG. 4.18 Shift in transition temperature for various steels and for strain rates in the range $10^{-3}\ s^{-1} \leq \dot{\varepsilon} \leq 10\ s^{-1}$.

strengths less than 140 ksi. The change in fracture-toughness values for loading rates varying from slow to impact rates is particularly important to those applications where the actual loading rates are slow or intermediate. Many types of structures such as buildings, bridges, ships, and pressure vessels fall into this classification.

For example, for a design stress of about 20 ksi (138 MN/m^2), a K_{Ic} of 40 ksi$\sqrt{\text{in.}}$ (44 MN/m$^{3/2}$) for a given steel would correspond to the tolerance of a through-thickness flaw of approximately 3 in. (76 mm). If a structure were loaded statically, this size flaw could be tolerated at extremely low temperatures. If the structure were loaded dynamically, however, the temperature at which this size flaw would cause failure would be much higher. That is, using the test results for an ABS-C ship steel presented in Figure 4.4 as an example, the static K_{Ic} value of 40 ksi $\sqrt{\text{in.}}$ occurs at −225°F, whereas the dynamic K_{Id} value of 40 ksi$\sqrt{\text{in.}}$ occurs at −50°F. Similar observations could be made for other steels as shown in Figures 4.5 and 4.6. This difference between the static and dynamic results decreases with increasing yield strength, as indicated in Equation 4.5 for T_{shift}.

If the loading rates of structures are closer to those of slow loading than to impact loading, a considerable difference in the behavior of these structures would be expected. Thus, not only should the effects of temperature and constraint (plate thickness) on structural steels be established, but more important, *the maximum loading rates that will occur in the actual structure being analyzed under operating conditions should be established.* As will be discussed in Part IV, the loading-rate shift has been used in establishing the AASHTO material fracture-toughness requirements for bridge steels [*12*].

The results presented in Figure 4.16 show that the largest shifts in transition temperature occurred for the low-strength steels and decreased with increasing yield strength up to strength levels of about 140 ksi. Above 140 ksi, no shift in transition temperature was observed. The shift in K_{Ic} behavior as a function of loading rate was presented in Figures 4.3, 4.4, and 4.5, for the low-strength steels. Figure 4.19 shows similar results for a 100-ksi yield strength steel, but the shift, as predicted, is smaller. Figure 4.20 shows test results for a 250-ksi yield strength steel, that, as predicted, has no temperature shift between slow and impact test results.

To determine whether the shift in CVN test values is the same as the shift in K_{Ic} test results, dynamic K_{Id} test results were shifted by an amount equal to the CVN transition-temperature shifts and compared with the actual slow-bend K_{Ic} test results. Figures 4.21 and 4.22 show examples of these predictions for steels with yield strengths of about 40 ksi and 100 ksi, respectively. In general, the measured values agreed quite well with the predicted values for these steels. It should be emphasized that a prediction can be made either from slow-bend K_{Ic} values to dynamic values (K_{Id}), or vice versa.

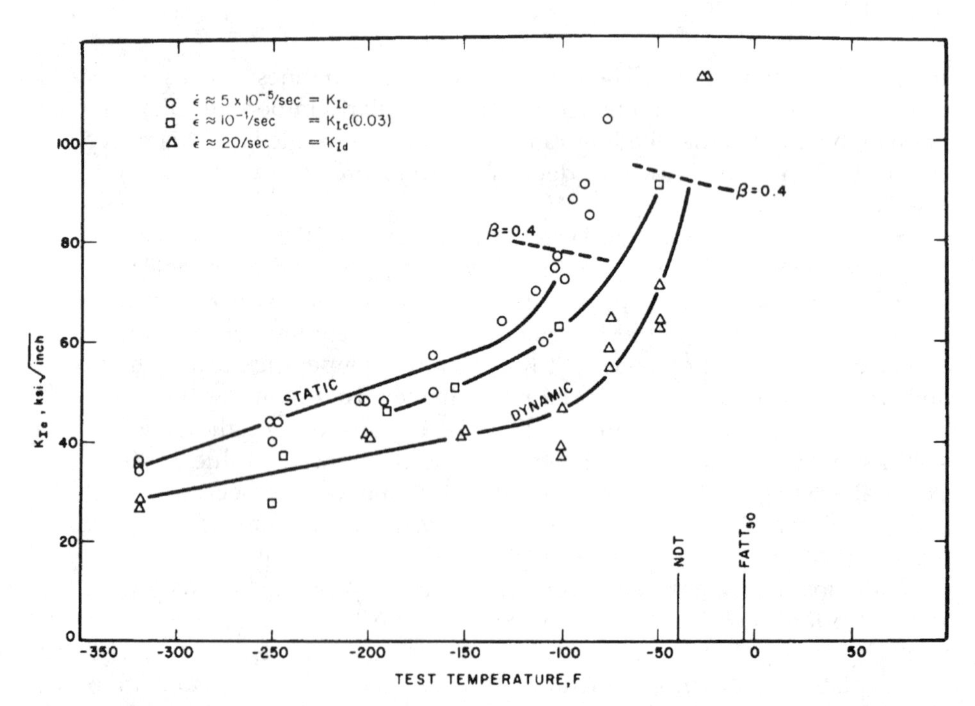

FIG. 4.19 Effect of temperature and loading rate on fracture toughness of an A517 Grade F steel.

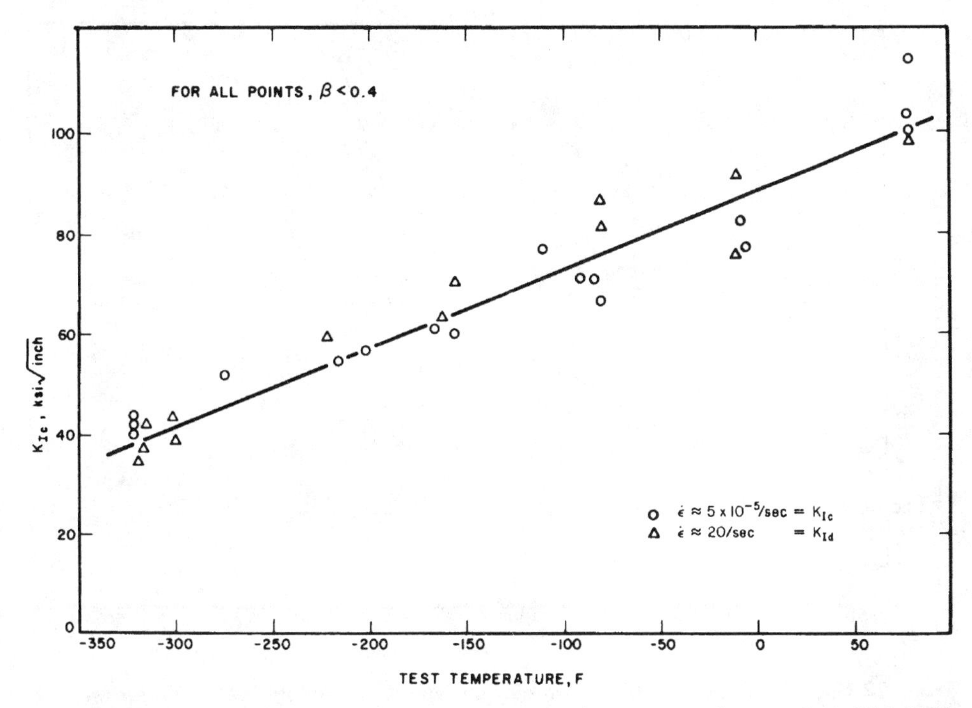

FIG. 4.20 Effect of temperature and loading rate on fracture toughness of an 18Ni (250) maraging steel.

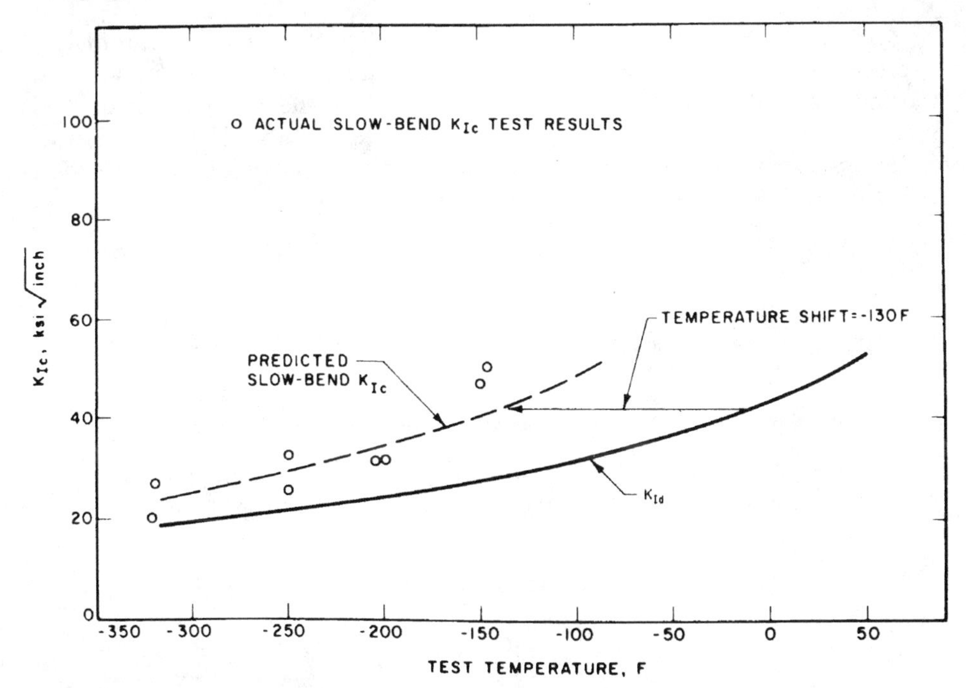

FIG. 4.21 Use of CVN test results to predict the effect of loading rate on K_{Ic} for an A302 Grade B steel.

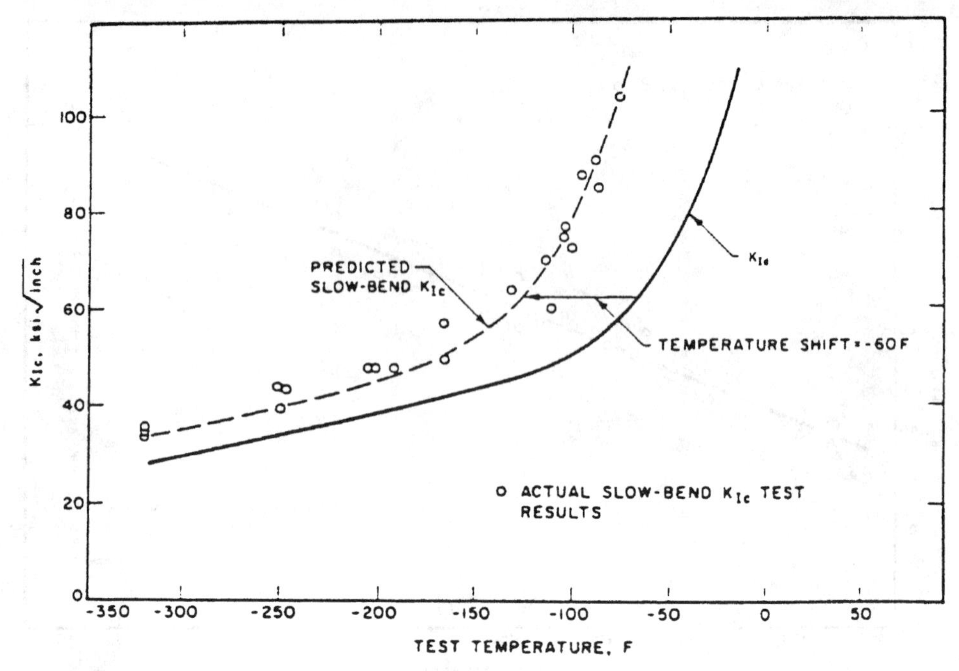

FIG. 4.22 Use of CVN test results to predict the effect of loading rate on K_{Ic} for an A517 Grade F steel.

Because dynamic K_{Id} tests are extremely difficult to conduct and analyze, the preceding developed prediction procedure is quite useful in obtaining a first-order approximation of the effects of loading rate on the K_{Id} behavior of steels by adjusting experimentally obtained slow-bend K_{Ic} values. Further application of this observation is described in Chapter 5 on Correlations Between Various K_{Id} Values and Fracture-Toughness Test Results.

4.6 References

[1] Pellini, W. S., "Design Options for Selection of Fracture Control Procedures in the Modernization of Codes, Rules, and Standards," *Proceedings: Joint United States—Japan Symposium on Application of Pressure Component Codes,* Tokyo, Japan, March 13–15, 1973.

[2] E 399-90 (Reapproved 1997). Standard Test Method for Plane-Strain Fracture Toughness of Metallic Materials, ASTM, Vol. 03.01.

[3] E 8-98. Standard Test Methods for Tension Testing on Metallic Materials, ASTM, Vol. 03.01.

[4] E 1221-96. Standard Test Method for Determining Plain-Strain Crack-Arrest Fracture Toughness, K_{Ia}, of Ferritic Steels, ASTM, Vol. 03.01.

[5] R., Krishna, G. V., and Yen, B. T., "Fracture Toughness of Bridge Steels—Phase II Final Report," Fritz Engineering Laboratory, Report No. 379.2, Lehigh University, Bethlehem, PA, June 1974.

[6] Barsom, J. M., "Relationship Between Plane-Strain Ductility and K_{Ic}," *Transactions of the ASME, Journal of Engineering for Industry,* Vol. 93, Series B, No. 4, November 1971, p. 1209.
[7] Barsom, J. M., Sovak, J. F., and Novak, S. R., "AISI Project 168—Toughness Criteria for Structural Steels: Fracture Toughness of A36 Steel," *Research Laboratory Report 97.021-001*(1), May 1, 1972 (available from the American Iron and Steel Institute).
[8] Barsom, J. M., Sovak, J. F., and Novak, S. R., "AISI Project 168—Toughness Criteria for Structural Steels: Fracture Toughness of A572 Steels," *Research Laboratory Report 97.021-001*(2), December 29, 1972 (available from the American Iron and Steel Institute).
[9] Barsom, J. M. and Rolfe, S. T., "The Correlations Between K_{Ic} and Charpy V-Notch Test Results in the Transition-Temperature Range," *Impact Testing of Metals, ASTM STP 466,* 1970, p. 281.
[10] Eftis, J. and Krafft, J. M., "A Comparison of the Initiation with the Rapid Propagation of Crack in a Mild Steel Plate," *Transactions of the ASME, Journal of Basic Engineering,* Vol. 87, Series D, March 1965, p. 257.
[11] Hahn, G. T., Hoagland, R. G., Rosenfield, A. R., and Sejnoha, R., "Rapid Crack Propagation in High Strength Steel," *Metallurgical Transactions,* Vol. 5, No. 2, February 1974, p. 321.
[12] Barsom, J. M., "Development of the AASHTO Fracture-Toughness Requirements for Bridge Steels," *Journal of Engineering Fracture Mechanics,* Vol. 7, No. 3, September 1975, p. 605.

5

CVN-K_{Id}-K_c Correlations

5.1 General

CHAPTER 3 DESCRIBED various fracture mechanics test methods that can be used to measure the resistance force of structural materials regardless of the fracture toughness level. That is, test methods ranging from linear elastic, K_{Ic} test methods to fully plastic *J-R* or δ_U test methods were described. Furthermore, the relationships between *K-J-δ* were described so that the engineer can estimate the resistance force in terms of a K_c value for comparison with the driving force, K_I, as described in Chapter 2. Thus, regardless of the inherent fracture toughness of a structural material or the service loading conditions, it is possible to use fracture mechanics to predict the fracture behavior of actual structures.

Unfortunately, fracture mechanics tests are complex and expensive to conduct, and very few laboratories are equipped to conduct all of the various types of tests. The cost of machining, fatigue precracking, and testing a K_c, J_c, or δ_U specimen and the size requirements necessary to ensure valid K_{Ic} or K_{Id} test results render the tests impractical as quality-control tests. Thus, although most codes and specifications were developed using principles of fracture mechanics, the *specific* fracture toughness tests specified for material purchase or quality control are in terms of *auxiliary* test specimens, such as the CVN impact test specimen rather than the expensive fracture mechanics types of tests. Consequently, the need exists to correlate K_c data with notch-toughness test results obtained with smaller and less costly specimens.

For example, the American Association of State Highway and Transportation Officials (AASHTO) material requirements for bridge steels [*1*] are based on concepts of fracture mechanics, but are specified in terms of Charpy V-notch impact test results. Fracture toughness requirements for thick-walled nuclear pressure-vessel steels are based on minimum dynamic fracture toughness values, K_{Id} (actually K_{IR} for critical reference values [2]). However, the actual material fracture toughness requirements for steels used in these pressure vessels are specified using NDT (nil-ductility transition) values and CVN impact values using lateral expansion measurements. Proposed fracture toughness requirements for welded ship hulls were developed using K_{Id}/σ_{ys} values, but the proposed material procurement values were in terms of NDT and CVN impact values. Thus empirical correlations as well as engineering judgement and experience are used to *translate* fracture-mechanics guidelines or controls into actual material fracture-toughness specifications.

5.2 Two-Stage CVN-K_{Id}-K_c Correlation

Many correlations have been developed using a wide variety of test specimens. However, because the CVN-impact test specimen is the most widely used quality control and specification specimen, the correlations presented in this chapter deal primarily with the CVN-impact test specimen as related to K_c results.

The Charpy V-notch impact specimen (ASTM E 23, Standard Methods for Notched Bar Impact Testing for Metallic Materials [3]) undergoes a transition in the same temperature zone as the impact plane-strain fracture toughness (K_{Id}), as shown in Chapter 4. Thus, it is not surprising that a correlation between these test results has been developed for the transition region. This correlation is given by the equation:

$$\frac{(K_{Id})^2}{E} = 5\ (\text{CVN}) \tag{5.1}$$

where K_{Id} is in psi $\sqrt{\text{in.}}$, E is in psi, and CVN is in ft-lb. The validity of this correlation is apparent from the data presented in Figure 5.1 for various grades of steel ranging in yield strength from about 36 to 140 ksi, and in Figure 5.2 for eight heats of SA 533B, Class 1 steel. Consequently, a given value of CVN impact energy absorption corresponds to a given K_{Id} value as predicted by Equation (5.1). The loading rates in both tests are the same. Thus, except for size, the only difference is notch acuity, and this is accounted for empirically by the factor of 5.

Recall that the temperature shift presented in Chapter 4, which described the loading rate shift between slow, K_{Ic}, and impact, K_{Id}, results are:

$$T_{\text{shift}} = 215 - 1.5\sigma_{ys} \tag{5.2}$$

Using these two relationships, the engineer can take CVN impact test results, which generally are widely available or fairly easy to obtain, and predict K_{Ic} values as follows and as shown schematically in Figure 5.3:

1. Obtain or test CVN impact specimens in the lower transition region for the material of interest.
2. At each temperature for which CVN values are available, calculate the corresponding K_{Id} values using Equation (5.1).

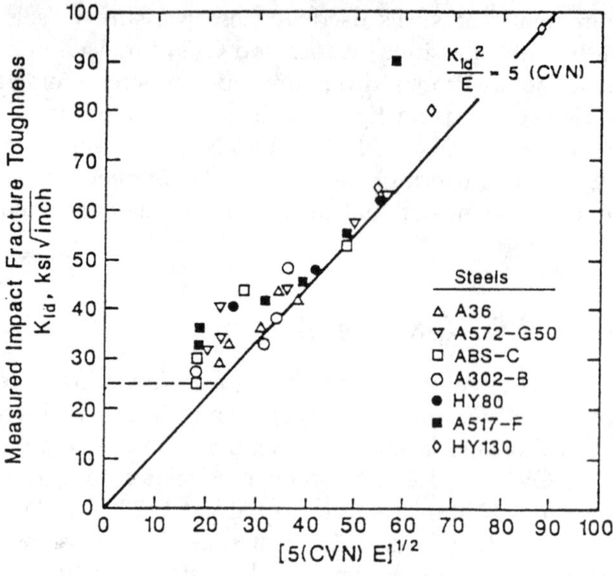

FIG. 5.1 Correlation of plane-strain impact fracture toughness and impact Charpy V-notch energy absorption for various grades of steel.

3. Shift the K_{Id} values to K_{Ic} values using Equation (5.2).
4. If intermediate loading rate K_{Ic} (1.0) values are required, use ¾ of the K_{Id} to K_{Ic} shift as shown in Figure 5.3.

5.3 K_{Ic}-CVN Upper-Shelf Correlation

The relationship between K_{Ic} and upper-shelf CVN test results is based on the results of various investigations by Clausing [4], Holloman [5], and Gross [6]. Clausing showed that the state of stress at fracture initiation in the CVN impact specimen is plane strain, which is the state of stress in a thick K_{Ic} specimen. Holloman has shown that for the dimensions used in the CVN specimen, the maximum possible lateral stress is obtained, indicating a condition approaching maximum constraint. Tests by Gross on CVN specimens of various thicknesses showed that the transition temperature for a standard CVN specimen is identical to the transition temperature for a CVN specimen of twice the standard width, substantiating the observation that the standard CVN test specimen has considerable constraint at the notch root.

The upper-shelf K_{Ic}-CVN correlation is shown in Figure 5.4. It was developed empirically from results obtained on 11 steels having yield strengths in the range

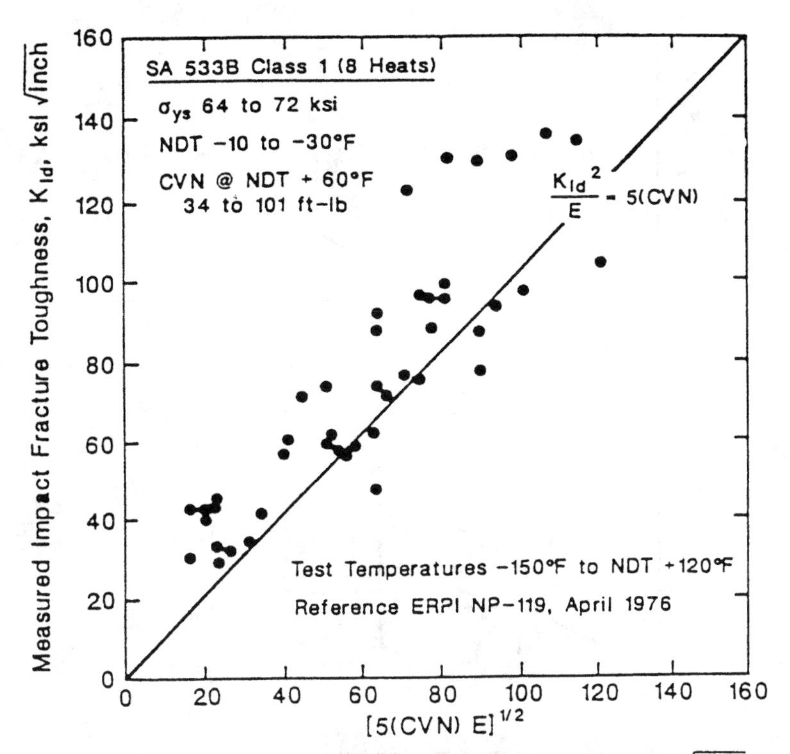

FIG. 5.2 Correlation of plane-strain impact fracture toughness and impact Charpy V-notch energy absorption for SA 533B Class 1 steel.

110–246 ksi (Table 5.1). The K_{Ic} values for these steels ranged from 87 to 246 ksi$\sqrt{\text{in.}}$, and the CVN impact values ranged from 16 to 89 ft-lb. Since the development of this correlation, additional test results have verified the correctness of this relation for upper-shelf values.

At the upper shelf, the effects of loading rate and notch acuity are not as critical as in the transition-temperature region. Thus, the differences in the K_{Ic} and CVN test specimens (namely, loading rate and notch acuity) are not that significant, and a reasonable correlation would be expected, as follows:

$$\left(\frac{K_{Ic}}{\sigma_{ys}}\right)^2 = \frac{5}{\sigma_{ys}}\left(CVN - \frac{\sigma_{ys}}{20}\right) \tag{5.3}$$

where K_{Ic} = critical plane-strain stress-intensity factor at slow loading rates, ksi$\sqrt{\text{in.}}$,

σ_{ys} = 0.2% offset yield strength at the upper shelf temperature, ksi, and

CVN = standard Charpy V-notch impact test value at upper-shelf, ft-lb.

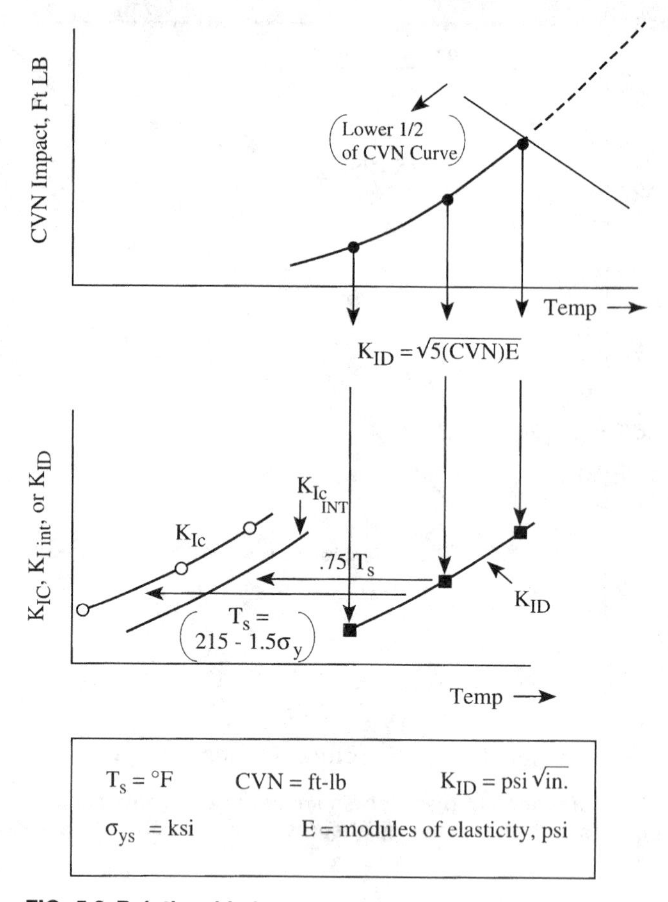

FIG. 5.3 Relationship between CVN-K_{ID}-K_{IC} (int) and K_{IC} test results.

In a discussion of a paper on the *J*-integral as a fracture criterion, P. C. Paris [7] states that, "This paper (*J*-integral) finally explains the reasonableness of the Rolfe-Novak-Barsom correlation of upper shelf Charpy values, CVN with K_{Ic} numbers. This equation relating CVN, a limit-load-relating energy parameter, to K_{Ic} is not only now acceptable but is, for us, in agreement with the *J* failure criteria."

Thus, in addition to empirically relating K_{Ic} to CVN test results over a wide range of strength and fracture toughness levels, the K_{Ic}-CVN relation shown in Figure 5.4 appears to have some theoretical basis as expressed in the development of the *J*-integral and has been substantiated by additional tests by other investigators.

Because K_{Ic} is a static test and the CVN impact test is an impact test, the relationship presented in Equation (5.3) is limited to steels having yield strengths

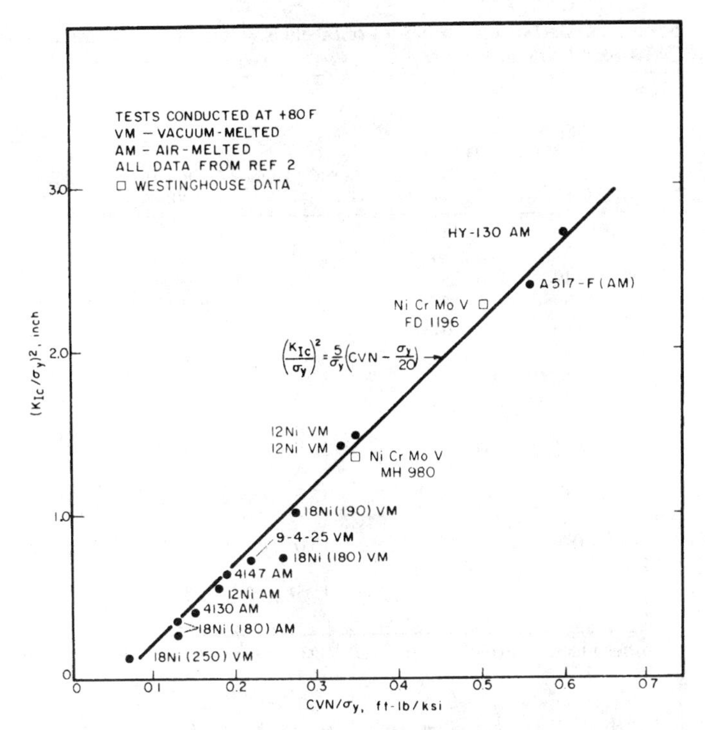

FIG. 5.4 Relation between K_{IC} and CVN values in the upper-shelf region.

greater than 100 ksi, where the effects of loading rate are small. However, because this correlation is an upper-shelf correlation, where the slow-bend and impact CVN values are constant for steels of various yield strengths, Equation (5.3) may be applicable to steels having yield strengths < 100 ksi. Because the upper-shelf CVN impact results are higher than the upper-shelf CVN slow-bend results, substituting the dynamic yield strength, σ_{yd}, into Equation (5.3) may give a better correlation for steels having a yield strength < 100 ksi. However, the extension of any correlation beyond the limits for which it was developed should be done with caution.

5.4 K_{Id} Value at NDT Temperature

In the drop weight NDT test (ASTM E 208, Standard Method for Conducting Drop Weight Test to Determine Nil-Ductility Transition Temperatures of Ferritic Steels [*8*]), a specimen is subjected to a crack initiated from a brittle weld bead under impact-loading conditions (Figure 5.5). After examining the crack shape obtained in this type of test, Irwin et al. [*9*] proposed an analysis for determining a dynamic fracture-toughness value, K_{Id}, from the drop weight test results. Irwin

TABLE 5.1. Longitudinal Mechanical Properties of Steels Investigated for Room Temperature K_{Ic}-CVN Upper-Shelf Correlation

STEEL & MELTING PRACTICE*	YIELD STRENGTH 0.2% OFFSET, ksi	TENSILE STRENGTH, ksi	ELONGATION IN 1 IN., %	REDUCTION OF AREA %	CHARPY V-NOTCH ENERGY ABSORPTION AT +80°F (ft-lb)	K_{Ic}, ksi$\sqrt{\text{in.}}$
A517-F, AM	110	121	20.0	66.0	62	170
4147, AM	137	154	15.0	49.0	26	109
HY-130, AM	149	159	20.0	68.4	89	246
4130, AM	158	167	14.0	49.2	23	100
12Ni-5Cr-3Mo, AM	175	181	14.0	62.2	32	130
12Ni-5Cr-3Mo, VIM	183	191	15.0	61.2	60	220
12Ni-5Cr-3Mo, VIM	186	192	17.0	67.1	65	226
18Ni-8Co-3Mo (200 Grade), AM	193	200	12.5	48.4	25	105
18Ni-8Co-3Mo (200 Grade), AM	190	196	12.0	53.7	25	112
18Ni-8Co-3Mo (190 Grade), VIM	187	195	15.0	65.7	49	160
18N-8Co-3Mo (250 Grade), VIM	246	257	11.5	53.9	16	87

NOTE: AM signifies electric-furnace air-melted; VIM signifies vacuum-induction-melted.

assumed that at the NDT temperature, the plate surface reached the dynamic yield stress, σ_{yd}, corresponding to the test temperature. Furthermore, analyses showed that the pop-in crack geometry had an $a/2c$ ratio of 1 to 4, where a is the crack length and c is half the crack width. These observations led to the following relationship for a part-through-thickness crack:

$$K_{Id} = 0.78\ (\sqrt{\text{inch}})\sigma_{yd} \tag{5.4}$$

Shoemaker [10] observed that the NDT temperature is close to the temperature at which a 1-in.-thick K_{Ic} specimen tested under impact loading ceases to satisfy the ASTM requirements for valid K_{Ic} tests. The thickness requirement for valid K_{Ic} tests is given by

$$B \geq 2.5 \left(\frac{K_{Ic}}{\sigma_{ys}}\right)^2 \tag{5.5}$$

where B = specimen thickness, and
σ_{ys} = yield strength.

This relationship can be used to represent this observation by Shoemaker concerning the dynamic K_{Id} value at NDT temperature. The resulting equation is

$$K_{Id} = 0.64\ (\sqrt{\text{inch}})\ \sigma_{yd} \tag{5.6}$$

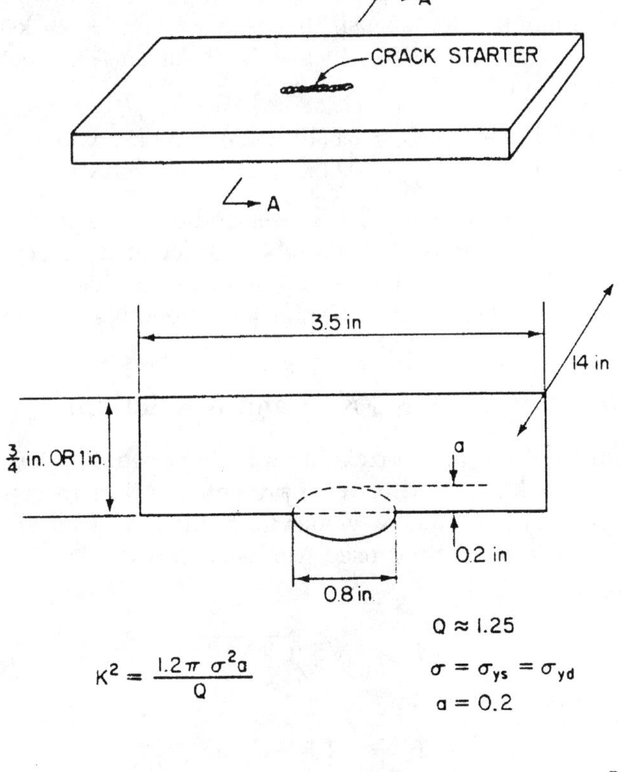

FIG. 5.5 Cross section of NDT test specimen showing initial crack and fracture-mechanics analysis.

where K_{Id} = critical plane-strain stress-intensity factor at NDT temperature and under dynamic loading ($\dot{\varepsilon} \approx 10\ s^{-1}$); and
σ_{yd} = dynamic yield strength at NDT temperature.

Pellini [*11*] estimated the factor relating K_{Id} and σ_{yd} to be 0.5. However, the differences in the various factors are slight, and the suggested relationship between K_{Id} and σ_{yd} at the NDT temperature is:

$$K_{Id} = 0.6\ (\sqrt{\text{inch}})\ \sigma_{yd} \tag{5.7}$$

where K_{Id} = dynamic critical plane-strain stress-intensity factor at the NDT temperature, ksi $\sqrt{\text{in.}}$, and
σ_{yd} = dynamic yield strength at the NDT temperature, ksi.

The values of the dynamic yield strength, σ_{yd}, are approximately equal to the static yield strength plus 25 ksi, that is $\sigma_{yd} \simeq \sigma_{ys} + 25$ ksi. Thus, using Equation (5.7), the calculated K_{Id} values at NDT for an A36 steel and an A572 Grade 50 steel would be

$$\text{A36:} \quad K_{Id} = 0.6\sigma_{yd} \simeq 0.6(40 + 25) \simeq 39 \text{ ksi}\sqrt{\text{in.}}$$
$$\text{A572 Grade 50:} \quad K_{Id} = 0.6\sigma_{yd} \simeq 0.6(55 + 25) \simeq 48 \text{ ksi}\sqrt{\text{in.}}$$

In Chapter 4, K_{Id} test results of these two steels indicate measured values of about 40 and 50 ksi$\sqrt{\text{in.}}$ compared with the above calculated values of 39 and 48 ksi$\sqrt{\text{in.}}$, respectively. Thus, Equation (5.7) appears to give a realistic approximation to K_{Id} at their NDT temperature for low-strength structural steels.

5.5 Comparison of CVN-K_{Id}-K_{Ic}-J and δ Relations

In Section 5.2, the CVN-K_{Id}-K_{Ic} correlation was described. Previously, in Chapter 3, relations between K_c, J_{Ic}, and δ_c were presented. All of these relations were used by Wellman [*12*] to predict K_c values for 5 different steels and compare the predictions. The various relations used were as follows:

Two-stage CVN-K_{Id}-K_{Ic}

$$K_{Id} = \sqrt{5\,(CVN)\,E} \tag{5.8}$$

Temperature shift between K_{Id} and K_c

$$T_{\text{shift}} = 215 - 1.5\sigma_{ys} \tag{5.9}$$

CTOD-K_c

$$K_c = \sqrt{1.7E\sigma_{\text{flow}}\delta_c} \tag{5.10}$$

J-integral

$$K_{Ic} = \sqrt{\frac{EJ_{Ic}}{1 - \nu^2}} \tag{5.11}$$

In addition, the Roberts-Newton [*13*] lower-bound CVN-K_{Ic} relation was also used. This relation is as follows:

$$K_c = 9.35\,(CVN)^{0.63} \tag{5.12}$$

Note that this relation does not account for the temperature shift and was developed as a conservative lower-bound relation between impact CVN test results and slow K_{Ic} test results. Also, unlike J and CTOD test results for the elastic-plastic and plastic behaviors, this correlation is based on initial crack extension and does not account for the increased fracture toughness accompanying subcritical ductile crack extension.

The steels studied were:

STEEL	σ_{ys}, ksi
ABS-B	38
A516	39
A533	62
A508	71
A517	108

The results of that study are presented in Figures 5.6 through 5.10 and show the benefits of using correlations to estimate K_c values.

In all cases, the Roberts-Newton lower-bound curve severely underestimates the true fracture toughness at elevated temperatures. The two-stage CVN-K_{Id}-K_{Ic} correlation appears to match the CTOD-K_c relation quite well, but its usefulness is limited to the lower region of the transition-temperature curve.

The K_c-CTOD and K_c-J relations yield results for the lower strength structural steels that increase with temperatures as might be expected. The high-strength steel (A517) behaved in a manner somewhat different from the lower-strength steels. Because of the higher yield strength, there is a much smaller loading-rate shift. Also because of the higher yield strength, the determination of stable crack-

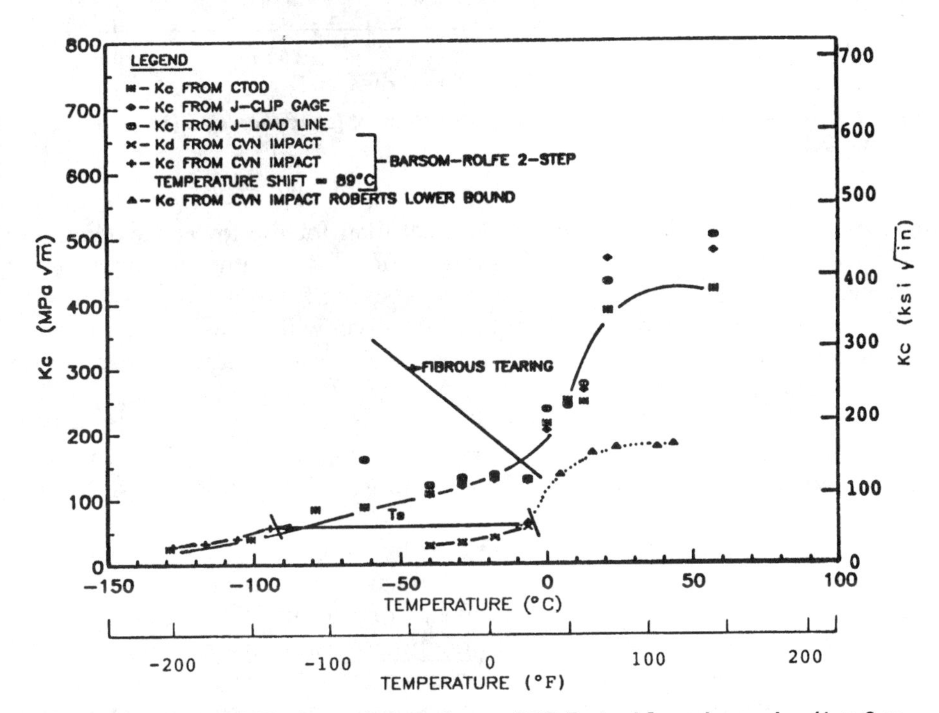

FIG. 5.6 K_c-CVN-CTOD-J correlations for an ABS-B steel [specimen size (1 × 2 × 8) lower-bound values].

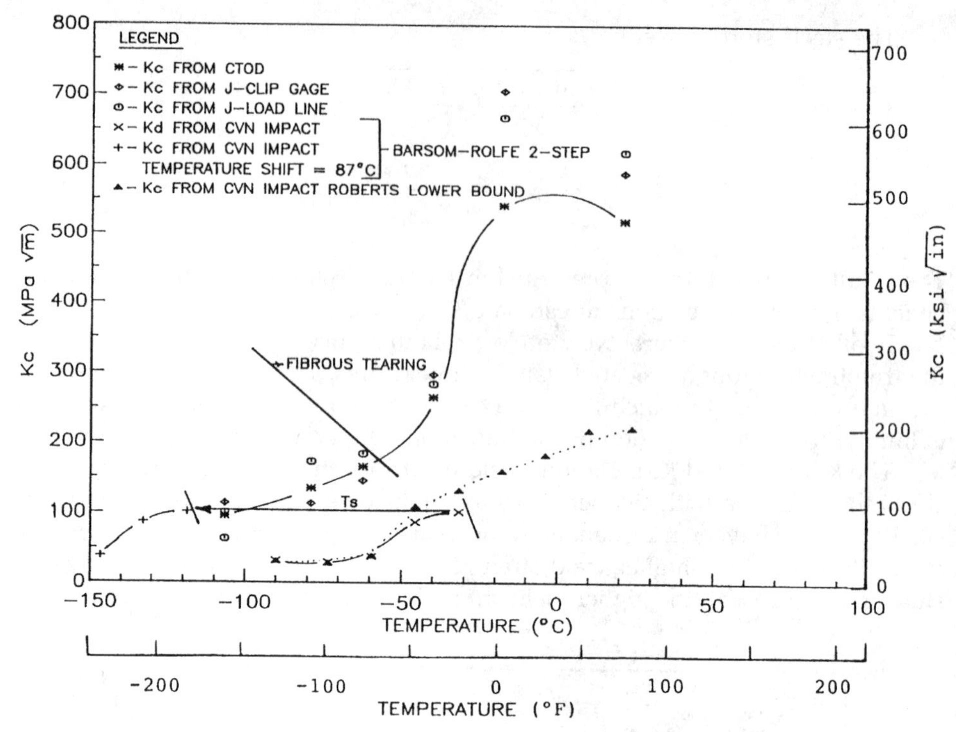

FIG. 5.7 K_c-CVN-CTOD-J correlations for an A516 steel [specimen size (1.6 × 3.2 × 12) lower-bound values].

ing is much more difficult for the A517 steel than for the lower-yield-strength steels. This is analogous to the difficulty encountered in rating the fracture appearance in Charpy V-notch specimens of high-strength steels. However, the general appearance of the temperature-transition curve, with a lower shelf, a lower- and upper-transition region, and an upper shelf, is maintained but at lower fracture toughness levels.

Figure 5.11 is a schematic showing the general regions of transition-temperature curves as described in Chapter 3. They can be divided into four regions, which are (1) lower shelf, (2) lower transition, (3) upper transition, and (4) upper shelf.

The lower-shelf region is characterized by little or no change in fracture toughness with changes in temperature. All fracture surfaces show brittle behavior both for initiation and propagation. Behavior is nearly linear-elastic, and K_{Ic} can be used to describe the behavior throughout most of this region.

The lower-transition region is characterized by a ductile behavior along the fatigue crack front followed by brittle crack propagation with no evidence of prior stable cracking. The fracture toughness increases steadily with increasing temperature from near the linear-elastic K_{Ic} values to the fracture toughness at

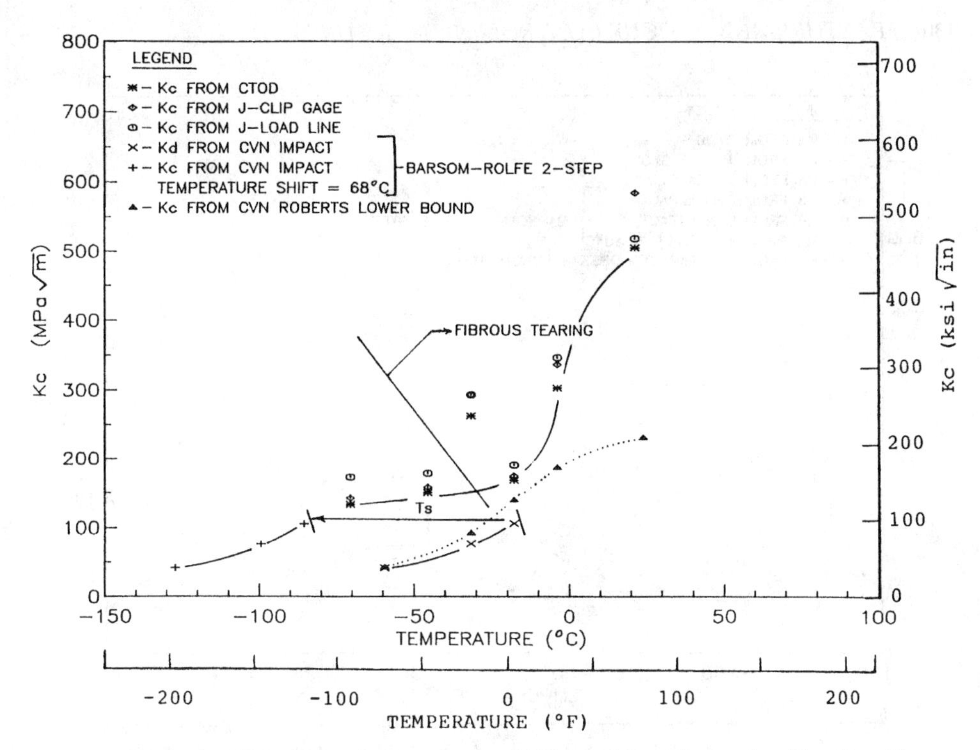

FIG. 5.8 K_c**-CVN-CTOD-*J* correlations for an A533 steel [specimen size (1 × 2 × 8) lower-bound values].**

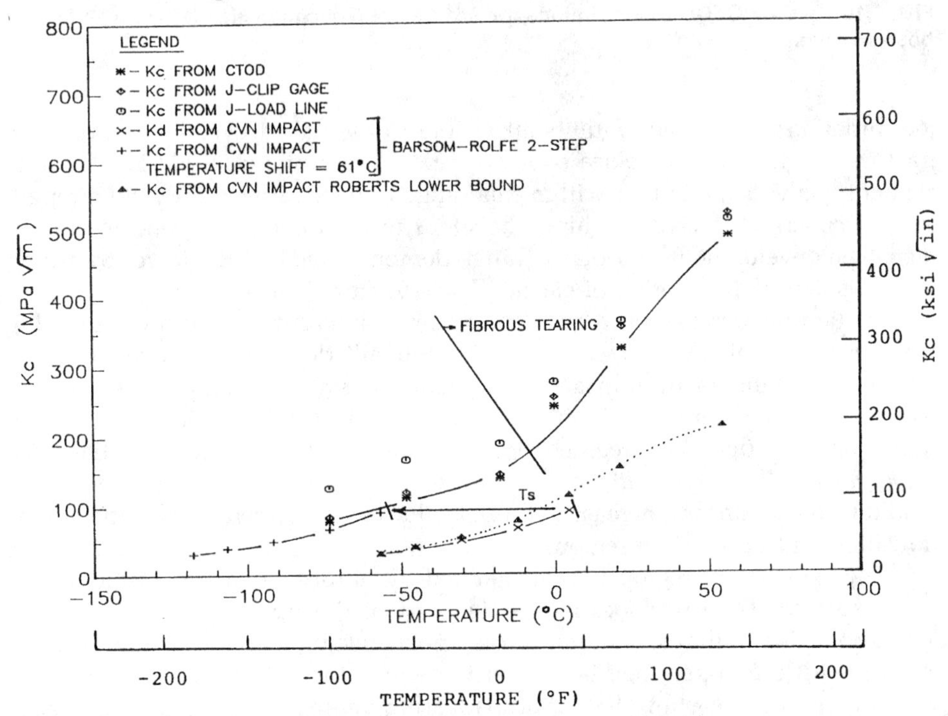

FIG. 5.9 K_c**-CVN-CTOD-*J* correlations for an A508 steel [specimen size (1 × 2 × 8) lower-bound values].**

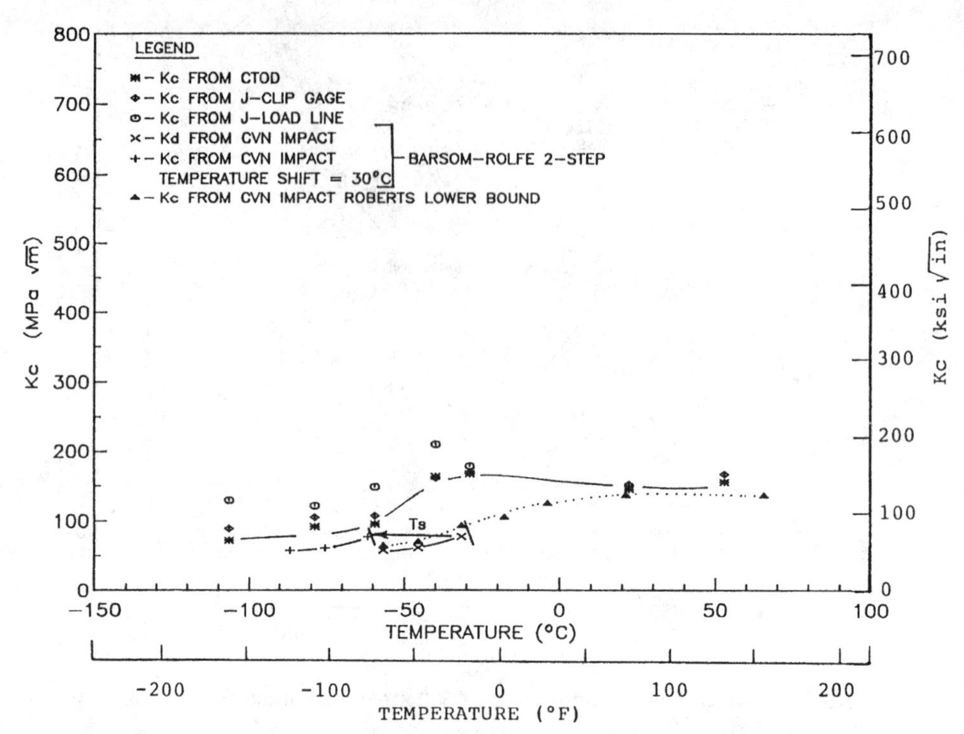

FIG. 5.10 K_c-CVN-CTOD-*J* correlations for A517 steel [specimen size (1 × 2 × 8) lower-bound values].

the initiation of a fibrous thumbnail visible to the naked eye. This increase in fracture toughness takes place over a temperature range of about 200°F. Finite element analyses show that within this range (which varies with yield strength and specimen size), a plastic hinge develops in the three-point bend test specimen. The development of a plastic hinge demonstrates that the lower transition region is definitely a region of elastic-plastic fracture behavior.

In the upper-transition region, the failure initiates by stable ductile tearing that is recognizable by a coarse fibrous "thumbnail" detectable by unaided visual observation of the fracture surface. This ductile "thumbnail" initiation is followed by brittle propagation. The significant increase in the CTOD values in the upper-transition and upper-shelf regions represents increased resistance to ductile crack *propagation.* If only the *initiation* of ductile crack propagation is plotted, there would be no significant increase in CTOD values with temperatures in the upper-transition and upper-shelf regions.

The upper-shelf region is characterized by fibrous ductile tearing over the entire surface. The exact location of the start of the upper shelf is somewhat ambiguous due to the data scatter in the upper-transition region. This ambiguity is avoided if the upper shelf is arbitrarily defined to start at the temperature at which all specimens show 100% fibrous tearing failures.

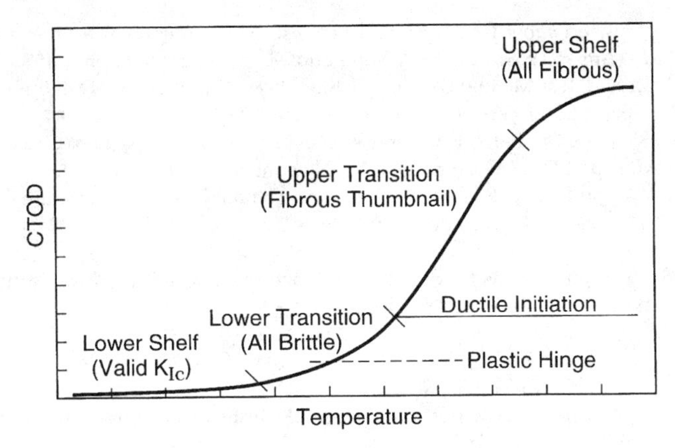

(a) CTOD versus temperature transition

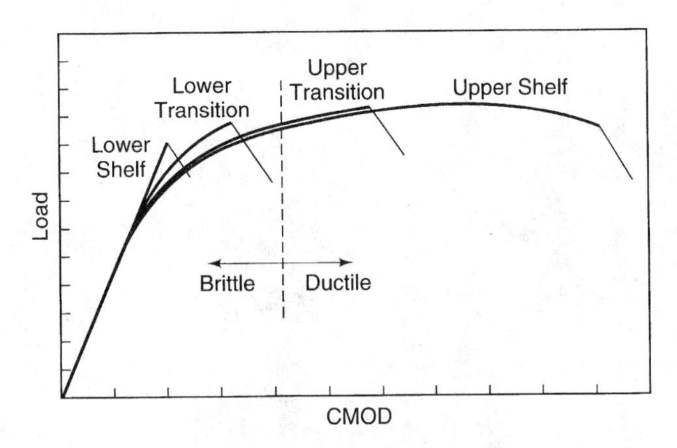

(b) Load versus crack-mouth opening displacement

FIG. 5.11 Schematic CTOD—temperature-transition curve showing four regions.

5.6 References

[1] Barsom, J. M., "The Development of the AASHTO Fracture-Toughness Requirements for Bridge Steels," American Iron and Steel Institute Paper, Washington, DC, February 1975.

[2] "PVRC Recommendations on Toughness Requirements for Ferritic Materials," *WRC Bulletin*, Vol. 175, New York, Aug 1972.

[3] E 23-96, Standard Test Methods for Notched Bar Impact Testing of Metallic Materials, ASTM, Vol. 03.01.

[4] Clausing, D. P., "Effect of Plane-Strain Sensitivity on the Charpy Toughness of Structural Steels," *International Journal of Fracture Mechanics*, Vol. 6, No. 1, March 1970.

[5] Holloman, J. H., "The Notched-Bar Impact Test," *Transactions, American Institute of Mining, Metallurgical, and Petroleum Engineers*, Vol. 158, 1944, pp. 310–322.

[6] Gross, J. H., "Effect of Strength and Thickness on Notch Ductility," *Impact Testing of Metals, ASTM STP 466*, American Society for Testing and Materials, Philadelphia, 1970, p. 21.

[7] Paris, P. C. discussion paper by Begley and Landes: "The J Integral as a Fracture Criterion," ASTM *STP 514,* American Society for Testing and Materials, Philadelphia, 1972.

[8] E 208-95a, Standard Test Method for Conducting Drop-Weight Test to Determine Nil-Ductility Transition Temperature of Ferritic Steels, ASTM, Vol. 03.01.

[9] Irwin, G. R., Krafft, J. M., Paris, P. C., and Wells, A. A., "Basic Aspects of Crack Growth and Fracture," *NRL Report 6598,* Washington, DC, November 21, 1967.

[10] Shoemaker, A. K. and Rolfe, S. T., "The Static and Dynamic Low-Temperature Crack-Toughness Performance of Seven Structural Steels," *Engineering Fracture Mechanics,* Vol. 2, No. 4, June 1971.

[11] Pellini, W. S., "Design Options for Selection of Fracture Control Procedures in the Modernization of Codes, Rules and Standards," *Proceedings of the Joint U.S.—Japan Symposium on Application of Pressure Component Codes,* Tokyo, March 13–15, 1973.

[12] Wellman, G. W. and Rolfe, S. T., "Engineering Aspects of CTOD Fracture Toughness Testing," *WRC Bulletin,* Vol. 299, November 1984.

[13] Roberts, R. and Newton, C., "Report on Small-Scale Test Correlations with K_{Ic} Data," *WRC Bulletin,* Vol. 299, November 1984.

6

Fracture-Mechanics Design

6.1 Introduction

"DESIGN" IS A TERM used in different ways by engineers. From a structural viewpoint, the term usually refers to the synthesis of various disciplines (statics, strength of materials, structural analysis, matrix algebra, etc.) to create a structure that is proportioned and then detailed into its final shape. When the word "design" is used in this sense, designing to prevent fracture usually refers to using an appropriate stress level as well as to the elimination (as much as possible) of those structural details that act as stress raisers and that can be potential fracture-initiation sites; for example, weld discontinuities, mismatch, and intersecting plates. Unfortunately, large complex structures (either welded or bolted) rarely are designed (or fabricated) without these discontinuities, although good design and fabrication practices can minimize their size and number.

From a materials viewpoint, the word "design" usually refers to the selection of a material and of the appropriate design stress level at the particular service temperature and loading rate to which the structure will be subjected. Thus, from this viewpoint, designing to prevent fracture refers to appropriate material selection as well as to selection of the appropriate allowable stress.

Both these approaches to design are valid, but they should not be confused. The first (and traditional) definition assumes that the designer starts with a given material and design stress level (often specified by codes). Thus, the design involves the process of detailing and proportioning members to carry the given loads without exceeding the allowable stress. The allowable stress is usually a certain percentage of the yield strength for tension members and a certain percentage of the buckling stress for compression members. These "allowable" design stress levels assume "perfect" fabrication in that the structures are usually assumed to have no crack-like discontinuities, defects, or cracks. It is realized that stress concentrations or mild discontinuities will be present, but the designer assumes that his structural materials have sufficient ductility (which is a different property than notch toughness) to yield locally and redistribute the load in the vicinity of these stress concentrations or mild discontinuities.

The second use of the term "design" refers to selection of materials and allowable stress levels based on the realization that discontinuities in large complex structures may be present or may initiate under cyclic loads or stress-corrosion cracking and that some level of notch toughness (which is a different property than ductility) may be desirable. This aspect of design has recently been made more quantitative by the development of fracture mechanics as an engineering science.

To provide a safe, fracture-resistant structure, both of these design approaches should be followed. The designer must properly proportion the structure to prevent failure by either tensile overload or compressive instability *and* by unstable crack growth. Historically, most design criteria have been established to prevent yielding (either in tension or compression) and to prevent buckling. The yielding and buckling modes of failure have received considerable attention because analytical design procedures were available and could be used in various theories of failure. The maximum shearing stress theory of failure and the Euler buckling analysis are among the most widely used analytical design principles to prevent failure by general yielding or buckling. The recent development of fracture mechanics as an analytical design tool finally "fills the gap" in the designer's various techniques for the safe design of structures.

Numerous textbooks describe the various design techniques to prevent either general yielding or buckling by proper proportioning of the various structural members, and the reader is referred to the large number of design textbooks in the particular field of structures (bridges, ships, pressure vessels, aircraft, etc.) for these well-established design procedures.

This book deals primarily with the selection of materials and appropriate design stress levels for fracture-resistant structures. This design approach assumes that discontinuities may be present in large complex structures. However, the incidence of fractures (as well as fatigue cracks) can be *reduced* and eliminated by good design and detailing practices and by minimizing discontinuities.

The remainder of this chapter deals with a description of how fracture mechanics can aid the engineer in the initial selection of:

1. Materials,
2. Design stress levels, and
3. Tolerable crack sizes for quality control or inspection

for fracture-resistant design of *any* large complex structures such as bridges, ships, pressure vessels, aircraft, and earth-moving equipment. Thus, this book should be used in conjunction with textbooks that describe the more traditional methods of designing to prevent either the buckling or yielding modes of failure to ensure that all possible failure modes are considered.

Regardless of the type of structure to be considered, fracture mechanics design assumes that the engineer has established the following general information:

1. Type and overall dimensions of the structure (bridge, pressure vessel, etc.).
2. General size of the tension members (length, diameter, etc.).
3. Additional performance criteria and service conditions (e.g., minimum weight, least cost, maximum resistance to fracture, specified design life, loading rate, operating temperature).
4. Applied stress and cyclic stress range, where crack growth can occur (as discussed in Chapters 7–11).

With this basic information, the designer can incorporate fracture mechanics values at the service temperature and loading rate in the "design" of a fracture-resistant structure. The fatigue life can be estimated as well.

If the designer knows or can measure the critical value of fracture toughness at the service conditions (that is, K_{Ic}, K_c, K_{Id}, etc., as described in Chapter 3) or can estimate these values using the correlations described in Chapter 5, the philosophy of design using fracture mechanics is fairly straightforward. Basically, the designer should make sure that the applied stress intensity factor, K_I, is always less than the critical stress intensity factor, K_c, that is appropriate for the particular structure in the same manner that σ is kept below σ_{ys} to prevent yielding.

Fracture-mechanics "design" can follow either a fail-safe or a safe-life principle, both of which have been used extensively in design. Fail-safe design assumes that if an individual member fails, the overall structure is still safe from total fracture. Conversely, safe-life principles assume that for the particular service loading the structure will last the entire design life of the structure and failure will not occur while the structure is in service. Structures with multiple-load paths are essentially fail-safe structures because of the structural redundancy. Safe-life structures may or may not have multiple-load paths, that is, redundancy. In either case, the objective is to keep the applied K_I below the critical K_c throughout the life of the structure.

Currently, most engineers recognize the fact that discontinuities may be present in large complex structures or that they may initiate and grow during the service life of the structure. The structural designer, therefore, should not limit his or her analysis to the traditional approaches of design whereby factors of safety or indices of reliability for unflawed structures are used but also should consider the possible presence of discontinuities in his or her structure.

It should be emphasized that fractures in structures are rare. Hence, fracture mechanics is not widely used, and this is actually a desirable situation. However,

1. As designs become more complex,
2. As the use of high-strength thick welded structural materials becomes more common compared with the use of lower-strength thinner bolted plates,

3. As the fabrication and construction become more complex,
4. As the magnitude of loading increases,
5. As actual factors of safety decrease because of the use of computer design,

the probability of fracture in large complex structures increases, and the designer should be aware of what can be done to prevent brittle fractures.

In this chapter, design procedures using fracture-mechanics concepts will be limited to design considerations related to the *terminal* failure by fracture after possible crack extension has occurred by fatigue or stress corrosion crack growth. Crack growth to terminal conditions is covered in Chapters 7–11.

Chapters 12, 13, and 14 in Part IV will focus on general methods to avoid fracture and fatigue failures in both new and existing structures. For existing structures, the options generally are more limited than in new design and the actual presence of discontinuities or fatigue cracks is more likely. Hence the concepts of fitness-for-service or life extension described in Chapter 14 clearly are pertinent for many structures that have seen extensive service. However, the principles described in this chapter serve as the basis for the application to new and existing structures described in Part IV.

6.2 General Fracture-Mechanics Design Procedure for Terminal Failure

The critical stress-intensity factor for a particular material at a given temperature and loading rate is related to the nominal stress and flaw size as follows:

$$K_{Ic}, K_c, K_{Ic}(t), K_{Id} = C\sigma\sqrt{a} \tag{6.1}$$

where K_{Ic}, K_c, etc. = critical fracture toughness of a material, ksi $\sqrt{\text{in.}}$ (Chapter 3), tested at a particular temperature and loading rate (Chapter 4). If J_c or δ_c values are obtained, convert these to K_c, as discussed in Chapters 3 and 5.

C = constant, function of crack and specimen geometries, Chapter 2.

σ = nominal applied stress, ksi, using tradition design methodologies.

a = flaw size as a critical dimension for a particular crack geometry, Chapter 2.

Thus, the maximum flaw size a structural member can tolerate at a particular stress level is:

$$a = \left[\frac{K_{Ic}, K_c, K_{Ic}\,(t), K_{Id}}{C \cdot \sigma}\right]^2 \tag{6.2}$$

Accordingly, the engineer can analyze the safety of a structure against failure by brittle fracture in the following manner:

1. Determine the values of the appropriate critical K value [K_{Ic}, $K_{Ic}(t)$, K_c, K_{Id}, etc.] and the corresponding yield strength at the service temperature and loading rate for the materials being considered for use in the structure. Note that for a complete analysis of welded structures, fracture toughness values for the weldment also should be determined.
2. Select the most probable type of flaw that can exist in the member being analyzed and the corresponding K_I equation. Figure 6.1 shows the fracture-mechanics models that describe three of the more common types of flaws occurring in structural members. Complex-shaped flaws often can be approximated by one of these models. Additional equations to analyze other crack geometries were described in Chapter 2.
3. Determine the stress–flaw-size relation at various possible design stress levels using the appropriate K_I expression and the appropriate critical K_c value.

As an example, the relation among stress, flaw size, and stress intensity factor, K_I, for the through-thickness crack geometry shown in Figure 6.1*a* is:

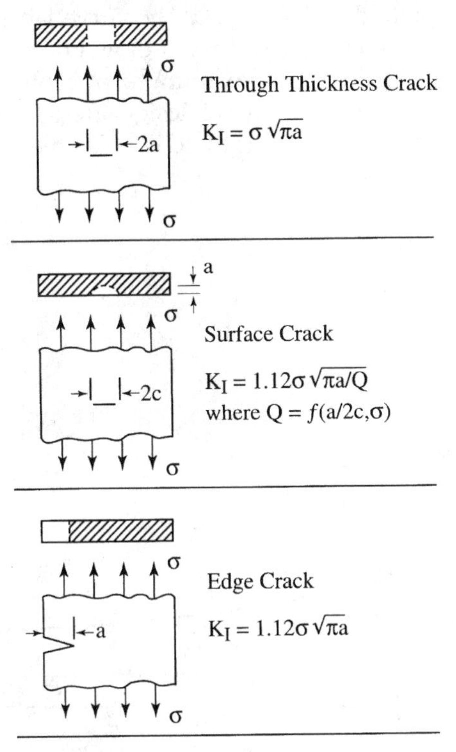

FIG. 6.1 K_I values for different crack geometries.

$$K_I = \sigma\sqrt{\pi a} \tag{6.3}$$

Assume that the material being analyzed has a K_{Ic} value at the service temperature of 50 ksi$\sqrt{\text{in.}}$ and a yield strength of 100 ksi. Substituting $K_I = K_{Ic}$, the possible combinations of stress and critical flaw size at failure are described by:

$$K_{Ic} = 50 \text{ ksi}\sqrt{\text{inch}} = \sigma\sqrt{\pi a}$$

Using this equation, values of the critical crack size (2*a*) for various stress levels are calculated as follows:

σ, ksi	2*a*, in.
10	16
20	4
30	1.8
40	1.0
50	0.64
60	0.44
70	0.32
80	0.24
90	0.20
100	0.16

These results are plotted in Figure 6.2. The curve labeled K_{Ic} is the locus of points at which unstable crack growth (fracture) will occur. Thus, if the nominal design stress level is 30 ksi, then the maximum tolerable flaw size, or *critical crack size,* 2*a*, is 1.8 in. Conversely, if the nominal design stress level is 60 ksi, the critical crack size is 0.44 in. It can be seen that there is no single critical crack size for a

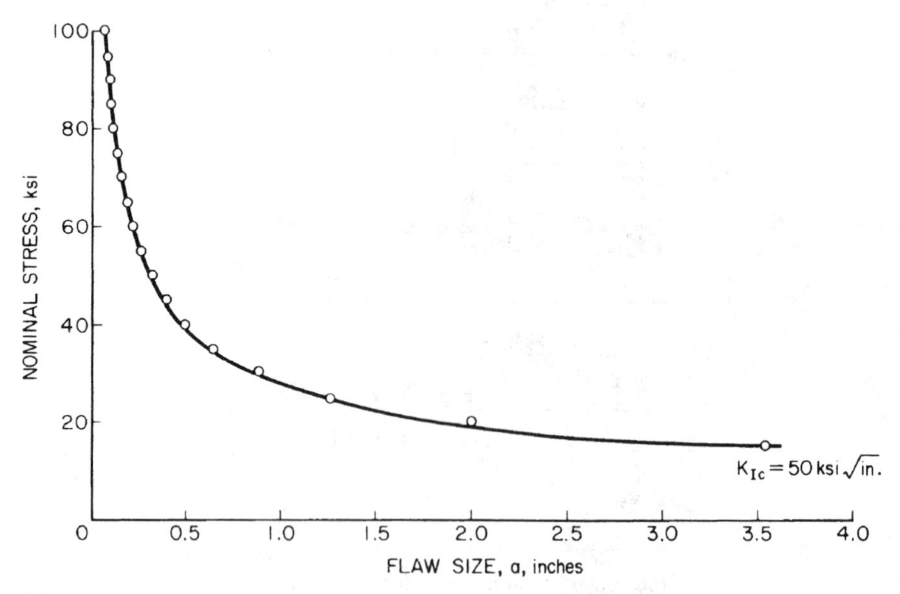

FIG. 6.2 Stress–flaw-size relation for a through-thickness crack in a material having K_{Ic} = 50 ksi$\sqrt{\text{in.}}$

particular structural material (at a given temperature and loading rate) but rather a "semi-infinite" number of critical crack sizes, depending on the nominal design stress level. Figure 6.3 shows the locus of values of stress and flaw size for a "design K_I" of $K_{Ic}/2$, that is, a factor of safety of two against fracture based on the critical stress-intensity factor using the traditional definition of factor of safety. In this case, a structure having a flaw size of 0.22 in. loaded to a design stress of 30 ksi has a factor of safety of two against fracture. Similarly the K_I = 25 ksi$\sqrt{\text{in.}}$ curve is the locus of all stress levels and corresponding flaw sizes with a factor of safety of two against fracture.

Obviously, for high fracture-toughness materials (compared with materials with lower fracture toughness), the possible combinations of design stress and allowable flaw sizes that will not lead to failure are larger. In Figure 6.4 the locus of failure points for a material with a K_{Ic} of 100 ksi$\sqrt{\text{in.}}$ is shown along with the locus of failure points of the material having a K_{Ic} of 50 ksi$\sqrt{\text{in.}}$ For the material with a K_{Ic} of 100 ksi$\sqrt{\text{in.}}$, the tolerable flaw sizes at all stress levels are considerably larger than are those for the material with the lower fracture toughness, and the possibility of fracture is reduced considerably.

Thus, it can be seen that to minimize the possibility of brittle fracture in a given structure, the designer has three *primary* factors that can be controlled:

1. Material fracture toughness at the particular service temperature and loading rate (critical K), ksi$\sqrt{\text{in.}}$
2. Nominal stress level (σ), ksi.

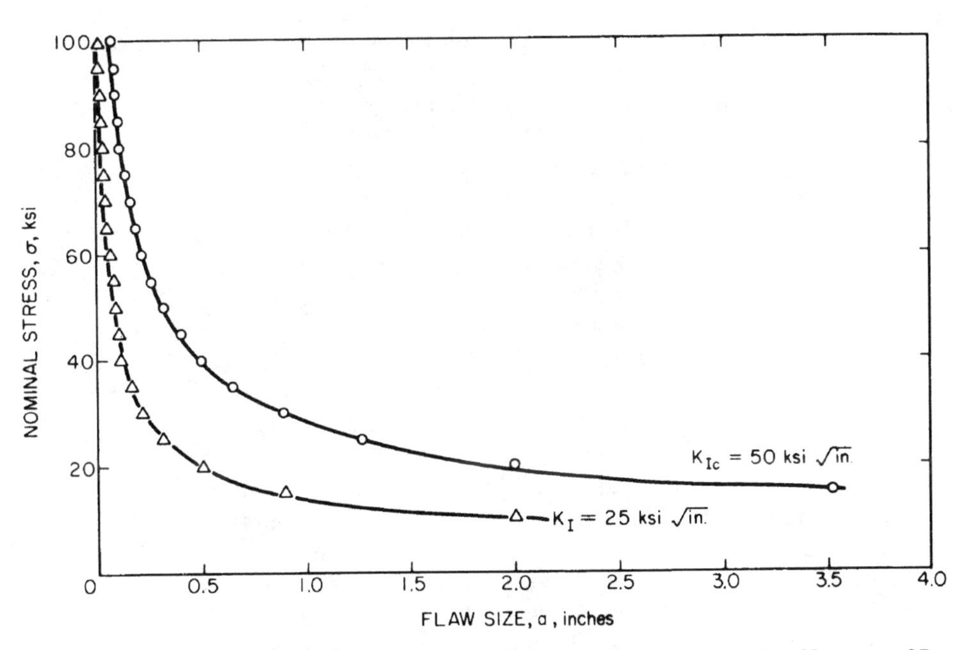

FIG. 6.3 Stress–flaw-size relation for a through-thickness crack showing $K_{I(design)}$ = 25 ksi$\sqrt{\text{in.}}$

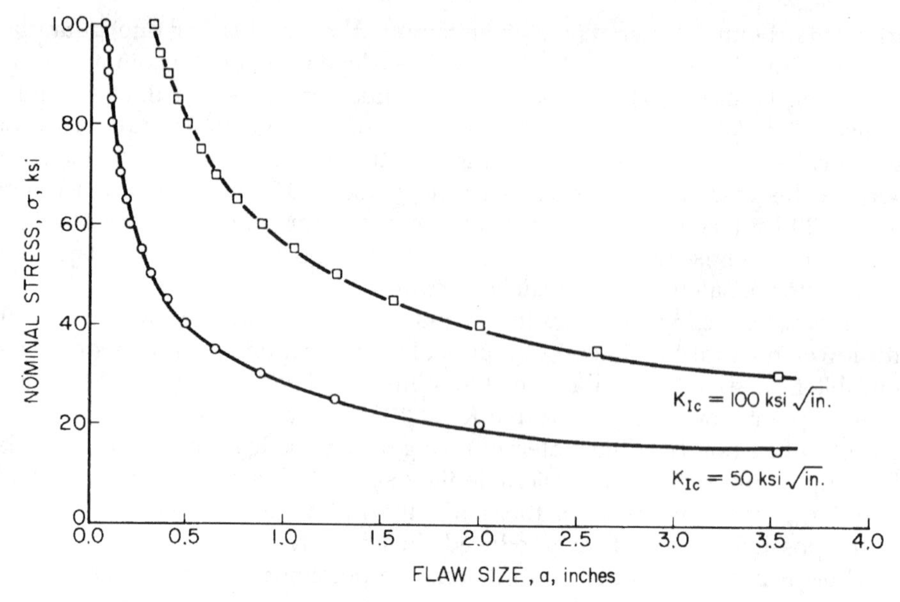

FIG. 6.4 Stress–flaw-size relation for a through-thickness crack showing effect of higher K_{Ic} (100 ksi$\sqrt{\text{in.}}$).

3. Flaw size present in the structure (a), in.

All three of these factors affect the possibility of a brittle fracture occurring in structures. Other factors such as temperature, loading rate, residual stresses, stress concentrations, and so on merely affect the three primary factors. It should be noted that flaws need not be present for brittle fractures to occur if the other factors are sufficiently severe and if the general level of constraint is high. An example of this behavior, i.e., high stresses at the toes of welds that join intersecting plates, will be described as a case study in Part V. Under these conditions, several factors other than the fracture toughness would be the controlling factors and the application of fracture mechanics technology is not appropriate for analyzing the fracture initiating event because there is no pre-existing crack.

Design engineers have known about these three primary factors for many years and have reduced the susceptibility of their structures to brittle fractures by applying these common sense principles to their structures qualitatively. The traditional use of "good design" practice, that is, by using appropriate stress levels and minimizing discontinuities, has led to the reduction of brittle fractures in many structures. In addition, the use of good fabrication practices, that is, eliminating or decreasing the flaw size by proper welding control and inspection, as well as the use of materials with good notch-toughness levels (for example, as specified in a Charpy V-notch impact test), has minimized the probability of brittle fractures in structures. This has been the traditional design approach to reducing brittle fractures in structures and has been quite successful. However, fracture mechanics offers the engineer the possibility of a *quantitative* approach to designing to prevent brittle fractures in structures containing crack-like imperfections.

In summary, the general relationship among material fracture toughness (K_c), nominal stress (σ), and flaw size (a) is shown schematically in Figure 6.5. If the particular combinations of stress and flaw size in a structure reach the K_c level, fracture can occur. Thus, there are many combinations of stress and flaw size, σ_f and a_f, that may cause fracture in a structure that is fabricated from a structural material having the particular K_c value at a particular service temperature and loading rate. Conversely, there are many combinations of stress and flaw size, for example, σ_o and a_o, that will not cause failure of a particular material.

As will be discussed later, K_I can increase throughout the life of a structure because of crack growth by fatigue. This behavior is shown schematically in Figure 6.6 for a crack, a, increasing from a_i to a_f by fatigue.

As discussed previously, a useful analogy for the designer is the relation among applied load (P), nominal stress (σ), and yield stress (σ_{ys}) in an unflawed structural member and among applied load (P), stress intensity factor (K_I), and critical stress intensity factor for fracture (K_c, K_{Ic}, K_{Id}, etc.) in a structural member with a flaw. In an unflawed structural member, as the load, P, is increased, the nominal stress increases until an instability (yielding at σ_{ys}) occurs. As the load is increased in a structural member with a flaw (or as the size of the flaw grows by fatigue or stress corrosion), the stress intensity factor, K_I, is increased until an

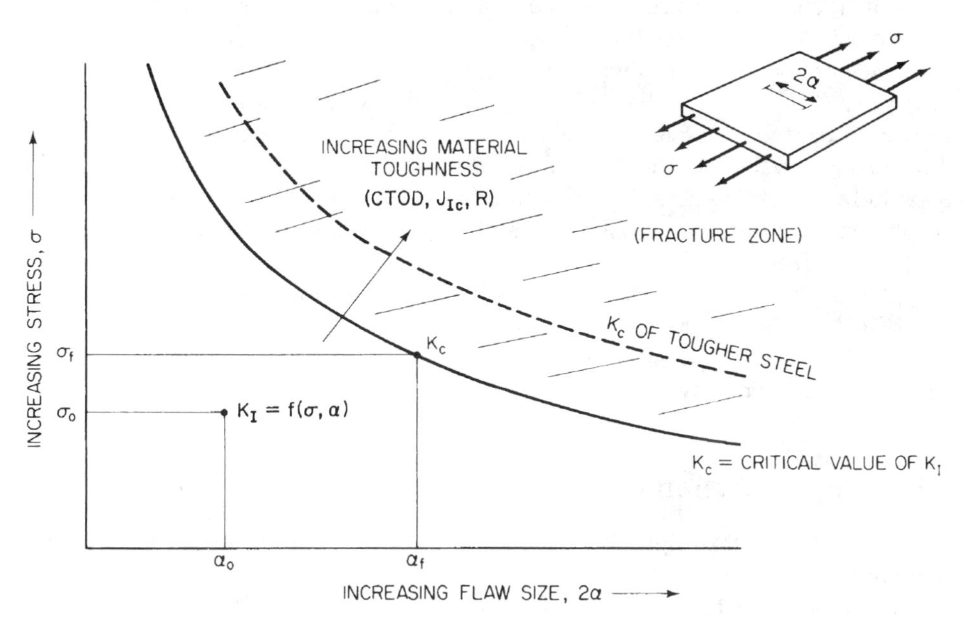

FIG. 6.5 Schematic relation among stress, flaw size, and material toughness.

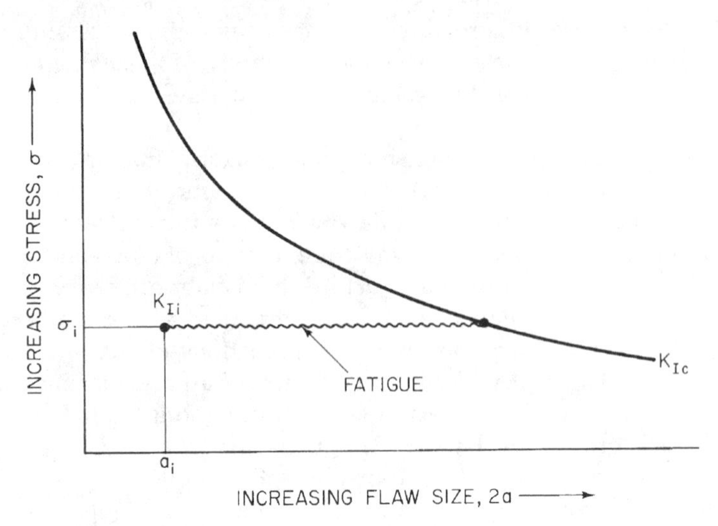

FIG. 6.6 Schematic showing how $K_{I(initial)}$ can increase by fatigue to K_{Ic}.

instability (fracture at K_c, K_{Ic}, K_{Id}, etc.) occurs. Thus, the K_I level in a structure should always be kept below the appropriate critical K_c value in the same manner that the nominal design stress (σ) is kept below the yield strength (σ_{ys}).

Another analogy that may be useful in understanding the fundamental aspects of fracture mechanics is the comparison with the Euler column instability (Figure 6.7). The stress level required to cause instability in a column (buckling) decreases as the L/r ratio increases. Similarly, the stress level required to cause instability (fracture) in a flawed tension member decreases as the flaw size (a) increases. As the stress level in either case approaches the yield strength, both the Euler analysis and the K_c analysis are invalidated because of yielding. To prevent buckling, the actual stress and L/r value must be below the Euler curve. To prevent fracture, the actual stress and flaw size, a, must be below the K_c line shown in Figure 6.7.

Obviously, using a material with a high level of fracture toughness (for example, a K_{Ic} level of 100 ksi$\sqrt{\text{in.}}$ compared with 50 ksi$\sqrt{\text{in.}}$ as shown in Figure 6.4) will increase the possible combinations of design stress and flaw size that a structure can tolerate without fracturing.

6.3 Design Selection of Materials

Current methods of design and fabrication of large complex structures are such that engineers expect these structures to be able to tolerate design stress loading in tension without failing. For yield stress loading, the critical crack size, a, is proportional to $(K_{Ic}/\sigma_{ys})^2$, $(K_{Id}/\sigma_{yd})^2$, or $(K_c/\sigma_{ys})^2$. Thus, these $(K_c/\sigma_{ys})^2$ ratios be-

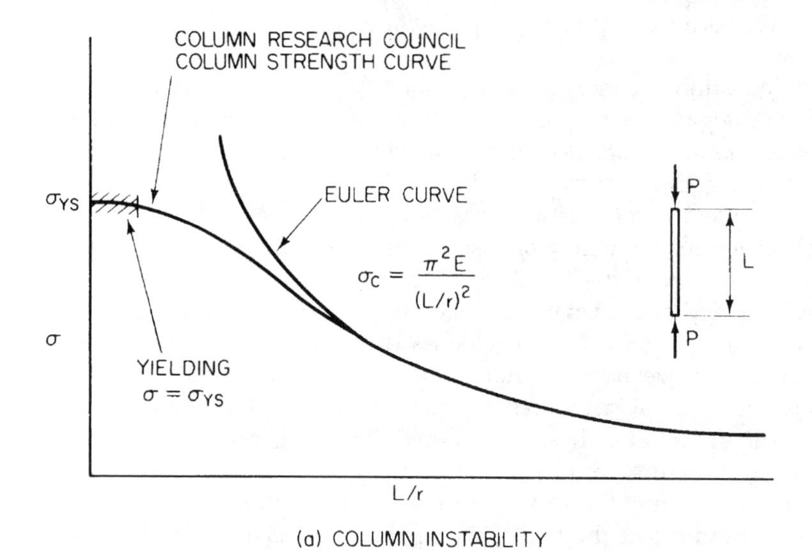

(a) COLUMN INSTABILITY

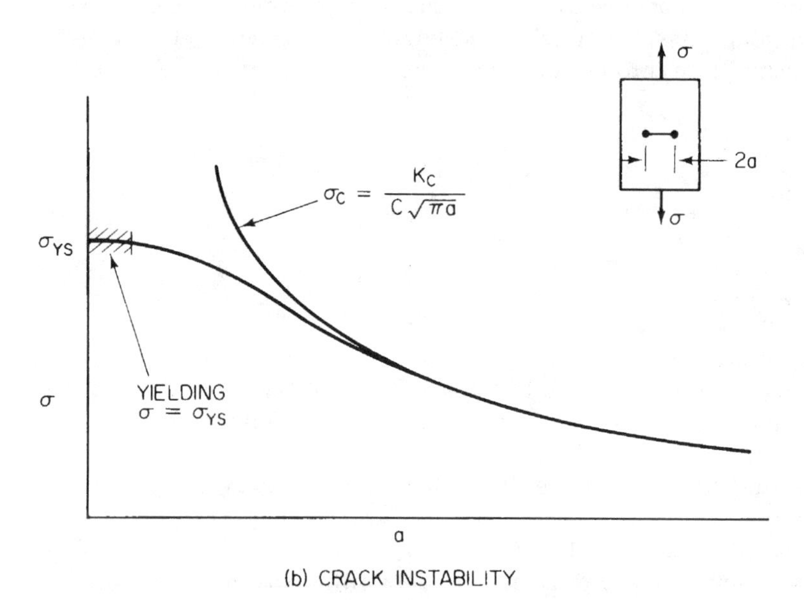

(b) CRACK INSTABILITY

FIG. 6.7 (*a*) Column instability and (*b*) crack instability.

come good indexes for measuring the relative fracture toughness of structural materials. Because for most structural applications it is desirable that the structure tolerate large flaws without fracture, the use of materials with as high a K_c/σ_{ys} ratio as is possible *consistent with economic considerations* is a desirable condition.

The question becomes, *How high must these ratios be to ensure satisfactory performance in large complex structures, where complete inspection for initial cracks and continuous monitoring of crack growth throughout the life of a structure may not always be possible, practical, or economical?*

No simple answer exists because the answer is obviously dependent on the type of structure, frequency of inspection, access for inspection, quality of fabrication, design life of the structure, consequences of failure for a structural member, redundancy of load path, probability of overload, fabrication and material costs, etc. However, fracture mechanics does provide an engineering approach to evaluate this question rationally. For example, conservative assumptions are that flaws do exist in structures and that yield stress loading is possible in parts of the structure. Under these conditions, the K_c/σ_{ys} ratios for materials used in a particular structure are one of the primary controlling design parameters that can be used to define the relative safety of a structure against fracture.

As an example of the use of the K_{Ic}/σ_{ys} ratio as a material selection parameter, the behavior of a wide plate with a through-thickness center crack of length $2a$, Figure 6.1, is analyzed for materials having assumed levels of strength and fracture toughness. The total crack length, $2a$, is calculated from the following relationship for a through-thickness crack in a wide plate, Figure 6.1,

$$K_{Ic} = \sigma_{design}\sqrt{\pi a} \tag{6.4}$$

Therefore,

$$a = \frac{1}{\pi}\left(\frac{K_{Ic}}{\sigma_{design}}\right)^2 \tag{6.5}$$

$$2a = \frac{2}{\pi}\left(\frac{K_{Ic}}{\sigma_{design}}\right)^2 \tag{6.6}$$

Table 6.1 presents assumed values of K_{Ic} for various steels having yield strengths that range from 40 to 260 ksi. It should be emphasized that there is no single unique critical stress-intensity factor for any one steel at a given test temperature and rate of loading because the various K_c, K_{Ic}, $K_{Ic}(t)$, or K_{Id} values for a given steel depend on the thermomechanical history, that is, heat treatment, rolling, etc. For each of these steels, the critical crack size, $2a$, is calculated for four design stress levels, that is, 100% σ_{ys}, 75% σ_{ys}, 50% σ_{ys}, and 25% σ_{ys}.

The results presented in Table 6.1 demonstrate the influence of strength level, fracture toughness, and design stress level on the critical crack size in a wide plate. For example, the critical flaw size for the 260-ksi yield strength steel loaded to 50% of the yield strength (design stress of 130 ksi) is 0.24 in. If a design stress of 130 ksi were required for a particular structure, it would be preferable to use

TABLE 6.1. Values of Critical Crack Size as a Function of Yield Strength and Fracture Toughness

	ASSUMED K_{Ic} VALUES, ksi$\sqrt{\text{in.}}$	CRITICAL FLAW SIZE, $2a$ (in.) (ACTUAL DESIGN STRESS LEVEL, ksi SHOWN IN PARENTHESES)			
σ_{ys}, ksi		$\sigma = 100\%\sigma_{ys}$	$\sigma = 75\%\sigma_{ys}$	$\sigma = 50\%\sigma_{ys}$	$\sigma = 25\%\sigma_{ys}$
260	80	0.06(260)	0.11(195)	0.24(130)	0.96(65)
220	110	0.16(220)	0.28(165)	0.64(110)	2.55(55)
180	140	0.39(180)	0.68(135)	1.54(90)	6.16(45)
180	220	0.95(180)	1.69(135)	3.80(90)	15.22(45)
140	260	2.20(140)	3.90(108)	8.78(70)	35.13(35)
110	170	1.52(110)	2.70(82.5)	6.08(55)	24.33(27.5)
80	200	3.98(80)	7.07(60)	15.92(40)	63.66(20)
40	100	3.98(40)	7.07(30)	15.92(20)	63.66(10)

a lower-strength, higher fracture-toughness material (e.g., the 180-ksi yield strength material) at a design stress of 75% σ_{ys} (135 ksi). The reason is that for either of the two 180-ksi yield-strength steels analyzed in Table 6.1, the critical crack sizes are much larger than 0.24 in. For example, the critical crack size is 0.68 in. for the material with a K_{Ic} of 140 ksi$\sqrt{\text{in.}}$ and 1.69 in. for the material with a K_{Ic} of 220 ksi$\sqrt{\text{in.}}$

Obviously, of these two materials, the one with higher fracture toughness would be a more fracture-resistant structural material, but may be a more expensive material also. This point illustrates one of the basic aspects of fracture-resistant design, namely economics, that is not as obvious in other more traditional modes of design. That is, structural materials that have very high levels of fracture toughness at service temperatures and loading rates are available. However, because the cost of these materials generally increases with their ability to perform satisfactorily under more severe operating conditions, the engineer usually does not want to specify more fracture toughness than is required for the particular application. Thus the problem of fracture-resistant design is one of optimizing structural performance consistent with safety and economic considerations. However, this is no different from the general definition of good engineering design, that is, an optimization of performance, safety, and cost.

Further analysis of Table 6.1 indicates that the traditional method of selecting a design stress as some percentage of the yield strength does not always give the same degree of safety and reliability against fracture as it is presumed to give for yielding. For example, assume that the design stress for the two steels having yield strengths of 220 ksi and 110 ksi is 50% σ_{ys}, or 110 ksi and 55 ksi, respectively. For the 220-ksi steel, the critical crack size is 0.64 in., whereas, for the 110-ksi yield-strength steel, the critical crack size is 6.08 in. If the design stress for the lower-yield-strength steel were increased to 100% σ_{ys} (110 ksi), the critical crack size would be 1.52 in. This critical crack size is still significantly larger than the 0.64-in. size for the 220-ksi yield-strength steel with a design stress of 50% σ_{ys}.

For the lower-strength steels shown in Table 6.1 that have higher values of K_{Ic}/σ_{ys}, the critical crack sizes become extremely large, indicating that fracture no longer controls the design in these cases. This is analogous to saying that for very low L/r ratios, the calculated Euler buckling stress is well above the yield stress and the Euler buckling analysis no longer controls the design.

In summary, there may be situations where the designer should specify a *lower*-yield-strength material at a *higher* design stress level (as a percentage of the yield strength), which will actually *improve* the overall safety and reliability of a structure from a fracture-resistant design viewpoint. Good design dictates that the engineer design to prevent failure of his or her structure against *all* possible modes of failure, including fracture.

6.4 Design Analysis of Failure of a 260-In.-Diameter Motor Case

An excellent example of the fact that specifying a percentage of yield strength is not always the best method of establishing the design stress level is the failure of a 260-in.-diameter motor case during hydrotest [*1,2*]. The motor case failed during hydrotest at a pressure of 542 psi, which was about 56% of the planned proof pressure. The motor case was constructed of 250 Grade maraging steel plate joined primarily by submerged arc automatic welding.

Gerberich [2] summarized the failure as follows:

> "Although this motor case was designed to withstand proof pressures of 960 psi, it failed during hydrotest at a pressure of 542 psi. In this 240 ksi yield strength material, the failure occurred at a very low membrane stress of 100 ksi. The fracture was both premature and brittle with crack velocities approaching 5000 feet/second. The result is this 65-ft-high chamber literally flying apart is shown in Figure 6.8. Post-failure examination revealed that the fracture had originated in an area of two defects that had probably been produced by manual-gas-tungsten arc weld repairs. After the crack initiated, it branched into multiple cracks [as shown in Figure 6.9], leading to complete catastrophic failure of the motor case.
>
> "The real lesson in this failure is the lack of design knowledge that went into the material selection. First, the chamber was 0.73 in. thick, which put it into the plane-strain regime for the high-strength material being considered. Grade 250 maraging steel which had a yield strength of about 240 ksi was chosen. For the plane-strain conditions this was not a particularly good choice since the base metal had a plane-strain fracture toughness, K_{Ic}, of only 79.6 ksi$\sqrt{\text{in.}}$ At the design stress of 160 ksi, a critical defect only 0.08 in. in depth could have caused catastrophic failure. Still, the chamber manufacturer thought that defects of this size could be detected. In retrospect, this was not a very judicious decision. Furthermore, the error was compounded by the fact that welding this material provided an even lower K_{Ic} value ranging from 39.4 to 78.0 ksi$\sqrt{\text{in.}}$, the toughness level depending on the location of the flaw in the weld."

FIG. 6.8 Failed motor case with pieces laid out in approximately the proper relation to each other. (Courtesy of J. E. Srawley, NASA Lewis Research Center.)

A post-failure analysis was run on several types of flaw configurations and weld positions which indicated the K_{Ic} value to range from 38.8 to 83.1 ksi$\sqrt{\text{in.}}$ with the average being 55.0 ksi$\sqrt{\text{in.}}$ Because the exact value of the fracture toughness at the failure origin in the chamber is not known, this average value is used as an estimate of K_{Ic}. Post-failure examination of the fracture origin indicated the responsible defect had an irregular banana shape that could best be approximated by an internal ellipse that was 0.22 in. in depth and 1.4 in. long. The critical dimension was the in-depth value of 0.22 in. since the crack would first propagate through the thickness from this dimension. To calculate the critical defect size from the fracture-mechanics principles, a solution for an internal ellipse gives

$$K_I = \sigma(a)^{1/2} f\left(\frac{a}{c}\right)$$

where

$$f\left(\frac{a}{c}\right) = \pi^{1/2}\left(\frac{1}{Q}\right)^{1/2}$$

and a is the half-crack depth, c is the half-crack length, and $f(a/c)$ is related to the complete elliptical integral of the second kind (Chapter 2). However, as a/c

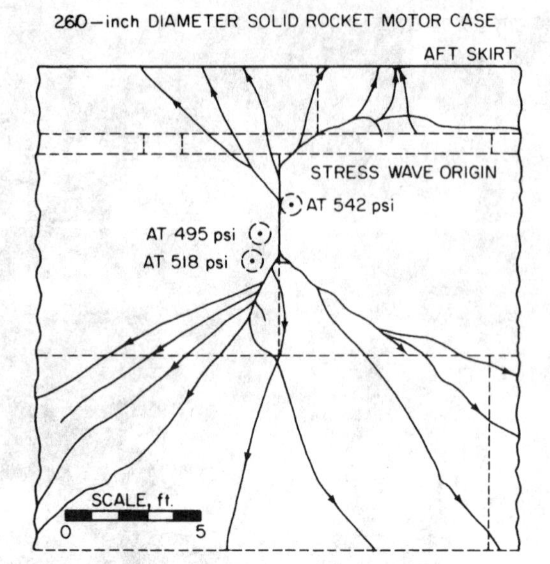

FIG. 6.9 Map of fracture path about failure origin; dashed lines indicate weld (Ref. 2).

approaches 0, $1/Q$ approaches 1.0 and $f(a/c)$ approaches $(\pi)^{1/2}$, and this equation approaches

$$K_I = \sigma(\pi a)^{1/2}$$

For the shape of flaw under consideration, the difference between Equations (6.5) and (6.6) is only 4%, and so for the sake of simplicity, the second equation is utilized. Based on the fracture toughness of 55.0 ksi $\sqrt{\text{in.}}$, the second equation was utilized to make a plot of membrane stress versus defect size for crack instability. This is shown in Figure 6.10 as the curve for Grade 250 maraging steel. An intercept of the 100-ksi failure stress for the chamber predicts a critical defect size, $2a$, of 0.2 in., which is very close to the observed value also indicated in Figure 6.10. Besides the Grade 250 data, Figure 6.10 includes a curve for Grade 200 maraging steel. Although this material has a yield strength that is 20% lower than a Grade 250, its fracture toughness is about triple the average K_{Ic} value for the Grade 250, with K_{Ic} being about 150 ksi$\sqrt{\text{in.}}$ for a member 0.7 in. thick. Using the 150-ksi$\sqrt{\text{in.}}$ value for K_{Ic} and Equation (6.6), a curve for Grade 200 maraging steel was constructed for Figure 6.10. Significantly, the defect that failed the Grade 250 chamber would not have failed a Grade 200 chamber, and, in fact, yield stresses could have been reached without failure. At the failure stress of 100 ksi, it would have taken a flaw 1.42 in. in depth to burst a Grade 200 chamber. This would not have been possible since the thickness was only 0.73 in. Even at the design stress of 160 ksi, it would take a flaw 0.56 in. in depth to cause plane-strain fracture, and this is a very large flaw.

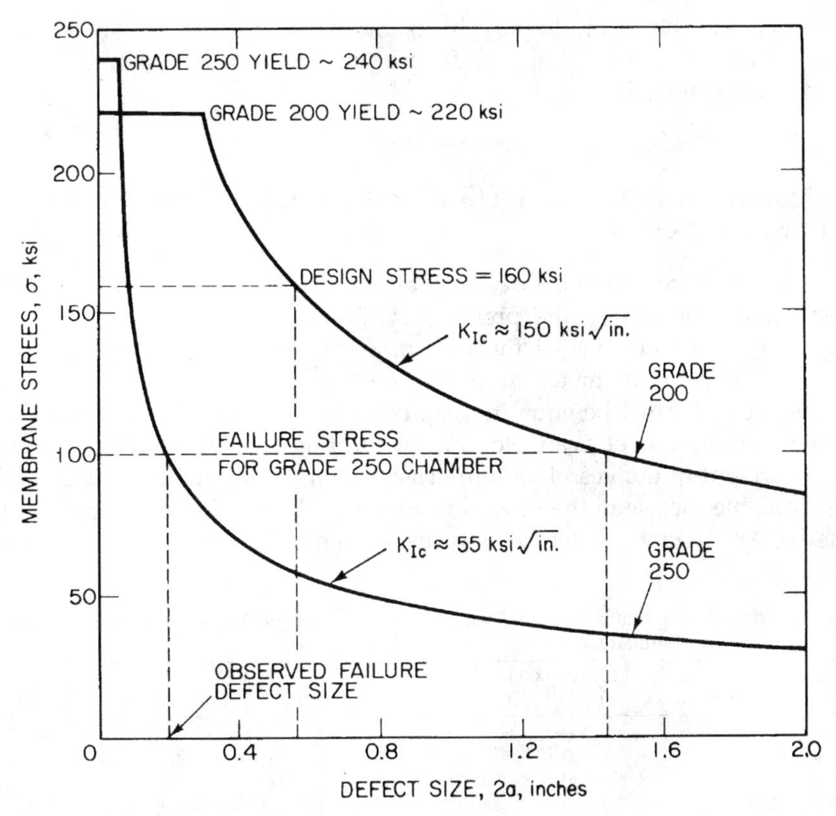

FIG. 6.10 Design curves for critical defect in 260-in. motor case (Ref. *2*).

Thus, the Grade 200 maraging steel would have been a more reliable material for this chamber than the Grade 250 maraging steel. The proof of this is that there was a competition to make the 260-in.-diameter chamber for NASA, and the competitor used Grade 200 maraging steel successfully. Not only were there two successful proof tests of Grade 200 chambers, but there were also two firings that developed thrusts of more than 6 million lb.

In summary, using a lower-strength steel with higher fracture toughness (a larger K_{Ic}/σ_{ys} ratio), even at a design stress which was a higher percentage of the yield strength, would have been better.

A failure in an F-111 Wing Pivot Fitting occurred in a D6-AC steel heat treated to obtain a high-strength level so that the design stress level of ⅔ σ_{ult} would lead to a weight reduction [3]. However, this also led to a lower fracture-toughness level so that the *actual* degree of safety was lowered. In this case, it also seems preferable to have used a heat treatment that gave a higher fracture toughness (but a lower σ_{ult}) and to have used a design stress somewhat higher than ⅔ σ_{ult} so that the *actual* design stresses were the same but the critical crack size was larger. This example plus the example of the 260-in. missile motor case illustrate that *both* the yielding and fracture modes of failure should be considered in structural design.

6.5 Design Example—Selection of a High-Strength Steel for a Pressure Vessel

As a simplified example of the desirability of using fracture-mechanics principles to *select* materials during the preliminary design stages, assume that a high-strength steel pressure vessel must be built to withstand 5000 psi of internal pressure, p, that the diameter, d, of the vessel is nominally 30 in., and that the wall thickness, t, must be equal to or greater than 0.5 in. The designer can use any yield-strength steel available, but in addition to satisfactory performance, cost and weight of the vessel are important factors that must be considered. The steels available for use in the vessel are shown in Table 6.2 along with their yield strengths, assumed K_{Ic} values at the service temperature and assumed cost of each steel.

TABLE 6.2. Yield-Strength and Fracture-Toughness Values of Steels Used in Example Problem as Well as Assumed Costs*

STEEL	YIELD STRENGTH, σ_{ys} (ksi)	K_{Ic} VALUES, ksi$\sqrt{\text{in.}}$	COST, \$/lb
A	260	80	1.40
B	220	110	1.40
C	180	140	1.00
D	180	220	1.20
E	140	260	0.50
F	110	170	0.15

*Assumed values for example only.

6.5.1 Case I—Traditional Design Approach

The traditional design approach is to assume "perfect" fabrication, and therefore there is no flaw. The factor of safety would be based on yielding only, e.g., $\sigma_{design} = \sigma_{ys}/2$ for a factor of safety of 2.

This design procedure is direct since:

$$\sigma_{design} = \frac{\sigma_{ys}}{2}$$

and

$$t = \frac{pd}{2\sigma_{design}}$$

Knowing t for each of the six steels studied in this example, the estimated weight per foot and cost per foot are:

$$\text{weight/ft} = \text{volume/ft} \times \text{density}$$

$$\text{volume/ft} = \pi dt \times 12 \text{ in.}$$

Therefore

$$\text{weight/ft} = \pi dt \times 12 \times \text{density}$$

and

$$\text{cost/ft} = \text{weight/ft} \times \text{cost/lb}$$

A typical calculation for Steel *D* with $\sigma_{ys} = 180$ ksi is as follows:

$$\sigma_{design} = \frac{\sigma_{ys}}{2} = \frac{180}{2} = 90 \text{ ksi}$$

$$t = \frac{pd}{2\sigma_{design}} = \frac{(5000)\,(30)}{2\,(90{,}000)} = 0.83 \text{ inches}$$

$$\text{volume/ft} = (3.14)(30)(0.83) \times 12 = 78.5(12) = 942 \text{ in.}^3$$
$$\text{weight/ft} = \text{volume/ft} \times \text{density} = 942 \times 0.283 = 267 \text{ lb/ft}$$
$$\text{cost/ft} = \text{weight/ft} \times \text{cost/ft} = 267 \times (\$1.20) = \$320/\text{ft}$$

The values for the other steels are calculated in a similar manner and are presented in Table 6.3. These results show the direct effect of reducing weight by using a higher-strength steel. The cost per foot is not necessarily directly related to yield strength because of the numerous factors involved in pricing of structural materials. However, weight is decreased directly by increasing the yield strength. And, if fracture is possible, the overall safety and reliability of the vessel may be decreased, as will be illustrated in the following examples. No K_{Ic} calculations are made in Case I because it is assumed there are no cracks.

TABLE 6.3. Comparison of Results for $\sigma_{design} = \sigma_{ys}/2$, Example Problem

STEEL	σ_{ys}, ksi	σ_{design}, ksi	t, in.	WEIGHT, lb/ft	COST, $/ft
A	260	130	0.58	185	259
B	220	110	0.68	218	306
C	180	90	0.83	267	267
D	180	90	0.83	267	267
E	140	70	1.07	343	172
F	110	55	1.36	437	66

6.5.2 *Case II—Fracture-Mechanics Design*

In this case, it is assumed that a flaw is present. As a first step, the designer should estimate the maximum possible flaw size that can exist in the vessel wall, based on fabrication, inspection, and service considerations. In this particular design example, assume that we want to prevent failures caused by a possible surface flaw of depth 0.5 in. with an $a/2c$ ratio of 0.25, as shown in Figure 6.11. This size flaw may be due to improper fabrication or crack growth by fatigue or

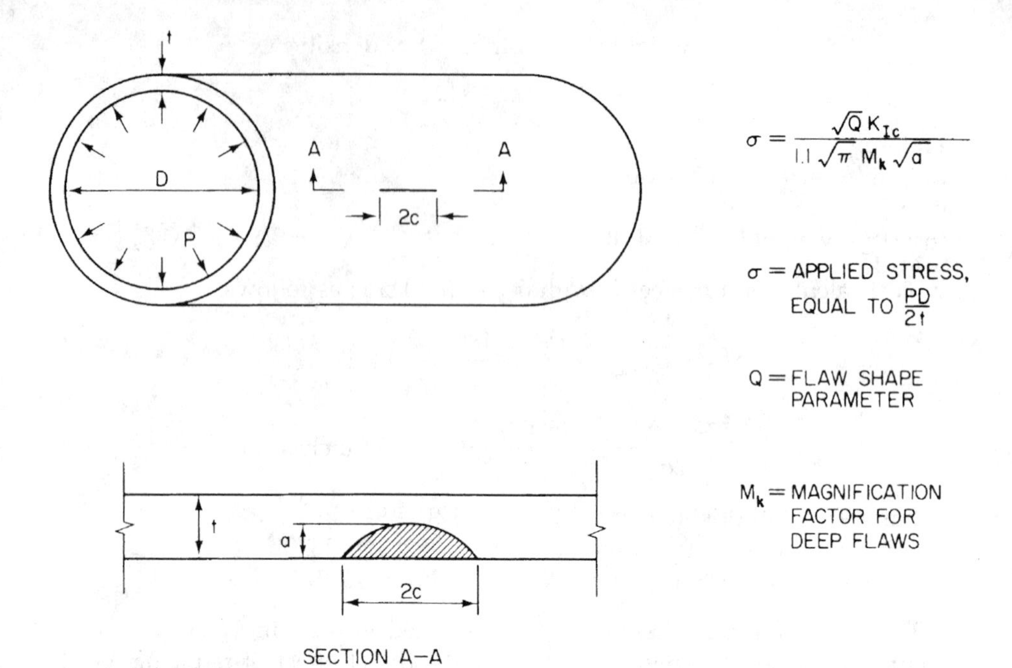

FIG. 6.11 Section of pressure vessel with surface flaw for example problem.

stress corrosion. Obviously, the decision regarding the maximum possible flaw size will depend on many factors related to the design, fabrication, and inspection for the particular structure, but the designer must use the best information available and make the decision. Too often, this step in the design process has been overlooked or, at best, casually handled by assuming that the fabrication will be inspected and "all injurious flaws removed," to quote one specification.

The general relation among K_I, σ, and a for a surface flaw as developed in Chapter 2 is

$$K_I = 1.12\sigma\sqrt{\frac{\pi a}{Q}} \cdot M_k$$

where M_k = magnification factor for deep flaws (similar to the tangent correction factor in Chapter 2), assumed in this example to vary linearly between 1.0 and 1.6 as a/t (crack depth/vessel thickness) varies from 0.5 to 1.0. Note that for a/t values less than 0.5, $M_k \simeq 1.0$.

Q = flaw shape parameter as shown in Figure 6.12 (also Figure 2.7—Chapter 2)

σ = applied hoop stress, ksi, equal to $pd/2t$ for the vessel section shown in Figure 6.11.

Rearranging the preceding equation yields

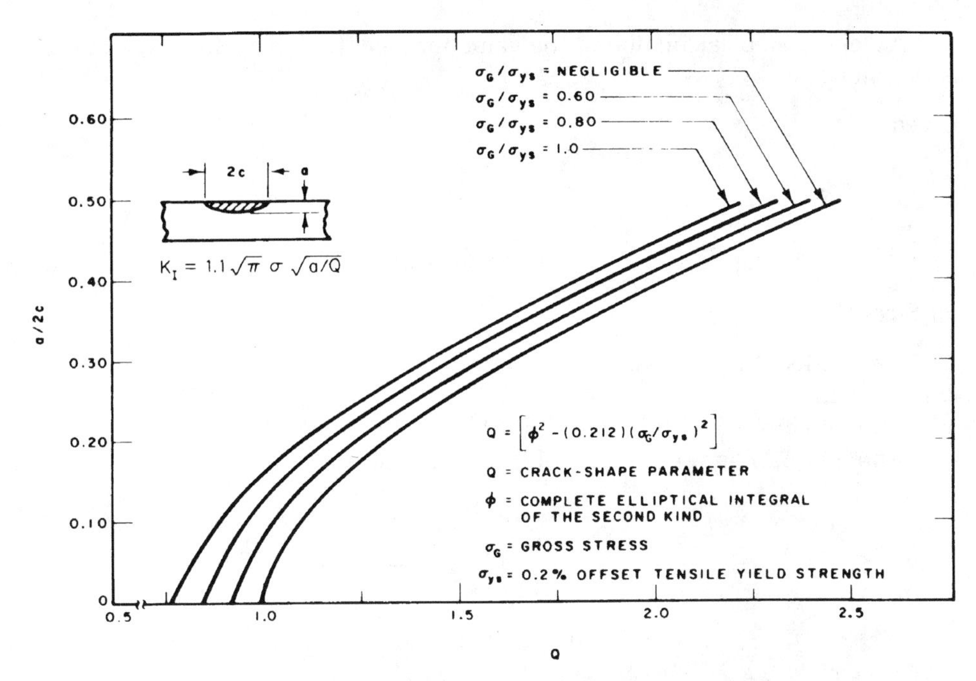

FIG. 6.12 Crack-shape parameter, Q, for surface flaw.

$$\sigma = \frac{K_I\sqrt{Q}}{1.12\sqrt{\pi a} \cdot M_k}$$

The design stress, σ_{design} for a factor of safety of 2.0 against fracture is calculated by first replacing K_I with $K_{Idesign} = K_{Ic}/2$ as follows:

$$\sigma_{design} = \frac{\left(\frac{K_{Ic}}{2}\right)\sqrt{Q}}{1.12\sqrt{\pi a} \cdot M_k}$$

and

$$t = \frac{pd}{2\sigma_{design}}$$

For $a = 0.5$, and the different values of K_{Ic} for the steels shown in Table 6.2, the calculated *design* stress values, σ, for each of the steels being studied are found by (1) calculating the design stress based on the fracture resistance needed for an 0.5-in.-deep crack and then (2) calculating the vessel thickness required at that stress level. Because M_k and Q are functions of the design stress, an iterative procedure must be used.

A step-by-step calculation of the value for Steel D (Table 6.2) ($\sigma_{ys} = 180$ ksi) is as follows:

Given

$$\sigma = \frac{\left(\frac{K_{Ic}}{2}\right)\sqrt{Q}}{1.12\sqrt{\pi a} \cdot M_k}$$

for Steel D

1. $K_{Ic} = 220$ ksi$\sqrt{\text{in.}}$
2. $a = 0.5$ in.
3. Assume that $\sigma/\sigma_{ys} = 0.55$ and thus $Q = 1.4$ (Figure 6.12).
4. Assume that $M_k = 1.0$ (Figure 6.13) for the first trial.

Thus

$$\sigma_{design} = \frac{\frac{220}{2}(\sqrt{1.4})}{1.12\sqrt{(3.14)(0.5)} \cdot 1.0}$$

$$\sigma_{design} = 95 \text{ ksi}$$

Using $\sigma = 95$ ksi, solve for the wall thickness, t, required to contain the design pressure of 5000 psi (5 ksi):

$$t = \frac{pd}{2\sigma}$$

$$= \frac{(5)(30)}{2(95)} = 0.8 \text{ inches}$$

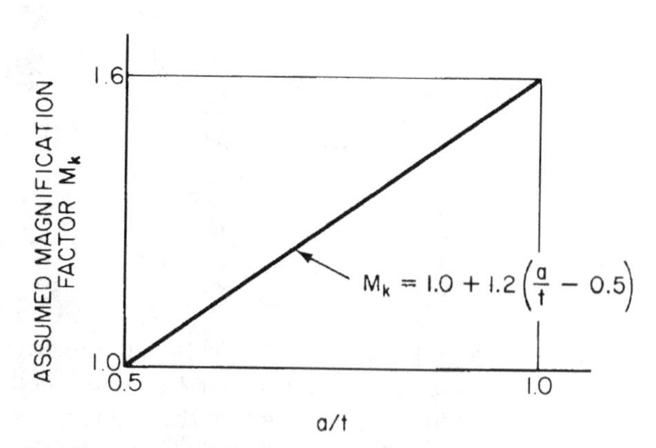

FIG. 6.13 Assumed magnification factor M_K, for example problem.

For $$t = 0.8 \text{ inches}$$

$$\frac{a}{t} = \frac{0.5}{0.8} = 0.63$$

therefore $$M_k = 1.15$$

also $$\frac{\sigma_{design}}{\sigma_{ys}} = \frac{95}{180} = 0.53$$

therefore $$Q = 1.4$$

Using $M_k = 1.15$ and $Q = 1.4$, a second iteration results in

$$\sigma_{design} = \frac{110\sqrt{1.4}}{1.12\sqrt{3.14(0.5)} \cdot 1.15}$$

$$\sigma_{design} = 82.2 \text{ ksi}$$

$$t = \frac{pd}{2\sigma} = \frac{(5)(30)}{2(82.2)}$$

$$t = 0.91 \text{ inches}$$

For ($t = 0.91$ inches)

$$\frac{a}{t} = \frac{0.5}{0.91} = 0.55$$

Therefore $$M_k = 1.06$$

and $$\frac{\sigma_{design}}{\sigma_{ys}} = \frac{82.2}{180} = 0.46$$

therefore $$Q = 1.42$$

Using $M_k = 1.06$ and $Q = 1.42$,

$$\sigma_{design} = \frac{110\sqrt{1.42}}{1.12\sqrt{3.14(0.5)} \cdot 1.06}$$

$$\sigma_{design} = 90 \text{ ksi}$$

$$t = \frac{pd}{2\sigma} = \frac{5(30)}{2(90)}$$

$$t = 0.83 \text{ inches}$$

Assuming that t will be $\simeq 0.85$ in.,

$$\frac{a}{t} = \frac{0.5}{0.85} = 0.59$$

thus

$$M_k = 1.11$$

also

$$\frac{\sigma_{design}}{\sigma_{ys}} = \frac{90}{180} = 0.5$$

then

$$Q \simeq 1.42$$

Therefore,

$$\sigma_{design} = \frac{110\sqrt{1.42}}{1.12\sqrt{3.14(0.5)} \cdot 1.11}$$

$$\sigma_{design} = 85.8 \text{ ksi}$$

$$t = \frac{(5)(30)}{2(85.8)}$$

$$t = 0.87 \text{ inches}$$

As the fourth iteration, for $\sigma = 86$ ksi and $t = 0.87$ in., assume that

1. $\sigma_{design}/\sigma_{ys} = 86/180 = 0.48$, and therefore $Q = 1.42$.
2. $a/t = 0.5/0.87$ and therefore $M_k = 1.10$.

$$\sigma_{design} = \frac{110\sqrt{1.42}}{1.12\sqrt{3.14(0.5)} \cdot 1.10}$$

$$\sigma_{design} = 87 \text{ ksi}$$

Thus, for a design stress of 87 ksi, the required wall thickness, t, is

$$t = \frac{pd}{2\sigma_{design}} = \frac{(5)(30)}{2(87)} = 0.86 \text{ inches}$$

This value agrees closely with the initial value of thickness of 0.86 in. for the assumed value, and thus further trials are not required. Note that the convergence is fairly rapid.

Wall thickness for the remaining steels in this example are calculated in a similar manner and are presented in Table 6.4. These results show that to withstand an internal pressure of 5000 psi in a 30-in.-diameter vessel having a 0.5-in.-deep surface flaw, the design stresses and wall thickness that should be used to give the same resistance to fracture vary considerably for the steels investigated. For example, the allowable design stress level for the 260-ksi yield-strength steel is only 35 ksi, and the required wall thickness is 2.14 in., whereas for a lower-strength steel having a 180-ksi yield strength and a K_{Ic} of 140 ksi$\sqrt{\text{in.}}$, the design stress is 61 ksi and the required wall thickness is 1.23 in. If a 180-ksi yield-strength steel with even higher fracture toughness is selected, that is, steel D

TABLE 6.4. Comparison of Results for $K_I = K_{Ic}/2$, Example Problem

STEEL	YIELD STRENGTH, σ_{ys} (ksi)	σ_{design}, ksi	K_{Ic}, ksi$\sqrt{in.}$	$K_I = K_{Ic}/2$, ksi$\sqrt{in.}$	t, in.	WEIGHT, lb/ft	COST, $/ft
A	260	35	80	40	2.14	685	959
B	220	48	110	55	1.56	499	699
C	180	61	140	70	1.23	394	394
D	180	87	220	110	0.86	272	330
E	140	95	260	130	0.79	253	127
F	110	72	170	85	1.04	333	50

with a K_{Ic} value of 220 ksi$\sqrt{in.}$, the design stress can be increased to 87 ksi and the wall thickness decreased to 0.86 in.

Because each of the vessels in this example is designed on the basis of *equivalent resistance to fracture in the presence of a 0.5-in.-deep surface flaw,* there would be an obvious savings in weight of the vessel by using a lower-strength steel with high fracture toughness compared with the 260-ksi yield-strength steel. That is, the weight is proportional to the wall thickness, and the required wall thicknesses for the lower-strength, higher-fracture toughness steels are *less* than for the higher-strength less fracture tough steels.

To show that the highest-strength steel may *not* yield the least weight or most economical vessel, estimates of the weight per foot and assumed cost per foot of the vessel (neglecting the costs of forming, fabrication, etc.) are presented in Table 6.4. The weights per foot of vessel were calculated by estimating the volume of material as the cross-sectional area (πdt) times a 12-in. length and then multiplying by the density of steel, 0.283 lb/in^3.

These results show the significant effect of fracture toughness on the allowable design stress and illustrate dramatically that high-strength materials with low values of fracture toughness do not necessarily yield the least weight vessel when fracture is a possible mode of failure. When cost is the primary criterion, the advantage of the lower-strength, higher-fracture toughness materials based on the assumed prices is obvious.

6.5.3 *General Analysis of Cases I and II*

Comparison of the required thicknesses tabulated in Tables 6.3 and 6.4 shows that for the two lowest-strength steels, yielding is the most likely mode of failure and the factor of safety against fracture will be larger than 2.0. Specifically for steel F, the required vessel thickness is 1.36 in., and the design stress ($\sigma_{ys}/2$) is 55 ksi; thus, the factor of safety against yielding is 2.0. The corresponding K_I value is 63 ksi$\sqrt{in.}$, which gives a factor of safety against fracture of 170/63 = 2.7. However, to have a factor of safety of at least 2 against *both* modes of failure, a wall thickness of 1.36 in. is required.

Conversely, the required thickness for the highest-strength steel is 2.14 in. based on a factor of safety against a fracture of 2.0. However, the corresponding

TABLE 6.5. Comparison of Factors of Safety Against Yielding and Fracture for Example Problem

STEEL	YIELD STRENGTH, ksi	THICKNESS REQUIRED FOR FACTOR OF SAFETY 2 AGAINST YIELDING, in.	THICKNESS REQUIRED FOR FACTOR OF SAFETY 2 AGAINST FRACTURE, in.	THICKNESS REQUIRED TO SATISFY BOTH CRITERIA, in.
A	260	0.58	2.14	2.14
B	220	0.68	1.56	1.56
C	180	0.83	1.23	1.23
D	180	0.83	0.86	0.86
E	140	1.07	0.79	1.07
F	110	1.36	1.04	1.36

TABLE 6.6. Weight and Cost of Steel for Factor of Safety of 2.0 or Greater Against Both Yielding and Fracture in Example Problem

STEEL	YIELD STRENGTH, σ_{ys} (ksi)	*t*, in.	WEIGHT, lb/ft	COST, $/ft
A	260	2.14	685	959
B	220	1.56	499	699
C	180	1.23	394	394
D	180	0.86	275	330
E	140	1.07	343	172
F	110	1.36	437	66

design stress value is 35 ksi, which gives a factor of safety against yielding of $260/35 = 7.43$.

The thickness required for factors of safety of 2 against yielding and fracture are tabulated in Table 6.5 and illustrate the necessity of considering all possible modes of failure prior to selecting a final geometry (thickness) as well as selecting a particular material.

As a last step in this example, the required thicknesses for a factor of safety of at least 2.0 against *both* yielding and fracture are listed in Table 6.6. In addition, the corresponding weight per foot and cost per foot are tabulated for each material. Analysis of the results shows that steel D would be the optimum selection on the basis of minimum weight and steel F would be the least expensive steel. Note that for both cases, the factors of safety against both yielding and fracture are 2.0 or greater. Thus, the vessels are compared on an equivalent performance basis and show the advantage of using fracture mechanics during the material selection process.

6.6 References

[1] Srawley, J. E. and Esgar, J. B., "Investigation of Hydrotest Failure of Thiokol Chemical Corporation 260-Inch-Diameter SL-1 Motor Case," *NASA TMX-1194,* Cleveland, January 1966.

[2] Gerberich, W. W., "Fracture Mechanics Approach to Design-Application," presented in a *Short Course on Offshore Structures,* University of California, Berkeley, CA, 1967.

[3] *Fracture Prevention and Control,* Proceedings ASM Symposium, March 13–16, 1972, Los Angeles, CA.

Part III: Fatigue and Environmental Behavior

7

Introduction to Fatigue

7.1 Introduction

THE DISCUSSION in the preceding chapters described the fracture behavior of components subjected to a monotonically increasing load. However, most equipment and structural components are subjected to repeated fluctuating loads whose magnitude is well below the fracture load under monotonic loading. Examples of equipment and structures subjected to fatigue loading include pumps, vehicles, earthmoving equipment, drilling rigs, aircraft, bridges, ships, and offshore structures.

Fatigue is the process of cumulative damage in a benign environment that is caused by repeated fluctuating loads and, in the presence of an aggressive environment, is known as corrosion fatigue. Fatigue damage of components subjected to normally elastic stress fluctuations occurs at regions of stress (strain) raisers where the localized stress exceeds the yield stress of the material. After a certain number of load fluctuations, the accumulated damage causes the initiation and subsequent propagation of a crack, or cracks, in the plastically damaged regions. This process can and in many cases does cause the fracture of components. The more severe the stress concentration, the shorter the time to initiate a fatigue crack.

The number of cycles required to initiate a fatigue crack is the fatigue-crack-initiation life, N_i. The number of cycles required to propagate a fatigue crack to a critical size is called the fatigue-crack-propagation life, N_p. The total fatigue life, N_t, is the sum of the initiation and propagation lives,

$$N_t = N_i + N_p \tag{7.1}$$

There is no simple or clear delineation of the boundary between fatigue-crack initiation and propagation. Furthermore, a pre-existing crack in a structural component can reduce or eliminate the fatigue-crack-initiation life and, thus, decrease the total fatigue life of the component.

Concern with fatigue damage was recognized in Europe early in the nineteenth century [1,2]. In 1852, Wohler conducted comprehensive experiments on axles [2] subjected to tensile, bending, and torsional repeated-load fluctuations. This work is important because it formed the basis for the Goodman diagram [2], which is the first developed methodology to predict the fatigue limit, at any stress ratio, given as a function of the ultimate tensile strength. Fatigue was incorporated into design criteria near the end of the nineteenth century and has been studied since. However, the most significant developments have occurred since the 1950s. At present, fatigue is part of design specification for many engineering structures.

7.2 Factors Affecting Fatigue Performance

Many parameters affect the fatigue performance of structural components. They include parameters related to stress (load), geometry and properties of the component, and the external environment. The stress parameters include state of stress, stress range, stress ratio, constant or variable loading, frequency, and maximum stress. The geometry and properties of the component include stress (strain) raisers, size, stress gradient, and metallurgical and mechanical properties of the base metal and weldments. The external environment parameters include temperature and aggressiveness of the environment. The effects of most of these parameters are discussed in the following chapters.

The primary factor that affects the fatigue behavior of structural components is the fluctuation in the localized stress or strain. Consequently, the most effective methods for increasing the fatigue life significantly are usually accomplished by decreasing the severity of the stress concentration and the magnitude of the applied nominal stress. In many cases, a decrease in the severity of the stress concentration can be easily accomplished by using transition radii in fillet regions, keyways, geometrical changes, and by minimizing the size of weld discontinuities.

The corrosion-fatigue behavior of components is affected by the same parameters that affect the fatigue behavior as well as factors that do not affect fatigue, for example, frequency and waveform of the stress cycle and the environment. Unfortunately, at present, the only means of determining the corrosion-fatigue behavior of materials is by conducting the tests on the actual material-environment system of interest. Environmental effects are discussed in Chapter 11.

7.3 Fatigue Loading

Structural components are subjected to a variety of load (stress) histories. The simplest of these histories is the constant-amplitude cyclic-stress fluctuation

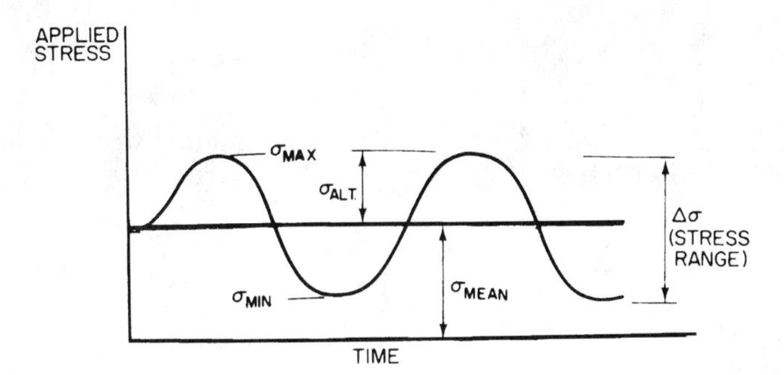

FIG. 7.1 Terminology used in constant-amplitude fatigue.

shown in Figure 7.1. This type of loading usually occurs in machinery parts such as shafts and rods during periods of steady-state rotation. The most complex fluctuating-load history is a variable-amplitude random sequence as shown in Figure 7.2. This type of loading is experienced by many structures, including offshore drilling rigs, ships, aircraft, bridges, and earthmoving equipment.

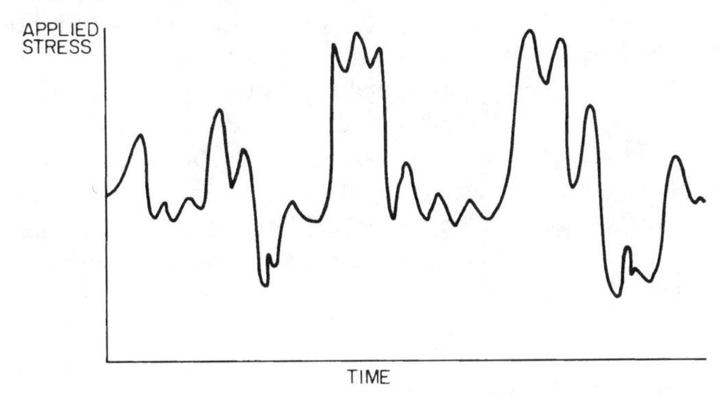

FIG. 7.2 Random-stress loading.

7.3.1 *Constant-Amplitude Loading*

Constant-amplitude load histories can be represented by a constant load (stress) range, $\Delta P(\Delta\sigma)$; a mean stress, σ_{mean}; an alternating stress or stress amplitude, σ_{amp}; and a stress ratio, R, Figure 7.1. The stress range is the algebraic difference between the maximum stress, σ_{max}, and the minimum stress, σ_{min}, in the cycle

$$\Delta\sigma = \sigma_{max} - \sigma_{min} \tag{7.2}$$

The mean stress is the algebraic mean of σ_{max} and σ_{min} in the cycle

$$\sigma_{mean} = \frac{\sigma_{max} + \sigma_{min}}{2} \tag{7.3}$$

The alternating stress or stress amplitude is half the stress range in a cycle

$$\sigma_{amp} = \frac{\Delta\sigma}{2} = \frac{\sigma_{max} + \sigma_{min}}{2} \tag{7.4}$$

The stress ratio, R, represents the relative magnitude of the minimum and maximum stresses in each cycle

$$R = \frac{\sigma_{min}}{\sigma_{max}} \tag{7.5}$$

Thus, a complete reversal of a load from a minimum compressive stress to an equal maximum tensile stress corresponds to $R = -1.0$, Figure 7.3*a*, and a stress fluctuation from a given minimum tensile load to a maximum tensile load would be characterized by a positive value for R that is larger than zero and is less than 1.0, $0 < R < 1.0$, as shown in Figure 7.3*c*.

7.3.2 Variable-Amplitude Loading

Variable-amplitude random-sequence load histories are very complex functions in which the probability of the same sequence and magnitude of stress ranges recurring during a particular time interval is very small. Such histories lack a describable pattern and cannot be represented by an analytical function. Examples of this type of fatigue-load histories include wind loading on aircraft, wave loading on ships and offshore platforms, and truck loading on bridges.

Between the extremes of constant-amplitude cyclic-stress histories and variable-amplitude random-sequence stress histories, there are a multitude of stress patterns of varying degrees of complexity. Many of these histories can be described by analytic functions and represented by various parameters. Some simple variable-amplitude stress histories are those corresponding to a single cycle

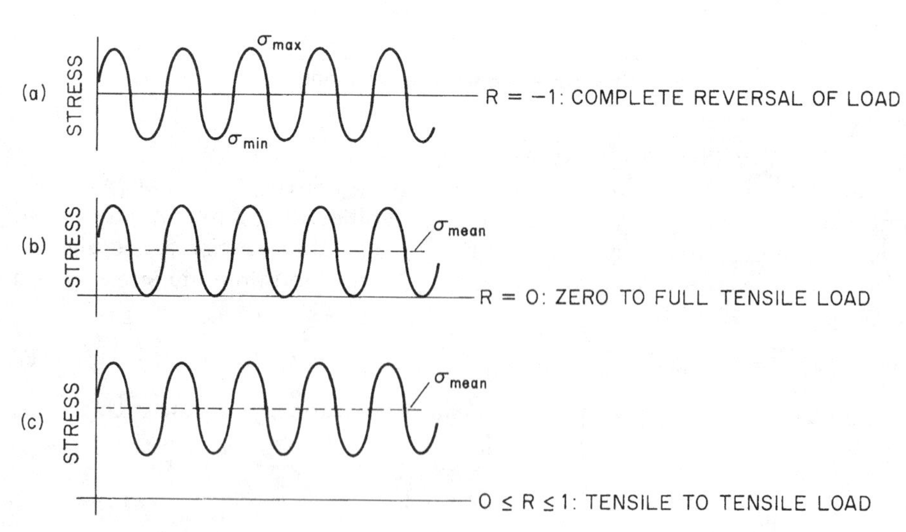

FIG. 7.3 Comparison of *R*-ratios for various loadings: $R = \sigma_{min}/\sigma_{max}$.

or multiple high-tensile-load cycles superimposed upon constant-amplitude cyclic-load fluctuations, Figure 7.4. Further discussion of variable-amplitude load histories is presented in Chapter 9.

7.4 Fatigue Testing

Fatigue tests are conducted on small laboratory specimens, specimens that simulate actual structural components, or on actual components. The objective of such tests is to develop information on the fatigue behavior of a particular ma-

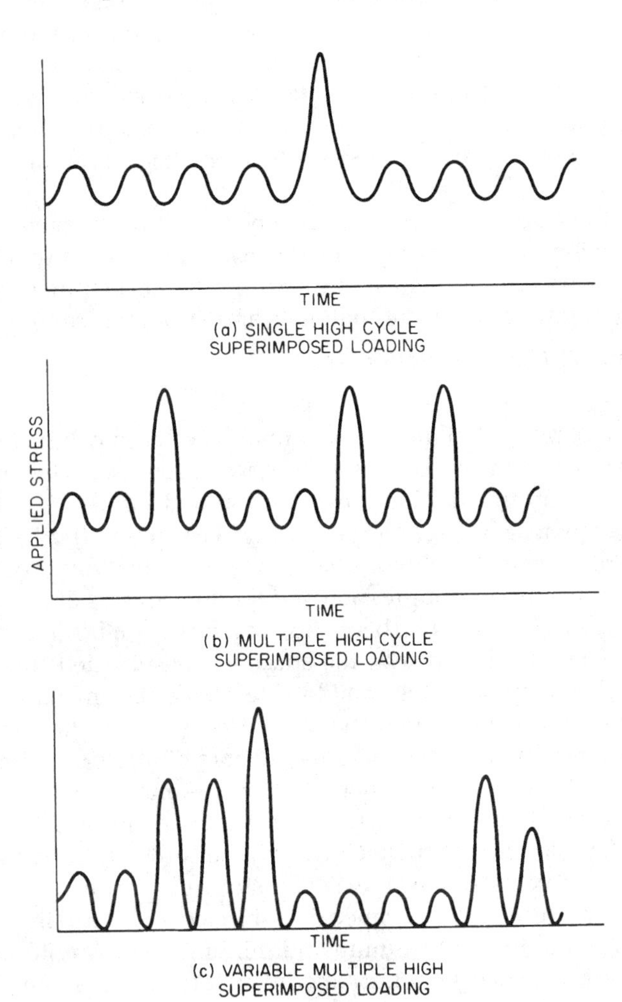

FIG. 7.4 Single or multiple high-cycle superimposed loading: (*a*) single high-cycle superimposed loading, (*b*) multiple high-cycle superimposed loading, and (*c*) variable multiple high-cycle superimposed loading.

terial, weldment, or geometry. This information can then be used to select the proper material, design, or both, for a particular application. Ideally, the material condition, stress history, and environment for the test closely simulate the actual service conditions for the structure under consideration.

7.4.1 Small Laboratory Tests

Small laboratory test specimens usually have simple geometries and are tested to obtain basic material properties for base metal and weldments. They can be used to study fatigue-crack-initiation or propagation life, but, in some cases, no distinction is made between the two life regions. Obtaining only the total life on small specimens complicates the use of the results to predict the behavior of actual structural components that have different size, shape, surface, and material conditions from those for the tested specimens. It is preferable to determine the initiation and propagation behaviors separately and then to combine them, as appropriate, to predict the behavior of an actual structural component.

Although there are many small specimens that have been used in laboratory tests, the following specimens and the corresponding test procedures will be discussed briefly because of their importance in the development of the present knowledge on fatigue behavior of materials and structural components.

7.4.1a. Fatigue-Crack-Initiation Tests

Stress-Life Tests

Extensive laboratory fatigue investigations have been conducted to evaluate the fatigue behavior of metals by testing a series of specimens in a rotating beam test, a flexure test, or an axial load test. For the rotating beam test, a polished, round specimen with a reduced cross section is supported as a beam and is subjected to a bending moment while being rotated so that fibers of the specimen are subjected alternately to compression and tension stresses of equal magnitude. For the flexure test, a specimen is bent back and forth as a beam instead of being rotated and, for the axial load test, the specimen is subjected to an alternating axial stress. In the flexure and the axial load tests, the specimen may be polished or may be cut from a plate so that the mill surface is left intact. In the rotating beam test, the stress ratio, R, which is the algebraic ratio of minimum to maximum stress, is -1.0, while in the other two tests the effect of various stress ratios may be investigated. In addition, there is a stress gradient over the cross section of the flexure or rotating beam specimen, whereas the stress is uniformly distributed over the cross section of the axially stressed specimen.

In each of the tests, the specimen is subjected to alternating stresses that vary between fixed limits of maximum and minimum stress until failure occurs. This procedure is repeated for other specimens at the same stress ratio but at different maximum stress values. The results of the tests are plotted to form an S-N diagram, where S represents the maximum stress, σ_{max}, in the cycle and N represents the number of cycles required to cause failure. An S-N diagram for polished specimens of A514 steel obtained from a series of rotating beam tests is shown in Figure 7.5 [3]. Because the number of cycles to failure span several orders of magnitude, and because most tests have been conducted under rotating bending, $R = -1.0$, where $\Delta\sigma$ is equal to $2\sigma_{max}$, the data are usually presented as a semilog plot of σ_{max} and the number of cycles to cause failure as shown in Figure 7.5. At any point, on the curve, the stress value is the "fatigue strength"—the value of maximum stress that will cause failure at a given number of stress cycles and at a given stress ratio—and the number of cycles is the "fatigue life"—the number of stress cycles that will cause failure at a given maximum stress and stress ratio. As shown in Figure 7.5, the fatigue strength of a structural steel decreases as the number of cycles increases until a "fatigue limit" is reached. If the maximum stress does not exceed the fatigue limit, an unlimited number of stress cycles can be applied at that stress ratio without causing failure. Tests on a large number of steels having tensile strengths up to 200,000 psi indicate that the fatigue limit of polished rotating beam specimens is about one-half the tensile strength. The fatigue limit of polished rotating beam specimens is about the same as that for axially loaded polished specimens, although the fatigue strength at a lower number of cycles is different.

Some materials, such as aluminum, do not exhibit a well-defined fatigue limit. Rather, the test results continue to decrease even beyond 10^7 cycles of loading, as shown in Figure 7.6 [4].

The influence of the stress ratio on fatigue strength is illustrated by the *S-N* curves for axially loaded polished specimens of A514 steel shown in Figure 7.7

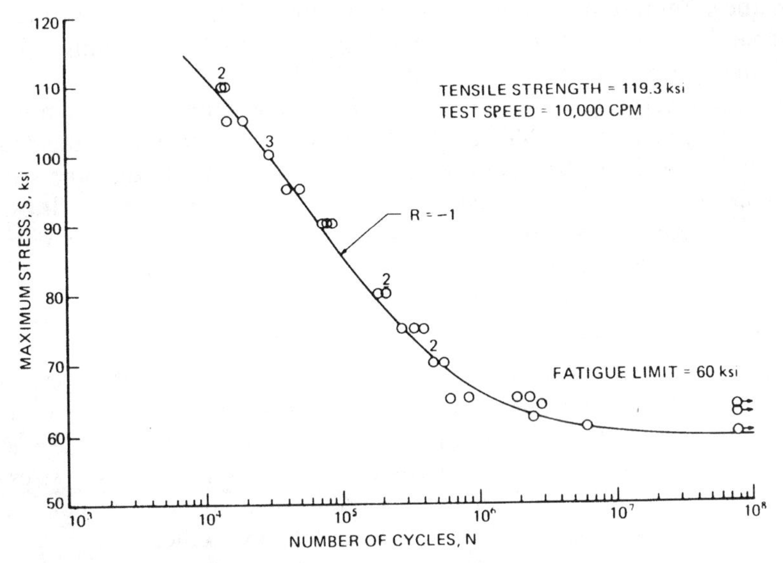

FIG. 7.5 *S-N* diagram for polished specimens of an A517 steel obtained from rotating-beam fatigue tests.

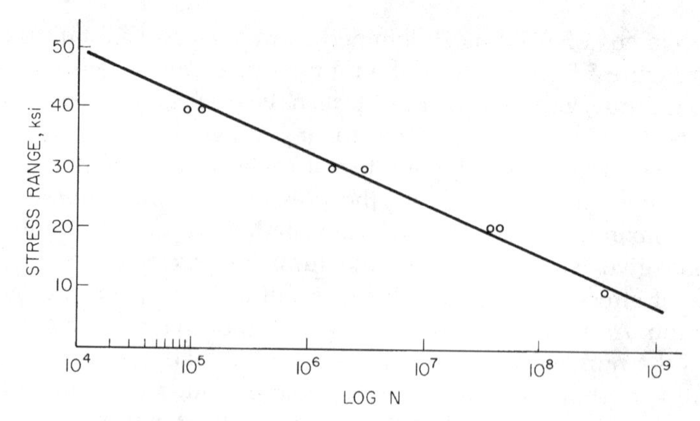

FIG. 7.6 Typical *S-N* fatigue test results for materials with no fatigue limit. (From N. Willems, J. T. Easley, and S. T. Rolfe, *Strength of Materials,* McGraw-Hill, New York, 1981, Figure 14.14).

[5]. Stress ratios of 0, $-1/2$, and -1 correspond to the following respective loading conditions: zero to tension, half compression to full tension, complete reversal of stress between equal compression, and tension stresses. The curves show that the fatigue strength decreases significantly as the stress ratio decreases. However, the apparent effect of stress ratio on the fatigue behavior is an artifact of the method adopted for presenting the *S-N* data in terms of σ_{max} rather than in terms of the stress range, $\Delta\sigma$, which is the primary stress parameter affecting fatigue performance. Representation of the data in Figure 7.7 in terms of $\Delta\sigma$ (rather than σ_{max}) versus *N* shows that stress ratio has less of an effect on the fatigue behavior of the tested specimens than would be predicted from Figure 7.7.

These *S-N* curves can be used to construct a fatigue chart shown in Figure 7.8 [3] for the polished A514 steel specimens. Each curved line in the chart represents the locus of all combinations of maximum and minimum stress at which failure will occur in the indicated number of cycles. Lines can be drawn from the origin to represent various stress ratios. Such charts are convenient for determining fatigue strength at a stress ratio different from those at which the tests were conducted.

Strain-Life Tests

Because the nominal stresses in most structures are elastic, the zone of plastically deformed metal in the vicinity of stress concentrations is surrounded by an elastic stress field. The true strains, ε, and true stresses, σ, of the plastic zones are limited by the elastic displacements of the surrounding elastic stress field. Therefore, even when the structure is stress controlled, the localized plastic zones are approximately strain controlled. Consequently, to predict the effects of stress concentration on the fatigue-crack-initiation behavior of structures, the fatigue behavior of the localized plastic zones has been simulated by testing smooth

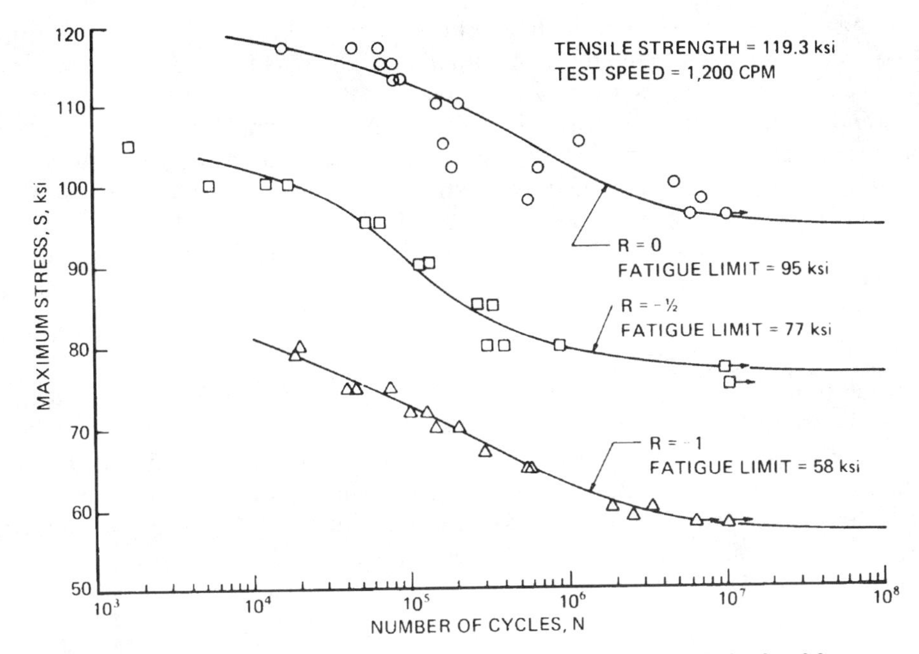

FIG. 7.7 *S-N* diagrams for polished specimens of an A517 steel obtained from axial-load fatigue tests.

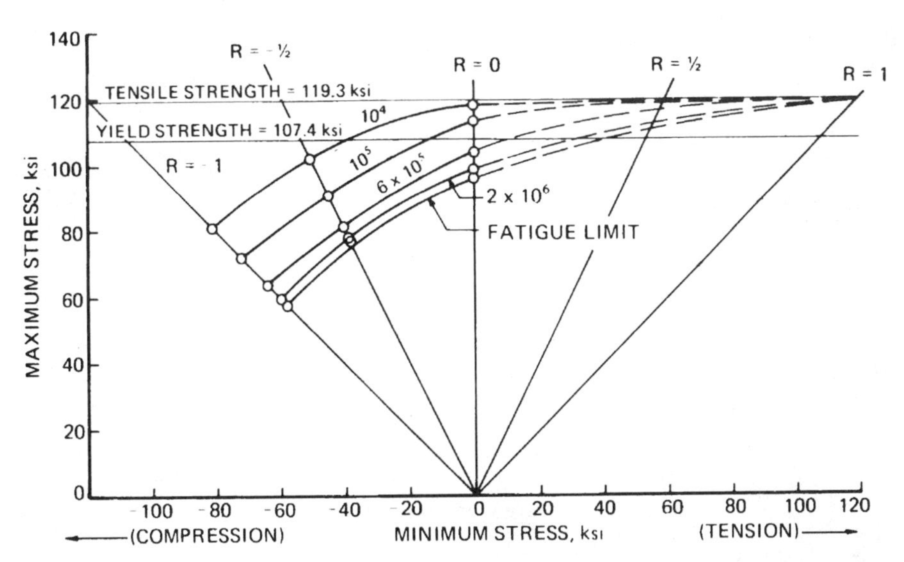

FIG. 7.8 Fatigue chart for axially loaded polished specimens of an A517 steel.

specimens under strain-controlled conditions, Figure 7.9, such that the minimum cross section for the specimen is some fraction of the plastic-zone size. However, suitable correction factors must be used to account for differences in stress state, size, and strain gradient between the smooth specimen and the plastic zone for the structural detail of interest [5]. Assuming that such correction factors are available, fatigue data obtained by testing smooth specimens under strain-controlled conditions can be used to predict the fatigue-crack-initiation behavior in structural components. This approach is not used in this book to analyze fatigue behavior of materials or components and, thus, will not be discussed further.

Fracture-Mechanics Tests

To predict the effects of stress concentration on the fatigue behavior of structures, the strain-life approach simulates the fatigue behavior of the localized plastic zone by testing smooth specimens under strain-controlled conditions, Figure 7.9. A better simulation of the fatigue behavior of this region is by testing notched specimens under stress-controlled conditions, Figure 7.9, because the applied stress can be related more directly to the applied loads. This approach is presented in detail in Chapter 8 on Fatigue-Crack Initiation.

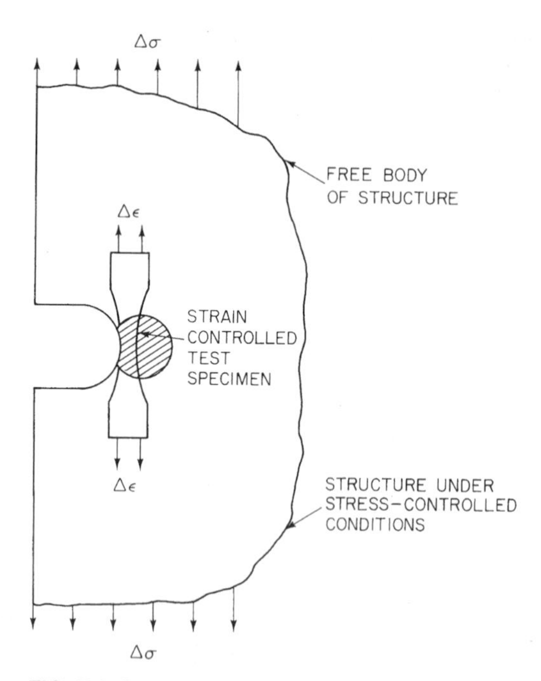

FIG. 7.9 Strain-controlled test specimen simulation for stress concentrations in structures.

7.4.1b. Fatigue-Crack-Propagation Tests

Most fatigue-crack-propagation tests are conducted by subjecting a fatigue-cracked specimen to cyclic-load fluctuations. Figure 7.10 shows a specimen that has been used extensively to obtain fatigue-crack-propagation data.

An incremental increase of crack length is measured visually at low magnification, ultrasonically, or by using electrical potential methods, and the corresponding number of elapsed load cycles are recorded. The data are presented as a linear plot of crack length, a, and the corresponding total number of load cycles, N. An increase in the magnitude of cyclic-load fluctuation results in a decrease of fatigue life of specimens having identical geometry (Figure 7.11). Furthermore, the fatigue life of specimens subjected to a fixed constant-amplitude cyclic-load fluctuation decreases as the length of the initial crack is increased (Figure 7.12). Consequently, under a given constant-amplitude stress fluctuation, most of the useful cyclic life is expended when the crack length is very small. Various a versus N curves can be generated by varying the magnitude of the cyclic-load fluctuation and/or the size of the initial crack. These curves reduce to a single curve when the data are represented in terms of crack-growth rate per cycle of loading, da/dN, and the fluctuation of the stress-intensity factor, ΔK_I, because ΔK_I is a single-term parameter that incorporates the effect of changing crack length and cyclic-load magnitude. The parameter ΔK_I is representative of the mechanical driving force and is independent of geometry. The most commonly used presentation of fatigue-crack-growth data is a log-log plot of the rate

FIG. 7.10 Compact-tension specimen used to determine fatigue crack propagation (specimen length is about 5 in.).

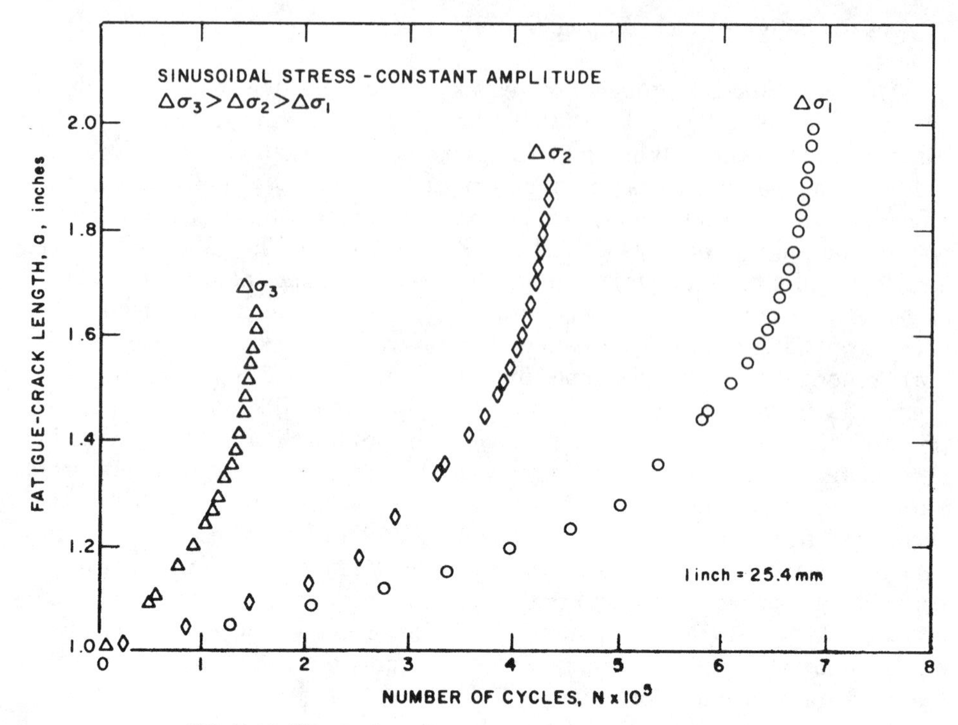

FIG. 7.11 Effect of cyclic-stress range on crack growth.

of fatigue-crack growth per cycle of load fluctuation, da/dN, and the fluctuation of the stress-intensity factor, ΔK_I.

7.4.2 *Tests of Actual or Simulated Structural Components*

Ideally, fatigue tests are conducted on actual structural components under conditions that closely simulate the loading and environment for the actual structure. Figure 7.13 is a photograph of a full-scale fatigue test of a 25-ft-long welded beam subjected to four-point loading. Data obtained from such tests can be used directly in design. However, these tests are difficult to conduct, time consuming, and usually very costly. Consequently, simple specimens that simulate structural components such as shown in Figure 7.14 are more commonly tested. Generally, these specimens are tested to failure to obtain information on the total fatigue life of the components. Test data from simulated and actual welded structural components are presented in Chapter 10.

7.5 Some Characteristics of Fatigue Cracks

Initiation and propagation of fatigue cracks are caused by localized cyclic-plastic deformation. A fatigue crack initiates more readily and propagates more rapidly as the magnitude of the local cyclic-plastic deformation increases. Thus, a smooth

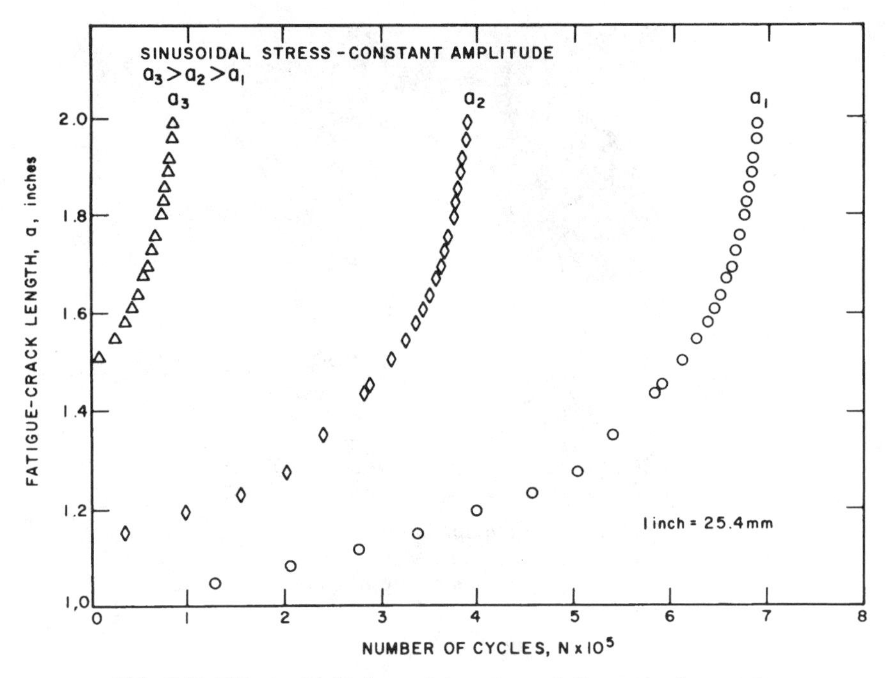

FIG. 7.12 Effect of initial crack length on fatigue crack growth.

FIG. 7.13 Full-scale fatigue test of welded beam used as structural component (specimen length is about 25 ft).

FIG. 7.14 Fatigue specimen used to determine fatigue behavior of welded attachment (specimen length is about 2 ft).

component free of imperfections and geometrical discontinuities does not fail by fatigue when subjected to elastic-stress (strain) fluctuations. However, when the same component is subjected to stress ranges approximately equal to or larger than the yield strength of the material, the plastic deformation causes the component to deform along slip planes that coincide with maximum shear stress—about 45 to 60° from the direction of the load in the case of axial loading, which results in slip steps on the surface. These steps correspond to stress raisers that become the nucleation sites for fatigue cracks which initiate along the maximum shear planes and propagate normal to the maximum tensile-stress component. Because the initial surface was free of stress raisers, the plastically induced stress raisers can occur at various locations on the surface, resulting in possible initiation of multiple fatigue cracks as shown in Figure 7.15. The stochastic character of this process results in the possible initiation of multiple cracks and their propagation at different rates until one reaches a critical size, causing failure of the component under the applied loads, Figure 7.15.

The probability for fatigue cracks to initiate at identical stress raisers subjected to the same stress range is equal. Consequently, multiple fatigue cracks may initiate in a uniformly loaded region of a component that contains stress raisers of equal severity.

The geometry of a component, (e.g., change in cross section), the type of loading, as well as other factors, can significantly affect the location of fatigue-crack initiation, the rate of crack propagation, and the shape of the propagating crack. Figure 7.16 [6] is a schematic representation of these observations for

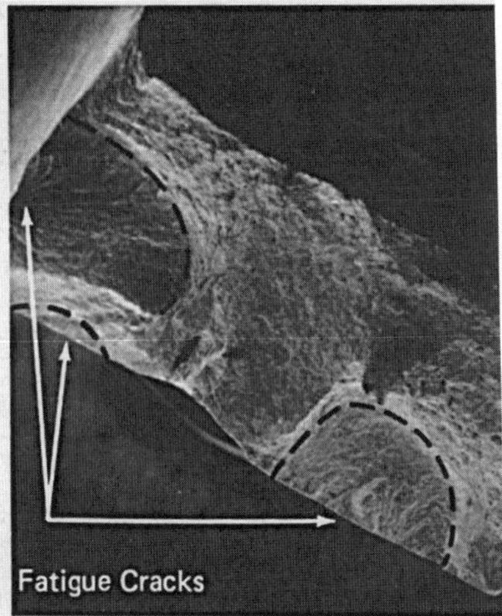

FIG. 7.15 Multiple fatigue-crack initiation in a smooth specimen.

smooth and notched components with round, square, and rectangular cross sections under various loading conditions.

In most structural components, fatigue cracks initiate and propagate from stress raisers. The stress raisers in unwelded components can be either surface imperfections or geometrical changes. In welded components, the stress raisers can be embedded imperfections, such as gas pockets, entrapped slag and lack of fusion, weld terminations and weld toes, or geometrical changes. The effects of these stress raisers on the fatigue behavior of weldments are discussed in Chapter 10.

A fatigue crack propagates normal to the primary tensile-stress component trying to assume a constant stress-intensity factor, K (or strain-energy-release rates, G), along the entire crack front. Thus, a fatigue crack that initiates from a point source in a uniform uniaxial tensile field propagates as a penny-shaped crack if the source is embedded and as a semicircular part-through crack if the source is on the surface, Figure 7.17. A crack that initiates from an irregularly shaped imperfection in a uniaxial tensile field propagates at different rates to achieve the circular crack-front shape, Figure 7.18.

A constant-K crack front in a pure bending stress field is a straight line as in a single-edge-notch specimen. Thus, a fatigue crack that initiates from a point source on the tension surface of a specimen in bending initially assumes a semicircular shape and then propagates more rapidly along the tension surface than in the depth direction until it reaches a constant-K straight-line crack front. The relative rate of crack propagation along the surface and in the depth direction depends on the stress gradient along the depth such that the higher the stress

FIG. 7.16 Schematic representation of marks on surfaces of fatigue fractures produced in smooth and notched components with round, square, and rectangular cross sections, and in thick plates, under various loading conditions at high and low nominal stress. (From *Metals Handbook,* Vol. 10, 8th ed., American Society for Metals, Metal Park, OH 44073, 1975, p. 102. With permission.)

FIG. 7.17 Semi-circular surface crack propagating from a single source.

gradient, the slower the relative propagation in the depth direction. Fatigue cracks in complex stress fields—like those possible when nominal, residual, and thermal stresses are superimposed—can exhibit complex fronts with widely varying propagation rates at different points and at the same relative point for different crack sizes.

Visually, a fatigue crack on a fracture surface appears smoother than the surrounding fracture regions, Figure 7.18. Microscopically, the fatigue region exhibits striations, Figure 7.19, which corresponds to crack extension with each cycle of a load. Usually, these striations are more distinct for aluminum than for steel.

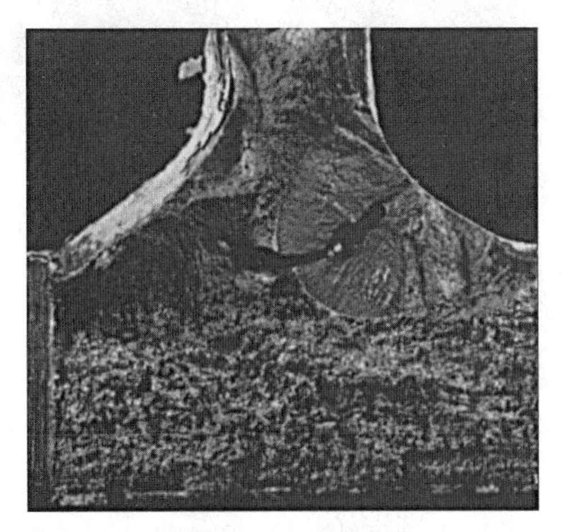

FIG. 7.18 Growth of a fatigue crack from irregularly shaped porosity maintaining a penny-shaped crack front.

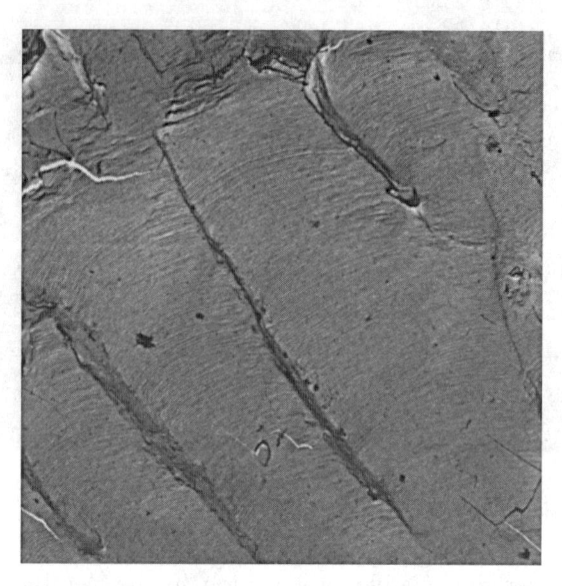

FIG. 7.19 Fatigue striations on a fracture surface of an aluminum alloy.

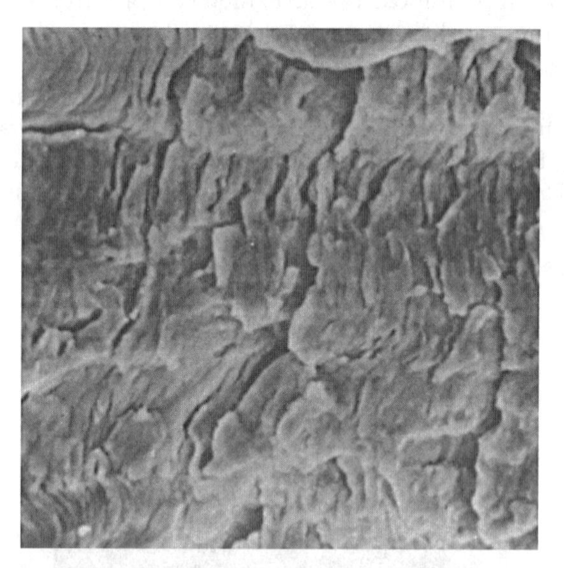

FIG. 7.20 Scanning electron micrograph of a fatigue crack surface in an A36 steel, X2500.

Figure 7.20 is a scanning electron micrograph of a fatigue surface in A36 steel. Although the surface indicates the presence of striations, some of the striation-like features are caused by the composite properties of ferrite-pearlite steels [7]. Ferrite-pearlite steels behave as particulate composites such that the

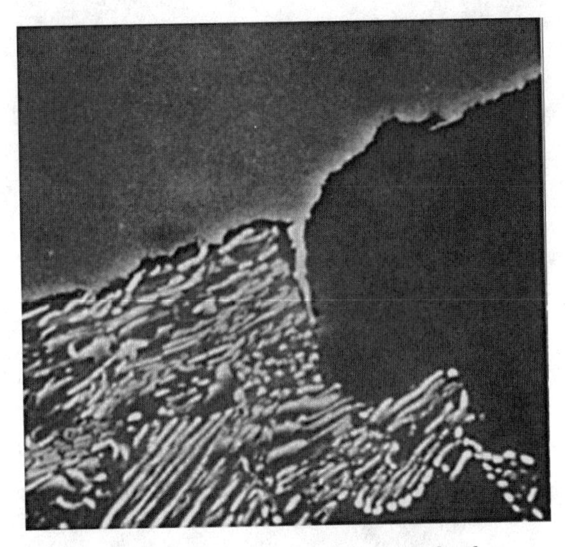

FIG. 7.21 Scanning electron micrograph of a fatigue crack profile in an A36 steel, X6000.

path of least resistance to fatigue-crack growth is through the ferrite matrix, and the pearlite colonies tend to retard crack growth. Figure 7.21 is a light micrograph of a cross-sectional portion of the fatigue surface shown in Figure 7.20. The micrograph shows that plastic deformation under cyclic loading is more extensive in the ferrite matrix than in the pearlite colonies. It also shows that secondary fatigue cracks preferentially seek to propagate around a pearlite colony rather than through it. Such secondary cracks make it very difficult to delineate the true striations on the fatigue surface.

Finally, an aggressive environment may eliminate the striation from the surface of a corrosion fatigue crack, thus making it very difficult to distinguish between corrosion-fatigue crack extension under cyclic loading and stress-corrosion crack extension under static loading.

7.6 References

[1] Cazaud, R., *Fatigue of Metals*, Chapman & Hall, London, 1953.

[2] *Structural Steel Design*, L. Tall, L. S. Beedle, and T. V. Galambos, Eds., Ronald Press, New York, 1964.

[3] Brockenbrough, R. L. and Johnston, B. G., *Steel Design Manual*, United States Steel Corporation, ADUSS 27-3400-04, January 1981.

[4] Willems, N., Easley, J. T., and Rolfe, S. T., *Strength of Materials*, McGraw-Hill Book Company, New York, 1981.

[5] Coffin, Jr., L. F., "Fatigue at High Temperature," *Fatigue at Elevated Temperatures, ASTM STP 520*, American Society for Testing and Materials, Philadelphia, 1973, pp. 5–34.

[6] *Metals Handbook*, Vol. 10, 8th ed., American Society for Metals, Metals Park, OH, 1975.

[7] Barsom, J. M., "Fatigue-Crack Propagation in Steels of Various Yield Strengths," *Transactions of the ASME, Journal of Engineering for Industry*, Series B, 93, No. 4, November 1971.

8

Fatigue-Crack Initiation

8.1 General Background

THE FATIGUE LIFE of structural components is determined by the sum of the elapsed cycles required to initiate a fatigue crack and to propagate the crack from subcritical dimensions to the critical size. Consequently, the fatigue life of structural components may be considered to be composed of three continuous stages: (1) fatigue-crack initiation, (2) fatigue-crack propagation, and (3) fracture. The fracture stage represents the terminal conditions (i.e., the particular combination of σ, a, and K_{Ic}) in the life of a structural component. The useful life of cyclically loaded structural components can be determined only when the three stages in the life of the component are evaluated individually and the cyclic behavior in each stage is thoroughly understood.

Conventional procedures used to design structural components subjected to fluctuating loads provide a design fatigue curve which characterizes the basic unnotched fatigue properties of the material and a fatigue-strength-reduction factor. The fatigue-strength-reduction factor incorporates the effects of all the different parameters characteristic of the specific structural component that make it more susceptible to fatigue failure than the unnotched specimen, such as surface finish, geometry, defects, as well as others. The design fatigue curves are based on the prediction of cyclic life from data on nominal stress (or strain) versus elapsed cycles to failure (*S-N* curves), as determined from laboratory specimens. Such data are usually obtained by testing unnotched specimens, and represent the number of cycles required to initiate a crack in the specimen plus the number of cycles required to propagate the crack from a subcritical size to a critical dimension. The dimensions of the critical crack required to cause failure depend on the magnitude of the applied stress and on the specimen size, as well as on the particular testing conditions used.

Figure 8.1 is a schematic *S-N* curve divided into an initiation component and a propagation component. The number of cycles corresponding to the endurance limit primarily represents initiation life, whereas the number of cycles expended in crack initiation at a high value of applied alternating stress is negligible. As the magnitude of the applied alternating stress increases, the total fatigue life decreases and the percent of the total fatigue life to crack initiation life decreases. Consequently, S-*N*–type data do not provide complete information regarding safe-life predictions in structural components, particularly in components having surface irregularities different from those of the test specimens and in components containing crack-like imperfections. This occurs because the existence of surface irregularities and crack-like imperfections reduces and may eliminate the crack-initiation portion of the fatigue life of structural components.

Many attempts have been made to characterize the fatigue behavior of metals [*1–3*]. The results of some of these attempts have proved invaluable in the evaluation and prediction of the fatigue strength of structural components. However, these fatigue-strength-evaluation procedures are subject to limitations, caused primarily by the failure to distinguish adequately between fatigue-crack initiation and fatigue-crack propagation.

Notches in structural components cause stress intensification in the vicinity of the notch tip. The material element at the tip of a notch in a cyclically loaded structural component is subjected to the maximum stress fluctuations, $\Delta\sigma_{max}$.

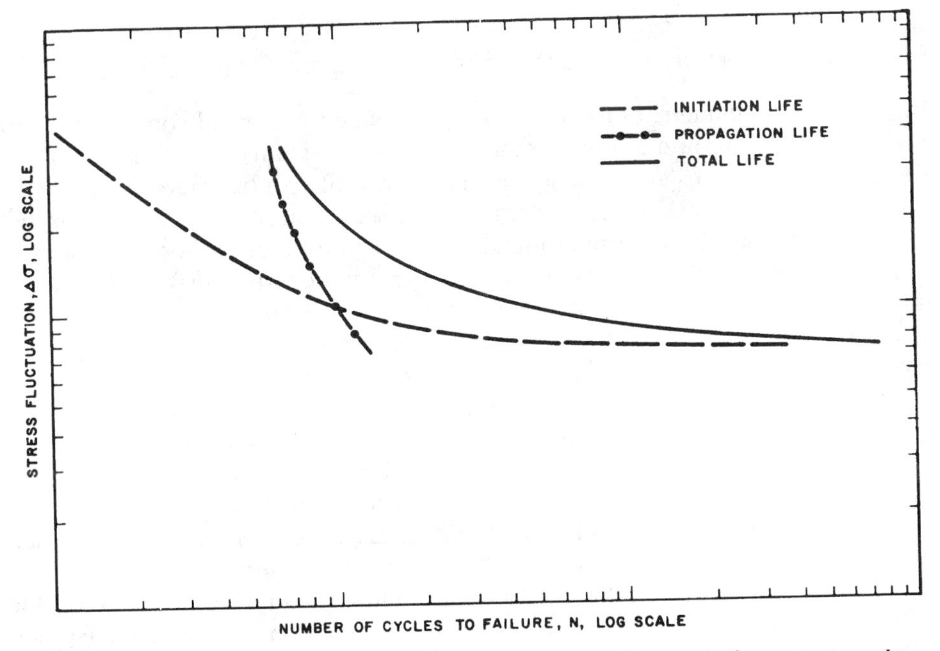

FIG. 8.1 Schematic *S-N* curve divided into initiation and propagation components.

Consequently, this material element is most susceptible to fatigue damage and is, in general, the site of fatigue-crack initiation. The maximum stress on this material element [4] is:

$$\sigma_{max} = k_t\sigma = \frac{2}{\sqrt{\pi}}\frac{K_I}{\sqrt{\rho}} \tag{8.1}$$

and the maximum stress range is:

$$\Delta\sigma_{max} = k_t\,(\Delta\sigma) = \frac{2}{\sqrt{\pi}}\frac{\Delta K_I}{\sqrt{\rho}} \tag{8.2}$$

where k_t is the stress-concentration factor for the notch and ρ is the notch-tip radius. Although these equations are considered exact only when ρ approaches zero, Wilson and Gabrielse [5] showed, by using finite element analysis of relatively blunt notches in compact-tensions specimens, where the notch length, a_n, was much larger than ρ, that these relationships are accurate to within 10% for notch radii up to 0.18 in.

Thorough understanding of fatigue-crack initiation requires the development of accurate predictions of the localized stress and strain behavior in the vicinity of stress concentrations. Developments in elastic-plastic finite-element stress analysis in the vicinity of notches contribute significantly to the development of quantitative predictions of fatigue-crack initiation behavior for structural components.

8.2 Effect of Stress Concentration on Fatigue-Crack Initiation

The effect of a geometrical discontinuity in a loaded structural component is to intensify the magnitude of the nominal stress in the vicinity of the discontinuity. The localized stresses may cause the metal in that neighborhood to undergo plastic deformation. Because the nominal stresses in most structures are elastic, the zone of plastically deformed metal in the vicinity of stress concentrations is surrounded by an elastic-stress field. The deformations (strains) of the plastic zone are governed by the elastic displacements of the surrounding elastic-stress field. In other words, when the structure is stress controlled, the localized plastic zones are strain controlled. Consequently, to predict the effects of stress concentrations on the fatigue behavior of structures, the fatigue behavior of the localized plastic zones has been simulated by testing smooth specimens under strain-controlled conditions, Figure 8.2*a*. A better simulation of the effects of stress concentrations on the fatigue behavior of structures is obtained by testing notched specimens under stress-controlled conditions, Figure 8.2*b*, because the applied stress can be more directly related to the structural loading.

The use of linear-elastic fracture-mechanics parameters to analyze the fatigue-crack-initiation behavior of notched specimens has been demonstrated in limited work by Forman [6], by Constable et al. [7], by Jack and Price [8], by

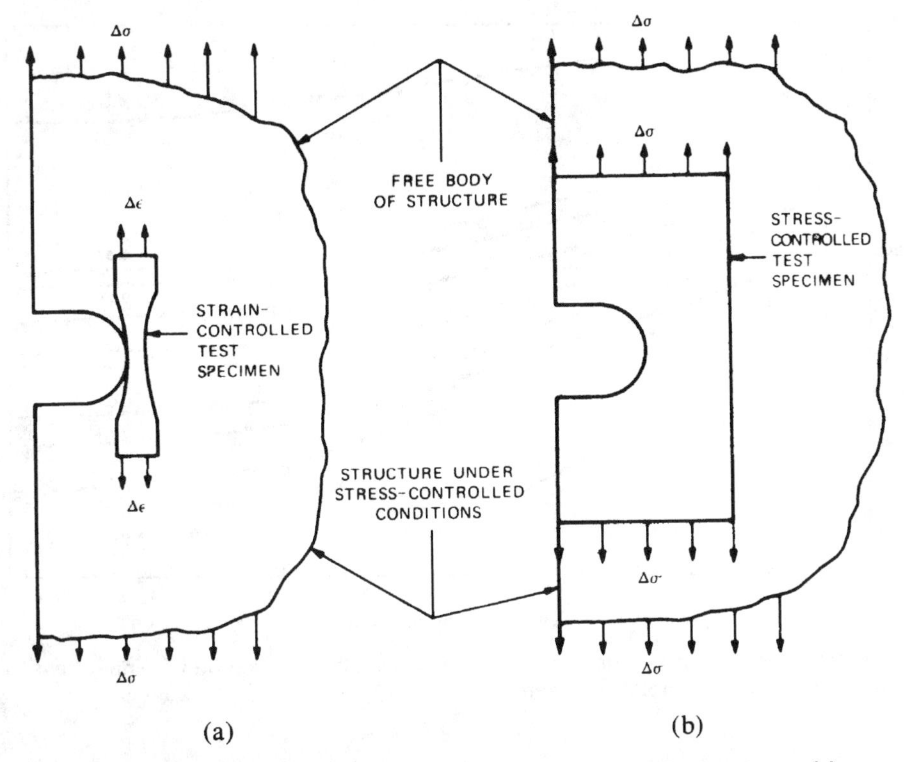

FIG. 8.2 Test specimen simulation of stress concentrations in structures: (*a*) strain-controlled specimen; (b) stress-controlled specimen.

Barsom and McNicol [*9*], and by Clark [*10*]. The fatigue-crack initiation behavior of an HY-130 steel was studied by Barsom and McNicol [*9*] under axial zero-to-tension cyclic load. The test data are presented in Figure 8.3 in terms of the stress range, $\Delta\sigma$, versus the number of cycles for fatigue crack initiation, and in Figure 8.4 in terms of the stress-intensity-factor range divided by the square root of the notch-tip radius, $\Delta K_I/\sqrt{\rho}$, versus the number of cycles for fatigue crack initiation. The stress-intensity-factor range was calculated as for a fatigue crack of length equal to the total notch depth. The data in Figure 8.3 demonstrate the severe detrimental effects of stress concentration whose magnitude increases as the notch-tip radius decreases and as the notch length increases. Figure 8.4 demonstrates that $\Delta K_I/\sqrt{\rho}$ may be used as a correlating parameter to account for the effects of notch geometry and stress range on the fatigue-crack initiation behavior of the steel.

The fatigue-crack-initiation behavior in various steels was investigated by Barsom and McNicol [*11*] in three-point bending at a stress ratio, *R* (ratio of nominal minimum applied stress to nominal maximum applied stress), equal to +0.1. The data were obtained by testing single-edge-notched specimens having

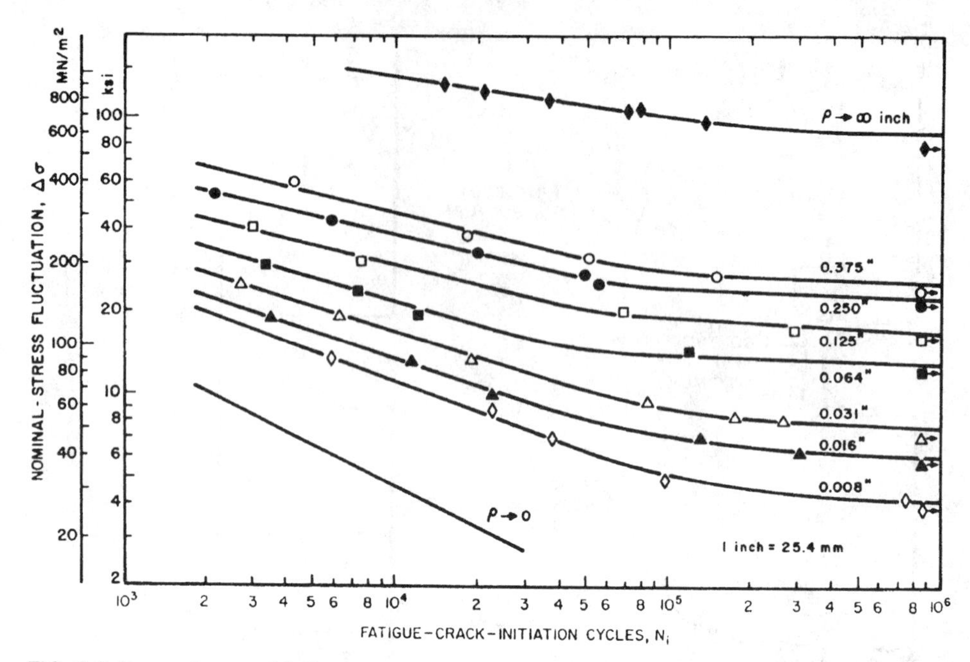

FIG. 8.3 Dependence of fatigue-crack initiation of an HY-130 steel on nominal-stress fluctuations for various notch geometries.

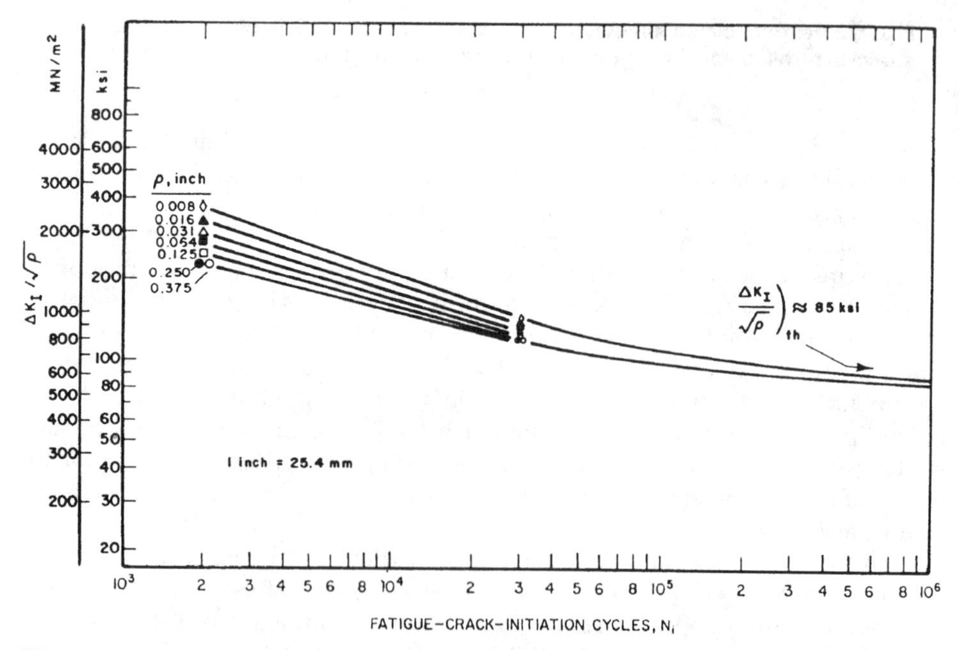

FIG. 8.4 Correlation of fatigue-crack-initiation life with the parameter $\Delta K_I/\sqrt{\rho}$ for an HY-130 steel.

a notch that resulted in a stress concentration of about 2.5. The fatigue-crack-initiation behavior of all the specimens tested is presented in Figure 8.5 in terms of the number of cycles for fatigue-crack initiation, N_i, versus the ratio of the stress-intensity-factor fluctuation to the square root of the notch-tip radius, $\Delta K_I/\sqrt{\rho}$. Because the same nominal notch length and tip radius were used for all specimens of the steels investigated, the differences in the fatigue-crack-initiation behavior shown in Figure 8.5 are related primarily to inherent differences in the fatigue-crack-initiation characteristics of the steels. The data show that fatigue cracks do not initiate in steel structural components when the body configuration, the notch geometry, and the nominal-stress fluctuations are such that the magnitude of the parameter $\Delta K_I/\sqrt{\rho}$ is less than a given value characteristic of the steel. In general, the value of this fatigue-crack-initiation threshold, $\Delta K_I/\sqrt{\rho})_{th}$, increased with increased yield strength or tensile strength of the steel.

8.3 Generalized Equation for Predicting the Fatigue-Crack-Initiation Threshold for Steels

To develop a generalized equation for predicting the fatigue-crack-initiation threshold for steels, the effect of stress ratio, R, on the threshold value must be established. Consequently, the fatigue-crack-initiation behavior for steels having yield strengths between 36 and 110 ksi was studied by Barsom at stress ratios of -1.0, 0.1, and 0.5 [12]. Analysis of the test results showed that stress ratio had a

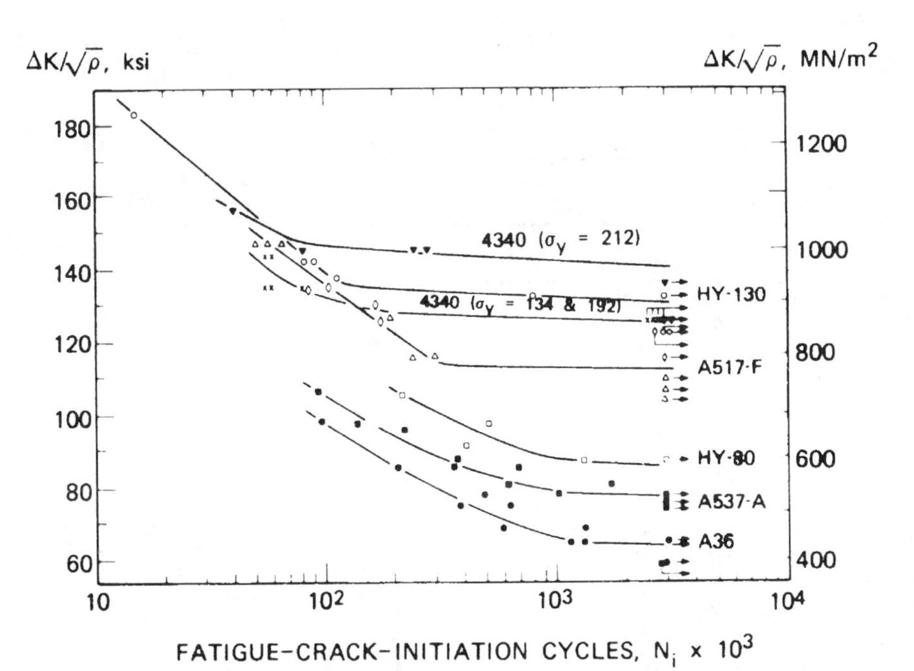

FIG. 8.5 Fatigue-crack initiation behavior of various steels.

negligible effect, if any, on the fatigue-crack-initiation threshold from notches when the ΔK in the $\Delta K/\sqrt{\rho}$ parameter are based on the total stress (tension plus compression) range. A plot of the data in terms of the $\Delta K_{total}/\sqrt{\rho}$ values at the threshold and R is shown in Figure 8.6 [*12,13*] where ΔK_{total} is calculated by using the total stress (tension plus compression) range. The data in Figure 8.6 indicate that the fatigue-crack-initiation damage caused by a compressive-stress range is equal to the damage caused by a tensile-stress range having the same magnitude. Although cracks can initiate under compressive cyclic loading, they would propagate only through the plastically deformed region where inelastic tensile stresses reside in the plastic zone at the tip of the notch and arrest in the vicinity of its boundary. However, under tensile-cyclic loading, once initiated, the crack would propagate beyond the plastically deformed region causing fracture of the specimen.

Analysis of the available data indicates that the fatigue-crack-initiation threshold for various steels subjected to stress ratios between −1.0 and 0.5 can be estimated best by using the relationship [*14*],

$$\left(\frac{\Delta K_{total}}{\sqrt{\rho}}\right)_{th} = 10\sqrt{\sigma_{ys}} \tag{8.3}$$

where ΔK_{total} is the stress-intensity factor range, calculated by using the sum of

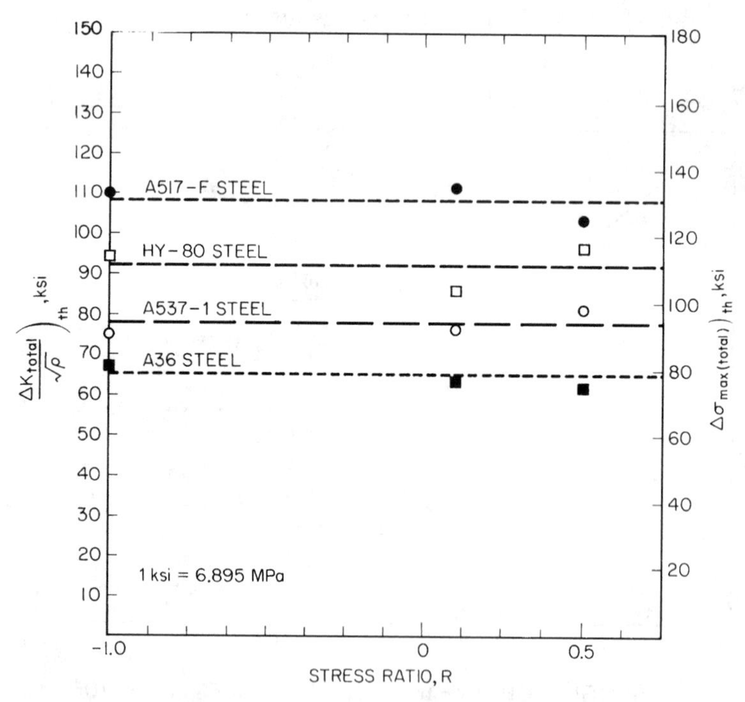

FIG. 8.6 Independence of fatigue-crack-initiation threshold from stress ratio.

the tensile- and compressive-stress ranges, and σ_{ys} is the yield strength of the steel, Figure 8.7 [14].

8.4 Methodology for Predicting Fatigue-Crack Initiation from Notches

Most structural components are subjected to a nominal elastic-stress (strain) field. However, the localized strains at regions of strain raisers embedded in the elastic field can be plastic. Under cyclic loading, these plastically deformed regions can become the nuclei of the fatigue-crack-initiation sites.

The fatigue-crack-initiation behavior for components that contain localized plastic deformation surrounded by an elastic-strain field can be analyzed accu-

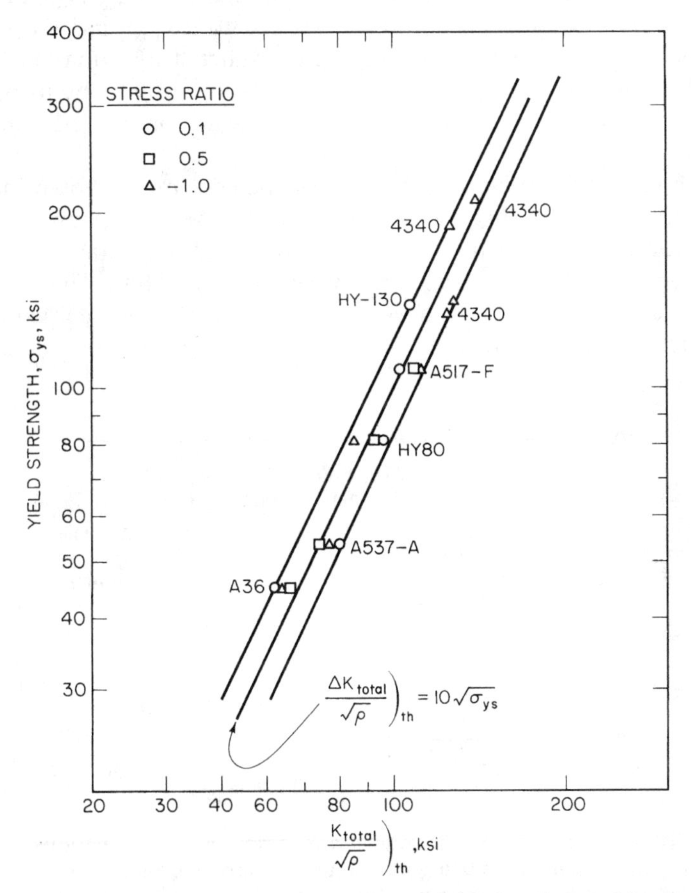

FIG. 8.7 Dependence of fatigue-crack-initiation threshold on yield strength. (From *Metals Handbook,* Desk Edition, American Society for Metals, Metals Park, OH 44073, 1985, pp. 32–33. With permission.)

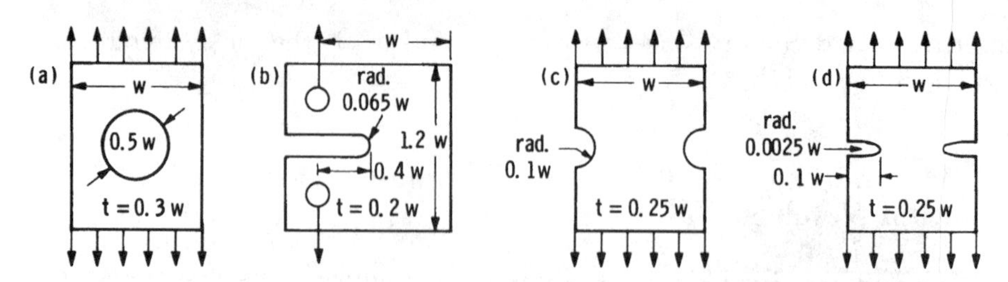

FIG. 8.8 Notched specimens: (*a*) center hole, (*b*) compact, (*c*) blunt double-edge notch, and (*d*) sharp double-edge notch. In each, W = 25.5 mm, t = thickness. (Reprinted with permission. © Society of Automotive Engineers, Inc., SAE Paper No. 820691, 1982.)

rately by using the maximum-stress-range parameter, $\Delta\sigma_{max}$. For notches with length larger than ρ, $\Delta K_I/\sqrt{\rho}$ is directly related to $\Delta\sigma_{max}$, and the relationship between these two parameters is accurately predicted by Equation (8.2). However, for blunt notches, $\Delta\sigma_{max}$ should be calculated accurately by using theory of elasticity and finite-element analyses when necessary or may be obtained from available handbooks [*15*].

The fatigue-crack-initiation behavior for the specimens shown in Figure 8.8 are presented in Figure 8.9 [*16*] as a function of the $\Delta K_I/\sqrt{\rho}$ parameter. The data separation for cycles larger than 10^3 is caused by the inaccuracy of the relationship between $\Delta K_I/\sqrt{\rho}$ and $\Delta\sigma_{max}$ for very blunt notches. The fatigue-crack-initiation data for lives equal to or less than 10^3 cycles was obtained under general yielding conditions, which caused further separation of the data.

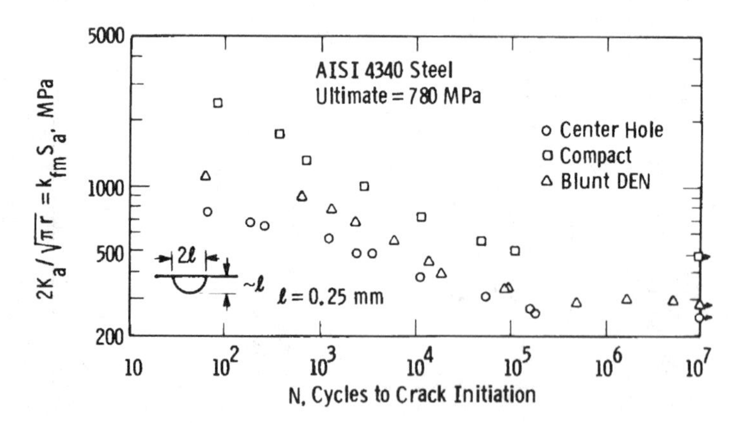

FIG. 8.9 Correlation of fatigue-crack-initiation data for various notched specimens using elastic stress-concentration factors estimated from fracture-mechanics relationships. (Reprinted with permission. © Society of Automotive Engineers, Inc., from SAE Paper No. 820691, 1982.)

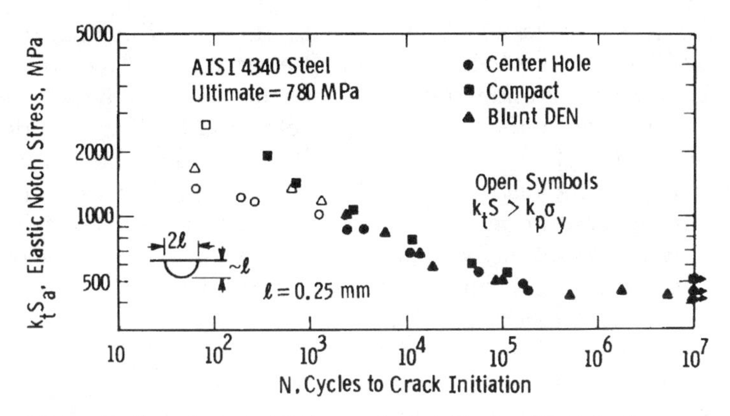

FIG. 8.10 Correlation of fatigue-crack-initiation-life data for various notched specimens using accurate elastic stress-concentration factors. (Reprinted with permission. © Society of Automotive Engineers, Inc. SAE Paper No. 820691, 1982.)

The fatigue-crack-initiation data for blunt and sharp notches embedded in an elastic field can be correlated better by using $\Delta\sigma_{max} = k_t(\Delta\sigma)$ rather than $\Delta\sigma_{max} = (2/\sqrt{\pi})(K_I/\sqrt{\rho})$. The use of an accurate value for k_t to calculate $\Delta\sigma_{max}$ results in an excellent correlation for notches whose plastic zone is fully surrounded by an elastic-stress field, Figure 8.10 [*16*].

The transition from a plastic zone embedded in an elastic field to general yielding of the specimen is gradual and is strongly influenced by the size of the

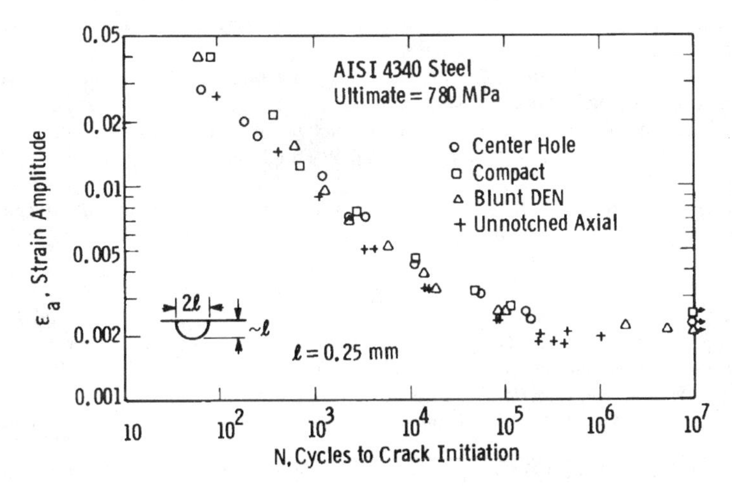

FIG. 8.11 Correlation of fatigue-crack-initiation-life data for various notched specimens using measured notched strains. (Reprinted with permission. © Society of Automotive Engineers, Inc. SAE Paper No. 820691, 1982.)

notch and its plastic zone relative to the size of the specimen. Thus, the data separation in Figure 8.10 was influenced by the large size of the hole relative to the specimen dimensions and would have been different had the ratio of the hole size and specimen width, the specimen geometry, and loading been different.

The most general correlation for fatigue-crack-initiation from various notches having plastic zones embedded in an elastic field or under general yielding conditions appears to be possible by using the measured stabilized cyclic strain range at the root of the notch, ϵ_a. Such a correlation is presented in Figure 8.11 [*16*] for the fatigue-crack-initiation data obtained by testing the specimens shown in Figure 8.8. Experimental determination of ϵ_a should be conducted for components that would be subjected to general yielding conditions and whose fatigue-crack-initiation lives are very short ($\leq 10^3$ cycles). However, must structural components are not subjected to such severe loading conditions, and, therefore, their fatigue-crack-initiation behavior can be determined accurately by using the $\Delta\sigma_{max}$ parameter.

The analyses and correlations discussed in this chapter can be used to evaluate quantitatively the fatigue-crack-initiation behavior of steels and can be incorporated in design considerations. However, more research is needed to determine the effects of various metallurgical and mechanical parameters on the fatigue-crack-initiation behavior of various metals.

8.5 References

[*1*] Manson, S. S., *Thermal Stress and Low-Cycle Fatigue,* McGraw-Hill, New York, 1966.

[*2*] Wundt, B. M., "Effect of Notches on Low-Cycle Fatigue," *ASTM STP 490,* American Society for Testing and Materials, Philadelphia, May 1972.

[*3*] Eeles, E. G. and Thurston, R. C. A., "Fatigue Properties of Materials," *Ocean Engineering,* Vol. 1, 1968.

[*4*] Creager, M., "The Elastic Stress-Field Near the Tip of a Blunt Crack," Master Science thesis, Lehigh University, Bethlehem, PA, 1966.

[*5*] Wilson, W. K. and Gabrielse, S. E., "Elasticity Analysis of Blunt Notched Compact Tension Specimens," *Research Report 71-1E7-LOWFA-R1,* Westinghouse Research Laboratory, Pittsburgh, February 5, 1971.

[*6*] Forman, R. G., "Study of Fatigue-Crack Initiation from Flaws Using Fracture Mechanics Theory," *Air Force Flight Dynamics Laboratory Technical Report AFFDL-TR-68-100,* Wright-Patterson Air Force Base, Dayton, OH, September 1968.

[*7*] Constable, I., Culber, L. E., and Williams, J. G., "Notch-Root-Radii Effects in the Fatigue of Polymers," *International Journal of Fracture Mechanics,* Vol. 6, No. 3, September 1970.

[*8*] Jack, A. R. and Price, A. T., "The Initiation of Fatigue Cracks from Notches in Mild-Steel Plates," *International Journal of Fracture Mechanics,* Vol. 6, No. 4, December 1970.

[*9*] Barsom, J. M., and McNicol, R. C., "Effect of Stress Concentration on Fatigue-Crack Initiation in HY-130 Steel," *ASTM STP 559,* American Society for Testing and Materials, Philadelphia, 1974.

[*10*] Clark, Jr., W. G., "An Evaluation of the Fatigue Crack Initiation Properties of Type 403 Stainless Steel in Air and Steam Environments," *ASTM STP 559,* American Society for Testing and Materials, Philadelphia, 1974.

[*11*] Barsom, J. M. and McNicol, R. C., unpublished data.

[*12*] Roberts, R., Barsom, J. M., Rolfe, S. T., and Fisher, J. W., *Fracture Mechanics for Bridge Design,* Report No. FHWA-RD-78-69, Federal Highway Administration, Washington, DC, 1977.

[13] Taylor, M. E. and Barsom, J. M., "Effect of Cyclic Frequency on the Corrosion-Fatigue Crack-Initiation Behavior of ASTM A517 Grade F Steel," *Fracture Mechanics: Thirteenth Conference, ASTM STP 743,* R. Roberts, Ed., American Society for Testing and Materials, Philadelphia, 1981.

[14] Barsom, J. M., "Fracture Mechanics-Fatigue and Fracture," *Metals Handbook—Desk Edition,* American Society for Metals, Metals Park, OH, 1985.

[15] Pilkey, W. D., *Petersen's Stress Concentration Factors,* 2nd ed., John Wiley, New York, 1997.

[16] Dowling, N. E., "A Discussion of Methods for Estimating Fatigue Life," *SEA Technical Paper Series, Paper No. 820691,* Society of Automotive Engineers, 1982.

9

Fatigue-Crack Propagation under Constant and Variable-Amplitude Load Fluctuation

9.1 General Background

THE LIFE OF STRUCTURAL components that contain cracks or that develop cracks early in their lives may be governed by the rate of subcritical crack propagation Proof-testing or nondestructive testing procedures or both may provide infor- mation regarding the relative size and distribution of possible pre-existing cracks prior to service. However, these inspection procedures are usually used to estab- lish upper limits on the size of undetectable discontinuities, rather than actua crack size. These upper limits are determined by the maximum resolution of th inspection procedure. Thus, to establish the minimum fatigue life of structura components, it is reasonable to assume that the component contains the larges discontinuity that cannot be detected by the inspection method. The useful lif of these structural components is determined by the fatigue-crack-growth be- havior of the material. Therefore, to predict the minimum fatigue life of structura components and to establish safe inspection intervals, an understanding of th rate of fatigue-crack propagation is required. The most successful approach t the study of fatigue crack propagation is based on fracture-mechanics concepts

The fatigue-crack-propagation behavior for metals can be divided into thre regions (Figure 9.1). The behavior in Region I exhibits a "fatigue-threshold" cycli stress-intensity-factor fluctuation, ΔK_{th}, below which cracks do not propagate un- der cyclic-stress fluctuations [1–15]. Region II represents the fatigue-crack prop- agation behavior above ΔK_{th} [16] which can be represented by:

$$\frac{da}{dN} = A\ (\Delta K)^m \quad (9.1$$

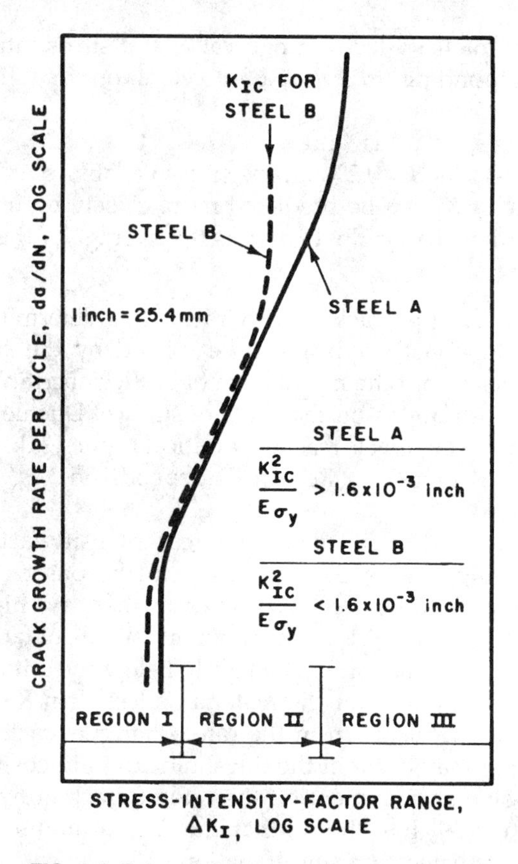

FIG. 9.1 Schematic representation of fatigue-crack growth in steel.

where a = crack length,
N = number of cycles,
ΔK = stress-intensity-factor fluctuation, and
A and m are constants.

In Region III the fatigue-crack growth per cycle is higher than that predicted for Region II. The data [16–21] show that the rate of fatigue-crack growth increases and that under zero-to-tension loading (that is, $\Delta K = K_{max}$) this increase occurs at a constant value of crack-tip opening displacement, δ_T, and at a corresponding stress-intensity-factor value K_T, given by:

$$\delta_T = \frac{K_T^2}{E\sigma_{ys}} = 1.6 \times 10^{-3} \text{ in. } (0.04 \text{ mm}) \tag{9.2}$$

where K_T = stress-intensity-factor-range value at a stress ratio, R, equals zero, corresponding to onset of acceleration in fatigue-crack-growth rates,
E = Young's modulus, and
σ_{ys} = yield strength (0.2% offset) (the available data indicate that the value of K_T can be predicted more closely by using a flow stress, σ_f, rather than σ_{ys}, where σ_f is the average of the yield and tensile strengths).

Acceleration of fatigue-crack-growth rates that determines the transition from Region II to Region III appears to be caused by the superposition of a ductile tear mechanism onto the mechanism of cyclic subcritical crack extension, which leaves fatigue striations on the fracture surface. Ductile tear occurs when the strain at the tip of the crack reaches a critical value [22]. Thus, the fatigue-rate transition from Region II to Region III depends on K_{max} and on the stress ratio, R.

Equation (9.2) is used to calculate the stress-intensity-factor, K_T (or ΔK_T for zero-to-tension loading), value corresponding to the onset of the fatigue-rate transition from Region II to Region III in materials that have high fracture toughness, such as Steel A in Figure 9.1. Acceleration in the rate of fatigue-crack growth occurs at a stress-intensity-factor value slightly below the critical-stress-intensity factor, K_{Ic}, when the K_{Ic} (or K_c) of the material is less than K_T, Steel B in Figure 9.1 [22]. Furthermore, acceleration in the rate of fatigue-crack growth in an aggressive environment may occur at the threshold stress-intensity factor, K_{Iscc}. The effect of an aggressive environment on the rate of crack growth is discussed in Chapter 11. Crooker [20] has shown that the ΔK_T of aluminum and titanium alloys can also be predicted by using Equation (9.2).

9.2 Fatigue-Crack-Propagation Threshold

The early investigations of fatigue-crack propagation by Frost and co-workers [2,3] indicated a significant deceleration in the fatigue-crack growth rates at low stresses. Their data suggested the possible existence of a threshold for fatigue-crack propagation below which fatigue cracks should not propagate. The existence of such a threshold was predicted by an elastic-plastic analysis conducted by McClintock [3]. The work of Linder [4] and Paris [5] showed that the fatigue-crack-propagation threshold can be established in the context of linear-elastic fracture mechanics and that a threshold stress-intensity-factor range, ΔK_{th}, can be determined below which fatigue cracks should not propagate.

In the late 1960s, Elber [6] noted that fatigue-crack surfaces interfere with each other through a closure mechanism. This plastically induced closure results from the presence of residual plastic deformation left behind in the wake of the growing fatigue crack. Soon after, Schmidt's [7] experimental studies indicated that crack closure may have a significant effect on the threshold behavior. The presently available data support this observation but indicate that crack closure can be related to four closure mechanisms [8], shown schematically in Figure 9.2: (1) plastically induced closure resulting from the presence of residual plastic deformation left in the wake of the growing fatigue crack, Figure 9.2*c*; (2) surface-roughness-induced closure caused by deviations of the crack trajectory associated with microstructural characteristics of the material (for example, grain size and interlamellar spacing), Figure 9.2*d*; (3) Mode II-induced closure caused by the displacement of the fatigue crack tip along shear planes, thus preventing a perfect match of the fatigue-crack surface features left behind the propagating crack, Figure 9.2*e*; and (4) environmentally induced closure resulting from corrosion products within the crack that wedge the crack surfaces, Figure 9.2*f*. Several of these mechanisms can be operative at the same time.

Several factors may influence the fatigue-crack-propagation threshold, including [8] yield strength, grain size and other microstructural elements, mean stress, stress history, residual stress, mode of crack-tip opening, Young's modu-

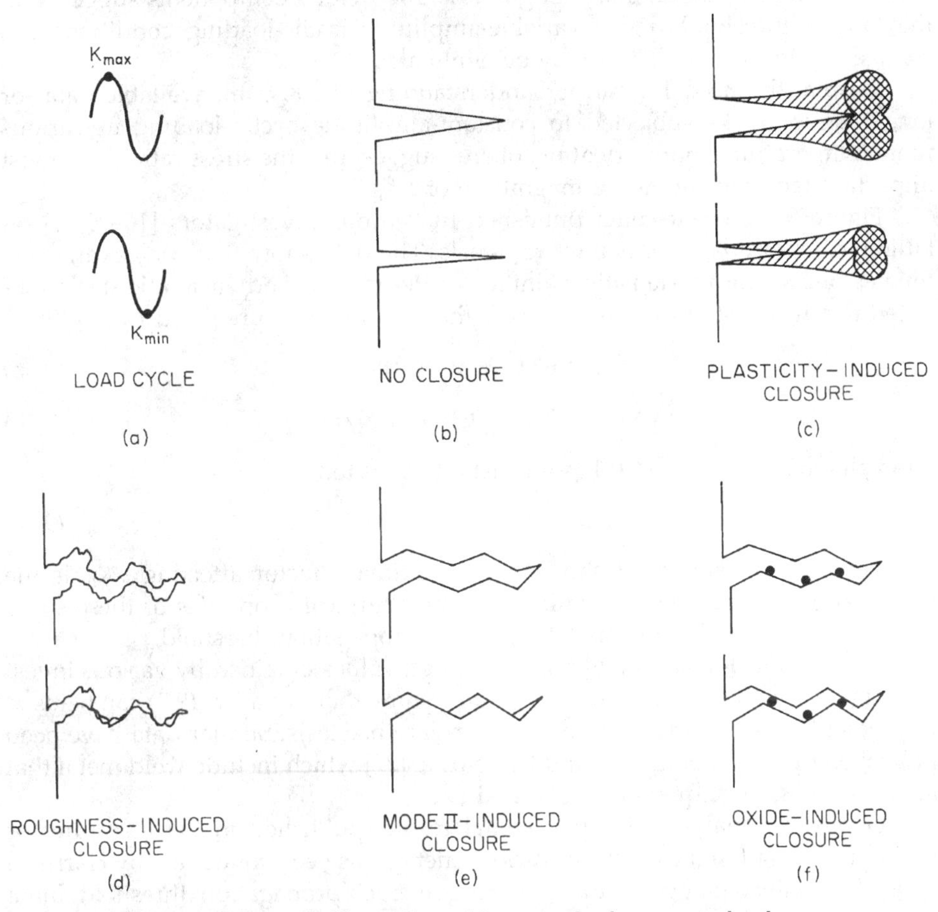

FIG. 9.2 Schematic illustration of four crack-closure mechanisms.

lus, temperature, and environment. The effects of most of these factors on ΔK_{th} can be explained by their relationship to the various mechanisms of crack closure.

Sufficient data are available to show the existence of a fatigue-crack propagation threshold, ΔK_{th}, below which existing fatigue cracks do not propagate under cyclic loading. However, more work is needed to understand better the effects of various factors on the magnitude and the existence of ΔK_{th} and to use it properly in design. Future work on ΔK_{th} should define better the effects of microstructure and material properties, various testing procedures, crack and specimen size and geometry, variable-amplitude loading, and various loadings (out-of-plane bending, torsion, and biaxial and triaxial loading.). Kitagawa and coworkers [*14*] found a constant ΔK_{th} value for crack lengths larger than 0.5 mm. However, below this crack size, a transfer occurred where the fatigue limit of an unnotched material, rather than ΔK_{th}, became the threshold condition for propagation. A similar trend for very small cracks was found by Haddad et al. [*15*]. These observations are extremely important for proper use of ΔK_{th} in structural design. Moreover, preliminary fatigue tests for welded components suggest that the fatigue threshold under variable-amplitude cyclic-loading conditions can change significantly [*26*] and may be eliminated.

Despite the need for further understanding of ΔK_{th}, the available data for long fatigue cracks subjected to constant-amplitude cyclic loading in various room temperature laboratory atmospheres suggest that the stress ratio is the most important factor affecting the magnitude of ΔK_{th}.

Figure 9.3 presents data published by various investigators [*10–12,24*] on fatigue-crack-propagation values for steels. The data show that conservative estimates of ΔK_{th} for martensitic, bainitic, ferrite-pearlite, and austenitic steels subjected to various stress ratios, R, larger than +0.1 can be predicted from [*20*]:

$$\Delta K_{th} = 6.4(1 - 0.85\,R)\text{ksi}\sqrt{\text{in.}} \tag{9.3a}$$

$$\Delta K_{th} = 7(1 - 0.85\,R)\text{MN/m}^{3/2} \tag{9.3b}$$

The value of ΔK_{th} for $R < 0.1$ is a constant equal to:

$$5.5\ \text{ksi}\sqrt{\text{in.}}(6\ \text{MN/m}^{3/2}) \tag{9.3c}$$

The data in Figure 9.3 show that the primary factor affecting ΔK_{th} is the stress ratio and that the mechanical and metallurgical properties of these steels have a negligible effect on the fatigue-crack-propagation threshold.

Lindley and Richards [*25*] have compiled values obtained by various investigators on a wide range of steels. The data show that Equation (9.3) can be used to predict lower-bound values of ΔK_{th} for various steels. Similar data have been published by Sasaki et al. [*26*] and by Priddle [*27*] which include weld metal that support the use of Equations (9.3a) and (9.3b).

Finally, an analysis of experimental results published in the literature for nonpropagating fatigue cracks in various metals has been conducted by Harrison [*10*]. His analysis suggests that the fatigue-crack-propagation threshold for a

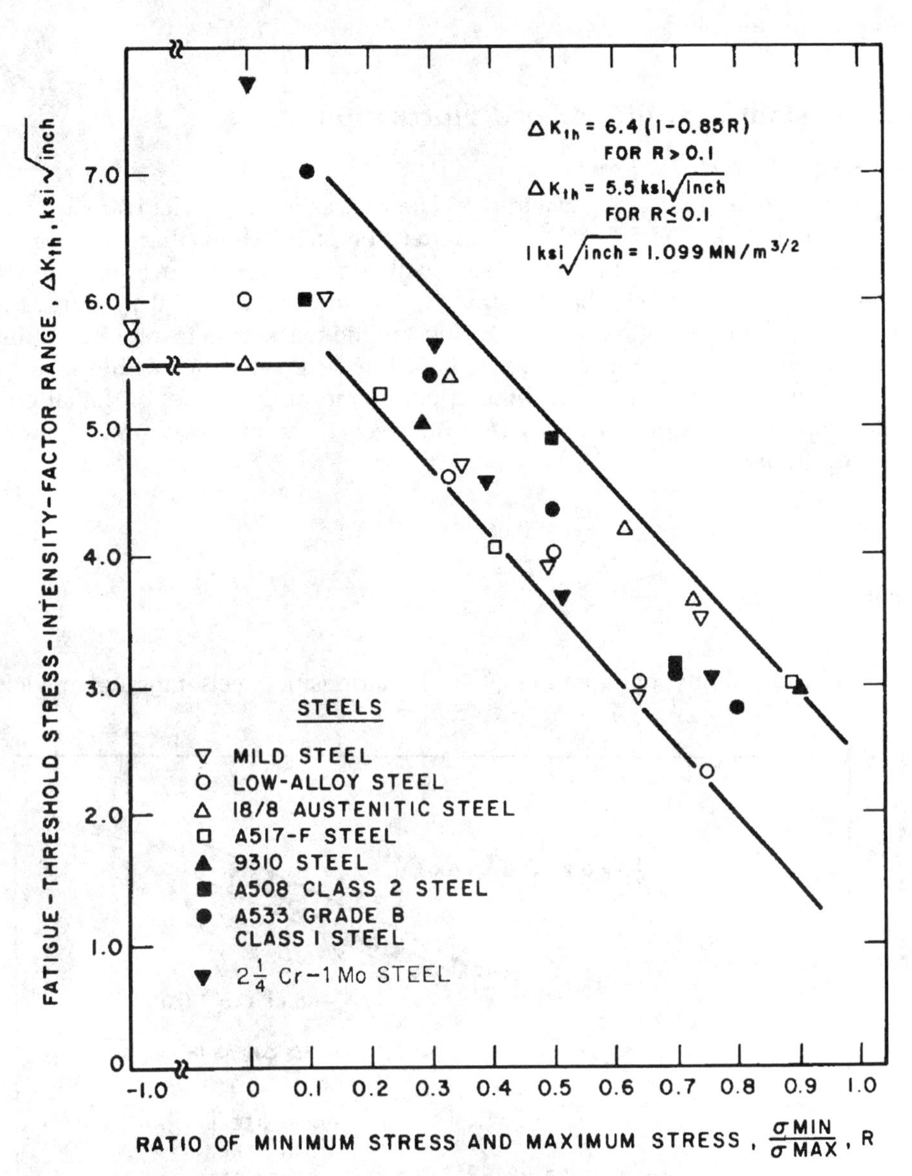

FIG. 9.3 Dependence of fatigue-thresholds stress-intensity-factor range on stress ratio.

number of materials can be superimposed by using the parameter $\Delta K_{th}/E$, where E is Young's modulus.

9.3 Constant Amplitude Load Fluctuation

9.3.1 Martensitic Steels

Extensive fatigue-crack-growth-rate data for various high-yield strength (σ_{ys} greater than 80 ksi or 552 MN/m^2) martensitic steels show that the primary parameter affecting growth rate in Region II is the range of fluctuation in the stress-intensity factor and that the mechanical and metallurgical properties of these steels have negligible effects on the fatigue-crack-growth rate in a room temperature air environment [*20*]. The data for these steels fall within a single band, as shown in Figure 9.4, and the upper bound of the scatter of the fatigue-crack-propagation-rate data for martensitic steels in an air environment can be obtained from:

$$\frac{da}{dN} = 0.66 \times 10^{-8}(\Delta K_I)^{2.25} \tag{9.4}$$

where a = in.
$\Delta K_I = \text{ksi}\sqrt{\text{in.}}$

The applicability of Equation (9.4) to martensitic steels ranging in yield

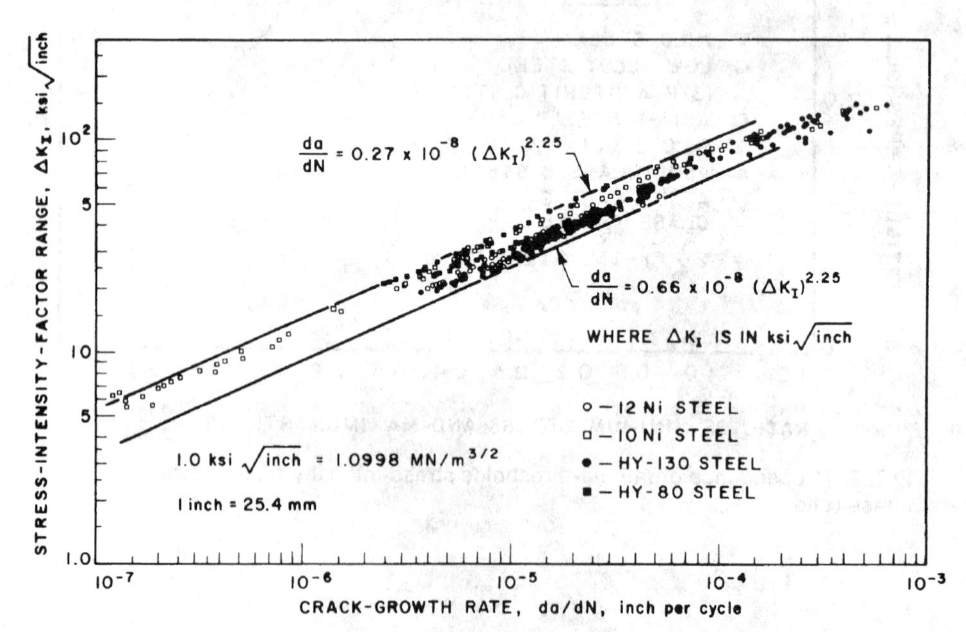

FIG. 9.4 Summary of fatigue-crack propagation for martensitic steels.

strength from 80 to 300 ksi (552 to 2068 Mn/m^2) has been established [*16,17*]. The validity of Equation (9.4) has been established further by using data obtained by testing various grades of ASTM A514 steels [*28,29*] (Figure 9.5) [*28*]; ASTM A517 Grade F steel [*9*]; ASTM A533 Grade B, Class 1 steel and ASTM A533 Grade A and ASTM A645 steels [*31*].

9.3.2 Ferrite-Pearlite Steels

The fatigue-crack-growth rate behavior in ferrite-pearlite steels [*16*] prior to the onset of fatigue-rate transition and above ΔK_{th} is presented in Figure 9.6. The data indicate that realistic estimates of the rate of fatigue-crack growth in these steels can be calculated from:

$$\frac{da}{dN} = 3.6 \times 10^{-10}(\Delta K_I)^{3.0} \tag{9.5}$$

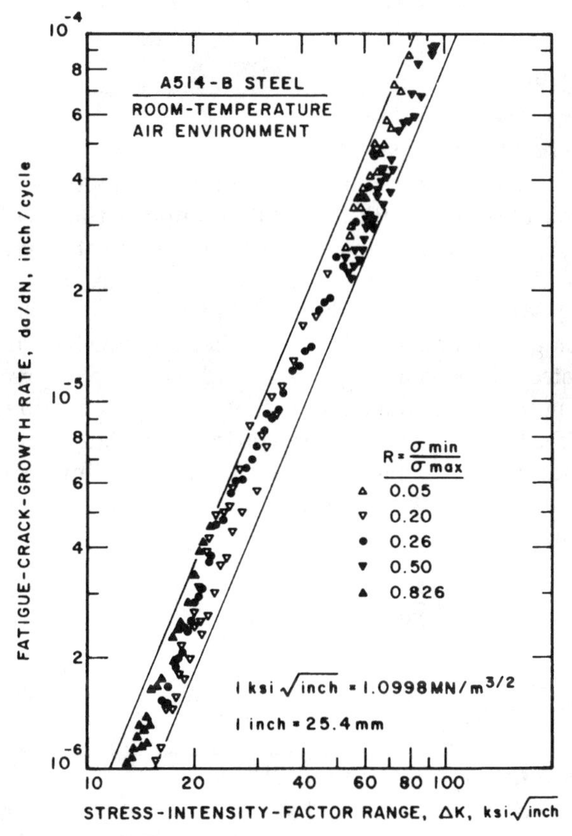

FIG. 9.5 Crack-growth rate as a function of stress-intensity-factor range for A514 Grade B steel.

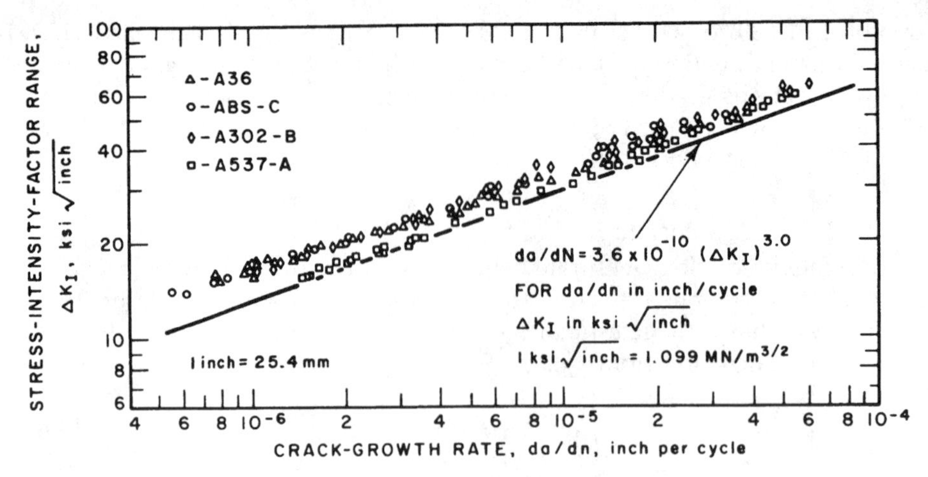

FIG. 9.6 Summary of fatigue-crack-growth data for ferrite-pearlite steels.

where a = in., and
$\Delta K_I = \text{ksi}\sqrt{\text{in.}}$

9.3.3 Austenitic Stainless Steels

Extensive fatigue-crack-growth-rate data for austenitic stainless steels in Region II have been obtained. Data for solution-annealed and cold-worked Type 316 stainless steel and for solution-annealed Type 304 stainless steel were obtained by James [32]. Data on these types of steels were also obtained by Shahinian et al. [33]. Weber and Hertzberg [34] examined the effect of thermomechanical processing on fatigue-crack propagation in Type 305 stainless steel. The steel was evaluated in the coarse-grained, cold-worked, and recrystallized conditions. A summary of their data shows that crack propagation was essentially similar irrespective of thermomechanical processing. Data from these and other investigations indicate that conservative and realistic estimates of fatigue-crack-propagation rates for austenitic steels in a room temperature air environment can be obtained from:

$$\frac{da}{dN} = 3.0 \times 10^{-10}(\Delta K_I)^{3.25} \tag{9.6}$$

where a = in.
$\Delta K_I = \text{ksi}\sqrt{\text{in.}}$

9.3.4 Aluminum and Titanium Alloys

The preceding discussion shows that the Region II fatigue-crack-growth-rate behavior of steels in a benign environment is essentially independent of mechanical and metallurgical properties of the material.

Figures 9.7 and 9.8 [*19*] present the fatigue-crack-growth behavior of aluminum and titanium alloys, respectively. The data indicate that, like steels, the fatigue-crack-growth rates for aluminum and titanium alloys fall within different but distinct scatter bands. However, the scatter bands for aluminum and titanium alloys are larger than for steels. Thus, it is apparent that for a given metal system subjected to a benign environment, the fatigue-crack-growth behavior in Region II is not a pertinent factor in material selection considerations within a given metal system.

The fatigue-rate transition from Region II to Region III for aluminum and titanium alloys has been investigated by Clark and Wessel [*35*] and by Crooker [*19,36*]. The data show conclusively that Equation (9.2) can be used to determine

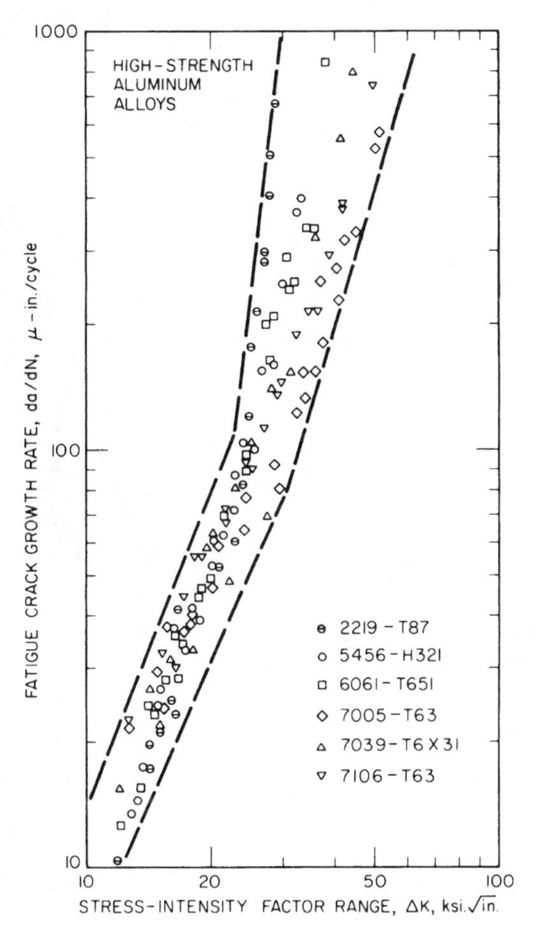

FIG. 9.7 Summary plot of *da/dN* versus Δ*K* for six aluminum alloys. The yield strengths of these alloys range from 34 to 55 ksi.

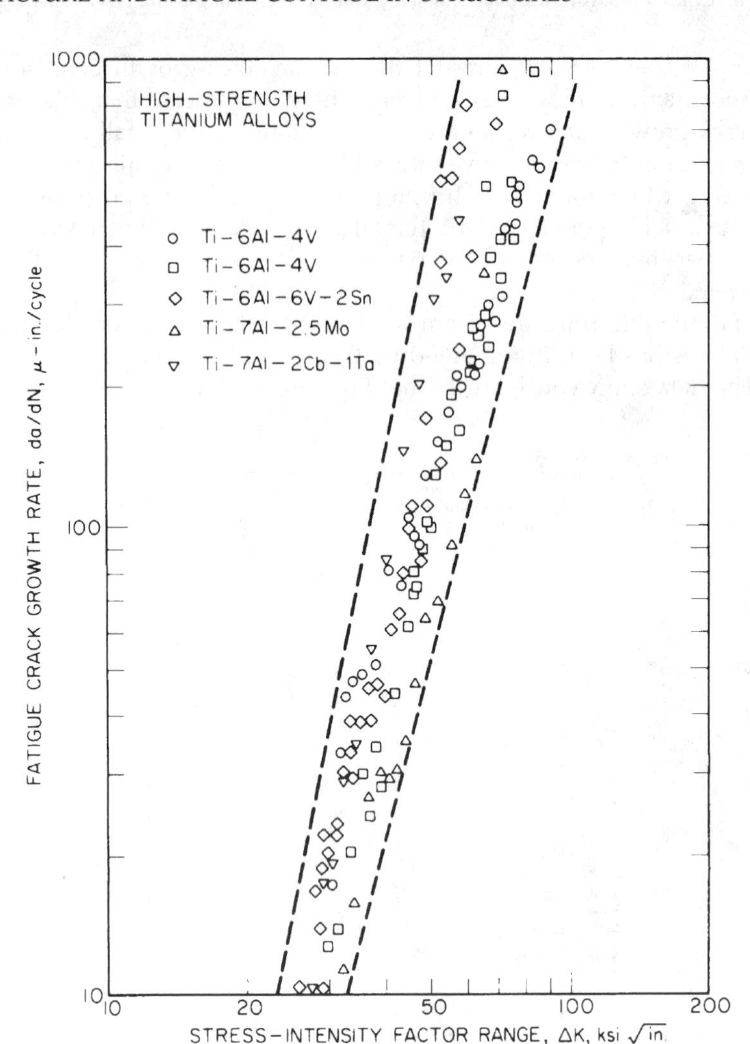

FIG. 9.8 Summary plot of *da/dN* versus Δ*K* data for five titanium alloys ranging in yield strength from 110 to 150 ksi.

the stress-intensity-factor value corresponding to the transition from Region II crack-growth-rate behavior to Region III in aluminum and titanium alloys.

9.4 Effect of Mean Stress on Fatigue-Crack Propagation Behavior

The effect of mean load on fatigue-crack-initiation and propagation behavior can be studied by using the load ratio parameter, R, where R is equal to P_{min}/P_{max} =

K_{min}/K_{max}, and $P_{min}(K_{min})$ and $P_{max}(K_{max})$ are the minimum and maximum loads (stress-intensity-factor values) in the cyclic process, respectively.

Several investigations have been conducted to study the effect of mean stress and stress ratio on fatigue-crack-propagation rate [*6,28,37–40*]. Available experimental data on ASTM A514 Grade B steel show no systematic change in fatigue-crack-growth rate with changes in R values from 0 to 0.82 (Figure 9.5) [*28*]. The data also show that this change in R values has negligible effects on the rate of crack propagation in Region II. On the other hand, Crooker [*38*] observed that the compression portion of fully reversed tension-compression cycling increased the growth rate in HY-80 steel by approximately 50% as compared with zero-tension cycling. Data on fatigue-crack growth in a 140-ksi (965-MN/m^2) yield-strength martensitic steel [*39*] showed a systematic increase in growth rate with increase in R values from 0 to 0.75 and with decrease in R values from 0 to -2, but that maximum nominal stress levels of 0.39 to 0.94 of the yield strength had no apparent effect (Figure 9.9). The maximum increase in fatigue-crack-growth rate as a function of variation of R from -2 to 0.75 was less than a factor of 2.

Crooker et al. [*38,39*] studied the effect of stress ratio, R, in the range $-2 \leq R \leq 0.75$ by using part-through-crack specimens subjected to axial loads. Because of the difficulties encountered in measuring crack depth in these specimens, the rate of fatigue-crack growth was calculated by measuring the rate of growth on the surface of the specimens and by assuming that the crack shape did not change (that is, a circular crack front remained circular) as the crack size increased. Despite reservations relating to the accuracy of this assumption and to the use of part-through-crack specimens for fatigue-crack-growth studies, analysis of the data presented in Figure 9.9 showed that the effect of positive stress ratio, $R \geq 0$, on the rate of fatigue-crack growth in the 140-ksi yield-strength steel investigated can be predicted by using:

$$\frac{da}{dN} = \frac{A(\Delta K)^{2.25}}{(1 - R)^{0.5}} \tag{9.7}$$

where A is a constant.

The upper bound of the scatter band of martensitic steels represented by Equation (9.4) is also presented in Figure 9.9. The data suggest that Equation (9.4) can be used to estimate conservative fatigue-crack-growth rates in martensitic steels subjected to various values of stress ratio.

The general effect of stress ratio, R, on fatigue-crack-growth rates in Regions I, II, and III are presented schematically in Figure 9.10 [*41*]. Data in support of this generalized behavior have been published by Wei [*41*].

9.5 Effects on Cyclic Frequency and Waveform

The effect of loading rate on fracture toughness was presented in Chapter 4. The available data show that the rate of loading can affect the fracture toughness

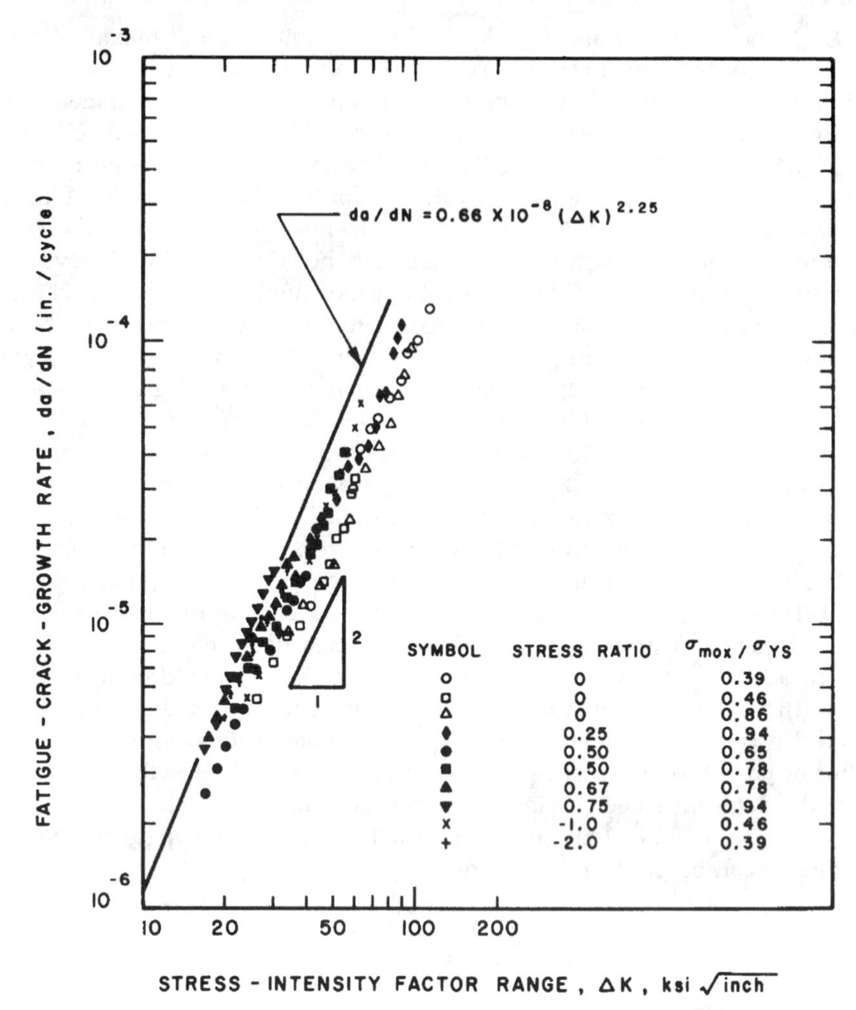

FIG. 9.9 Effect of stress ratio on the fatigue-crack-propagation rate in a 140-ksi yield strength martensitic steel.

significantly and that the magnitude of this effect increases with decrease in yield strength. Because a change in cyclic frequency corresponds to a change in the rate of loading, the effect of cyclic frequency on fatigue-crack-growth rate was investigated by Barsom for a 36-ksi and a 180-ksi yield strength steel.

The fatigue-crack-growth rate for A36 steel (σ_{ys} = 36 ksi) under cyclic frequencies of 6 to 3000 cycles per minute (cpm) is presented in Figure 9.11. Similar data for a 12 Ni-5Cr-3Mo maraging steel (σ_{ys} = 180 ksi) were obtained under cyclic frequencies of 6 to 600 cpm [*21,42,43*]. The data show that the rates of fatigue-crack growth in a room temperature benign environment were not affected by the frequency of the cyclic-stress fluctuations.

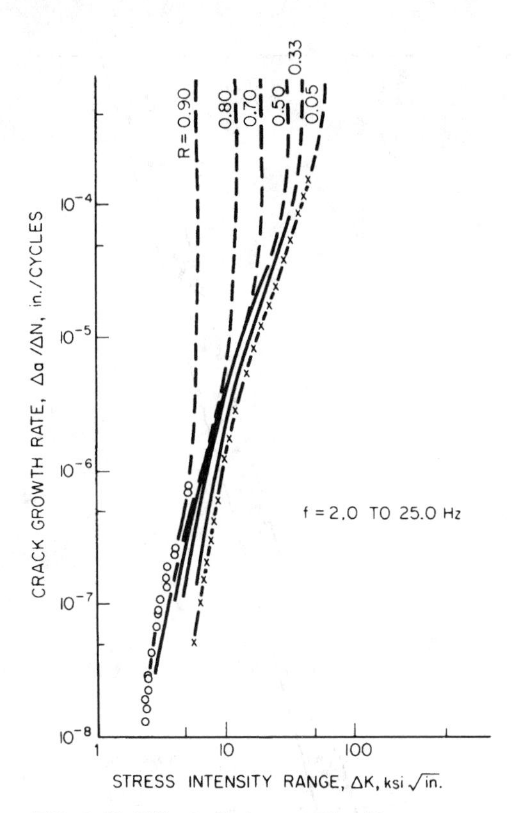

FIG. 9.10 Effect of stress ratio, *R*, on fatigue-crack-growth behavior.

The discussions in the preceding sections show that the stress-intensity-factor fluctuation, ΔK_I, is related to the magnitude of the stress fluctuation and to the square root of crack length. This parameter does not account for possible differences in growth rates that may exist between various cyclic waveforms such as sine and square waves. Consequently, data on the fatigue-crack-growth rate of 12Ni-5Cr-3Mo maraging steel in a room temperature air environment were obtained under sinusoidal, triangular, square, positive sawtooth (/|), and negative sawtooth (|\) cyclic-stress fluctuations. The combined data, presented in Figure 9.12 [42], show that the rates of fatigue-crack growth in a room temperature air environment were not affected by the form of the cyclic-stress fluctuations.

The effects of cyclic frequency and waveform on corrosion-fatigue crack-growth rates are presented in Chapter 11.

9.6 Effects of Stress Concentration on Fatigue-Crack Growth

The effects of stress-concentration factors on fatigue-crack initiation was discussed in Chapter 8. Because the fluctuation of the stress-intensity factor is the

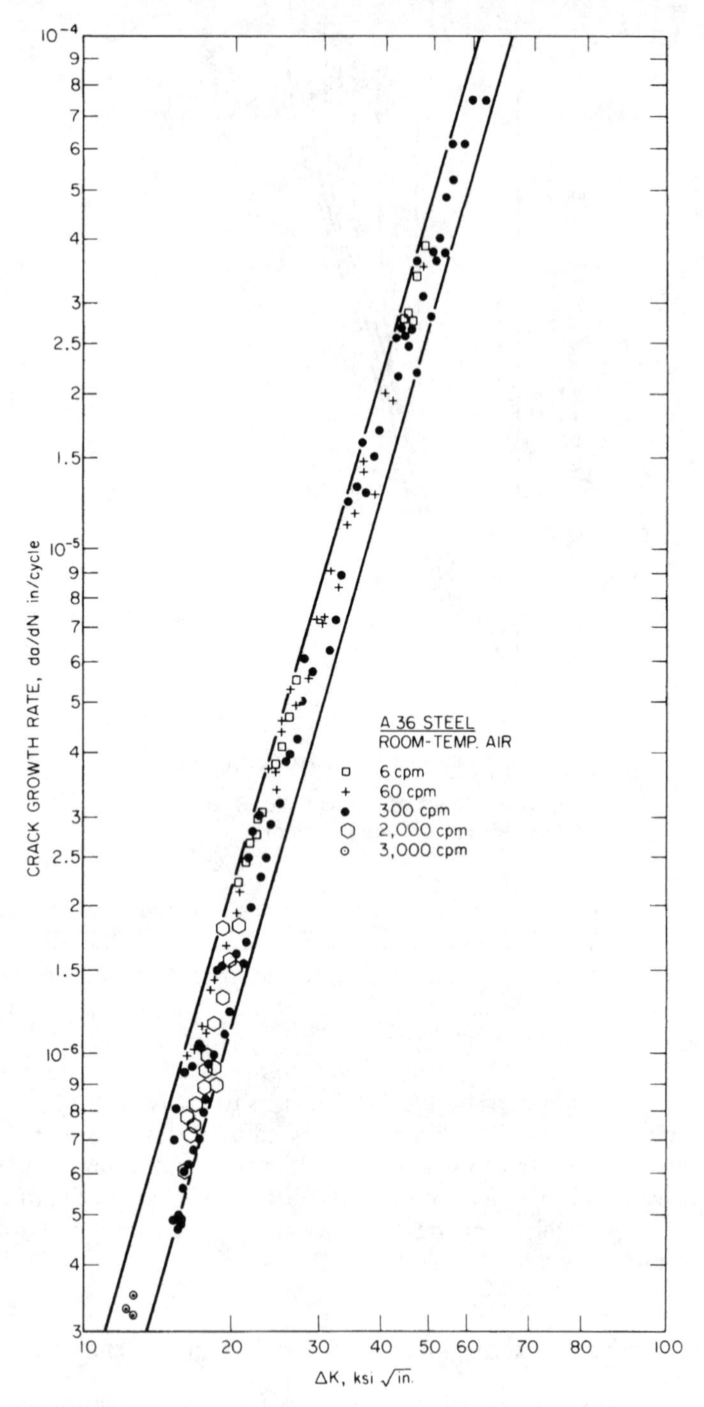

FIG. 9.11 Effect of cyclic frequency on crack-growth rates in a benign environment.

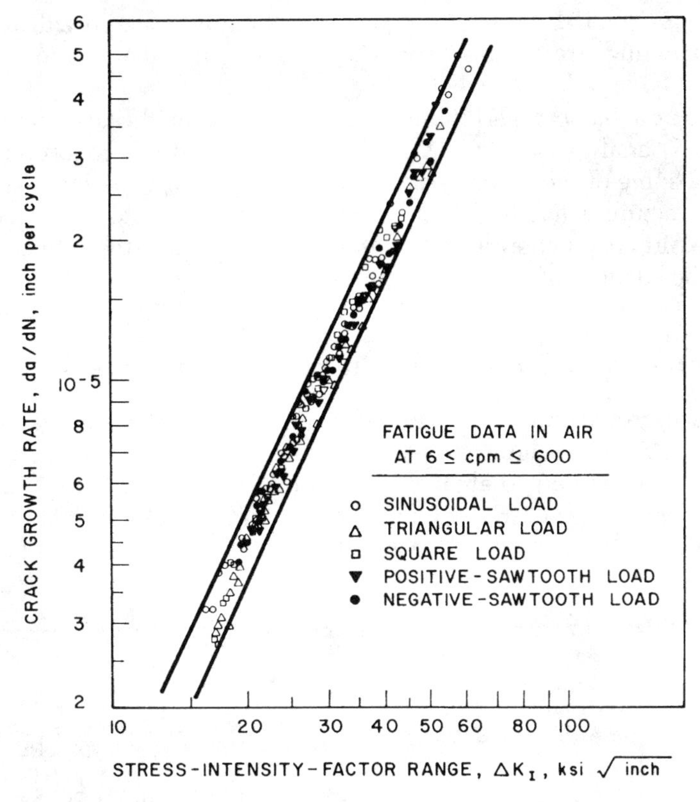

FIG. 9.12 Fatigue-crack-growth rates in a 12Ni-5Cr-3Mo steel under various cyclic-stress functions with different stress-time profiles.

primary fatigue-crack-propagation force, it can be surmised that the rate of fatigue-crack propagation in the vicinity of a stress raiser, such as a notch, would be governed by the local stress-intensity-factor fluctuation. That is, the fatigue-crack-propagation rate per cycle, da/dN, in the shadow of a notch must be governed by the relationship:

$$\frac{da}{dN} = A(\Delta K_{\text{eff}})^m \tag{9.8}$$

where A and m = constants,

$\Delta K_{\text{eff}} \cong k_t(a)\Delta\sigma\sqrt{a}$

$\Delta\sigma$ = nominal stress fluctuation,

a = crack length, and

$k_t(a)$ = stress-concentration factor; this factor is a function of crack length such that as the crack propagates outside the field of influence of the notch, $k_t(a)$ approaches unity (see Chapter 2 for cracks emanating from notches).

Equation (9.8) could be used to analyze the fatigue-crack-growth behavior of cracks emanating from holes, nozzle corners, or other regions of stress concentrations.

Novak and Barsom [*44*] summarized the state of the art of all available theoretical K_I analyses for cracks in the vicinity of stress concentrations. The accuracy of some of these analyses was verified by their experimental results on the brittle-fracture behavior for cracks emanating from notches. These analyses can be used in conjunction with Equation (9.8) to predict the rate of growth of cracks in the shadow of stress concentrations.

9.7 Fatigue-Crack Propagation in Steel Weldments

Fatigue-crack-growth rates in weldments of various steels are available in the literature [*31,45–48*]. Fatigue-crack-growth rates in the weld metal and the heat-affected zone of submerged arc weldments of ASTM A533 Grade B, Class 1 steel have been obtained by Clark [*45*]. In general, fatigue cracks initiated in the heat-affected zone grew into the adjacent weld metal. Figures 9.13 and 9.14 present

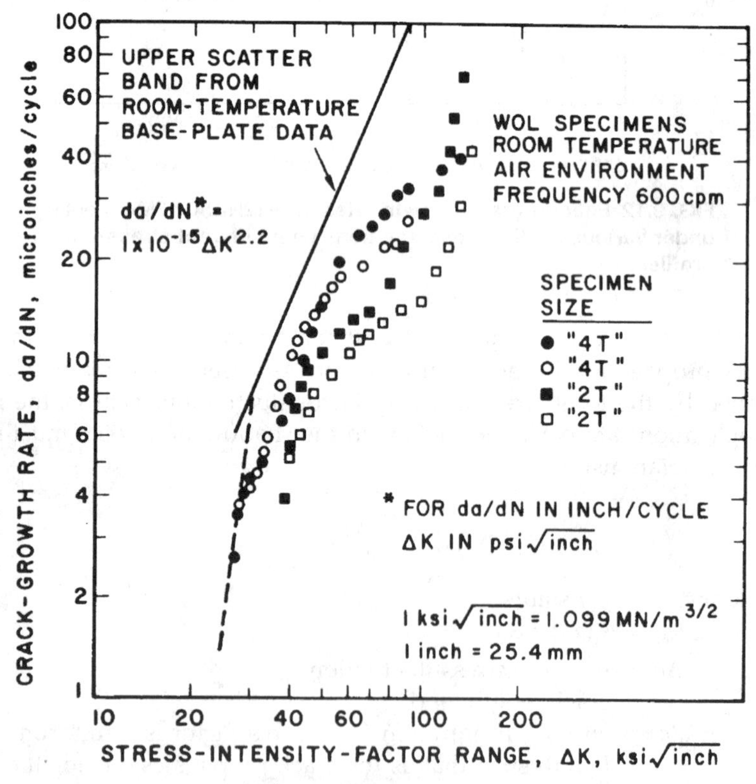

FIG. 9.13 Effect of specimen size on the crack-growth-rate behavior of an A533 Grade B, Class 1 steel weld metal.

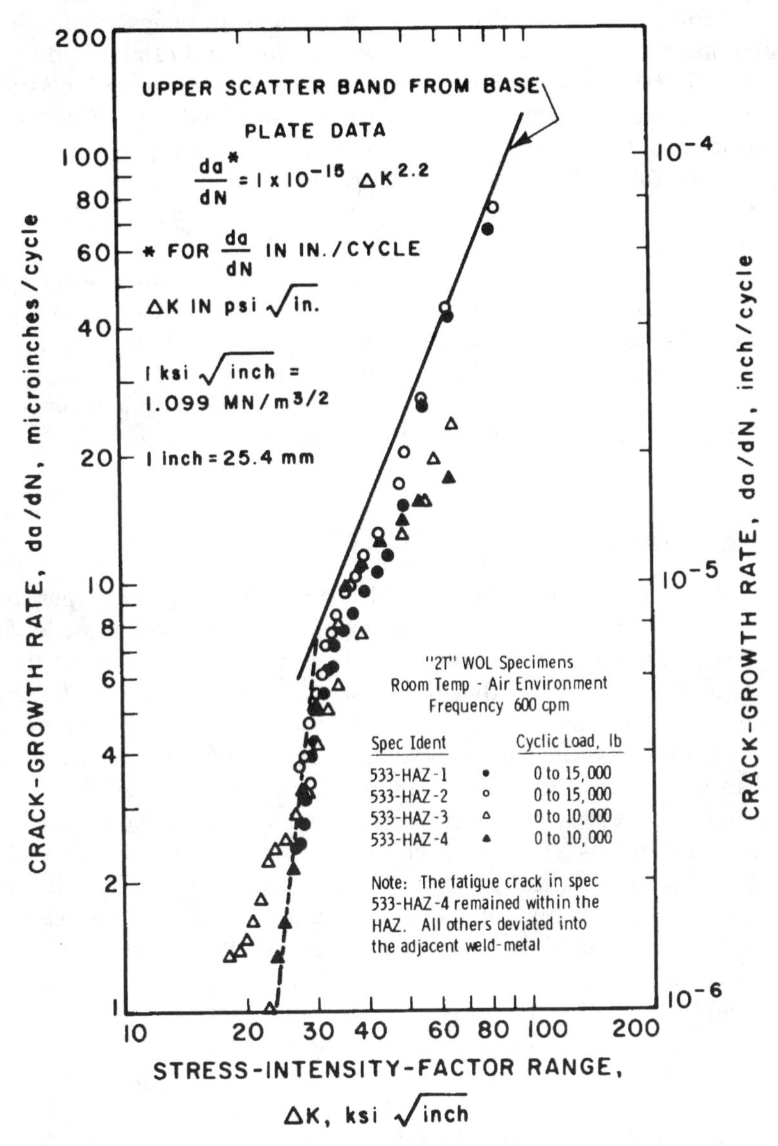

FIG. 9.14 Fatigue-crack-growth-rate behavior observed in the vicinity of the heat-affected zone of an ASTM A533 Grade B, Class 1 weldment.

the fatigue-crack-growth behavior in the weld metal and in the heat-affected zone, respectively. The data show that the rate of fatigue-crack growth in the weld metal and in the heat-affected zone was equal to or less than that in the base metal and that the upper bound of the scatter band established by testing the base metal at room temperature may be used as a conservative estimate of fatigue properties of the weld metal and the heat-affected zone. This conclusion is confirmed by data obtained on HY-140 steel weldments [*46*], on a 5% nickel steel (ASTM A645) weldment [*31*], on a Type 308 weld metal [*47*], on a Type 316 weld metal [*47*], and for structural C-Mn steel weld metals, heat-affected zones, and base metals [*48*].

As-welded structural components usually contain residual tensile stresses often of yield stress magnitude [*49*]. Consequently, fatigue cracks in regions of tensile-residual stresses propagate under high stress ratios. Preceding discussions showed that high stress ratios affect the magnitude of ΔK_{th} and the magnitude of K_T but have negligible effect on the fatigue-crack-growth-rate behavior in Region II. This observation is supported further by Maddox [*48*].

9.8 Design Example

For most structural materials, the tolerable flaw sizes are much larger than any initial undetected flaws. However, for structures subjected to fatigue loading (or stress-corrosion cracking), these initial cracks may grow throughout the life of the structure. Thus, to ensure that the structural component does not fracture, the calculated critical crack size, a_{cr}, at design load must be large, and the number of cycles of loading required to grow a small crack to the critical crack must be greater than the design life of the structure.

Although *S–N* curves have been widely used to analyze the fatigue behavior of steels and weldments, closer inspection of the overall fatigue process in complex welded structures indicates that a more rational analysis of fatigue behavior is possible by using concepts of fracture mechanics. Specifically, fabrication flaws may be present in welded structures, even though the structure has been inspected. Accordingly, a conservative approach to designing to prevent fatigue failure would be to assume the presence of an initial flaw and analyze the fatigue-crack-growth behavior of the structural member. The size of the initial flaw is obviously dependent on the detail geometry, quality of fabrication, and inspection.

A schematic diagram of the general relation between fatigue-crack initiation and propagation is shown in Figure 9.15. The question of when does a crack "initiate" to become a "propagating" crack is somewhat philosophical and depends on the level of observation of a crack, that is, crystal imperfection, dislocation, microcrack, lack of penetration, and so on. A conservative approach to fatigue would be to assume an initial flaw size on the basis of the quality of inspection used and then to calculate the number of cycles it would take for this crack to grow to a size critical for fracture.

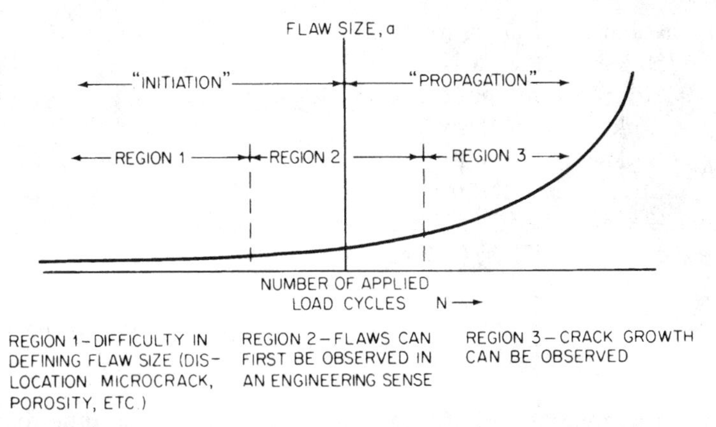

FIG. 9.15 Schematic showing relation between "initiation" life and "propagation" life.

The procedure to analyze the crack-growth behavior in steels and weld metals using fracture-mechanics concepts is as follows:

1. On the basis of quality of inspection, estimate the maximum initial flaw size, a_0, that may be present in the structure and the associated K_I relation (Chapter 2) for the member being analyzed.
2. Knowing K_c or K_{Ic} and the nominal maximum design stress, calculate the critical flaw size, a_{cr}, that would cause fracture.
3. Obtain an expression relating the fatigue-crack-growth rate of the steel or weld metal being analyzed. The following conservative estimates of the fatigue-crack growth per cycle of loading, da/dN, have been determined for martensitic steels (for example, A514/517) as well as ferrite-pearlite steels (for example, A36) in a room temperature air environment and were discussed previously,

$$\textit{martensitic steels:}\quad \frac{da}{dN} = 0.66 \times 10^{-8}(\Delta K_I)^{2.25} \tag{9.4}$$

$$\textit{ferrite-pearlite steels:}\quad \frac{da}{dN} = 3.6 \times 10^{-10}(\Delta K_I)^{3} \tag{9.5}$$

where da/dN = fatigue-crack growth per cycle of loading, in./cycle, and
ΔK_I = stress-intensity-factor range, ksi$\sqrt{\text{in.}}$

4. Determine ΔK using the appropriate expression for K_I, the estimated initial flaw size, a_0, and the range of live-load stress (cyclic-stress range).
5. Integrate the crack-growth-rate expression between the limits of a_0 (at the initial K_I) and a_{cr} (at K_{Ic}) to obtain the life of the structure prior to failure.

A numerical example of this procedure is as follows:

1. Assume the following conditions:

a. A514 steel, $\sigma_{ys} = 100$ ksi (689 MN/m^2).
b. $K_{Ic} = 150$ ksi$\sqrt{\text{in.}}$(165 MN/mn$^{3/2}$).
c. $a_0 = 0.3$ in. (7.7 mm), edge crack in tension (Chapter 2).
d. $\sigma_{max} = 45$ ksi (310 MN/m^2);
$\sigma_{min} = 25$ ksi (172 MN/m^2);
$\Delta\sigma = 20$ ksi (138 MN/m^2) (live-load stress range).
e. $K_I = 1.12\sqrt{\pi}\sigma\sqrt{a}$ edge crack in tension (Chapter 2).

2. Calculate a_{cr} at $\sigma = 45$ ksi (310 MN/m^2):

$$a_{cr} = \left(\frac{K_{Ic}}{1.12\sqrt{\pi} \cdot \sigma_{max}}\right)^2 = \left(\frac{150}{1.12(1.77)(45)}\right)^2 = 2.8 \text{ in. (71.1 nm)}$$

3. Assume an increment of crack growth, Δa. In this case assume that $\Delta a = 0.1$ in. (2.5 mm). If smaller increments of crack growth were assumed, the accuracy would be increased slightly.
4. Determine the expression for ΔK_I for a_{avg}, which is the average crack size between the two crack increments a_i and $a_{i+0.1 \text{ in.}}$

$$\Delta K_I = 1.12\sqrt{\pi}\Delta\sigma\sqrt{a_{avg}}$$
$$= 1.98\ (20)\sqrt{a_{avg}}$$

5. Using the appropriate expression for crack-growth rate,

$$da/dN = 0.66 \times 10^{-8}(\Delta K_I)^{2.25}$$

solve for ΔN for each increment of crack growth, replacing da/dN by $\Delta a/\Delta N$:

$$\Delta N = \frac{\Delta a}{0.66 \times 10^{-8}[1.98(20)\sqrt{a_{avg}}]^{2.25}}$$

$$\Delta N = 12{,}500 \text{ cycles}$$

for the first increment of crack extension from $a_0 = 0.30$ in. to $a = 0.40$ in.

6. Repeat for $a = 0.4$ to 0.5 in. (10.2 to 12.7 mm), and so on, by numerical integration, as shown in Table 9.1. The flaw size—life results for this example are presented in Figure 9.16. If only the desired total life is required, the expression for ΔN can be integrated directly. In this example, direct integration yielded a life of 87,600 cycles, while the numerical technique gave a life of 86,700 cycles.

Note that the total life to propagate a crack from 0.3 to 2.8 in. (7.6 to 71.1 mm) in this example is 86,700 cycles. If the required life were 100,000 cycles, this design would be inadequate, and one or more of the following changes should be made:

TABLE 9.1. Fatigue-Crack Growth Calculations.

$$\Delta N = \frac{\delta a}{0.66 \times 10^{-8}[1.98(\Delta\sigma)\sqrt{a_{avg}}]^{2.25}}$$

where $\Delta a = 0.10$ in. (2.54 mm), and

$\Delta\sigma = 20$ ksi (138 MN/m²)

a_0, in.	a_f in.	a_{avg}, in.	ΔK, ksi$\sqrt{in.}$	ΔN, cycles	ΣN, cycles
0.3	0.4	0.35	23.5	12,500	12,500
0.4	0.5	0.45	26.7	9,750	22,250
0.5	0.6	0.55	29.4	7,550	29,800
0.6	0.7	0.65	32.2	6,150	35,950
0.7	0.8	0.75	3.6	5,200	41,150
0.8	0.9	0.85	36.6	4,600	45,750
0.9	1.0	0.95	38.8	4,100	49,850
1.0	1.1	1.05	40.5	3,700	53,550
1.1	1.2	1.15	42.5	3,300	56,850
1.2	1.3	1.25	44.5	2,950	59,800
1.3	1.4	1.35	46.1	2,700	62,500
1.4	1.5	1.45	47.7	2,550	65,050
1.5	1.6	1.55	49.3	2,350	67,400
1.6	1.7	1.65	51.0	2,200	69,600
1.7	1.8	1.75	52.5	2,050	71,650
1.8	1.9	1.85	54.0	1,900	73,550
1.9	2.0	1.95	55.6	1,800	75,350
2.0	2.1	2.05	56.8	1,700	77,050
2.1	2.2	2.15	58.5	1,600	78,650
2.2	2.3	2.25	59.6	1,500	80,150
2.3	2.4	2.35	60.8	1,450	81,600
2.4	2.5	2.45	62.5	1,400	83,000
2.5	2.6	2.55	63.5	1,350	84,350
2.6	2.7	2.65	64.8	1,200	85,550
2.7	2.8	2.75	66.0	1,150	86,700

1 in. = 25.4 mm

1 ksi$\sqrt{in.}$ = 1.11 MN m$^{3/2}$

1. Increase the critical crack size at failure [a_{cr} = 2.8 in. (71.1 mm)] by using a material with a higher K_{Ic} value.
2. Lower the design stress, σ_{max}, to increase the critical crack size at failure.
3. Lower the stress range ($\Delta\sigma$) to decrease the *rate* of crack growth, thereby increasing the number of cycles required for the crack to grow to the critical size. Note that because the rate of crack growth is a power function of $\Delta\sigma$, or actually ΔK, lowering the stress range slightly has a significant beneficial effect on the life.
4. Improve the fabrication quality and inspection capability so that the initial flaw size (a_0) is reduced. It is clear from Table 9.1 and Figure 9.16 that most of the life is taken up in the early stages of crack propagation. In fact, to double the initial crack size during the early stages of propagation requires almost one third the total number of cycles. Therefore, any decrease in

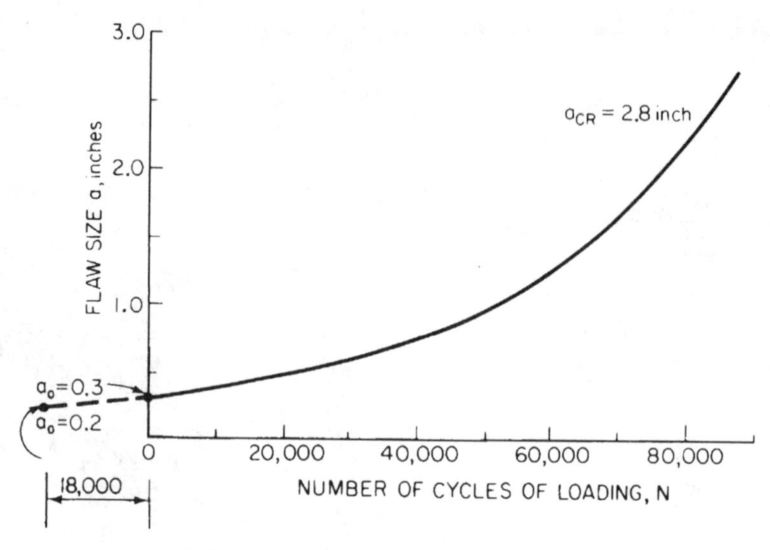

FIG. 9.16 Fatigue-crack-growth curve.

initial flaw size has a very significant effect on the fatigue life of a structural member.

In this example, if a_0 were only 0.2 in. (5.1 mm), the design would be satisfactory. That is, the number of cycles to grow a crack from 0.2 to 0.3 in. (5.1 to 7.6 mm) is about 18,000 cycles, as indicated in Figure 9.16, which [added to the 86,700 cycles required to grow the crack from 0.3 to 2.8 in. (7.6 to 71.1 mm)] would make the total life equal to 104,700 cycles. It should be noted that for steels with high fracture toughness levels, the state of stress ahead of large cracks may be plane stress, and thus larger cracks could be tolerated than are calculated on the basis of plane-strain behavior. However, because the crack-growth rate increases rapidly for large cracks, a significant increase in the fracture toughness of the material (that is, in the size of the tolerable crack at failure) may result in a negligible increase in the fatigue life of the structural member.

9.9 Variable-Amplitude Load Fluctuation

9.9.1 Probability-Density Distribution

Many engineering structures such as bridges, ships, and others are subjected to variable-amplitude random-sequence load fluctuations. The probability of occurrence of the same sequence of stress fluctuations for a given detail in such structures obtained during a given time interval is very small. Consequently, the magnitude of stress fluctuations must be characterized to study the fatigue behavior of components subjected to variable-amplitude random-sequence stress fluctuations. The magnitude of the stress fluctuations should be characterized and described by analytic functions. The use of probability-density curves to characterize variable-amplitude cyclic-stress fluctuations appears to be very useful [28,50–52].

Stress history, or stress spectrum, for a particular location in a structure subjected to variable-amplitude stress fluctuation can be defined in terms of the frequency of occurrence of maximum (peak) stresses. Usually, frequency-of-occurrence data are presented as a histogram, or a bar graph (Figure 9.17), in which the height of the bar represents the percentage of recorded maximum

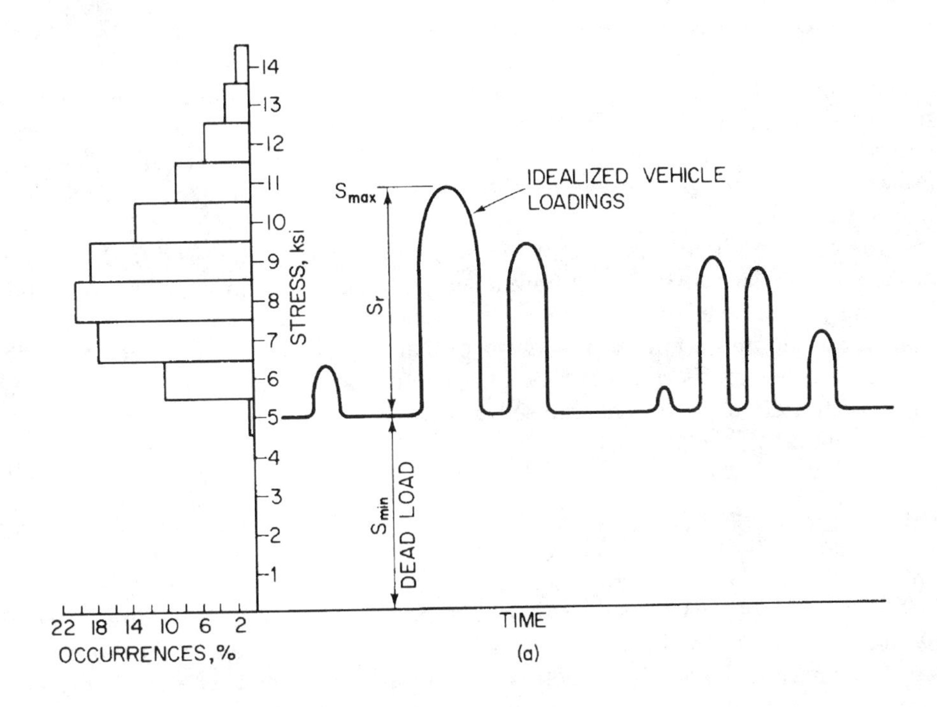

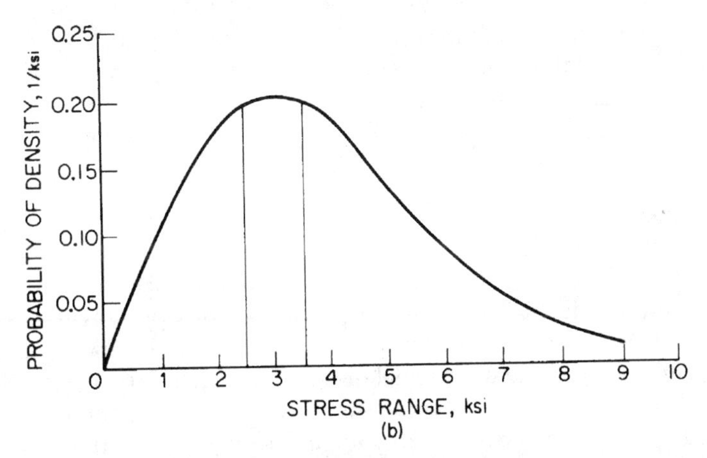

FIG. 9.17 Frequency-of-occurrence data.

stresses that fall within a certain stress interval represented by the width of the bar. For example, 20.2% of the maximum stresses in Figure 9.17*a* fall within the interval between 7.5 and 8.5 ksi (52 to 59 MN/m^2). The frequency of occurrence of stress ranges can be represented by similar plots with the vertical scale changed according to the relationship between σ_{max},σ_{min}, and stress range, σ_r, or $\Delta\sigma$. Since stress range is the most important stress parameter controlling the fatigue life of structural components, stress range is used to define the major stress cycles in the following discussion.

The frequency-of-occurrence data can be presented in a more general form by dividing the percent of occurrence for each interval, that is, the height of each bar, in Figure 9.17*a* by the interval width to obtain a probability-density curve such as shown in Figure 9.17*b*. Thus, data from sources that use different stress-range intervals can be compared by using the probability-density curve. The area under the curve between any two values of $\Delta\sigma$ represents the percent occurrence within that interval.

A single nondimensional mathematical expression can be used to define the probability-density curves for different sets of data. For example, Klippstein and Schilling [61] showed that the following nondimensional mathematical expression, which defines a family of skewed probability-density curves referred to as Rayleigh curves or distribution functions, can be used to fit a probability-density curve accurately to each available set of field data for bridges:

$$p' = 1.011\, x'\, e^{-1/2(x')^2} \tag{9.9}$$

where $x' = (\sigma_r - \sigma_{r\,min})/\sigma_{rd}$, σ_r, (i.e., $\Delta\sigma$) is the stress range, and $\sigma_{r\,min}$ (i.e., $\Delta\sigma_{min}$) and σ_{rd} (that is, $\Delta\sigma_d$) are constant parameters that define any particular probability-density curve from the family of curves represented by Equation (9.9). Equation (9.9) is plotted in Figure 9.18*a*. As illustrated in Figure 9.18*b*, a particular curve from the family is defined by two parameters: (1) the modal stress range, σ_{rm}, which corresponds to the peak of the curve; and (2) the parameter σ_{rd}, which is a measure of the width of the curve or the dispersion of the data. The curve could be shifted sideways by changing σ_{rm}. Mathematical expressions for the modal, median, mean, and root-mean-square (rms) values of the spectrum are given in Figure 9.18. The root-mean-square (rms) value is defined as the square root of the mean of the squares of the individual values of x' or σ_r. Stress (σ) is represented as *S* in Figure 9.18.

9.9.2 *Fatigue-Crack Growth Under Variable-Amplitude Loading*

Many attempts have been made to predict fatigue-crack-growth behavior under variable-amplitude loading. The following sections present the behavior under simple and complex variable-amplitude loading. The simple loadings correspond to a single or multiple high-tensile-load fluctuations superimposed upon constant-amplitude cyclic-load fluctuations (Figure 7.4). Complex variable-amplitude loadings correspond to multivalue cyclic-load fluctuations.

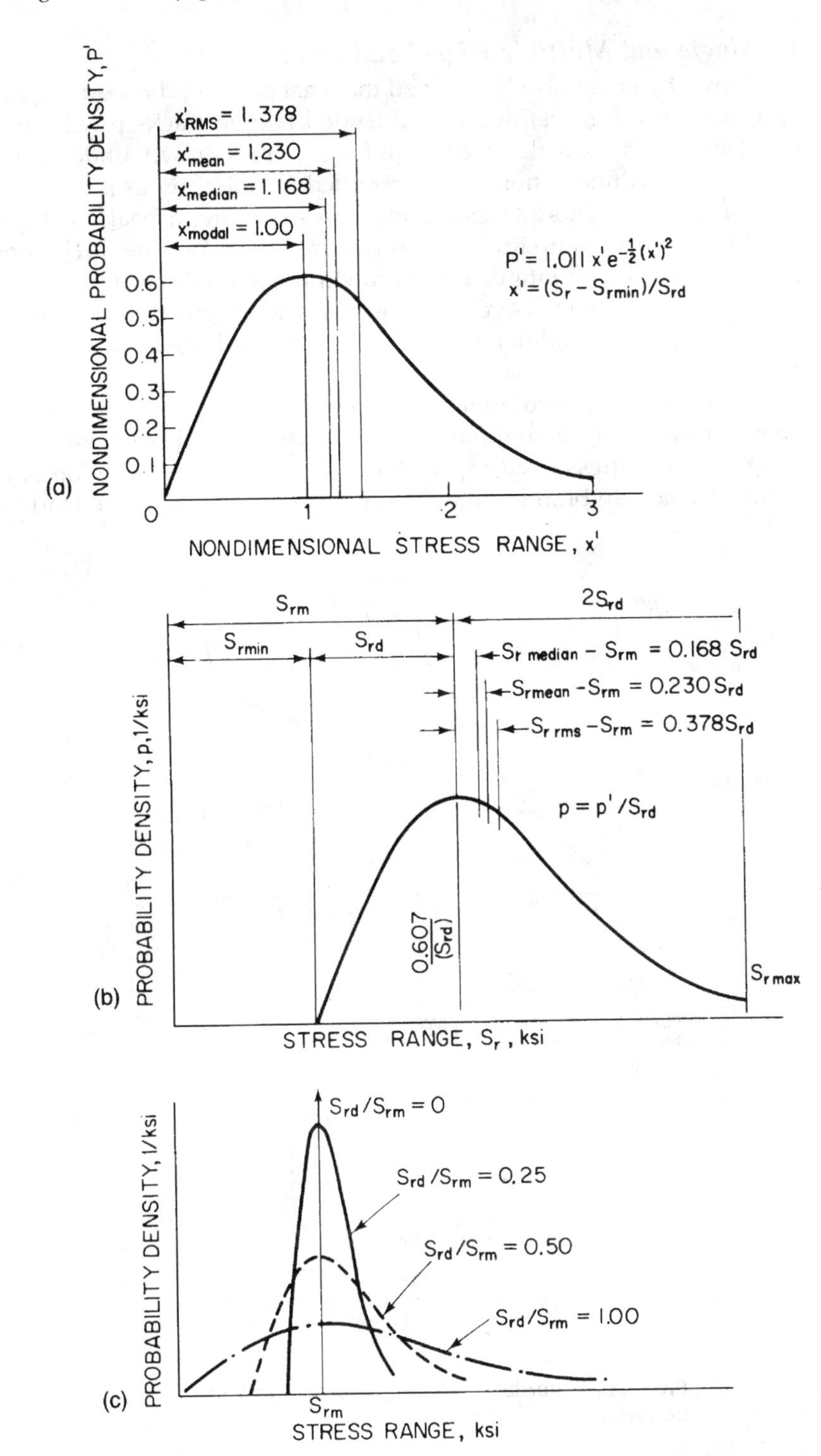

FIG. 9.18 Charactristics of Rayleigh probability curves.

9.9.3 Single and Multiple High-Load Fluctuations

Several investigators [*53–60*] observed that changes in cyclic-load magnitude (Figure 7.4) may result in retarded or accelerated fatigue-crack-growth rate. Extensive published data show that the rate of fatigue-crack growth under constant-amplitude cyclic-load fluctuation can be retarded significantly as a result of application of a single or multiple tensile-load cycles having a peak load greater than that of the constant-amplitude cycles (Figure 9.19). Von Euw [*61*] observed that the minimum value of fatigue-crack-growth rate did not occur immediately following the high-tensile-load cycle but that the rate of growth decelerated to a minimum value. This deceleration region has been termed "delayed retardation," Figure 9.19*b*.

Several models have been advanced to explain the phenomenon of crack-growth delay. In general, these models attribute the delayed behavior to crack-tip blunting, residual stresses [*62,63*] crack closure [*6*], or a combination of these mechanisms. A crack-tip-blunting model advocates that high-tensile-load cycles

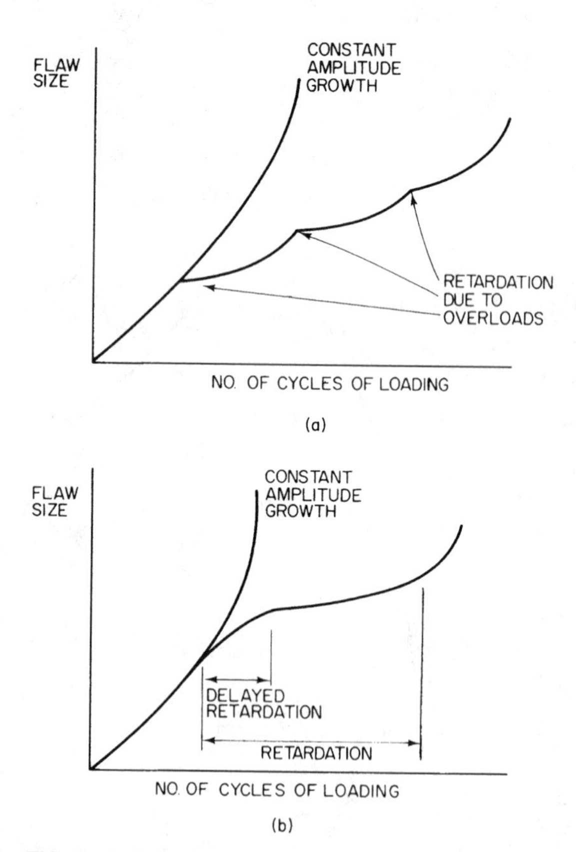

FIG. 9.19 Retardation in fatigue-crack-growth behavior.

cause crack-tip blunting, which in turn causes retardation in fatigue-crack growth at the lower cyclic-load fluctuations until the crack is resharpened. The residual-stress model suggests that the application of a high-tensile-load cycle forms residual compressive stresses in the vicinity of the crack tip that reduce the rate of fatigue-crack growth. Finally, the crack-closure model postulates that the delay in fatigue-crack growth is caused by the formation of a zone of residual tensile deformation left in the wake of a propagating crack that causes the crack to remain closed during a portion of the applied tensile-load cycle. Consequently, fatigue-crack growth delay occurs because only the portion of the tensile-load cycles above the crack-opening level is effective in extending the crack. These models are useful to describe trends in fatigue-crack-growth-rate behavior caused by single or multiple high-tensile-load cycles but are of little value to predict fatigue lives under these conditions. Fatigue-crack-growth delay has been shown to be strongly dependent on all the loading variables, such as the stress-intensity-factor-fluctuation, ΔK_I, of the high tensile-load cycle, the ΔK_I for the constant-amplitude cycles (Fig. 9.20) [57], the stress ratios of these ΔK_I values and the number of constant-amplitude cycles between the high-tensile-load cycles [55,57,64,65]. Extensive research is necessary to further our understanding of the significance of these variables in order to develop equations that can be used to predict accurately the fatigue life of components subjected to single or multiple high-tensile-load cycles.

9.9.4 Variable-Amplitude Load Fluctuations

Extensive investigations have been conducted to develop methods to predict fatigue lives under variable-amplitude load fluctuations. Some of these investigations resulted in models that can be used to describe trends in fatigue-crack-propagation rates but with varying degrees of success in predicting fatigue lives under variable-amplitude loads [28,51,62,63]. This section presents the root mean square (RMS) model advanced by Barsom [28,50,66], which relates fatigue-crack-

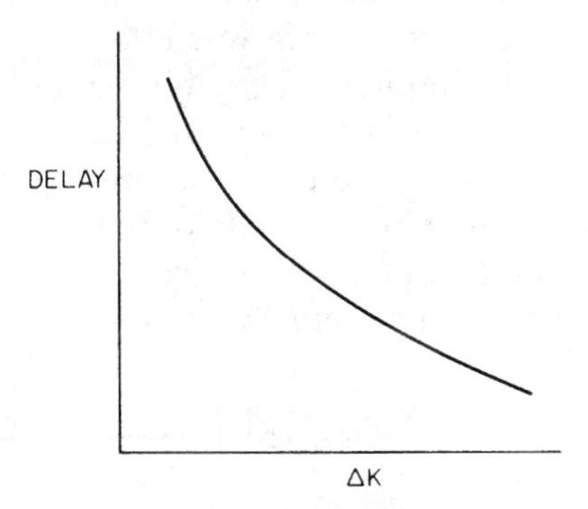

FIG. 9.20 Schematic showing effect of ΔK on fatigue-crack-growth delay.

growth rate per cycle to an effective stress-intensity factor characteristic of the probability-density curve.

9.9.4.1 The Root-Mean-Square (RMS) Model Incremental increase of crack length and the corresponding number of elapsed load cycles can be measured under variable-amplitude random-sequence load spectra. However, unlike constant-amplitude cyclic-load data, the magnitude of ΔK_I changes for each cycle. Reduction of data in terms of fracture-mechanics concepts requires the establishment of a correlation parameter that incorporates the effects of crack length, cyclic-load amplitude, and cyclic-load sequence.

Barsom [*28*] attempted to determine the magnitude of constant-amplitude cyclic-load fluctuation, which results in the same *a* versus *N* curve obtained under variable-amplitude cyclic-load fluctuation when both spectra are applied to identical specimens (including initial crack length). In other words, one of the objectives of his investigation was to find a single stress-intensity parameter, such as mean, modal, or root mean square, that can be used to define the crack-growth rate under both constant and variable-amplitude loadings. The selected parameter must be a characteristic of the probability-density curve.

This methodology for analyzing the fatigue behavior under variable-amplitude loading should be applicable to loading conditions that result in relatively smooth, continuous crack-length (*a*) versus number-of-cycles (*N*) curves, as shown in Figure 9.21. Such curves can be obtained from many variable-amplitude loadings and from frequently applied overloads but not from single or infrequent overloads.

Variable-amplitude random-sequence load spectra having Rayleigh probability-density curves of P_{rd}/P_{rm} (or σ_{rd}/σ_{rm}) values equal to 0.5 and 1.0 were investigated as part of Project 12-12 of the National Cooperative Highway Research Program (NCHRP) of the National Academy of Sciences [*28,50,52*]. A typical portions of the 500-cycles loading block for each is shown in Figure 9.22.

A correlation between data obtained under constant-amplitude and variable-amplitude random-sequence load spectra was obtained on the basis of the root mean square of the load distribution, where the root mean square is the square root of the mean of the squares of the individual load cycles in a spectrum. The combined crack-growth-rate data are presented in Figure 9.23 as a function of ΔK_{rms}. The data show that, within the limits of the experimental work, the average fatigue-crack-growth rates per cycle, da/dN, under variable-amplitude random-sequence stress spectra can be represented by:

$$\frac{da}{dN} = A(\Delta K_{rms})^m \tag{9.10}$$

where *A* and *m* are constant and,

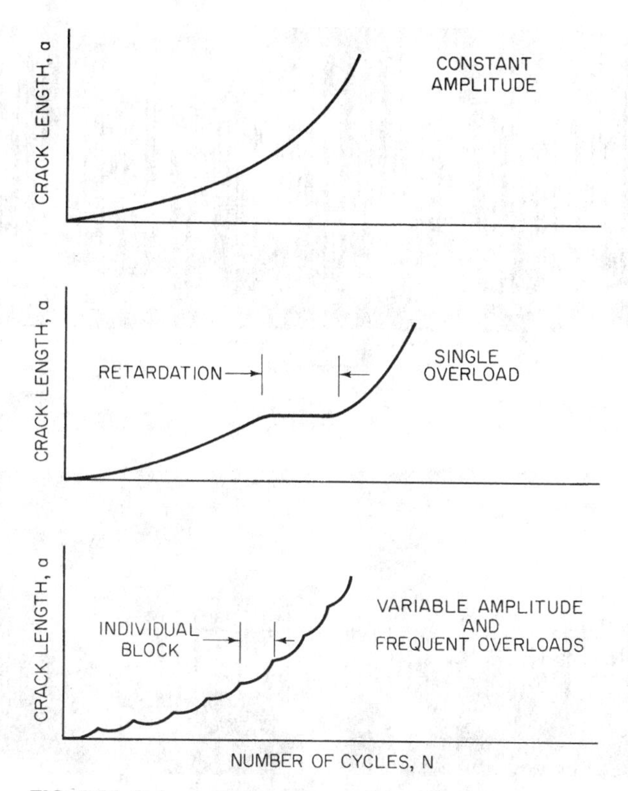

FIG. 9.21 Schematic representation of crack extension under different loading conditions.

$$\Delta K_{\text{rms}} = \sqrt{\frac{\sum_{i=1}^{k} \Delta K_i^2}{n}} \tag{9.11}$$

The root-mean-square value of the stress-intensity factor under constant-amplitude cyclic-load fluctuations is equal to the stress-intensity-factor fluctuation. Consequently, the average fatigue-crack-growth rate can be predicted for constant-amplitude and variable-amplitude random-sequence stress fluctuations by using Equation (9.10),

9.9.4.2 Fatigue-Crack Growth Under Variable-Amplitude Ordered-Sequence Cyclic Load The root-mean-square stress-intensity factor ΔK_{rms} is characteristic of the load-distribution curve and is independent of the order of the cyclic-load fluctuations. To determine whether the order of load fluctuations affects the average rate of crack growth, fatigue tests were performed on identical specimens under random and ordered variable-amplitude cyclic-load fluctua-

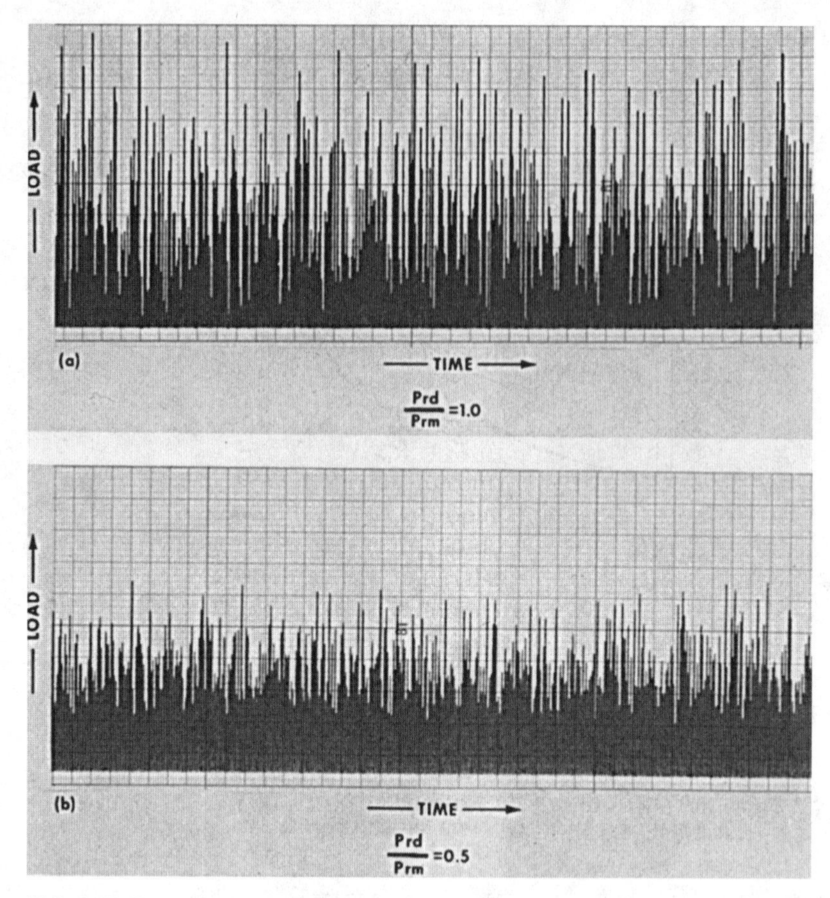

FIG. 9.22 Two variable-amplitude random-sequence load fluctuations investigated.

tions that represent the same load-distribution curve with $P_{rd}/P_{rm} = 1.0$ [28,50]. The tests were conducted at a constant minimum load, P_{min}, with $P_{min}/P_{rm} = 0.25$.

Fatigue-crack-growth rate tests were conducted by using the variable-amplitude random-sequence load fluctuations shown in Figure 9.24*a*. In other tests these same load fluctuations were arranged in descending magnitudes, Figure 9.24*b*, ascending magnitudes, Figure 9.24*c*, and combined ascending-descending magnitudes, Figure 9.24*d*. The fatigue-crack-growth data obtained under these various conditions and under a constant-amplitude cyclic-load fluctuation of $P_r = \Delta P_{rms}$, Figure 9.24*e*, are presented in Figure 9.25. The data show that the average rate of fatigue-crack growth is represented accurately by Equation (9.10) regardless of the order of occurrence of the cyclic-load fluctuations.

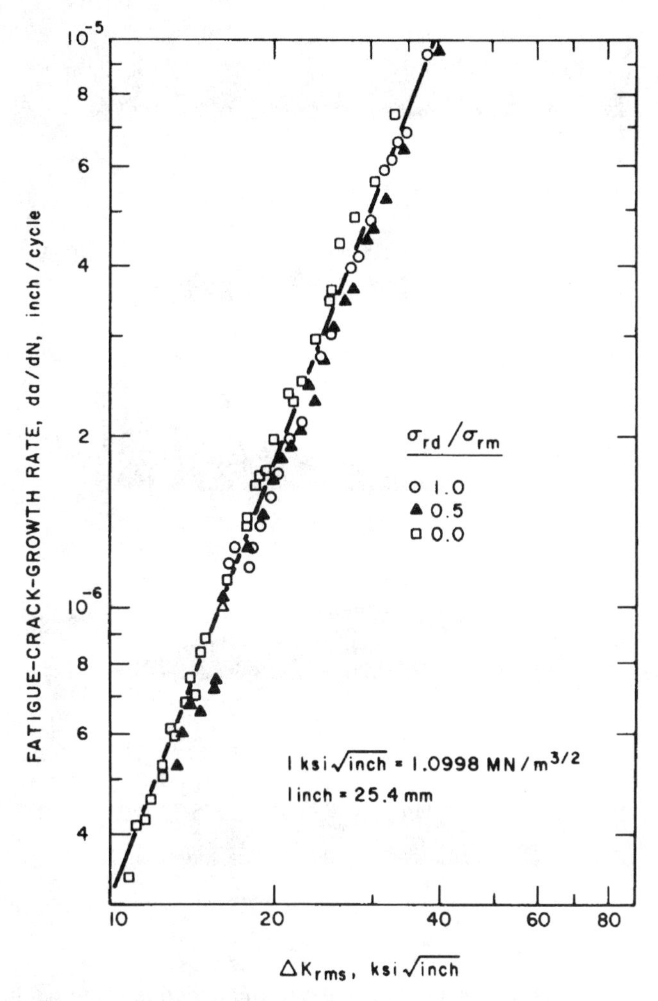

FIG. 9.23 Crack-growth rate as a function of the root-mean-square stress-intensity-factor range.

9.10 Fatigue-Crack Growth in Various Steels

The preceding results were obtained by testing A514 Grade B steel under variable-amplitude random and ordered-sequence cyclic-load fluctuations. Because several investigators [*53–60*] have noted that changes in cyclic-load magnitude can lead to accelerated or retarded rates of fatigue-crack growth in various metals, the applicability of the RMS model to steels of various yield strengths was investigated under NCHRP Project 12–14 [*66*]. Fatigue-crack-growth rates under constant-amplitude and variable-amplitude random-sequence load fluctuations

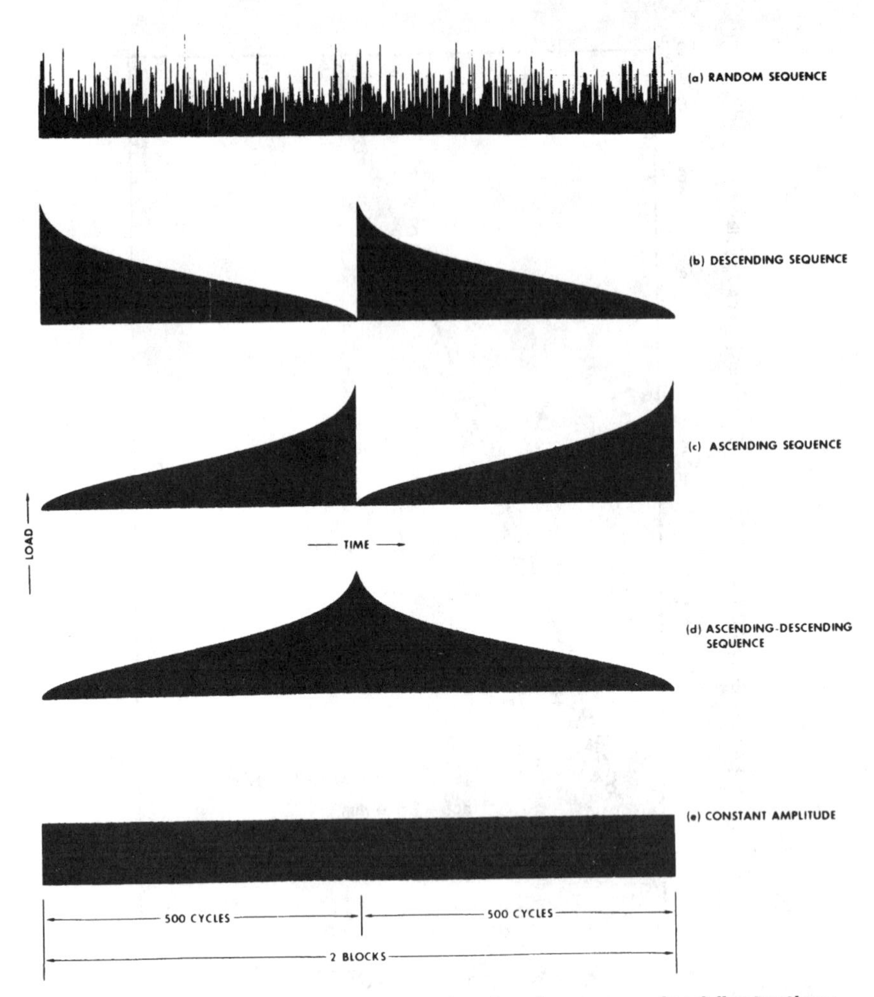

FIG. 9.24 Various random-sequence and ordered-sequence load fluctuations studied to establish the ΔK_{rms} analysis.

were investigated for A36, A588 Grade A, A588 Grade B, A514 Grade E, and A514 Grade F steels. All loadings followed a Rayleigh probability-density curve, with the ratio of the load-range deviation to the model (peak) load (P_{rd}/P_{rm}) equal to either 0 or 1.0. The data presented in Figures 9.26 and 9.27 show that, within the limits of the experimental work, the average fatigue-crack-growth rates per cycle, da/dN, in various steels subjected to variable-amplitude load spectra can be represented by Equation (9.10). Moreover, the average fatigue-crack-growth rates for the various steels studied under variable-amplitude random-sequence load fluctuations are equal to the average fatigue-crack-growth rates obtained under constant-amplitude load fluctuations when the stress-

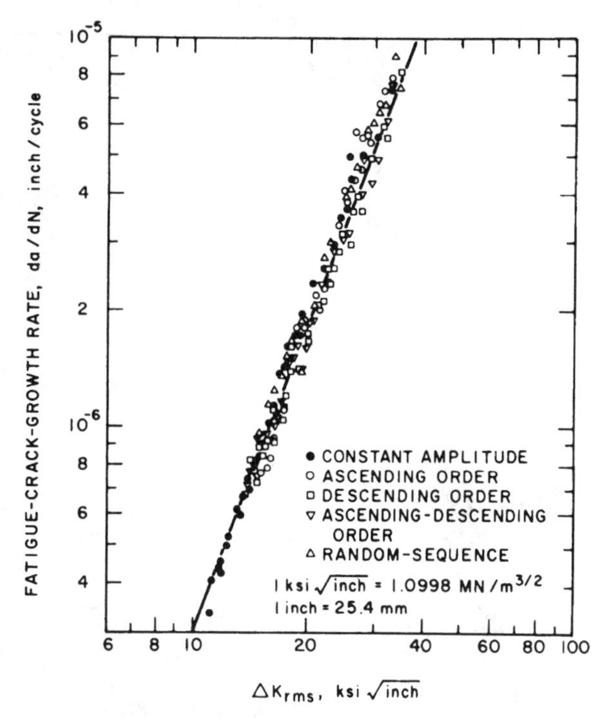

FIG. 9.25 Summary of crack-growth-rate data under random-sequence and ordered-sequence load fluctuations.

intensity-factor range, ΔK_I, under constant-amplitude load fluctuations is equal in magnitude to the ΔK_{rms} of the variable-amplitude spectra.

9.11 Fatigue-Crack Growth Under Various Unimodal Distribution Curves

The applicability of the RMS model for correlating crack-growth rates under variable-amplitude random-sequence load fluctuations that follow unimodal distribution curves different from the Rayleigh type have been studied. Fatigue-crack-growth rates under constant-amplitude load fluctuations and under four unimodal variable-amplitude random-sequence load fluctuations were investigated by Barsom [67]. A block of variable-amplitude random-sequence load fluctuations of each unimodal distribution curve and of constant-amplitude load fluctuations is presented in Figure 9.28. Each block was applied repeatedly to a single specimen until the test was terminated. The four distribution curves corresponding to the various blocks shown in Figure 9.28 are presented in Figure 9.29. These curves cover a wide variation of unimodal distribution curves.

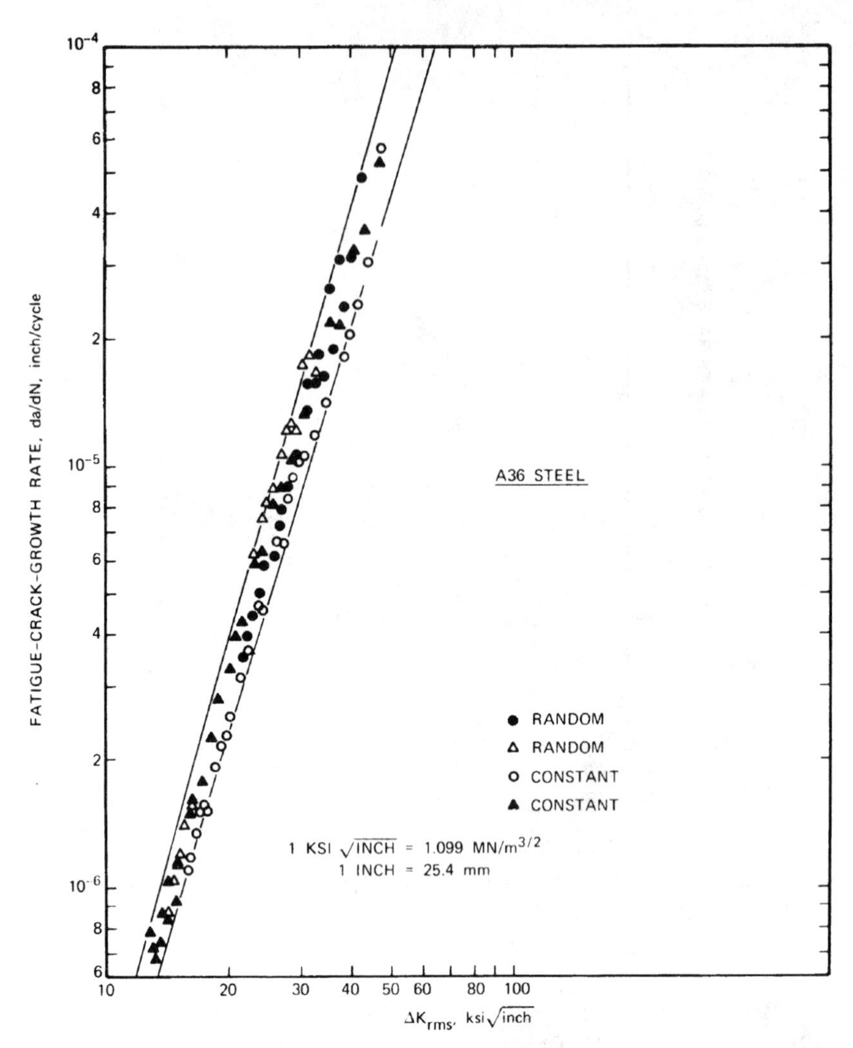

FIG. 9.26 Crack-growth rate as a function of the root-mean-square stress-intensity factor for an A36 steel.

Data of fatigue-crack-growth rate per cycle and the corresponding root-mean-square stress-intensity-factor fluctuation, ΔK_{rms}, obtained by subjecting identical specimens to the load fluctuations shown in Figure 9.28 are presented in Figure 9.30. The data show that the average fatigue-crack-growth rate, da/dN, under constant-amplitude and variable-amplitude random-sequence load fluctuations that follow various unimodal distribution curves can be predicted by using the RMS model, Equation (9.10), of fatigue-crack growth. Further investigations are needed to establish the effects of various parameters, such as variable minimum load, on the rate of fatigue-crack growth under variable-amplitude

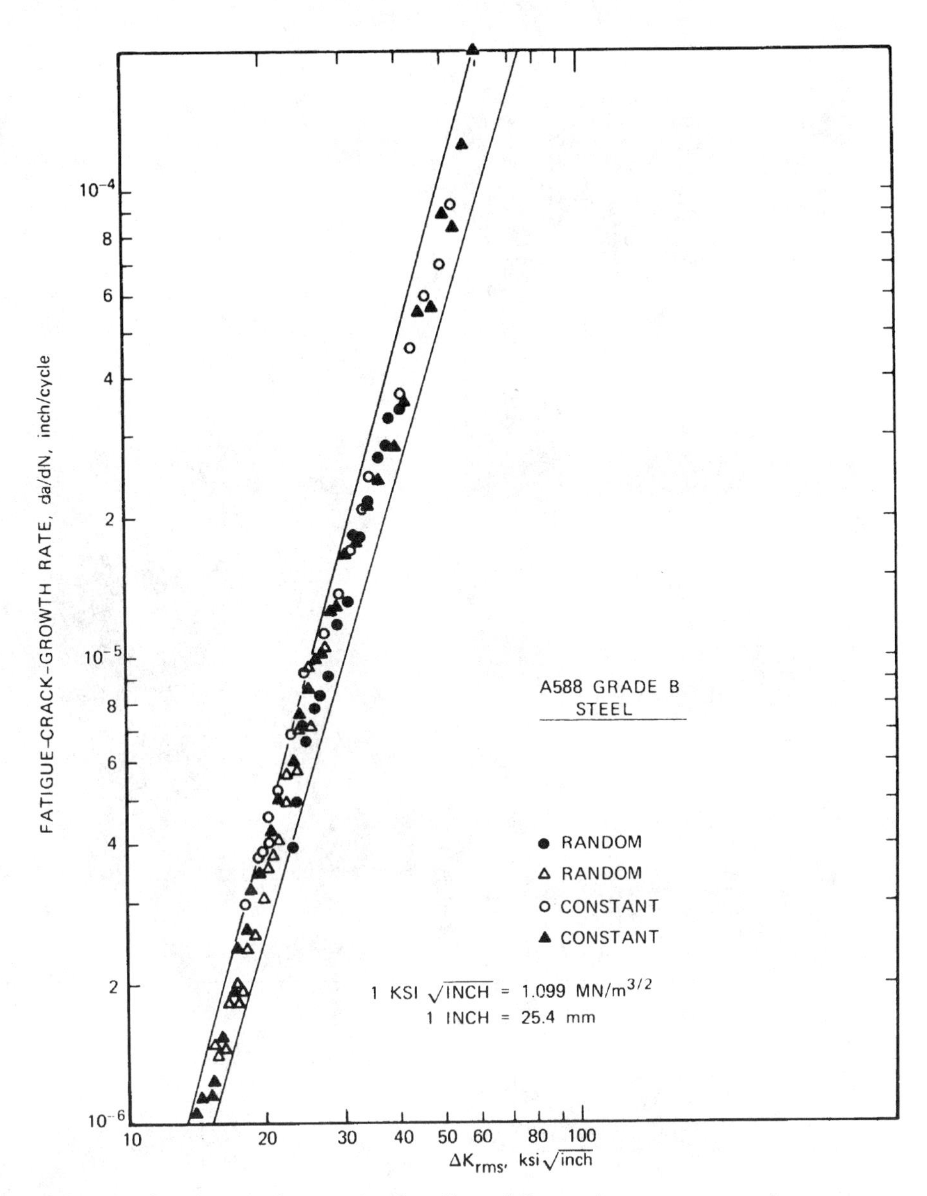

FIG. 9.27 Crack-growth rate as a function of the root-mean-square stress-intensity factor for an A588 Grade B steel.

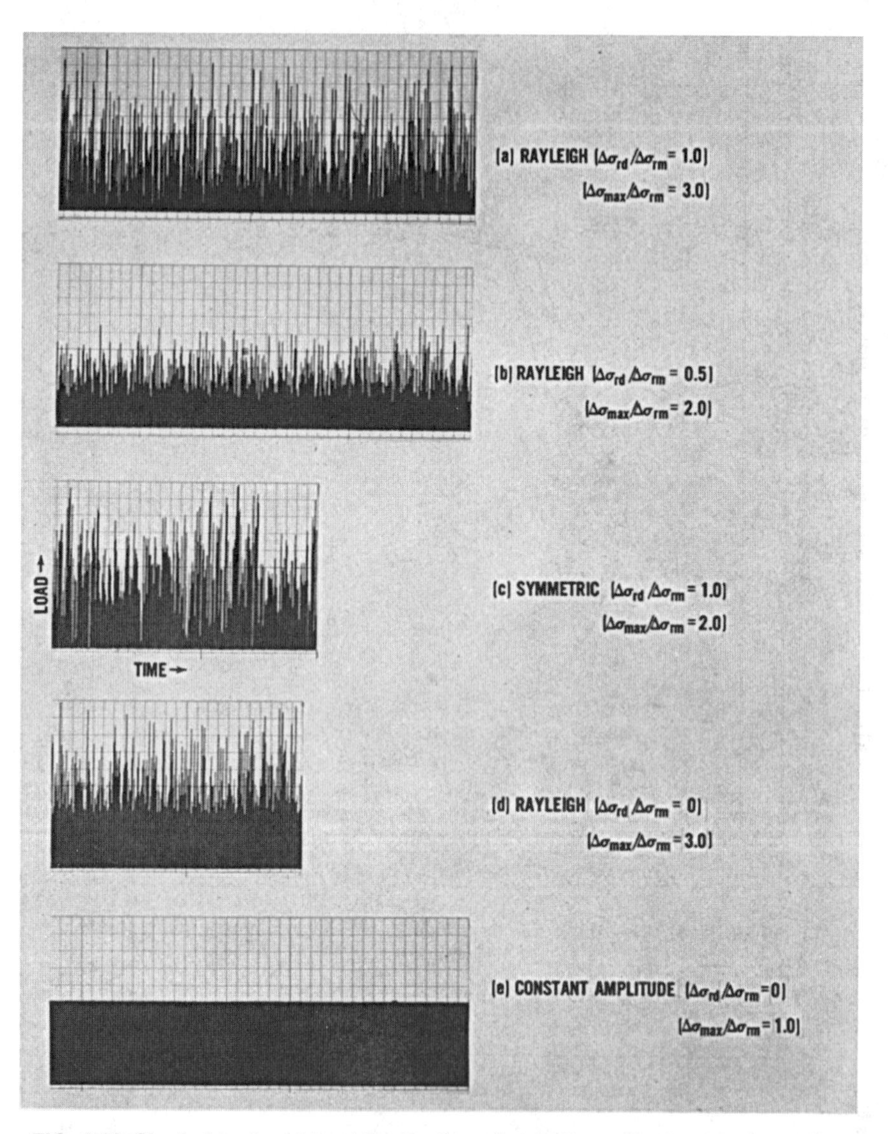

FIG. 9.28 Single blocks of load fluctuations for various distribution functions.

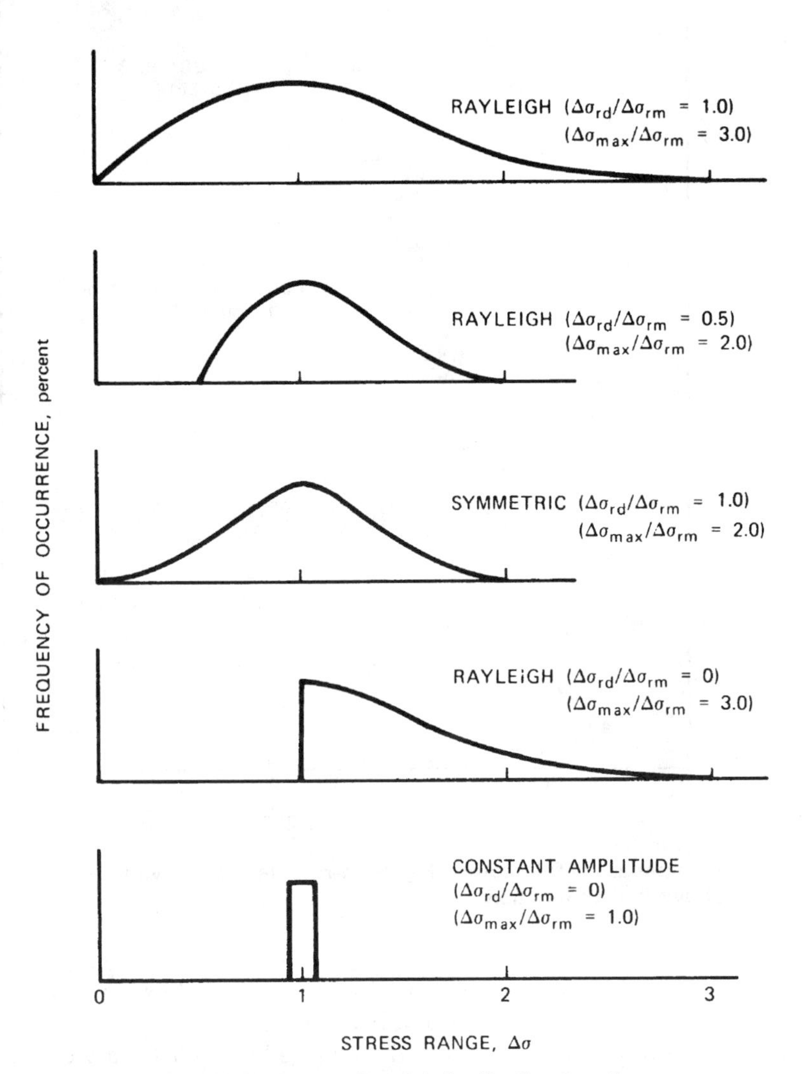

FIG. 9.29 Various unimodal distribution functions.

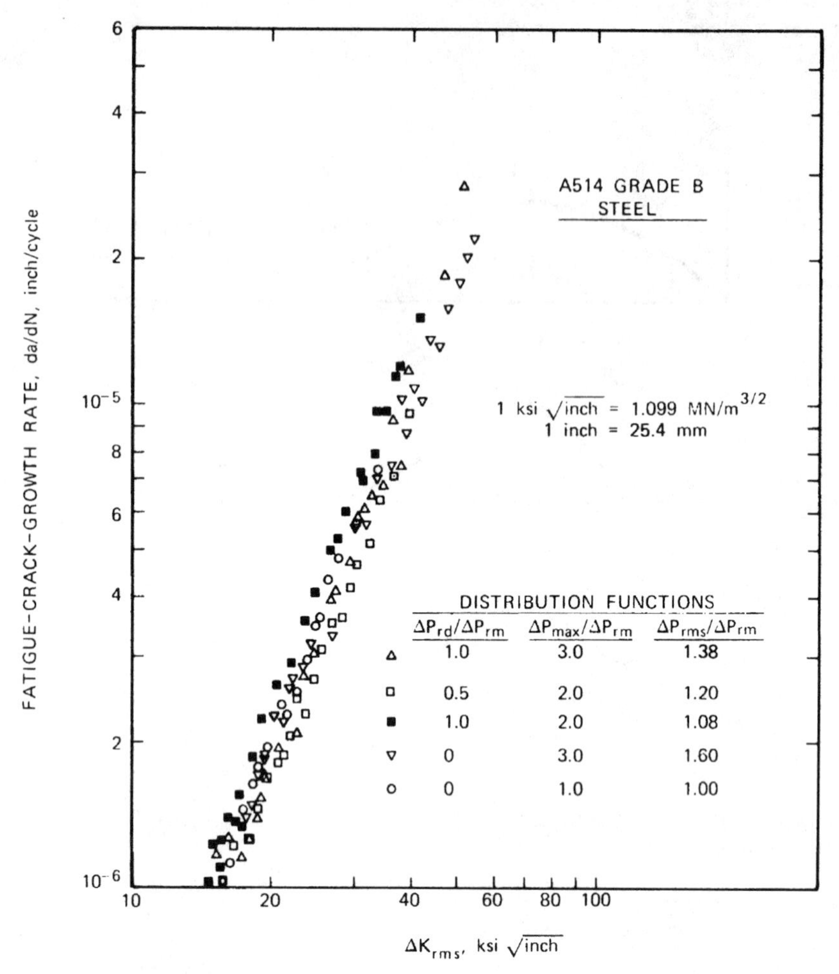

FIG. 9.30 Summary of fatigue-crack growth-rate data under various unimodal distribution functions.

load fluctuations and the necessary modifications to the RMS model to account for these effects or the development of a better model.

Finally, an example for analysis of the fatigue behavior for a welded structural component subjected to variable-amplitude cyclic loading is presented in Chapter 10.

9.12 References

[1] Frost, N., "Notch Effects and the Critical Alternating Stress Required to Propagate a Crack in an Aluminum Alloy Subjected to Fatigue Loading," *Journal of the Mechanical Engineering Society*, Vol. 2, No. 2, 1960, pp. 109–119.

[2] Frost, N. E., "The Growth of Fatigue Cracks," *Proceedings of the First International Conference on Fracture,* Sendai, Japan, 1966, p. 1433.
[3] McClintock, F. A., "On the Plasticity of the Growth of Fatigue Cracks," *Fracture of Solids,* Wiley-Interscience, New York, 1963, p. 65.
[4] Lindner, B. M., "Extremely Slow Crack Growth Rates in Aluminum Alloy 7075-T6," M. S. thesis, Lehigh University, Bethlehem, PA, 1965.
[5] Paris, P. C., "Testing for Very Slow Growth of Fatigue Cracks," *MTS Closed Loop Magazine,* Vol. 2, No. 5, 1970.
[6] Elber, W., "The Significance of Crack Closure," *Damage Tolerance in Aircraft Structures, ASTM STP 486,* American Society for Testing and Materials, Philadelphia, 1971, pp. 230–242.
[7] Schmidt, R. A. and Paris, P. C., "Threshold for Fatigue Crack Propagation and Effects of Load Ratio and Frequency," *Progress in Flaw Growth and Fracture, ASTM STP 536,* American Society for Testing and Materials, Philadelphia, 1973, p. 79.
[8] *Fatigue Threshold—Fundamentals and Engineering Applications,* Vols. I and II, J. Backlund, A. F. Blom and C. J. Beevers, Eds., Engineering Materials Advisory Services Ltd., United Kingdom, 1982.
[9] Bucci, R. J., Clark, Jr., W. G., and Paris, P. C., "Fatigue-Crack-Propagation Growth Rates Under a Wide Variation of ΔK for an ASTM A517 Grade F (T-1) Steel," *ASTM STP 513,* American Society for Testing and Materials, Philadelphia, 1972.
[10] Harrison, J. D., "An Analysis of Data on Non-Propagating Fatigue Cracks on a Fracture Mechanics Basis," *British Welding Journal,* Vol. 2, No. 3, March 1970.
[11] Pook, L. P., "Fatigue Crack Growth Data for Various Materials Deduced from the Fatigue Lives of Precracked Plates," *ASTM STP 513,* American Society for Testing and Materials, Philadelphia, 1972.
[12] Paris, P. C., Bucci, R. J., Wessel, E. T., Clark, W. G., and Mager, T. R., "Extensive Study of Low Fatigue Crack Growth Rates in A533 and A508 Steels," *ASTM STP 513,* American Society for Testing and Materials, Philadelphia, 1972.
[13] Yoder, G. R., Cooley, L. A., and Crooker, T. W., "A Critical Analysis of Grain Size and Yield Strength Dependence of Near-Threshold Fatigue-Crack Growth in Steels," *Naval Research Laboratory Report,* Naval Research Laboratory, Washington, DC, July 15, 1981.
[14] Kitagawa, H. and Takahashi, S., "Applicability of Fracture Mechanics to Very Small Cracks or the Cracks in the Early Stages," *Proceedings of the Second International Conference on Mechanical Behavior of Materials,* American Society for Metals, Metals Park, OH, 1976, p. 627.
[15] El Haddad, M. H., Dowling, N. E., Topper, T. H., and Smith, K. N., "J Integral Applications for Short Fatigue Cracks at Notches," *International Journal of Fracture,* Vol. 16, 1980, p. 15.
[16] Barsom, J. M., "Fatigue-Crack Propagation in Steels of Various Yield Strengths," *Transactions of the ASME, Journal of Engineering for Industry,* Series B, Vol. 93, No. 4, November 1971.
[17] Barsom, J. M., Imhof, Jr., E. J., and Rolfe, S. T., "Fatigue-Crack Propagation in High-Yield-Strength Steels," *Engineering Fracture Mechanics,* Vol. 2, No. 4, June 1971.
[18] Barsom, J. M., "The Dependence of Fatigue Crack Propagation on Strain Energy Release Rate and Crack Opening Displacement," *ASTM STP 486,* American Society for Testing and Materials, Philadelphia, 1971.
[19] Crooker, T. W., "Crack Propagation in Aluminum Alloys Under High-Amplitude Cyclic Load," *Naval Research Laboratory Report 7286,* Washington, DC, July 12, 1971.
[20] Barsom, J. M., "Fatigue Behavior of Pressure-Vessel Steels," *WRC Bulletin,* No. 194, Welding Research Council, New York, May 1974.
[21] Barsom, J. M., "Investigation of Subcritical Crack Propagation," Doctor of Philosophy dissertation, University of Pittsburgh, 1969.
[22] Imhof, E. J. and Barsom, J. M., "Fatigue and Corrosion-Fatigue Crack Growth of 4340 Steel at Various Yield Strengths," *ASTM STP 536,* American Society for Testing and Materials, Philadelphia, 1973.
[23] Barsom, J. M. and Vecchio, R. S., "Fatigue of Welded Structures," Welding Research Council, Bulletin 422, June 1997.

[24] Suresh, S. and Ritchie, R. O., "Closure Mechanisms for the Influence of Load Ratio on Fatigue Crack Propagation in Steels," U.S. Department of Energy, Contract No. DE-ACO3-76SF00098, April 1982.
[25] Lindley, T. C. and Richards, C. E., "Near-Threshold Fatigue Crack Growth in Materials Used in the Electricity Supply Industry," *Fatigue Thresholds—Fundamentals and Engineering Applications,* Engineering Materials Advisory Services Ltd., United Kingdom, 1982, pp. 1087–1113.
[26] Sasaki, E., Ohta, A., and Kosuge, M., "Fatigue Crack Propagation Rate and Stress Intensity Threshold Level of Several Structural Materials at Varying Stress Ratios ($-1 \sim 0.8$)," *Transactions of the National Research Institute for Metals (Japan),* Vol. 19, No. 4, 1977, pp. 183–199.
[27] Priddle, E. K., "The Threshold Stress Intensity Factor for Fatigue Crack Growth in Mild Steel Plate and Weld Metal: Some Effects of Temperature and Environment," *Fatigue Thresholds—Fundamentals and Engineering Applications,* Engineering Materials Advisory Services Ltd., United Kingdom, 1982, pp. 581–600.
[28] Barsom, J. M., "Fatigue-Crack Growth Under Variable-Amplitude Loading in ASTM A514 Grade B Steel," *ASTM STP 536,* American Society for Testing and Materials, Philadelphia, 1973.
[29] Parry, M., Nordberg, H., and Hertzberg, R. W., "Fatigue Crack Propagation in A514 Base Plate and Welded Joints," *Welding Journal,* Vol. 51, No. 10, Oct. 1972.
[30] Clark, Jr., W. G., "Effect of Temperature and Section Size on Fatigue Crack Growth in A533 Grade B, Class 1 Pressure Vessel Steel," *Journal of Materials,* Vol. 6, No. 1, March 1971.
[31] Bucci, R. J., Greene, B. N., and Paris, P. C., "Fatigue Crack Propagation and Fracture Toughness of 5Ni Steel at Cryogenic Temperatures," *ASTM STP 536,* American Society for Testing and Materials, Philadelphia, 1973.
[32] James, L. A., "The Effect of Elevated Temperature upon the Fatigue-Crack-Propagation Behavior of Two Austenitic Stainless Steels," *Mechanical Behavior of Materials,* Vol. III, The Society of Materials Science, Tokyo, 1972.
[33] Shahinian, P., Watson, H. E., and Smith, H. H., "Fatigue Crack Growth in Selected Alloys for Reactor Applications," *Journal of Materials,* Vol. 7, No. 4, 1972.
[34] Weber, J. H. and Hertzberg, R. W., "Effect of Thermomechanical Processing on Fatigue Crack Propagation," *Metallurgical Transactions,* Vol. 4, Feb. 1973.
[35] Clark, Jr., W. G. and Wessell, E. T., "Interpretation of the Fracture Behavior of 5456-H321 Aluminum with WOL Toughness Specimens," *Scientific Paper 67-1D6-BTLFR-P4,* Westinghouse Research Laboratory, Pittsburgh, September 1967.
[36] Crooker, T. W., "Factors Determining the Performances of High Strength Structural Metals (Slope Transition Behavior of Fatigue Crack Growth Rate Curves)," *NRL Report of Progress,* Naval Research Laboratory, Washington, DC, December 1970.
[37] Walker, K., "The Effect of Stress Ratio During Crack Propagation and Fatigue for 2024-T3 and 7075-T6 Aluminum," *ASTM STP 462,* American Society for Testing and Materials, Philadelphia, 1970.
[38] Crooker, T. W., "Effect of Tension-Compression Cycling on Fatigue Crack Growth in High-Strength Alloys," Naval Research Laboratory Report 7220, Naval Research Laboratory, Washington, DC, January 1971.
[39] Crooker, T. W. and Krause, D. J., "The Influence of Stress Ratio and Stress Level on Fatigue Crack Growth Rates in 140-ksiYS Steel," *Report of NRL Progress,* Naval Research Laboratory, Washington, DC, December 1972.
[40] Krause, D. J. and Crooker, T. W., "Effect of Constant-Amplitude Loading Parameters on Low-Cycle Fatigue-Crack Growth in a 140- to 150-ksi Yield Strength Steel," *Report of NRL Progress,* Naval Research Laboratory, Washington DC, March 1973.
[41] Wei, R. P., "Fracture Mechanics Approach to Fatigue Analysis in Design," *ASME Paper No. 73-DE-22,* New York, April 1973.

[42] Barsom, J. M., "Effect of Cyclic-Stress Form on Corrosion-Fatigue Crack Propagation Below K_{Iscc} in a High-Yield-Strength Steel," in *Corrosion Fatigue: Chemistry, Mechanics, and Micro-Structure,* International Corrosion Conference Series, Vol. NACE-2, National Association of Corrosion Engineers, Houston, 1972.
[43] Barsom, J. M., "Corrosion Fatigue Crack Propagation Below K_{Iscc}," *Journal of Engineering Fracture Mechanics,* Vol. 3, No. 1, July 1971.
[44] Novak, S. R. and Barsom, J. M., "Brittle Fracture (K_{Ic}) Behavior for Cracks Emanating from Notches," *Cracks and Fracture, ASTM STP 601,* American Society for Testing and Materials, Philadelphia, 1976.
[45] Clark, Jr., W. G. "Fatigue Crack Growth Characteristics of Heavy Section ASTM A533 Grade B Class 1 Steel Weldments," *ASME Paper No. 70-PVP-24,* American Society of Mechanical Engineers, New York, 1970.
[46] Clark, Jr., W. G. and Kim, D. S., "Effect of Synthetic Seawater on the Crack Growth Properties of HY-140 Steel Weldments," *Engineering Fracture Mechanics,* Vol. 4, 1972.
[47] Shahinian, P., Smith, H. H., and Hawthorne, J. R., "Fatigue Crack Propagation in Stainless Steel Weldments at High Temperature," *Welding Journal,* Vol. 51, No. 11, November 1972.
[48] Maddox, S. J., "Assessing the Significance of Flaws in Welds Subject to Fatigue," *Welding Journal,* Vol. 53, No. 9, September 1974.
[49] Gurney, T. R., *Fatigue of Welded Structures,* Cambridge University Press, New York, 1968.
[50] Schilling, C. G., Klippstein, K. H., Barsom, J. M., and Blake, G. T., "Fatigue of Welded Steel Bridge Members Under Variable-Amplitude Loadings," NCHRP Report 188, Transportation Research Board, Washington, DC, 1978.
[51] Klippstein, K. H. and Schilling, C. G., "Stress Spectrum for Short-Span Steel Bridges," *Fatigue Crack Growth Under Spectrum Loads, ASTM STP 595,* American Society for Testing and Materials, Philadelphia, 1976.
[52] Schilling, C. G., Klippstein, K. H., Barsom, J. M., and Blake, G. T., "Fatigue of Welded Steels Bridge Members Under Variable-Amplitude Loadings," *Research Results Digest,* Vol. 60, National Cooperative Highway Research Program, April 1974.
[53] Schijve, J., "Significance of Fatigue Cracks in Micro-Range and Macro-Range," *Fatigue Crack Propagation, ASTM STP 415,* American Society for Testing and Materials, Philadelphia, 1967.
[54] McMillan, J. C. and Pelloux, R. M. N., "Fatigue Crack Propagation Under Program and Random Loads," *ASTM STP 415,* American Society for Testing and Materials, Philadelphia, 1967.
[55] Von Euw, E. F. J., Hertzberg, R. W., and Roberts, R., "Delay Effects in Fatigue Crack Propagation," *ASTM STP 513,* American Society for Testing and Materials, Philadelphia, 1973.
[56] Jones, R. E., "Fatigue Crack Growth Retardation After Single-Cycle Peak Overload in Ti6Al-4V Titanium Alloy," *Engineering Fracture Mechanics,* Vol. 5, 1973.
[57] Wei, R. P. and Shih, T. T., "Delay in Fatigue Crack Growth," *International Journal of Fracture,* Vol. 10, No. 1, March 1974.
[58] Hardrath, H. F. and McEvily, A. T., "Engineering Aspects of Fatigue-Crack Propagation," *Proceedings of the Crack Propagation Symposium,* Vol. 1, Cranfield, England, October 1961.
[59] Schije, J., Jacobs, F. A., and Tromp, P. J., "Crack Propagation in Clad 2024-T3Al Under Flight Simulation Loading. Effect of Truncating High Gust Loads," *NNLR TR-69050-U,* National Lucht-En Ruimtevaart-Laboratorium (National Aerospace Laboratory NLR—The Netherlands), June 1969.
[60] Hudson, C. M. and Hardrath, H. F., "Effects of Changing Stress Amplitude on the Rate of Fatigue-Crack Propagation of Two Aluminum Alloys," *NASA Technical Note D-960,* NASA, Cleveland, September 1961.
[61] Von Euw, E. F. J., "Effect of Single Peak Overloading on Fatigue Crack Propagation," Master's dissertation, Lehigh University, Bethlehem, PA, 1968.
[62] Wheeler, O. E., "Spectrum Loading and Crack Growth," *General Dynamics Report FZM-5602,* Fort Worth, June 30, 1970.

[63] Willenborg, J., Engle, R. M., and Wood, H. A., "A Crack Growth Retardation Model Using an Effective Stress Concept," *Technical Memorandum 71-1-FBR,* Air Force Flight Dynamics Laboratory, January 1971.
[64] Gardner, F. H. and Stephens, R. I., "Subcritical Crack Growth Under Single and Multiple Periodic Overloads in Cold-Rolled Steel," *ASTM STP 559,* American Society for Testing and Materials, Philadelphia, 1974.
[65] Trebules, Jr., V. W., Roberts, R., and Hertzberg, R. W., "Effect of Multiple Overloads on Fatigue Crack Propagation in 2024-T3 Aluminum Alloy," *ASTM STP 536,* American Society for Testing and Materials, Philadelphia, 1973.
[66] Barsom, J. M. and Novak, S. R., "Subcritical Crack Growth and Fracture of Bridge Steels," *NCHRP Report 181,* Transportation Research Board, Washington, DC, 1977.
[67] Barsom, J. M., unpublished data.

10

Fatigue and Fracture Behavior of Welded Components

10.1 Introduction

WELDING TECHNOLOGY has had a significant impact on industrial developments. Fabrication by welding is an effective method to reduce production and fabrication costs and can be mechanized, computer controlled, and incorporated in assembly lines. Welding fabrication has revolutionized many industries, including shipbuilding and automotive production, and has resulted in the development of various products, such as pressure vessels, that could not otherwise have achieved their present functions.

Welding technology is complex and fabrication by welding encompasses characteristics that should be understood to different levels by the design engineer, the fabricator, and the welder. Some of these characteristics pertinent to the present discussion are residual stresses, imperfections, and stress concentrations.

Failures in engineering structures occur predominately at component connections, even in those structures which have been designed, fabricated, and inspected according to Code. "Connections" refers to those locations in a structure where elements are joined to reconcile changes in geometry and/or accommodate fabrication or service requirements. For example, fatigue cracking in bridges, ships, offshore structures, pressure vessels, and buildings occurs, almost without exception, at the welded or bolted connections and attachments such as cover plate fillet weld terminations, stiffeners, backing bars, and seam and girth weld toes [*1–3*].

Recent reviews by the ASME Section XI Task Group on Fatigue in Operating Plants and others [*4–6*], as well as examination of in-house files spanning nearly 40 years of failure investigations [*7,8*], revealed the vast majority of pressure boundary component fatigue failures to have occurred at the welded connections. In fact, none of the investigated fatigue failures occurred in a base metal that did not contain a weld, weld repair, or significant stress concentration associated with the fatigue crack initiation site.

Given the preponderance of weld-related fatigue failures, it is reasonable to expect that state-of-the-art design codes would incorporate rules or procedures for addressing the fatigue life of welded components. However, several design codes, including the ASME Boiler and Pressure Vessel Code, Section III, "Rules for Construction of Nuclear Power Plant components" [9], and Section VIII, Division 2 do not incorporate explicit fatigue life curves for welded components. Rather, the Section III and Section VIII fatigue rules are currently based on stress versus number of cycles (*S-N*) fatigue curves developed nearly 30 years ago from smooth, base metal specimens tested in air at room temperature. Welds, as well as other structural discontinuities, are evaluated in Section III by calculating the peak stress at the detail of concern, incorporating an appropriate fatigue strength reduction factor (FSRF) and comparing this stress to the smooth specimen *S-N* curve.

Although the applied service stresses may be within Code design allowable for "fatigue rated" components, welded and bolted connections can sufficiently elevate local stresses to initiate and propagate cracks. Fatigue cracks typically initiate from discontinuities within the weld or base metal. Such discontinuities can be volumetric (e.g., porosity, slag inclusions) or planar (e.g., lack of fusion, undercut) in nature and in either case elevate local stresses sufficiently to reduce weld joint fatigue strength. In addition, the weld geometry itself can induce stress concentrations higher than those associated with the aforementioned weld discontinuities. For instance, one of the most fatigue-sensitive weld details is a fillet weld termination oriented perpendicular to the applied cyclic stress field. In this case, fatigue cracking initiates from the toe of the fillet weld and propagates through the adjacent base metal. In fact, the majority of weld-related fatigue failures initiate at the surface, generally at the weld toe. Consequently, the overall fatigue strength of a welded joint will be governed by the combined severity of the weld discontinuity and geometry-induced stress concentration. Other parameters also known to affect the behavior of welded components include residual stress, distortion, heat treatment, environment, as well as others. Although these parameters tend to have a secondary influence on weldment fatigue strength, all of these topics, except for environmental effects, are discussed. Due to the breadth of possible environmental effects, this topic is not considered herein and is deserving of its own discourse. Most of the information presented in Chapter 11 on environmental effects applies to weldments.

10.2 Residual Stresses

Residual stresses are those that exist in a component free from externally applied forces. They are caused by nonuniform plastic deformations in neighboring regions. These regions can be small, as occurs within weldments, or large, as may occur in curving or straightening a beam or a shell during fabrication. Furthermore, residual stresses are always balanced so that the stress field is in static equilibrium. Consequently, wherever tensile residual stresses occur, compressive residual stresses exist in neighboring regions.

Residual stresses can be either beneficial or detrimental to the behavior of components. For example, controlled thermal or mechanical residual stresses are used to curve or straighten various components. Also, compressive residual stresses are used to minimize environmental effects on surfaces of components and to improve their fatigue initiation performance. Because fatigue life is governed by the stress range rather than the magnitude of the static or steady-state (applied or residual) stresses, tensile residual stresses usually have only a secondary effect on the fatigue behavior of components. On the other hand, excessive tensile residual stress can also initiate unstable fracture in materials with low-fracture toughness. Consequently, the magnitude of unfavorable residual stresses should be controlled, especially in thick, highly constrained weldments with low fracture toughness.

In addition, residual stresses can be induced by thermal, mechanical, or metallurgical processes. Thermal residual stresses are caused by nonuniform permanent (plastic) deformations when a metal is heated, then cooled under restraint. Unrestrained expansion and contraction do not generate residual stresses. However, restrained expansion and contraction induce permanent deformation (strains) and corresponding residual stresses. Generally, metal that cools last is in tension.

Mechanically induced residual stresses are caused by nonuniform permanent deformation when a metal is mechanically stretched or compressed under restraint. Therefore, the occurrence of mechanically induced residual stresses requires the presence of both permanent mechanical deformation and restraint that prevents the deformed metal from contracting or expanding to its new unrestrained equilibrium dimension. In general, the sign (tension or compression) of mechanically induced residual stress is opposite to the sign of nonuniform plastic strain that produced the residual stress. This process is used to mechanically curve or straighten components and, as in peening, to produce a compressive stress layer on the surface of a component to improve its fatigue or corrosion behavior.

Fabrication by welding usually results in stresses that are locked into the fabricated assembly. These stresses are either residual stresses or reaction stresses or both. Residual stresses are caused by the inability of the deposited molten weld metal to shrink freely as it cools and solidifies. Reaction stresses are caused by the inability of assembly components to move relative to each other during thermal expansion and contraction of the deposited weld metal, or the molten base metal for autogenous welds and surrounding base metal. Contraction of solidifying weld metal is restricted by adjacent materials, resulting in complex three-dimensional residual stresses. The magnitude of these stresses depends on several factors, including size of the deposited weld beads, weld sequence, total volume of the deposited weld metal, weld geometry, and strength of the deposited weld metal and of the adjoining base metal as well as other factors. Often, the magnitude of these stresses exceeds the elastic limit of the lowest strength region in the weldment.

Methods for measuring residual stresses include sectioning, hole drilling, and X-ray diffraction. The sectioning method is a destructive test in which residual stresses are determined by removing slices from the member and measuring the resulting strain. Hole drilling is a semi-destructive test for measuring residual stresses near the surface of the material; it involves placing strain gages on the surface and measuring strain relaxation as a hole is drilled in the vicinity of the gages. The X-ray diffraction method is a nondestructive test in which surface residual stresses are determined by measuring the change in the lattice spacing of the material; only surface residual stresses are measured in a very localized area.

Development of residual stresses in weldments may be demonstrated by considering the following simplified example of a groove weld shown in Figure 10.1. As the deposited molten filler metal cools, it contracts longitudinally along the weld and transversely across the weld. This contraction is resisted by the surrounding base metal, resulting in residual tensile (+) and compressive (−) stresses as indicated in Figure 10.1. The combined tensile and compressive stresses are balanced in neighboring areas to achieve equilibrium. These stresses may cause distortion, deformation during post-weld machining, and stress corrosion cracking and fracture.

10.3 Distortion

Thermal contraction of deposited weld metal may cause distortion when the assembly components are free to move, as shown in Figure 10.2. If the assembly components are not free to move, distortion decreases and the magnitude of residual reaction stresses increases. Distortion of butt welds may result in out of plane bending, angular distortion, and dimensional changes, Figure 10.2. More complex geometries may exhibit complex distortion patterns. Distortion produces secondary stresses and, under cyclic loading, secondary stress fluctuations, which, when superimposed onto the primary stress fluctuations, decrease the fatigue life of components.

Distortion can be minimized and controlled by proper design and fabrication practices. These practices include:

1. Minimize amount of deposited weld metal: use smallest acceptable weld size; use groove geometries that require the least amount of weld metal per length; use double-sided joints in place of single-sided joints; use square edge or narrow gap procedures where possible; avoid excessive reinforcement.
2. Obtain good fit-up.
3. Preset and pre-camber members prior to welding.

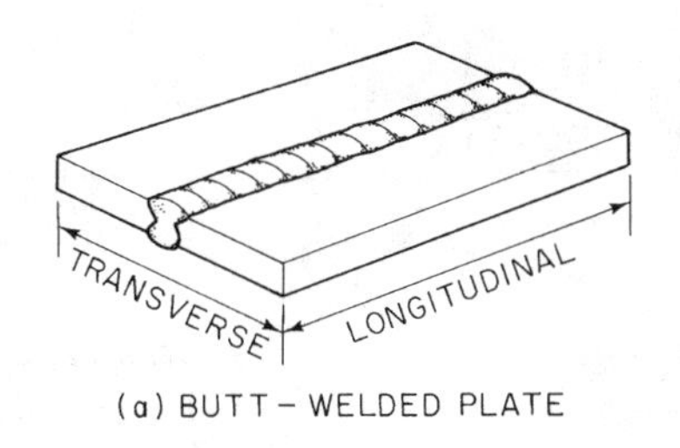

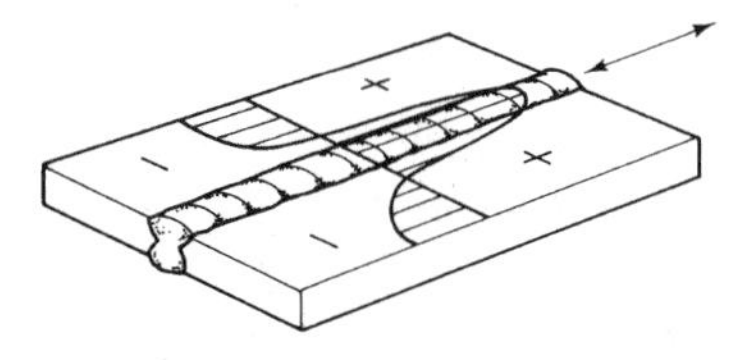

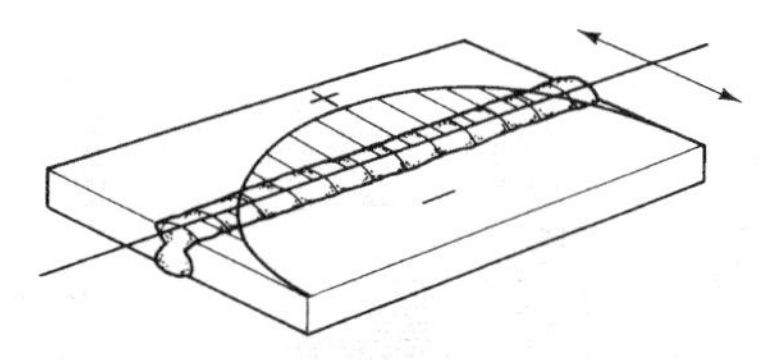

FIG. 10.1 Residual stresses for a butt-welded plate: (*a*) butt-welded plate, (*b*) longitudinal residual stress, and (*c*) transverse residual stress.

4. Use a well-planned welding sequence.
5. Place welds near the neutral axis.
6. Balance welds about the neutral axis.
7. Use high travel speeds and low heat input processes.

10.4 Stress Concentration

Stress fields in components having a uniform cross section can be visualized as evenly spaced parallel lines that traverse the component from one loaded end to the other, Figure 10.3*a*. Changes in geometry of the component that intersect the stress-field flow lines cause an increase in stress at locations where the flow lines

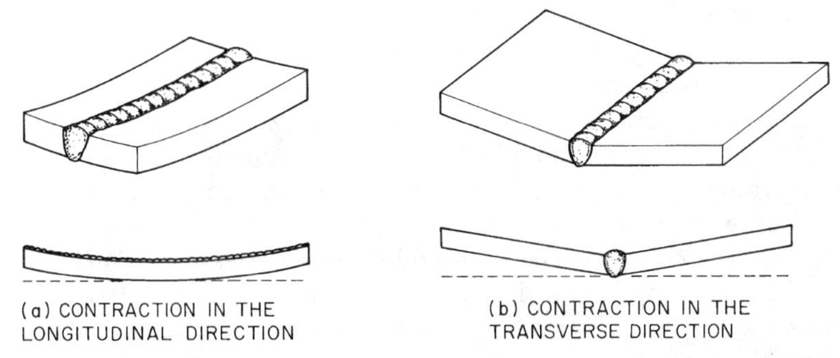

FIG. 10.2 Exaggerated distortions for a single-pass butt-welded plate: (*a*) contraction in the longitudinal direction; (*b*) contraction in the transverse direction.

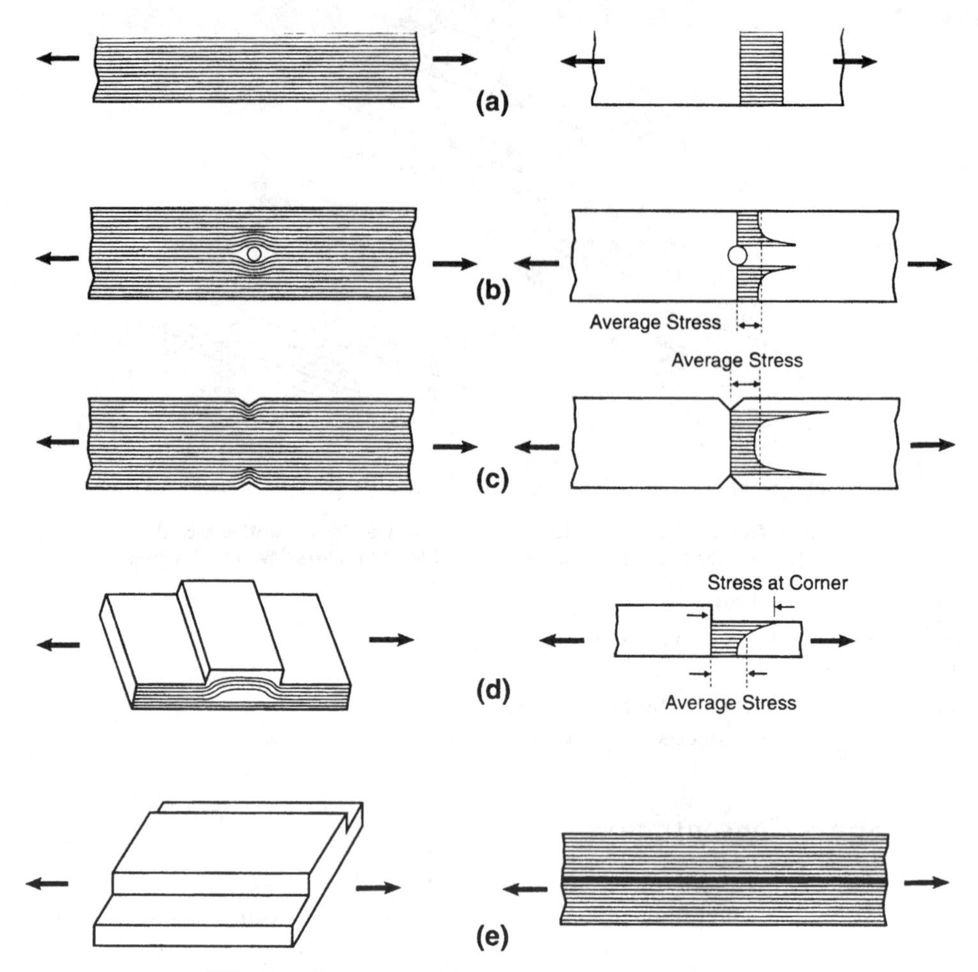

FIG. 10.3 Stress flow lines and stress concentrations.

are disturbed most. Figures 10.3*b*, 10.3*c*, and 10.3*d* present three examples that demonstrate the effect of geometric discontinuities on stress concentration. Figure 10.3*e* shows a geometric change that does not intersect but is in the direction of stress-field flow lines and, consequently, does not result in stress intensification.

Examples 10.3*b*, 10.3*c*, and 10.3*d* show that geometric discontinuities concentrate and intensify the stresses in a very local area. These stresses decay rapidly until they equal the applied nominal stresses away from the "shadow" of the geometric discontinuity. The magnitude of the stress concentration, k_t, depends on the geometry of the discontinuity and is defined as the ratio of the maximum local stress, σ_{max}, and the nominal stress, σ, remote from the influence of the discontinuity so that:

$$k_t = \frac{\sigma_{max}}{\sigma} \tag{10.1}$$

The magnitude of the stress intensification at the edge of a given planar discontinuity in a unidirectional tensile stress field depends on its projected size and shape on the plane perpendicular to the direction of the principal tensile stress. The maximum stress intensity for a given planar discontinuity occurs when the plane of the discontinuity is perpendicular to the direction of the tensile stress and approaches zero as the plane of the discontinuity becomes parallel to the direction of the tensile stress. Thus, planar discontinuities, such as plate laminations, whose plane is parallel to the surfaces of a plate subjected to in-plane tensile stress fluctuations, rarely cause a degradation in the fatigue life of the plate but can be detrimental to the fatigue life when the plate is subjected to tensile stress fluctuations in the through-thickness direction.

The stress intensification caused by a surface discontinuity is about twice the stress intensification caused by an embedded discontinuity of equal size and shape. Thus, for a given shape, an embedded discontinuity must be about twice as large as a surface discontinuity to cause the same stress intensification.

The stress intensification for planar discontinuities whose plane is perpendicular to the direction of the tensile stress is higher than the stress intensification for a volumetric discontinuity having equal planar size and shape projected on the plane perpendicular to the direction of the stress. Fabrication by welding may result in stress concentration in the fabricated joint. Stress concentration may be caused by the geometry of the welded component or by various imperfections and cracklike discontinuities in the base metal, weld metal, or heat-affected zone. Examples of stress concentration regions caused by geometric discontinuities in fabricated joints are shown in Figure 10.4. These locations usually correspond to initiation sites for fatigue. This figure shows that the location of the stress concentration is a function of the direction of loading. Also, unwelded regions loaded perpendicular to their plane in partial penetration groove welds or when lack of fusion occurs behave like cracks and, depending on the geometry of the weldment, can cause crack initiation and propagation from the weld root.

Figure 10.5 presents various types of weld discontinuities and cracks that result in stress concentration and may cause the initiation and propagation of fatigue cracks. The following section presents a brief discussion of discontinuities and their effects on the fatigue behavior of welded components.

10.5 Weld Discontinuities and Their Effects

Fabrication by welding may result in various discontinuities and cracks in the deposited weld metal or in the heat-affected zone of the base metal [*10–12*]. Codes and specifications define acceptance levels for discontinuities in terms of their type, size, orientation, and distribution. Usually, cracks and crack-like dis-

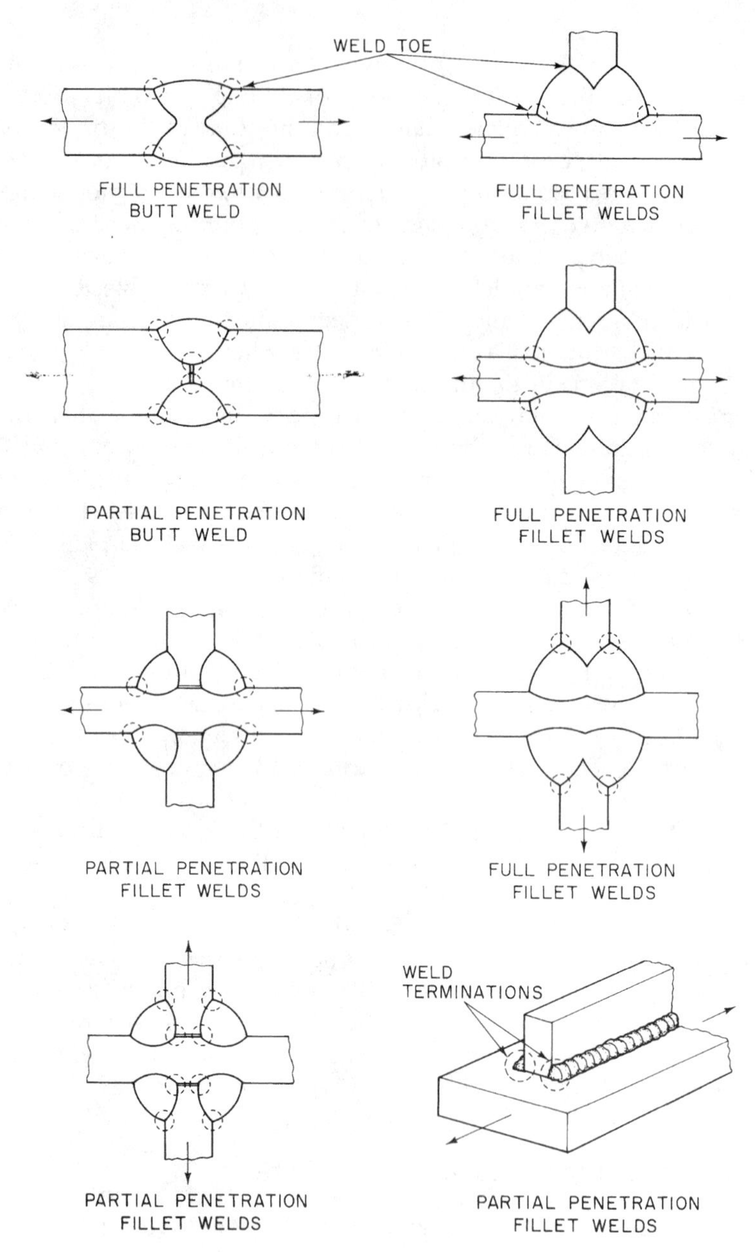

FIG. 10.4 Stress-concentration regions (indicated by dashed circles) for weldments.

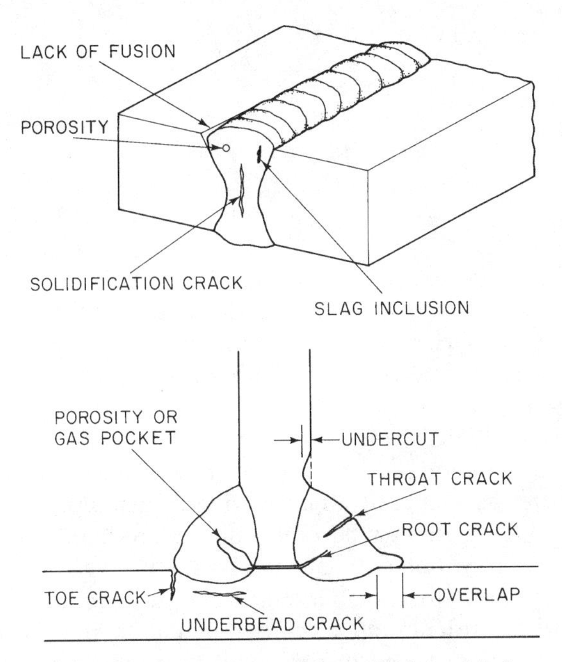

FIG. 10.5 Imperfections and cracks in welded joints.

continuities are prohibited. Discontinuities are designated as defects only when their size, orientation, and distribution exceed specification limits and their presence affects the integrity of the component and renders it unfit for its intended application.

Various types of weld discontinuities, cracks, and crack-like imperfections, their causes and methods to eliminate them have been the subject of many publications. In general, these weld discontinuities may be caused by: (1) improper design that restricts accessibility for welding; (2) incorrect selection of a welding process or welding parameters; (3) improper care of the electrode or flux, or both, and (4) other causes including welder performance. These generalized observations are presented to emphasize that assurance of good-quality fabrication by welding requires considerations and decisions that start in the design stage and continue throughout the entire fabrication process.

Weld discontinuities may be divided into three categories that correspond to different characteristics. These three categories are:

1. Crack-like discontinuities
   - Cracks
   - Lack of fusion
   - Lack of penetration
   - Overlap

2. Volumetric discontinuities
   - Porosity
   - Slag inclusions
3. Geometric discontinuities
   - Undercut
   - Incorrect profile
   - Misalignment

Some of these discontinuities are shown in Figure 10.5.

The mere existence of discontinuities does not indicate a product's unsuitability for a given application. Product suitability for service is based on the severity of the discontinuity measured in accordance with applicable specifications and analyses.

As described previously, the severity of a discontinuity is governed by its size, shape, and orientation, and by the magnitude and direction of the design and fabrication stresses. Generally, the severity of discontinuities increases as the size increases, and the geometry becomes more planar and the orientation more perpendicular to the direction of tensile stresses. Thus, volumetric discontinuities are usually less injurious than planar, crack-like discontinuities. Also, crack-like discontinuities whose orientation is perpendicular to the tensile stress can be injurious, while the same or larger discontinuities whose orientation is parallel to the tensile stress would be innocuous. Furthermore, for a given size and shape, a surface discontinuity whose plane is perpendicular to the tensile stresses is more severe than if it were embedded.

In summary, geometric discontinuities are caused by the welder or the weld procedure. They are stress raisers that intensify the local stresses in their immediate vicinity. Their effect on the fatigue performance of a component is directly related to their severity as stress raisers.

10.5.1 Fatigue Crack Initiation Sites

Fatigue cracks in fabricated components originate at the location where the localized stress range is maximum. This location may not correspond to the location where the stress concentration factor is maximum because the stress raiser may be in a low-stress fluctuation region. For example, fatigue cracks would not initiate at a weld attachment, resulting in severe stress concentration if the attachment is placed at the neutral plane of a beam in bending where the nominal stress range and, therefore, the localized stress range is low.

Stress concentration in weldments occurs at weld discontinuities, Figure 10.5, and at geometric discontinuities, Figure 10.4. These locations usually correspond to initiation sites for fatigue. Thus, fatigue cracks in weldments may originate either at internal discontinuities such as inclusions and porosity or at weld toes and weld terminations [*13–16*]. The majority of fatigue cracks in welded structures originate at a weld toe or at a weld termination rather than from internal discontinuities. This behavior is attributed to the fact that for a given fatigue life, a much larger embedded discontinuity can be tolerated than a surface discontinuity.

The effect of weld discontinuities on the fatigue behavior of welded components depends on their size, shape, distribution, orientation, resistance to deformation, and on the strength and ductility of the surrounding material. In general, fatigue cracks initiate at larger inclusions rather than at small ones, at angular and planar inclusions rather than at spherical ones, at refractory-type inclusions rather than at sulfide inclusions or at porosity, and at closely spaced inclusions rather than at widely separated ones. The size and frequency of discontinuities depend on the welding process, geometry of the weldment including ease of access for welding, the knowledge and care exercised in making the weld, and on the applied nondestructive examination procedures (i.e., radiography, ultrasonic, etc. or a combination thereof) and required sensitivities.

Fatigue cracks that initiate at the root of web-to-flange fillet welds are a good example of fatigue cracks that originate from internal discontinuities, Figure 10.6 [17]. The crack initiated at a gas pocket and propagated as an embedded crack that continually changed its shape until it intersected the fillet-weld surface as a penny-shaped crack. These cracks continued to propagate in all directions in a

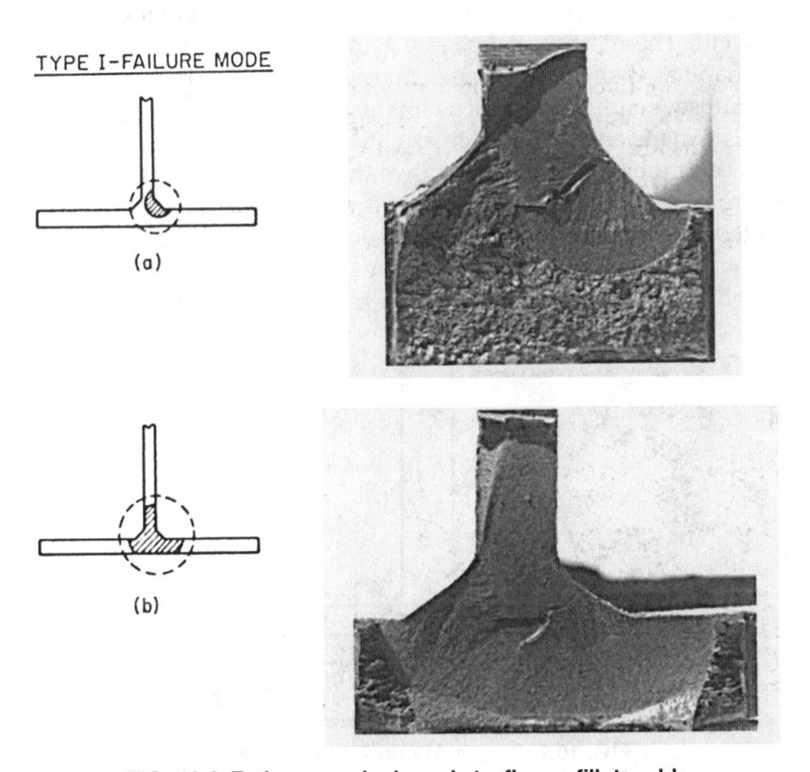

FIG. 10.6 Fatigue cracks in web-to-flange fillet welds.

plane perpendicular to the direction of maximum tensile stress until they became three-corner cracks, Figure 10.6. The crack-like discontinuity formed between the flange surface and the edge of the web by this partial penetration weld did not contribute to the fatigue damage because the plane of the discontinuity was parallel to the direction of the applied fluctuating load. Figure 10.7 [*17*] shows that most of the fatigue life was exhausted when the crack was small.

The majority of fatigue cracks in welded members initiate at a weld toe or at a weld termination near a stiffener, or other attachments such as a gusset plate, or end of a cover plate. These are regions of high stress concentration, high residual stresses, and may contain small weld discontinuities. The residual stresses become redistributed under cyclic loading. Moreover, because the surface of the deposited weld metal is invariable rippled, the toe angle between the weld metal and the base metal can vary significantly at neighboring points along the weld toe, resulting in variations in the stress concentration and the initiation of fatigue cracks at multiple sites along the weld toe. Figure 10.8 [*17*] shows crack formation at the termination of a longitudinal fillet weld and at the toe of a transverse fillet weld in cover-plate details. For the longitudinal fillet welds, the fatigue crack initiates at the termination of the weld and propagates as a single crack changing its shape as it propagates, as shown in Figure 10.9 [*17*]. For the cover plate with a transverse fillet weld, multiple fatigue cracks initiate at the toe of the weld. These cracks propagate as part-through cracks that continually change in shape as their size increases and as they approach adjacent propagating cracks. Subsequently, these cracks link up to form a single crack. Most of the fatigue life is exhausted when the cracks are still small and difficult to detect.

The preceding brief discussion shows that fatigue cracks in weldments can (1) initiate from small discontinuities that are either embedded or on the surface, (2) be located in regions of high stress concentration where the level of stress concentration may vary appreciably in small neighboring locations along the weld toe, and (3) reside in regions of high residual stress that become redistrib-

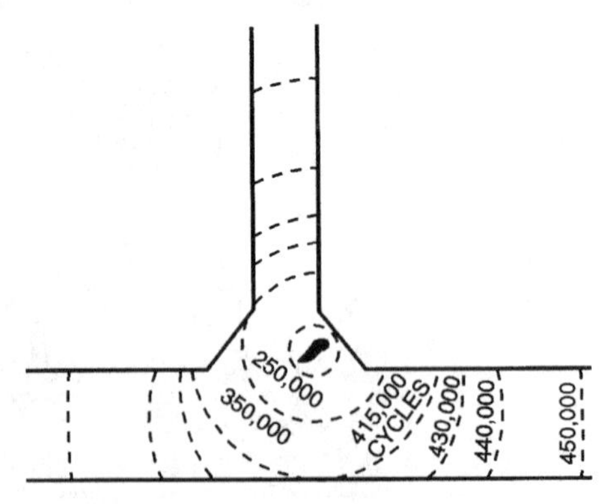

FIG. 10.7 Stages of crack propagation for a web-to-flange fillet weld.

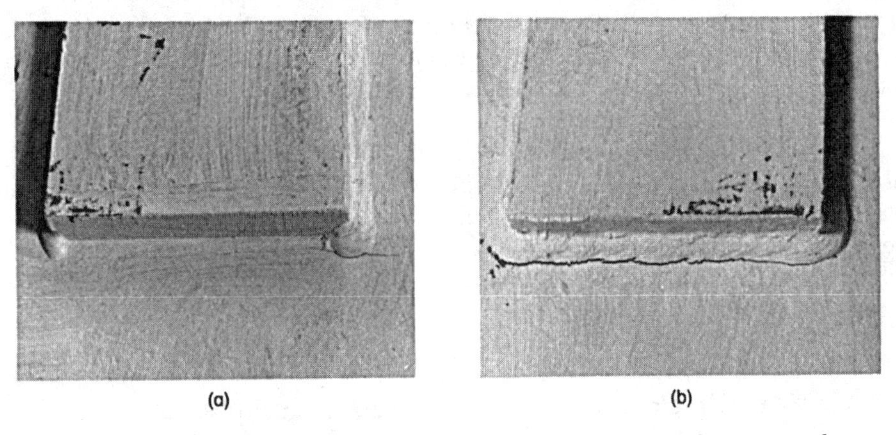

FIG. 10.8 Fatigue crack at ends of cover plates: (*a*) crack formation at toe of longitudinal fillet weld; (*b*) crack formation at toe of transverse fillet weld.

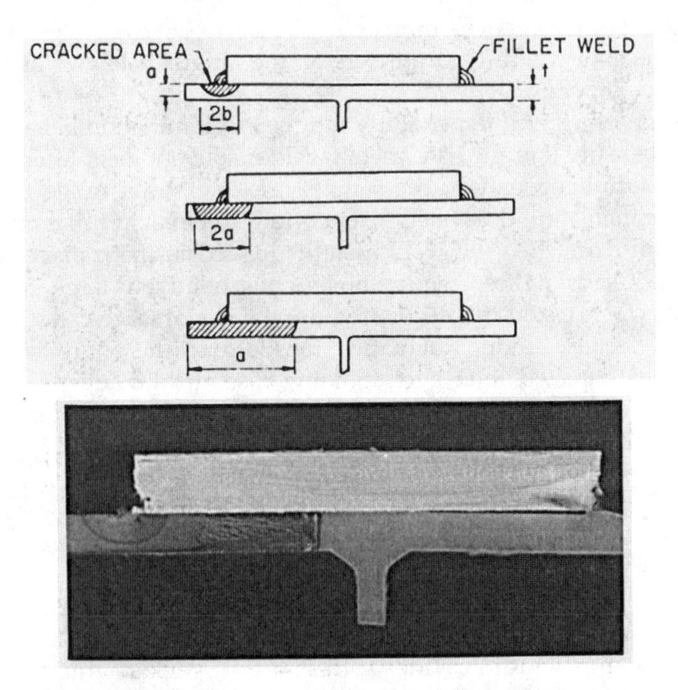

FIG. 10.9 Stages of crack propagation at the end of a longitudinally welded cover-plate detail.

uted under cyclic loading. The discontinuities from which fatigue cracks initiate have different characteristics, sizes, and shapes and, in most cases, are very difficult and costly to locate and define nondestructively. Moreover, the fatigue crack changes its shape throughout most of its propagation life. The magnitude of the change depends on the shape and location of the fatigue-crack initiating discontinuity, the stress-field distribution, and the physical shape of the weld joint configuration. Finally, the rate of fatigue-crack propagation generally increases exponentially with increased crack length in components subjected to a given tensile load fluctuation. Consequently, the fatigue life of most welded components is expended when the fatigue crack is small.

10.6 Fatigue Crack Behavior of Welded Components

10.6.1 Fatigue Behavior of Smooth Welded Components

10.6.1.1 Specimen Geometries and Test Methods Smooth-machined welded specimens can be used to establish the fatigue crack initiation behavior of weldments subjected to cyclic loads. Fatigue crack initiation sites can be determined either during testing or by postmortem analysis of fracture surfaces. In the absence of weld discontinuities, fatigue cracks initiate on the specimen surface in the least resistant region of the base metal, weld metal, or heat-affected zone. Thus, smooth specimens may be used also to develop filler metals and to determine effects of welding parameters on the fatigue behavior of weld metals and heat-affected zones.

The design of a specimen to study fatigue crack initiation behavior of weldments should ensure that the base metal, weld metal, and heat-affected zone will be subjected simultaneously to the same stress and strain range in each cycle. Also, the specimen should have sufficient width to develop full constraint along the weld metal axis when the weld metal is loaded in the transverse direction. Specimens that satisfy these requirements have relatively large cross-sectional areas, which, when tested in tension, require the use of large-capacity machines. Furthermore, because under fluctuating tensile load the entire volume of the weld metal, heat-affected zone, and adjoining base metal is subjected to the same tensile stress range, fatigue crack initiation usually occurs at weld discontinuities rather than from the least fatigue-resistant metallurgical region. These difficulties can be minimized by fatigue testing smooth-machined weldments in bending. The following discussion describes a cantilever-beam specimen developed and used by U.S. Steel Research in the mid-1960s to study the fatigue behavior of weldments.

The weld specimen, Figure 10.10, has a tapered geometry that subjects the weld metal, heat-affected zone, and adjoining base metal to the same strain range. The specimen width provides sufficient transverse restraint to initiate the fatigue crack at mid-width of the specimen. Such specimens were subjected to

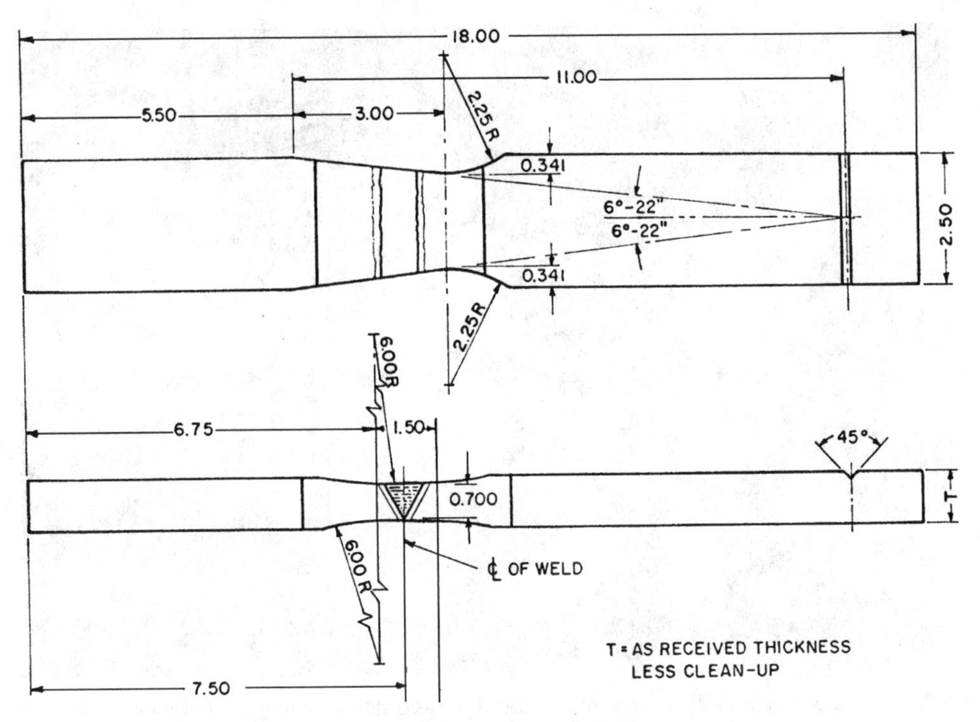

FIG. 10.10 Tapered welded low-cycle fatigue specimen.

cyclic strain in bending using hydraulically operated machines, as shown in Figure 10.11. Strain gages were mounted on the reduced section of the specimens and the output fed into an X-Y recorder. The output from strain gages on weigh bars, which measured the load applied to the test specimen, was also fed into the X-Y recorder to obtain a complete applied moment-strain curve. The total strain range at 10 cycles was used as the total strain range for any particular test. Failure was defined as a 3/16-in.-long surface crack, Figure 10.12.

Tests conducted on weldments of HY-80, HY-130, and higher-strength steels showed that fatigue cracks may initiate in the weld metal, heat-affected zone, or base metal. The initiation site is determined primarily by imperfections in the weld metal or heat-affected zone and to a lesser extent by properties of the various zones in the weldments.

10.6.1.2 Effects of Surface Roughness Fatigue crack initiation and fatigue crack propagation of very shallow cracks are affected by the surface condition of the component. Fatigue cracks initiate more readily from rough surfaces than from smooth surfaces. Component surfaces may be in the as-produced condition such as hot rolled, cold rolled, forged or cast, or may be prepared by various processes such as machining, grinding, electric discharge machining, and elec-

FIG. 10.11 Low-cycle fatigue machines and X-Y recorder showing typical hysteresis loop.

tropolishing. The beneficial or harmful effects of these processes on fatigue strength depend on surface roughness, the magnitude and depth of surface residual stress, and the size of crack-like imperfections induced in the surface.

Machining is a widely used process to produce the desired shape of a component. The machining tool tears the metal by shearing, producing a compressively stressed work-hardened surface layer. Rough machining can produce a deformed surface layer that contains crack-like tears and lapped metal. Also, rough machining can produce grooves and scratches, which are stress raisers. Isolated or widely spaced grooves or scratches are more severe stress raisers than closely spaced parallel ones of identical geometry because the latter provide mutual relaxation of stress concentration.

Stress concentration caused by grooves, scratches, and crack-like surface irregularities adversely affect the fatigue strength of components. The magnitude of this effect is directly related to the severity of the stress raiser. The necessary and sufficient surface condition to ensure the needed fatigue life of a component depends on the magnitude and number of service stress cycles. Improved fatigue life may be achieved when the surface is finished with fine cut or ground and the direction of the finishing operation is parallel to the principal fluctuating stresses. Weldments are often ground to improve their profile, assure quality NDE, and increase their fatigue life.

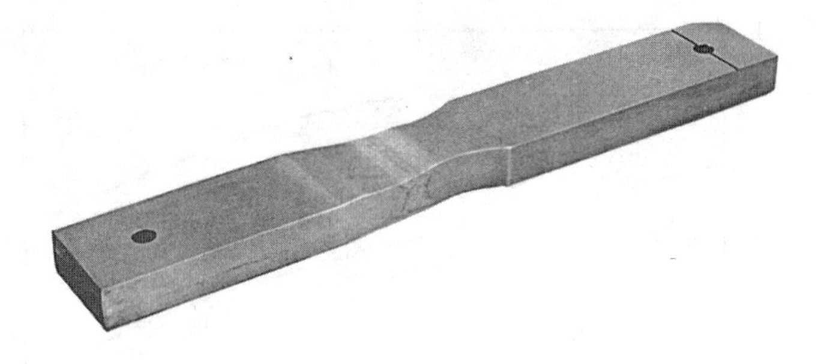

A. Over-all specimen (1 x 2-1/2 x 18 inches).

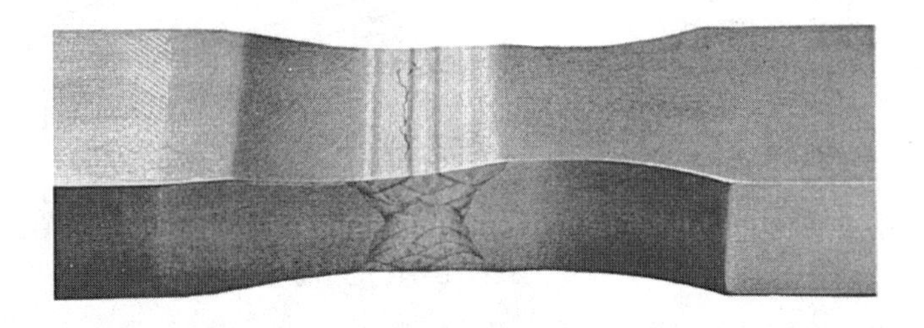

B. Close-up of weld showing crack location.

FIG. 10.12 Tapered welded low-cycle fatigue specimen.

The reduction in fatigue life caused by surface roughness increases as the strength of the steel increases, as shown in Figure 10.13 [*18*]. This behavior is related to a reduction in stress concentration caused by plastic deformation, which occurs more easily as strength decreases. Similarly, the effect of surface roughness is more pronounced at low-stress, high-cycle life than at high-stress, low-cycle life.

Surface roughness is defined by the distance between adjacent peaks and the depth between peaks and adjacent valleys. Usually, these measurements are recorded with a stylus along a selected line. Unfortunately, such traces do not adequately describe the three-dimensional topography of the surface irregularities such as lapped metal in rough machining or lack of fusion in welds. Consequently, the use of surface roughness to evaluate fatigue life of components should be done with care.

10.6.2 Fatigue Behavior of As-Welded Components

Ideally, the fatigue life of components would be established from the number of cycles to initiate a crack and to propagate the crack to terminal size. The

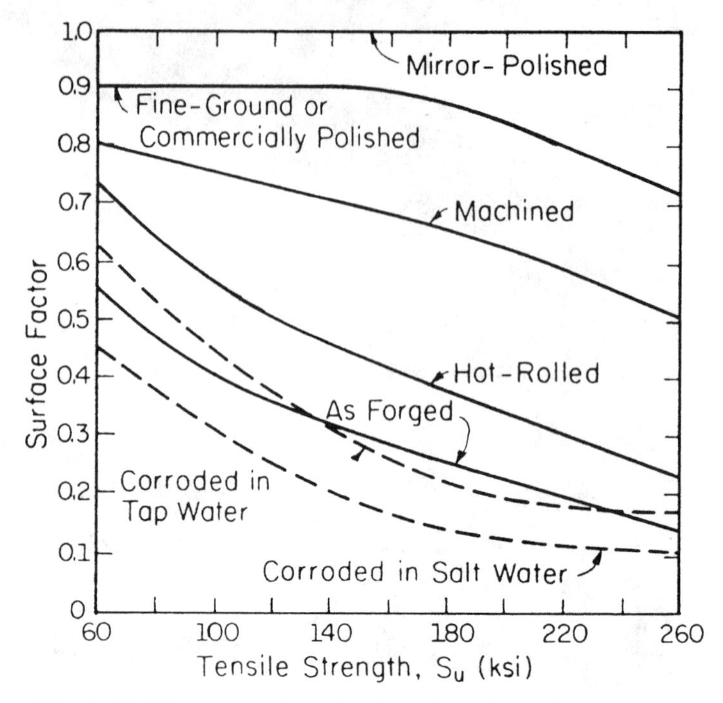

FIG. 10.13 Effect of surface finish on the fatigue limit of steels.

number of cycles to initiate a fatigue crack depends on the definition and identity of the initiated crack, the stress range, the stress concentration factor, the size, shape, and orientation of the initial discontinuity, and the material properties at the initiation site. The fatigue crack propagation life depends on the size, shape, and orientation of the initial crack-like discontinuity or the fatigue crack initiated from the pre-existing discontinuity. The higher the stress range, the more severe the stress raiser, and the more planar the initial discontinuity, the shorter the initiation life. Because geometrically identical components may contain different discontinuities all within permissible code requirements, the minimum fatigue life obtained by testing a set of identical specimens is exhibited by the specimen containing the most severe discontinuity. The fatigue crack initiation life of such a specimen is decreased or eliminated, and, therefore, the total fatigue life may be governed primarily by the rate of fatigue crack propagation.

Fatigue crack propagation can be analyzed best by using fracture mechanics methodology. The usefulness of fracture mechanics depends on the accuracy of the input data for stress range, stress ratio, size and shape of the initial discontinuity and terminal crack, material properties, and the stress intensity factor solution for the particular problem under investigation. Difficulties may be encountered in the application of fracture mechanics analyses to accurately predict fatigue lives of welded components. This is attributed to: (1) inadequacies in the input data, (2) the extreme difficulties in nondestructively characterizing small weld discontinuities, (3) the stress concentration factors along weld toes and at weld terminations, and (4) the need to accurately predict fatigue crack propagation of very small cracks residing in regions of high residual stress and stress concentration. Also, fracture mechanics analysis shows that, under a constant stress range, the fatigue crack propagation rate accelerates exponentially as the crack length increases. Consequently, most of the fatigue crack propagation life is consumed when the crack is small and the contribution of long cracks to the total fatigue life is usually minimal. This behavior is observed also for welded details, as shown in Figures 10.7 and 10.14 [*17*]. Thus, small differences in the size of the initial crack or crack-like discontinuity result in large differences in the total fatigue life. Because of the difficulties in nondestructively characterizing the size and shape of very small crack-like discontinuities, fracture mechanics analyses of weldment fatigue behavior are usually based on the assumption of a large pre-existing crack having the most severe shape and orientation. Consequently, fracture mechanics predictions of welded component fatigue life are usually overly conservative.

Based, in part, on the proceeding observations, a pragmatic stress range versus fatigue life approach derived from extensive tests of fabricated components is used widely in codes and standards. Figure 10.15 [*14*] presents an example of this approach for welded beams fabricated from carbon-manganese steels having yield strength between 36 and 100 ksi (248 and 690 MPa) and subjected to various minimum loads. The large scatter observed for identical geometries tested at a given stress range is caused by differences in the number of cycles to initiate a fatigue crack, the severity (e.g., size and shape) of the initial discontinuity, or both.

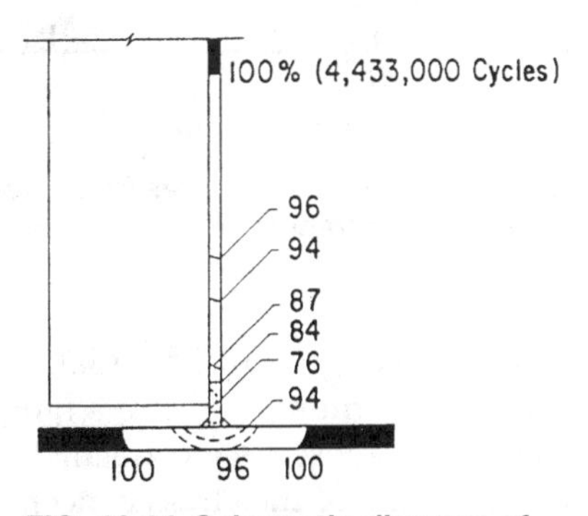

FIG. 10.14 Schematic diagram of phases of crack growth for a stiffener attached to the web alone.

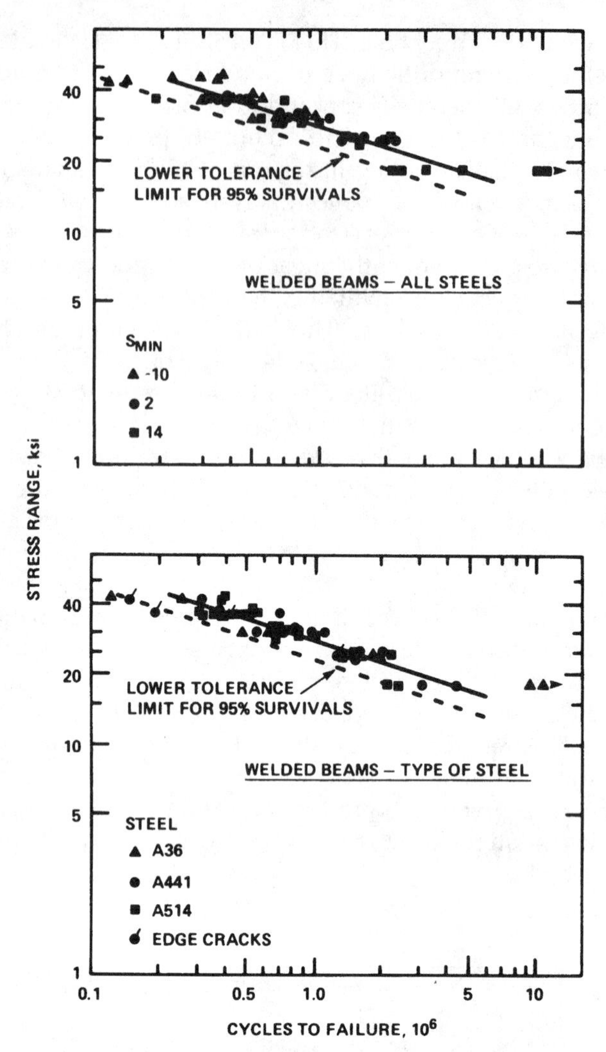

FIG. 10.15 Effects of minimum stress and steel grade on the fatigue strength of welded beams.

10.6.2.1 Effect of Geometry Fatigue cracks initiate and propagate from material elements subjected to the maximum local stress range. Under uniform loading, these elements reside at the most severe detail having the largest stress concentration factor. The magnitude of the stress concentration factor is dictated by the local geometry of the structural detail.

Welded components invariably contain geometric discontinuities caused by a localized change in section. The change in section may be caused by the deposited weld metals as in the case of a butt weld with weld reinforcement left in place, or by welding intersecting components, attachments, and components having a different section, as shown in Figure 10.16 [1]. These changes in geometry cause a concentration of stress at the toe of the weld where the weld metal and base metal surfaces meet. The magnitude of the stress concentration factor depends on several factors, including the angle formed at the juncture of the weld metal and base metal surfaces, by the radius at this juncture, and by the dimension (length) of the attachment in the direction of loading.

The dimensions of deposited weld metal dictate the magnitude of stress concentration. The larger the length in the direction of loading for a transverse

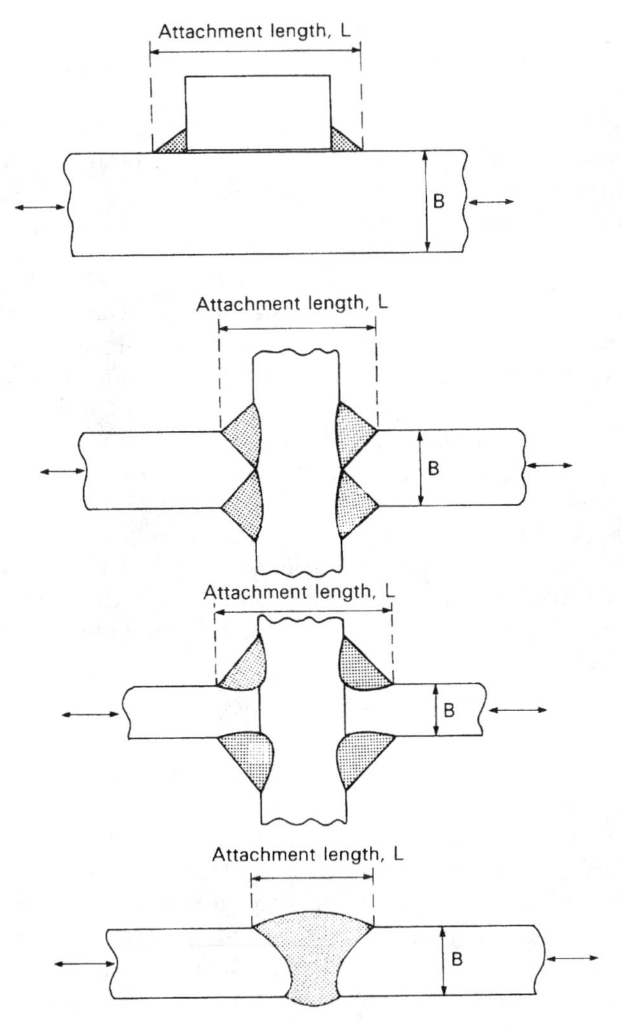

FIG. 10.16 Dimensions relevant to size effects in transverse fillet and butt-welded joints.

butt weld, the higher the magnitude of stress concentration. Also, the larger the height of this weld, the higher the stress concentration. Grinding the excess weld metal to a uniform surface eliminates the geometric stress concentration.

The effects of dimensions on the magnitude of stress concentration are usually much more pronounced for components having different sections, attachments, or intersecting members than from weld reinforcement of a transverse butt weld between plates of equal thickness. Frequently, plates of unequal thickness are butt welded. The larger the difference between the two thicknesses, the larger the angle at the weld toe and the larger the stress concentration. The magnitude of the stress concentration can be reduced by:

1. *Welding the plates with their centerlines aligned.* However, in many cases this arrangement is not possible or practical and the plates are welded with only one of the surfaces aligned.
2. *Tapering the thicker plate.* The smaller the taper angle, the lower the stress concentration. However, the stress concentration cannot be eliminated completely because the difference in plate thickness cannot be eliminated and because the offset centerlines form a secondary moment at the transition region.

Welded attachments, such as gusset plates, stiffeners, and lifting lugs become an integral part of a fabricated component, resulting in stress concentration when the component is stressed. The angle at the weld toe is inherently large compared with the angle formed by weld reinforcement for a butt weld joining plates of equal thickness and, depending on the thickness difference, is equal to or larger than the angle formed between two plates of unequal thickness. The height of these attachments is very large and, therefore, the effect of the height dimension on stress concentration is constant. Consequently, the primary dimensional factor that affects the stress concentration is the length of the attachment in the direction of loading such that the longer the length, the larger the stress concentration factor, as shown in Figure 10.17 [*1*], and the lower the fatigue strength.

10.6.2.2 Effect of Composition Fatigue crack propagation of various steels indicated that the difference in microstructure between ferrite-pearlite and martensitic steels can have an effect on the rate of crack propagation. However, the differences are small enough such that, for all practical purposes, one may conclude that the rate of crack propagation is independent of steel composition, microstructure, strength, or other properties. Because the fatigue life of most welded components is governed primarily by fatigue crack propagation, the fatigue life of welded components is considered independent of steel metallurgical and mechanical properties. This observation is supported by extensive test results of welded components including the data presented in Figure 10.15 [*14*]. Differences between fatigue crack propagation for ferrite-pearlite and martensitic steels are masked by inherent scatter observed in testing weldments. Furthermore, because fatigue crack propagation is relatively insensitive to microstructure, fatigue

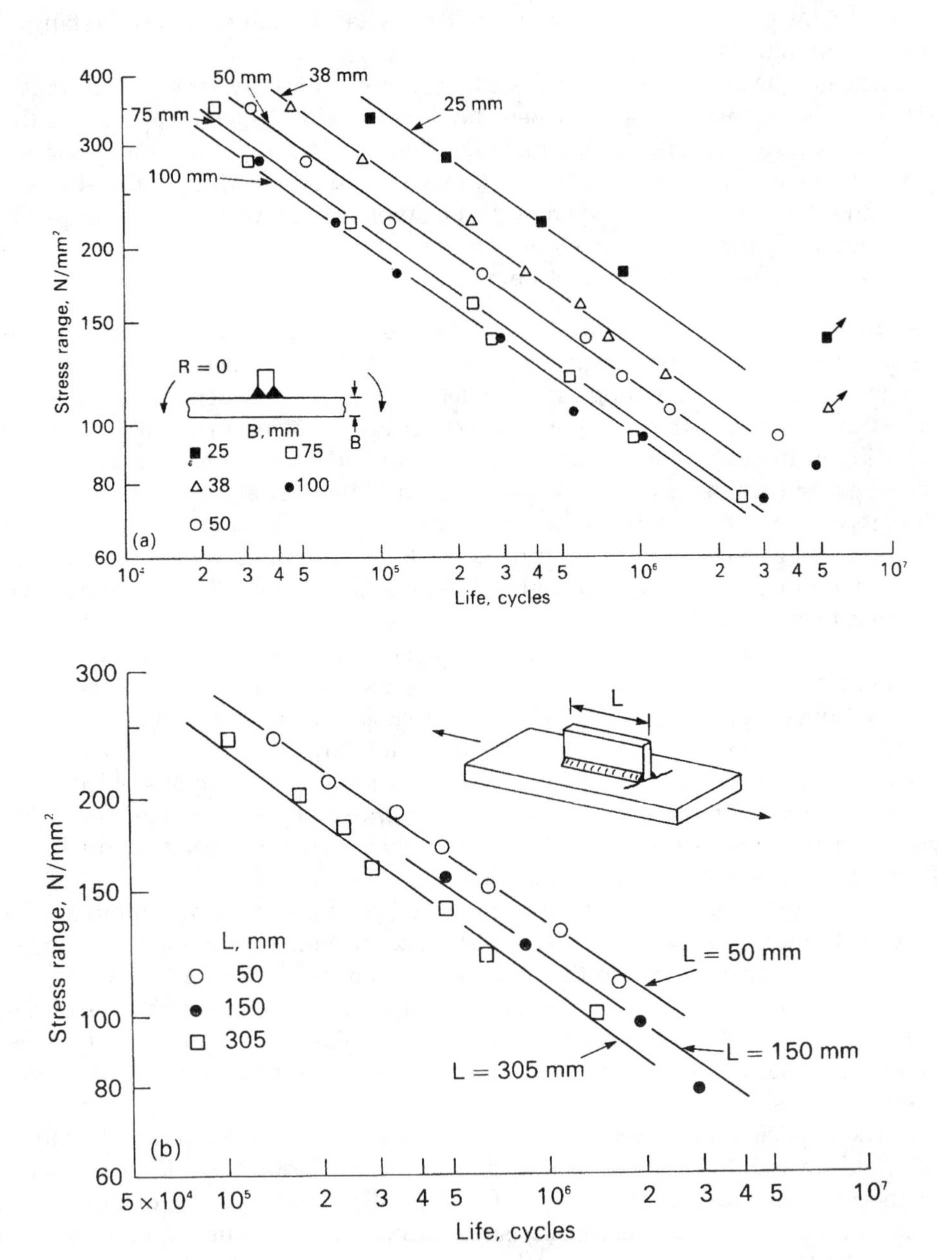

FIG. 10.17 Fatigue test results showing influence of: (*a*) plate thickness; (*b*) attachment length.

crack propagation in welded components through changing microstructures of base metal, weld metal, and heat-affected zone has a small effect on the fatigue life of components.

The fatigue crack propagation rate increases significantly at very large stress-intensity-factor ranges, ΔK, or when the stress intensity factor approaches the critical fracture toughness of the material. This acceleration is caused by the superposition of a ductile or brittle crack extension mechanism onto the striation mechanism for normal fatigue crack propagation. Consequently, in the presence of a very large initial crack-like discontinuity, the fatigue life of welded components can be affected by the fracture toughness of the material.

10.6.2.3 Effect of Residual Stress Stresses that open a crack and separate its surfaces in a direction normal to each other stretch the crack tip and generate a new surface, thus extending the crack length a small increment. Stress reversal closes the crack and the newly generated surface is folded upon itself. Further stressing in the unloading direction by a compressive force forces the crack surfaces against each other without generating a driving force at the crack tip. Therefore, the compression portion of a stress cycle is usually subtracted from the total stress range, and only the tensile portion of the stress range is used to calculate the driving force for fatigue crack propagation and the fatigue life of components without tensile residual stress.

Welded joints may be subjected to tensile residual stress of yield stress magnitude. Under these conditions a crack within the weld joint is forced open and remains open even when the weldment is subjected to an externally applied nominal compressive stress. Consequently, externally applied tensile and compressive stresses cause fatigue damage, and the total fatigue life of welded components is related to the total (tensile plus compressive) stress range. This principle can be demonstrated by analyzing the interaction between residual and fluctuating stresses for a butt-welded joint.

The longitudinal stresses for a butt-welded component are maximum tensile stresses in the weld, then decrease rapidly and become compressive. The magnitude of the tensile stress in the weld may be equal to the yield stress, σ_{ys}, of the weld metal. The magnitude and distribution of the compressive stresses depend on plate dimensions and must satisfy the equilibrium requirement that the area under the tensile residual stress must be equal to the area under the compressive stress.

The application of a nominal tensile stress, σ_n, whose magnitude is less than the yield stress of the weld, in the direction of the weld, results in plastic straining of the weld metal. The magnitude of this plastic strain is equal to the elastic strain in the surrounding elastically stressed material to ensure equilibrium between the two regions. Consequently, in most practical applications, the plastic straining of the weld metal is small.

The stress-strain curve for the weld metal can be either of the two types shown in Figure 10.18. The type of curve that represents a given weld metal

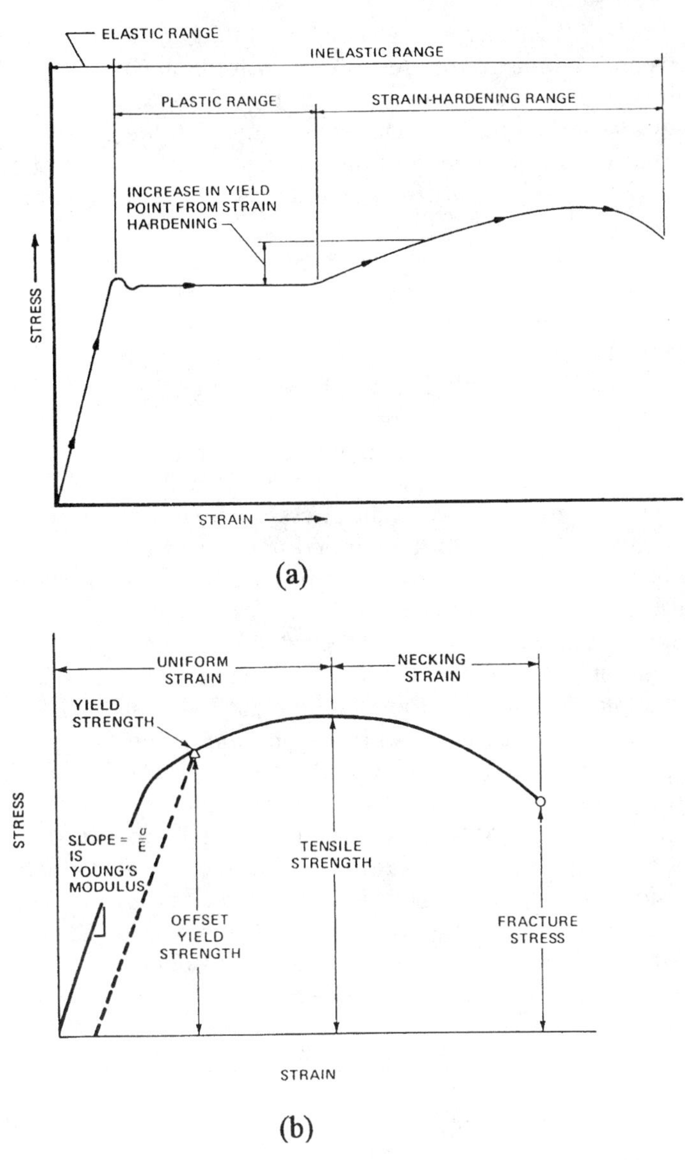

FIG. 10.18 Dependence of fatigue crack initiation threshold on yield strength.

depends on weld metal composition and amount of base metal dilution. When the inelastic strain due to the applied nominal stress is within the plastic range, Figure 10.18*a*, the maximum (residual plus applied) stress, σ_{max}, in the weld metal remains equal to the yield stress of the weld metal, σ_{ys}. However, when the applied inelastic strain is beyond the plastic range or when the stress-strain curve for the weld metal is represented by the curve in Figure 10.18*b*, the weld metal strain hardens and the maximum (residual plus applied) stress, σ_{max}, in the weld metal is slightly higher than σ_{ys}, as shown in Figure 10.18. The behavior of weld metal with strain-hardening characteristics, Figure 10.18*b*, is considered later. Weld metal that does not undergo strain hardening becomes a special case of this generalized discussion with the maximum stress, σ_{max}, remaining equal to σ_{ys}.

The application of a nominal tensile stress, σ_n, increases the weld metal stress slightly and increases all the surrounding elastic stresses by an increment equal to σ_n, Figure 10.19. Removing the applied stress unloads the plastically strained weld metal and the surrounding stress field elastically, causing redistribution of the residual stress, Figure 10.19. Because the weld metal unloads elastically, the magnitude of the decrease in weld metal stress, $\Delta\sigma$ (i.e., maximum initial stress under load, $\sigma_{max(i)}$, minus maximum final stress after unloading, $\sigma_{max(f)}$, is equal to the magnitude of the applied stress, Figure 10.19. Subsequent application and removal of σ_n results in an elastic cyclic stress range, $\Delta\sigma$, of the weld metal and surrounding stress field equal to the applied nominal stress, σ_n.

The application of a compressive nominal stress, $-\sigma_n$, in the direction of the weld decreases the tensile residual stress in the weld metal and the elastic stress in the surrounding stress field by a value equal to $-\sigma_n$, Figure 10.20. Compressive applied nominal stresses unload the weld metal into the tensile elastic stress region without plastic straining the weld metal. Removal of the compressive stress returns the stress magnitude and distribution to their original condition. Repeated application of the same compressive stress results in an elastic stress range, $\Delta\sigma$, in the weld metal equal to σ_n, Figure 10.20.

The preceding discussion shows that the repeated application of a nominal tensile stress or compressive stress results in tensile stress fluctuation in weld metal with residual tensile stress. Consequently, in the presence of tensile residual stress, repeated application of compressive stress causes the same fatigue damage as repeated application of tensile stress of equal absolute magnitude. Furthermore, the fatigue life of as-welded components is governed by the total (tension

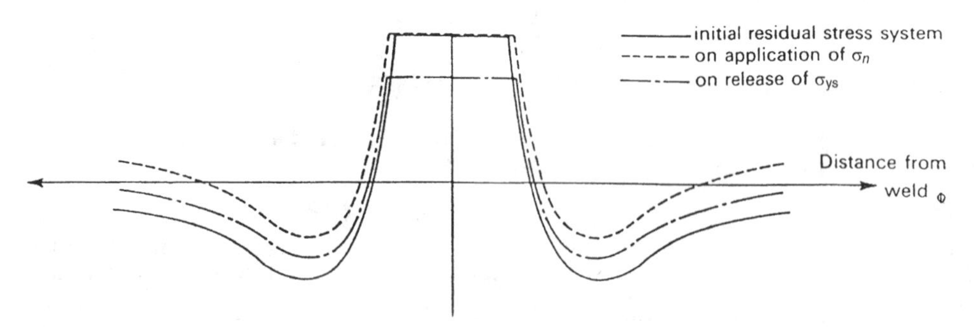

FIG. 10.19 Superposition of an applied tensile stress, σ_n, on a residual stress, σ_{ys}, of the weld metal.

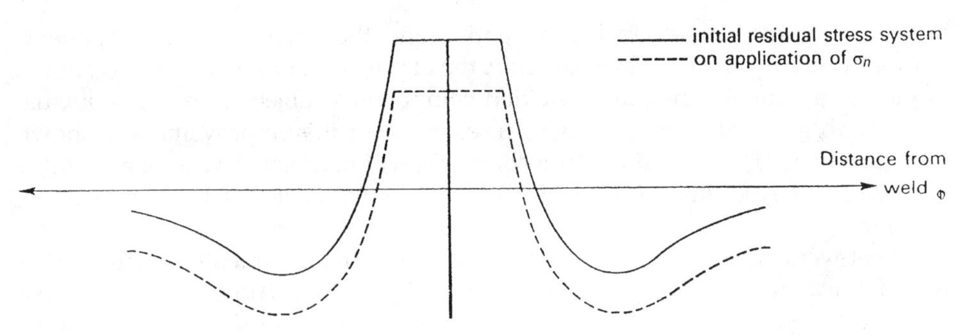

FIG. 10.20 Superposition of an applied compressive stress, $-\sigma_n$, on a residual stress equal to the yield stress, σ_{ys}, of the weld metal.

and compressive) stress range and, for all practical purposes, is independent of the stress ratio, R ($R = \sigma_{min}/\sigma_{max}$). Therefore, several fatigue design codes are based on stress range versus number of cycles to failure curves that are independent of stress ratio. Finally, although fatigue cracks can initiate under nominal tensile or compressive stress fluctuations, fatigue cracks under compressive cyclic loading, unlike tensile cyclic loading, propagate across the residual tensile stress region only. The nominal compressive stress prevents the propagation of fatigue cracks beyond the region where the residual tensile stresses reside. Therefore, stress range versus number of cycles to failure curves are usually applied to tension members and not to compression members.

10.6.2.4 Effect of Postweld Heat Treatment Various postweld heat treatments are available that produce metallurgical or mechanical effects, or both effects in steel base metal, weld metal, and heat-affected zone. These effects can be either beneficial or detrimental to the performance of the treated component. The effects of postweld heat treatments depend on (1) the composition, microstructure, and thermomechanical history including welding of the steels, (2) the maximum temperature imposed and its duration, and (3) the rate and uniformity of the cooling rate throughout the component from the maximum temperature to room temperature. Postweld heat treatment of steels may affect residual stress magnitude and distribution, microstructure, hardness, fracture toughness, stress corrosion cracking susceptibility, hydrogen-induced cracking (HIC) susceptibility, and dimensional stability. The effects of postweld heat treatments on various steels have been discussed thoroughly in Refs *19* and *20*.

Postweld heat treatments, such as thermal stress relief, may minimize residual stress in weldments, resulting in possible fatigue strength improvements. In the absence of residual stress, the application of a nominal compressive stress results in a uniform compressive stress throughout the weldment. Because fatigue cracks do not propagate in a compressive stress field, compressive stress fluctuations are not considered in analyzing fatigue strength of components without residual stresses. Thus, only the tensile portion of a stress range is considered in predicting the rate of crack propagation and the fatigue life of components without residual stresses. Consequently, the elimination of residual stress should improve the fatigue strength of welded components subjected to stress fluctuations partly or totally compressive. An example of this improvement is shown in Figure 10.21 [*1*], which presents a comparison of fatigue behavior of as-welded and stress-relieved small-fillet welded components subjected to various stress ratios.

Post-weld heat treatments redistribute rather than eliminate residual stress from fabricated components of any practical significance. At best, residual stress equal to the yield strength at the stress relief temperature should be expected. Higher residual stresses usually occur due to nonuniform cooling of the component and from reaction stresses during assembly.

10.7 Methodologies of Various Codes and Standards

10.7.1 General

Pressure vessels and piping, as well as most modern structures, are fabricated using welded construction techniques in accordance with one or more codes or standards. The intent of these design and construction codes is to ensure that safe and reliable structures are produced at reasonable cost. To this end,

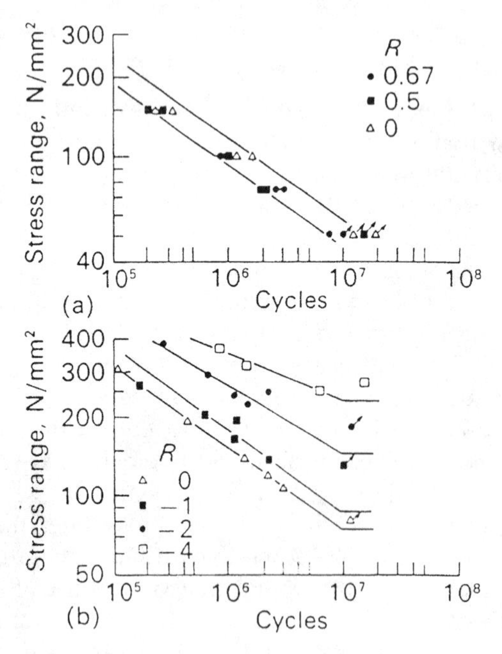

FIG. 10.21 Effect of stress relief on fatigue strength of fillet welds in steel in the as-welded (*a*) and stress relieved (*b*) conditions.

such codes as the ASME Boiler and Pressure Vessel (B & PV) Code, Section III, AASHTO Standard Specification for Highway Bridges, and British Standards BS5500 and BS7608, specifications for fusion-welded pressure vessels and structures, have incorporated procedures or rules for assessing the fatigue performance of their respective structures.

Various codes differ in the methodologies employed for fatigue life assessment. For example, the ASME B & PV Code, Sections III and VIII, utilize an analysis-dominated approach, whereas AASHTO and BS7608 incorporate empirical assessment procedures based on "full-scale" fatigue test results. In this context, full-scale refers to full-size components such as a welded girder or pressure vessel nozzle, rather than an entire reactor vessel. BS5500 utilizes an analytical approach nearly identical to Section III, except that the fatigue design curve is based on small-welded test specimen data, rather than the smooth-specimen, base-metal data upon which Section III is based.

In an analysis approach, stresses or strains are calculated for a given component using a variety of techniques including the finite element method. The calculated stress fluctuations are compared to the appropriate material fatigue curve derived from small, smooth specimen test results (see Chapter 7), and fatigue life is determined at that stress value from either a mean data curve or "adjusted" design curve. The burden in the analytical approach is developing a sufficiently accurate assessment of the local peak stress. In general, this approach is reasonable and conservative for component geometries in which the stresses can be determined accurately such as elbows, ground and inspected butt welds, and moderately complex structural geometries.

In contrast, empirically based approaches such as AASHTO and BS7608 do not rely on highly complex stress analyses. Rather, AASHTO and other empirical approaches rely on statistically significant fatigue test data from actual or full-size weldment test specimens. The importance of such statistical data is that design curves can be specified with known confidence levels with respect to the mean. Of equal or greater importance is that the empirical approaches are generally based on the nominal or net applied stress rather than highly complex local stresses. That is, the stress concentrating effects of the weld detail and/or possible discontinuity, which are not easily captured in a stress analysis, are incorporated in the fatigue curve detail classification, as described in the next section.

A brief description of the AASHTO fatigue evaluation procedure is presented in the following section. The reader is referred to the specific codes and their accompanying commentaries for detailed evaluation procedures and code basis.

10.7.2 AASHTO Fatigue Design Curves for Welded Bridge Components

Bridge engineers have recognized for a long time the effect of cyclic loading on the structural integrity of welded bridge components. The American Welding

Society (AWS) *Specifications for Welded Highway and Railway Bridges* [21] was based on fatigue tests of welded details conducted in the 1940s. In the late 1950s, the observation of fatigue cracks at welded details in the American Association of State Highway Officials (AASHO) Road Test [22] bridges indicated the need for further study of the fatigue behavior of welded details and for modifications of the existing specifications. The present AASHTO fatigue design specifications [23,24] are based on extensive fatigue-test results and field experience that have been accumulated since the early 1960s.

The present AASHTO fatigue design specifications are based on experimental curves that relate the fatigue life, N, of a welded detail to the total (tension plus compression) applied nominal stress range, $\Delta\sigma$ [*14*]. A large number of tests for a given detail have been conducted to generate a statistically significant stress-range–fatigue-life relationship. The design curves represent the 95% survival for a given detail.

Figures 10.15 and 10.22 [*24*] present fatigue-test results for welded beams and cover-plated beams, respectively, fabricated from bridge steels having yield strengths between 36 and 100 ksi (248 and 690 MPa) and subjected to various minimum loads. Statistical analysis of the available data indicates that the stress range is the primary parameter controlling the fatigue life and that the minimum stress, the maximum stress, and the grade of steel have secondary influence on the fatigue behavior of welded components [*25*].

Other fatigue tests were conducted on beams and girders with welded attachments and with transverse stiffeners [*16*]. The available fatigue data for various attachments show that the fatigue strength of a girder with welded attachments is strongly governed by the length of the attachment in the stress direction [*16,24*]. The longer the attachment, the higher the stress concentration at the toe of the weld and the lower the fatigue strength. Welded attachments longer than 4 in. (101 mm) have fatigue strengths equivalent to welded partial length cover plates. Transverse stiffeners are similar to very short attachments and have fatigue strengths equivalent to those of welded attachments that are 2 in. (50 mm) long or shorter.

The extensive fatigue data that have been obtained by testing welded bridge details have been used to establish allowable stress ranges for various categories of steel bridge details, Figure 10.23. Each category represents welded-bridge details that have equivalent fatigue strengths. For example, all welded attachments having a length, L, in the direction of stress equal to or less than 2 in. (Category C) are considered to have equivalent fatigue strength. In reality, under identical fabrication and geometrical conditions, a 2-in.-long attachment results in a higher stress concentration than does a shorter attachment and, therefore, would have a shorter fatigue life. Because the curve for each category corresponds to the 95% confidence limit for 95% survival of all the details in a given category, the fatigue-design curves correspond to approximately the shortest lives obtained for details in each category and are, therefore, governed by the details in that category that have the most severe geometrical or weld stress concentration.

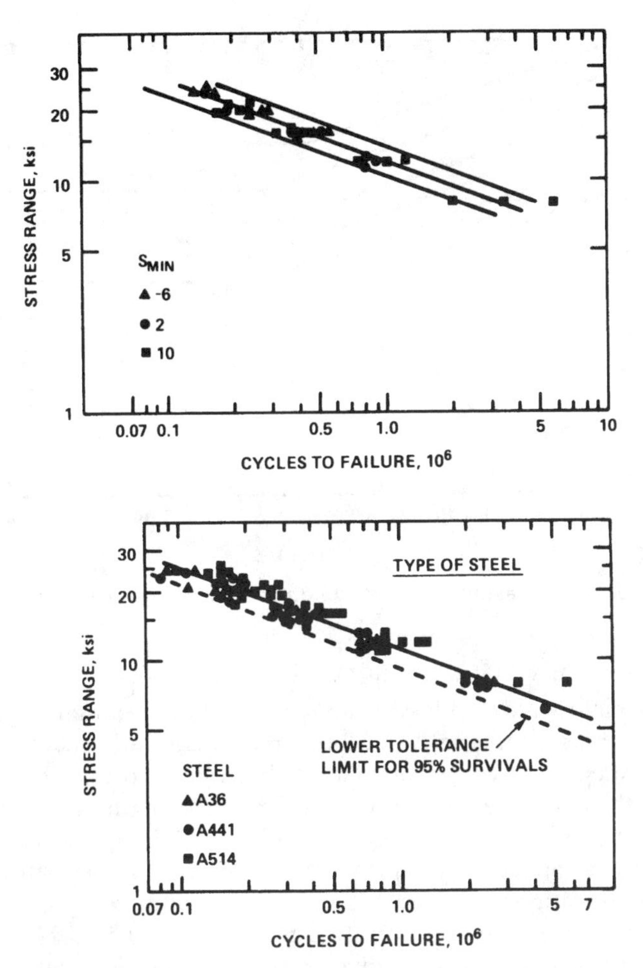

FIG. 10.22 Effects of minimum stress and steel grade on the fatigue strength of beams with transverse end-welded cover plates.

The existence of gouges and weld-imperfection stress raisers in a structural detail of a given geometry decrease the fatigue life of the detail. Consequently, significant variability (scatter) in fatigue-life data can be obtained by testing many details of identical geometry but containing different size imperfections. This variability in the data is very apparent in the database used to establish the AASHTO fatigue categories. For example, the longest life obtained for a Category C detail (stiffener) that was tested at a stress range, $\Delta\sigma$, of 25 ksi was about four times longer than the same detail that exhibited the shortest life. The difference in fatigue life for these two specimens was caused primarily by the difference in the size of the initial imperfections that existed in the specimens.

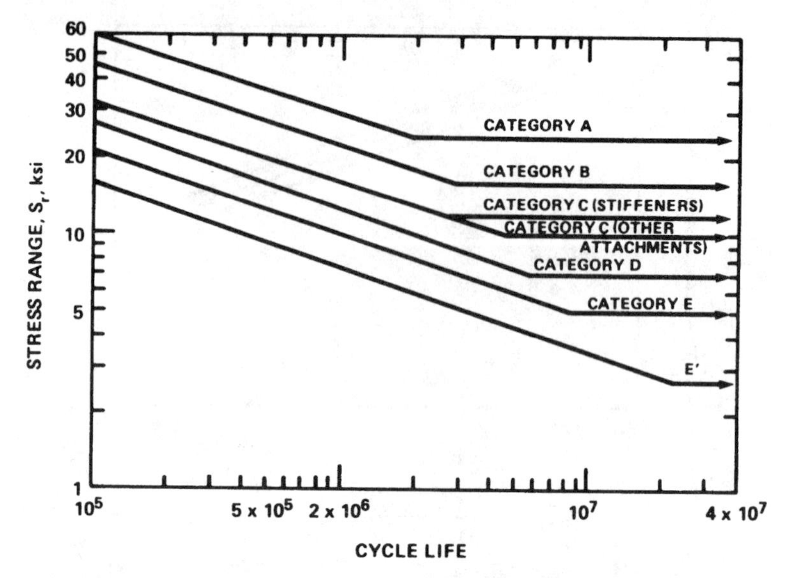

FIG. 10.23 Design stress range curves for categories A to E′.

Categories, A, B, C, D, and E in Figure 10.23 correspond to plain plate and rolled beams, plain welds, welded beams, plate girders, stiffeners, and short attachments (less than 2 in. long), 4-in.-long attachments, and cover-plated beams, respectively. Category E′ corresponds to thick flanges and thick cover plates and suggests that thickness may also affect the fatigue strength of welded girders as has been observed from highway bridges [26] and laboratory tests [27]. The horizontal lines for each category represent the applied nominal stress range corresponding to the fatigue limit (over 2×10^6 cycles) and are extremely important for highway bridges located on heavily traveled roads. The stress-range threshold corresponding to long life for a given category is related to either the fatigue-crack-initiation threshold or the fatigue-crack-propagation threshold.

The AASHTO fatigue design curves represent the 95% confidence limit for 95% survival of all the details in a given category and are governed primarily by the details in a given category that have the most severe geometrical discontinuities, imperfections, or both. Because these discontinuities and imperfections minimize or eliminate the fatigue-crack-initiation life, the fatigue life for these details is governed by the fatigue-crack-propagation behavior for the particular geometry and steel. Fatigue-crack propagation was shown in Chapter 9 to be independent of the strength of the steels. Thus, the AASHTO fatigue design curves should be essentially independent of the strength of the steel. However, it is important to realize that, unlike the fatigue design stress range, the static design stress usually is increased as the strength of the steel is increased.

The fatigue life for a given structural component is determined by the most severe detail in that component. Thus, it is essential to identify that detail and to design the component by using the fatigue category appropriate to the most severe detail it contains.

10.8 Variable Amplitude Cyclic Loads

Most structures are subjected to time-varying sequences of stress fluctuations. The resulting stress histories are very complex because they do not exhibit a describable pattern and cannot be represented by analytical functions. In recent years, actual service load histories have been defined for several structural applications. In many cases a representative load history or block loading is used to simulate actual service stresses. Defining actual service stress histories accurately is essential, especially because fatigue crack propagation and fatigue life of most weldments are proportional to the stress range raised to the third power. Thus, small errors in estimating stress range may lead to large errors in measured or predicted fatigue life. Although stress histories may be approximated, the most reliable method to determine stress histories is by actual field measurements.

Understanding fatigue behavior under variable amplitude stress history is accomplished, in part, by breaking down the history into discrete cycles with corresponding stress ranges. This process of cycle counting in combination with cumulative damage rules is used to estimate the fatigue life of components from the sum of fatigue damage caused by these cycles.

Extensive tests were made of simulated-steel highway-bridge members under variable-amplitude random-sequence loading, such as occurs in actual highway bridges [25]. Welded beams with and without partial-length cover plates and fabricated from both A36 and A514 steels were tested. These details represent the approximate upper and lower bounds, respectively, for fabricated bridge members. The results showed that, like fatigue-crack initiation and fatigue-crack propagation under variable-amplitude loading, Chapters 8 and 9, respectively, the fatigue life for welded components that are subjected to variable-amplitude loading spectra can be predicted by using a single constant-amplitude effective parameter that is a characteristic of the stress-range distribution function.

Figure 10.24 [25] compares the cover-plated beam data obtained under variable-amplitude random-sequence cyclic loading and the AASHTO fatigue-design curve for cover-plate ends (Category E) on the basis of the root-mean-square stress range. Similarly, Figure 10.25 [25] compares the data obtained for welded beams and the AASHTO fatigue-design curve for Category B. For both types of details, the appropriate fatigue-design curve closely approximated the lower limit (95% tolerance limit) of previous constant-amplitude test results where almost all the data points are above these curves. This shows that the AASHTO fatigue-design curves provide an approximate lower limit for variable-amplitude test results when plotted on the basis of the root-mean-square stress range. The scatter of data in Figures 10.24 and 10.25 is reasonable considering that data for

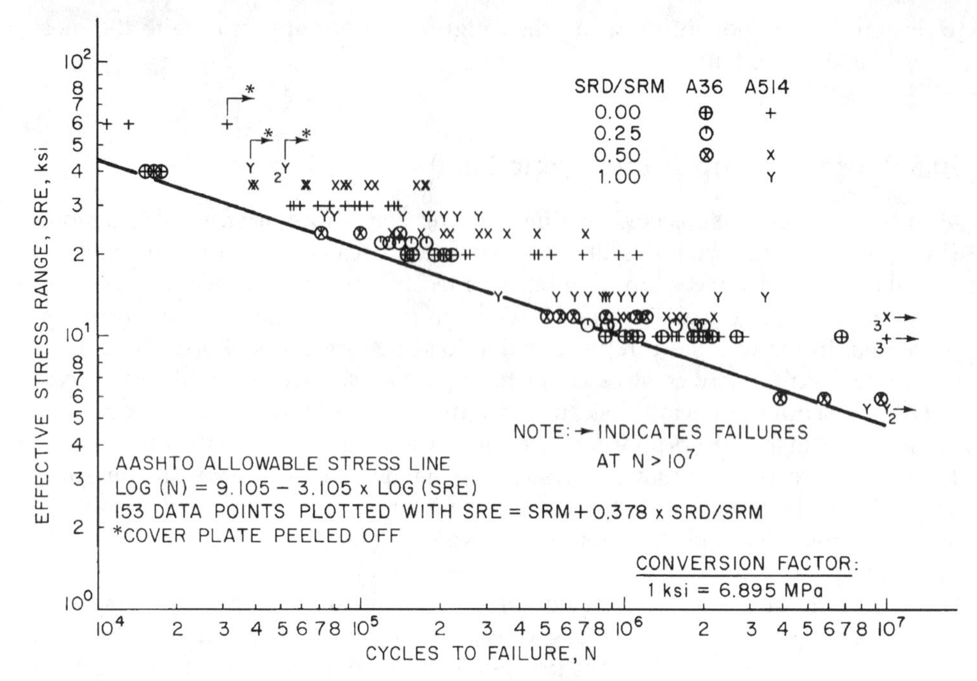

FIG. 10.24 Comparison of cover-plate beam results with AASHTO allowable stress for category E. (→ indicates failure at $N > 10^7$.)

different steels, minimum stress levels, and welding sequences are included in these plots. The effects of various secondary parameters, such as minimum stress and type of steel, on the fatigue life under variable-amplitude cyclic loading were similar to the effects of these parameters on the fatigue life under constant-amplitude cyclic loading.

Preliminary data suggest that the fatigue limit for over 2×10^6 cycles under constant-amplitude loading may not exist under variable amplitude loading if a few (>10%) of the stress ranges exceed the stress range corresponding to the fatigue limit under constant-amplitude loading. The data also suggest that, under these conditions, the fatigue behavior for a given structural component can be analyzed by extrapolating the fatigue-design curve for that category to lower stress ranges.

10.8.1 Example Problem

The following is a simple example that illustrates the procedure for predicting the fatigue life of a fabricated component subjected to variable-amplitude loading. The calculations are based on the assumption that the fatigue limits for constant-amplitude loading, Figure 10.23, do not exist under variable-amplitude loading and that the long-life behavior is represented by an extrapolation of the finite-life fatigue-design curves.

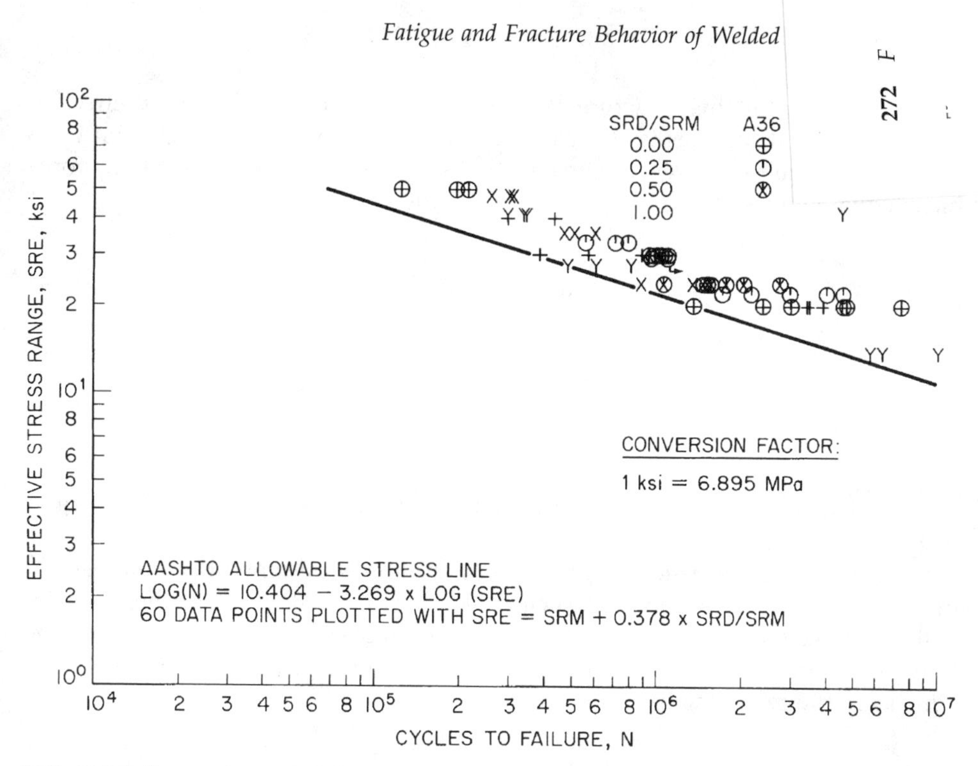

FIG. 10.25 Comparison of welded-beam results with AASHTO allowable stress for Category B.

The fatigue life of a structure is usually governed by the localized behavior of the structural detail, which in combination with the applied load fluctuations results in the shortest fatigue life. Thus, a detail with a severe geometrical discontinuity subjected to stress fluctuations of small magnitude may exhibit a longer fatigue life than a detail with a moderate geometrical discontinuity subjected to large stress fluctuations.

Consider a welded structure expected to be subjected each year to the cycles, N_i, and their corresponding stress ranges, $\Delta\sigma_i$, that are tabulated in Table 10.1.

TABLE 10.1 Example Problem: Data and Calculations for Fatigue-Life Determination.

i	NUMBER OF CYCLES PER YEAR, N_i	STRESS RANGE, $\Delta\sigma_i$ (ksi)	$\alpha_i = \frac{N_i}{\Sigma_i N_i}$	$\alpha_i(\Delta\sigma_i)^2$
1	4,500,000	0.1	0.792	0.008
2	800,000	0.6	0.141	0.051
3	140,000	5.6	0.025	0.784
4	232,000	7.8	0.041	2.490
5	3,900	10.2	0.0007	0.073
6	113	14.0	0.00002	0.004
7	2,300	15.0	0.0004	0.090

NOTE: $\Sigma_i N_i = 5{,}678{,}313$.
$\Sigma_i\, \alpha_i(\Delta\sigma_i)^2 = 3.50$.
$\Delta\sigma_{rms} = \sqrt{\Sigma_i\, \alpha_i\, (\Delta\sigma_i)^2} = 1.87$ ksi.

ne fatigue life for this structure can be determined by calculating an effective stress range that represents the various loadings shown in Table 10.1 and the frequency of their occurrence. The effective stress range represented by the root-mean-square stress range, $\Delta\sigma_{rms}$, is given by the equation

$$\Delta\sigma_{rms} = \sqrt[2]{\sum_{i} \alpha_i(\Delta\sigma_i)^2} \quad (10.2)$$

where

$$\alpha_i = \frac{N_i}{\sum_{i} N_i} \quad (10.3)$$

Table 10.1 presents the details for calculating $\Delta\sigma_{rms}$, which is shown to be equal to 1.87 ksi.

Assume that the structure under consideration is designed to contain structural details that correspond to Category E′, Figure 10.23, and that these details will be subjected to the loadings presented in Table 10.1. The fatigue lives for Category E′ details are represented by the E′ fatigue-design curve, Figure 10.23, which has the relationship

$$\log N_f = 8.59 - 3 \log \Delta\sigma \quad (10.4)$$

where N_f is the total fatigue life in cycles of loading. Substituting $\Delta\sigma = \Delta\sigma_{rms} = 1.87$ ksi, one obtains $N_f = \sim 60{,}000{,}000$ cycles. Thus, the fatigue life in years is given by

$$\frac{N_f}{\sum_{i} N_i} \cong 10.5 \text{ years}$$

If this life is too short, the structure should be redesigned to eliminate the category E′ details in this section of the structure, or to decrease the stress ranges, or both.

10.9 Fracture-Toughness Behavior of Welded Components

10.9.1 General Discussion

The fracture toughness and other mechanical properties of steels and weldments depend on several metallurgical factors, including composition, microstructure, and cleanliness. The effects of these factors are complicated by synergisms and, especially for weldments, by a heterogeneous microstructure. A discussion of these factors and their effects is available elsewhere [*11,28*] and is beyond the scope of this text. However, the following general statements may help to clarify some of the observations related to fracture toughness of weldments.

Chemical elements are added to steel products to obtain certain desired properties such as higher strength, hardenability, toughness, and corrosion resistance. Similarly, chemical elements may be added to filler metals and fluxes to obtain the desired weld metal properties. The effect of a given element on fracture toughness can depend on many factors, including the amount of the addition, the interaction of the element with other elements, and the thermomechanical processing of steel.

The microstructure of a given steel has a significant effect on its fracture-toughness behavior. The microstructural constituents present in structural steels can be ferrite, pearlite, bainite, or martensite. The prominence of each of these constituents depends on the steel composition, processing, and heat treatment. Steels with low hardenability, such as A36, that are subjected to relatively slow cooling rates have ferrite-pearlite microstructures. Quenched and tempered steels, such as A514, have a tempered bainite and tempered martensite microstructure.

Fine-grain microstructures for any of these constituents improve the fracture toughness of the steel. The size of the prior austenite or ferrite grains has been recognized as one of the most important factors that controls the fracture toughness of steels. In general, strength decreases and impact fracture-toughness transition temperature increases as the microstructure changes from tempered martensite, to tempered bainite, to ferrite-pearlite.

The elevated-temperature (≥1650°F) microstructure of all structural steels is austenite. Rapid cooling rates may transform this microstructure for hardenable steels to as-quenched (untempered) martensite. Untempered martensitic structures contain trapped carbon atoms resulting in high hardness and low fracture toughness. Subsequent heating (tempering) to a temperature where the trapped carbon atoms have mobility defuses the carbon atoms to form carbides. The resulting tempered martensitic structure has lower strength and higher fracture toughness than the untempered martensitic structure. However, tempering followed by slow cooling may embrittle the steel and reduce its fracture toughness, thus offsetting some of the benefits gained by tempering. This temper embrittlement is one of several embrittling mechanisms that can degrade the fracture toughness of steels and weldments [*11,28,29*].

10.9.2 Weldments

Many welding processes are available that can produce satisfactory joints in steels. The selection of a particular process is based on many factors that include the thickness and size of the parts to be joined, the position of the weldment, the desired properties and appearance of the finished weldment, the particular application, and the cost of fabrication as well as other factors. No single process can be used to produce satisfactory weldments for all steels, thicknesses, and positions. The most suitable process is the one that produces the desired properties in the final product at the lowest possible cost.

In arc welding, which is the most widely used welding process for structural steels, filler metal is melted and used to fill a weld groove. The arc-welding process, welding procedure, and joint geometry influence weld penetration and admixture of the filler metal with the base metal. Because of this admixture, the chemical composition of the base metal can have significant influence on the microstructural and mechanical properties of the weld metal. This influence is significant, especially for electroslag and electrogas welds, because they are high-heat-input single-weld-pass processes. The final properties of the weld metal depend on many factors, especially the composition of the weld metal and the conditions governing its solidification and subsequent cooling. Because the heat flow in the weld metal is highly directional toward the adjacent cooler metal, the weld metal develops distinctly columnar grains. Furthermore, the rapid cooling of the weld metal may not allow sufficient time for diffusion of the chemical constituents, resulting in microstructural heterogeneities. This segregation and the directional solidification of the weld metal may result in weld-metal properties having pronounced directionality.

For the arc welding processes, the maximum temperature of the weld metal is above the melting temperature of the base metal joined. This temperature decreases as the distance from the weld increases. Thus, partial melting of the base metal occurs at the weld-metal–base-metal interface, and microstructural changes occur in the base metal in the immediate vicinity of the weld, forming a heat-affected zone. The size of this zone is determined by the rate of heating, the volume and temperature of the weld metal, and the rate of cooling of the weld metal and surrounding base metal. These factors as well as the composition and microstructure of the base metal determine the grain size, the grain-size gradient, the microstructure, and therefore the fracture toughness of the heat-affected zone. Because of the high temperatures and the large variations in temperature gradient and cooling rate, adjacent regions in the heat-affected zone can exhibit large differences in microstructure and properties. In general, for carbon and low-alloy steels, the closer the distance to the weld, the coarser the microstructure. Coarse-grain regions adjacent to the weld interface generally exhibit the poorest toughness.

Grotke [*30*] divided the heat-affected zone associated with arc welds in plain-carbon and low-alloy steels into five general regions:

1. A partially spherodized region adjacent to the unaffected base metal where the steel underwent a modest alteration in microstructure.
2. A transition region that includes all the microstructures that have undergone partial reaustenitization.
3. A grain-refined region where the steel was completely transformed to austenite but at too low of a temperature and for too short of a time to permit significant grain growth.
4. A grain-coarsened region, resulting from exposure to extremely high austenitizing temperatures.

5. A partially melted region in which incomplete liquidation has occurred, located between the unmelted grain-coarsened region and the entirely fused weld metal.

Illustrations of intermediate heat-affected-zone microstructure associated with bead-on-plate deposits on hot-rolled plain-carbon steels and on a low-alloy quenched-and-tempered steel are shown in Figures 10.26 and 10.27, respectively. Both welds were made on plates of the same thickness, using identical heat-input conditions; but, the nominal carbon contents were slightly different. They were 0.20% for the carbon steel and 0.16% for the low-alloy quenched-and-tempered steel.

The heat-affected zone in a single-pass weld forms under the influence of a single thermal cycle. The temperature and temperature distribution from the weld metal into the base metal in a direction perpendicular to the weld is essentially identical at different locations along the weld groove of a simple butt joint for two constant-thickness plates. Consequently, the various microstructural regions in the heat-affected zone can be continuous. However, the weld metal in a multipass weld is built up by the deposition of successive weld beads. The structure and properties of deposited weld beads and existing heat-affected zones are usually altered by the heating effects of subsequent weld-bead deposits. The heat from subsequent weld passes may refine the grain size of the deposited weld metal and existing heat-affected zone, may change the columnar structure of the weld metal to an equiaxed structure, and may temper the microstucture of the existing heat-affected zone. In multipass welds, unlike single-pass welds, the heat-affected zone regions that exhibit relatively low toughness occur intermittently adjacent to the weld interface. In either case, these lower-toughness regions are surrounded by heat-affected zone regions of higher toughness.

10.9.3 Fracture-Toughness Tests for Weldments

Weldability is a complex property affected by many interrelated factors. Consequently, many fracture-toughness tests have been developed to determine the effects of these factors [*11*]. Some of these tests are related to the fabrication qualities for the weldment, while others are related to the service performance. The tests for fabrication "must be designed to measure the susceptibility of the weld-metal–base-metal system to such conditions as cracks, porosity or inclusions under realistic and properly controlled conditions of welding [*31*]. A discussion of these tests is beyond the scope of this text. The service performance tests include yield and tensile strengths, ductility, fracture toughness, stress rupture, stress-corrosion cracking, fatigue, and corrosion fatigue. Some of the fracture-toughness characteristics for weldments are discussed in this section.

A large number of tests have been developed or adapted to determine the fracture toughness for weldments [*11*]. These tests include the Charpy V-notch test, various bend tests, the drop weight test, the explosion bulge test, the drop weight tear test, the wide-plate test, and the fracture-mechanics-type tests. Each

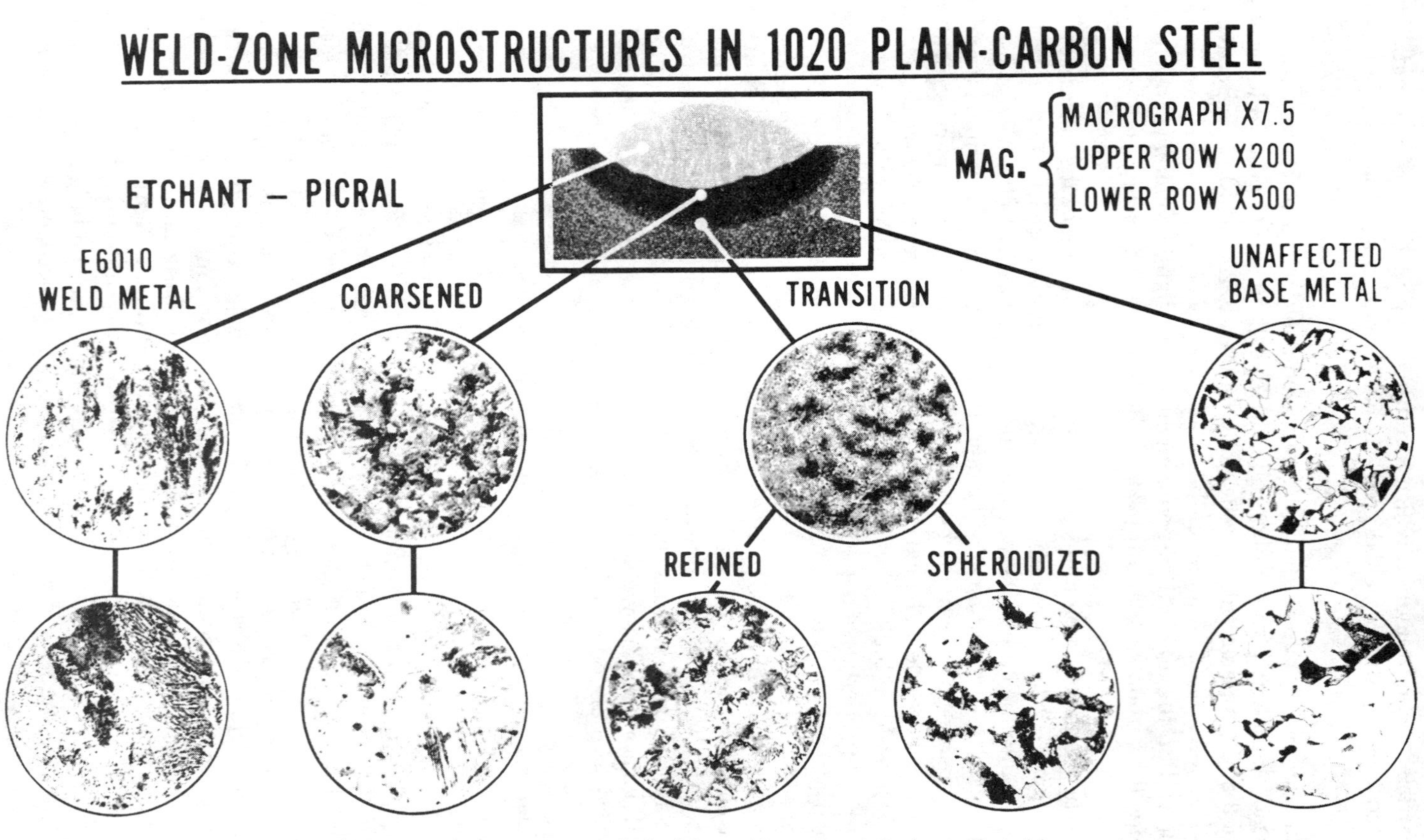

Figure 14.16 Microstructures typical of the weld metal and the heat-affected base metal in a mild-steel weld.

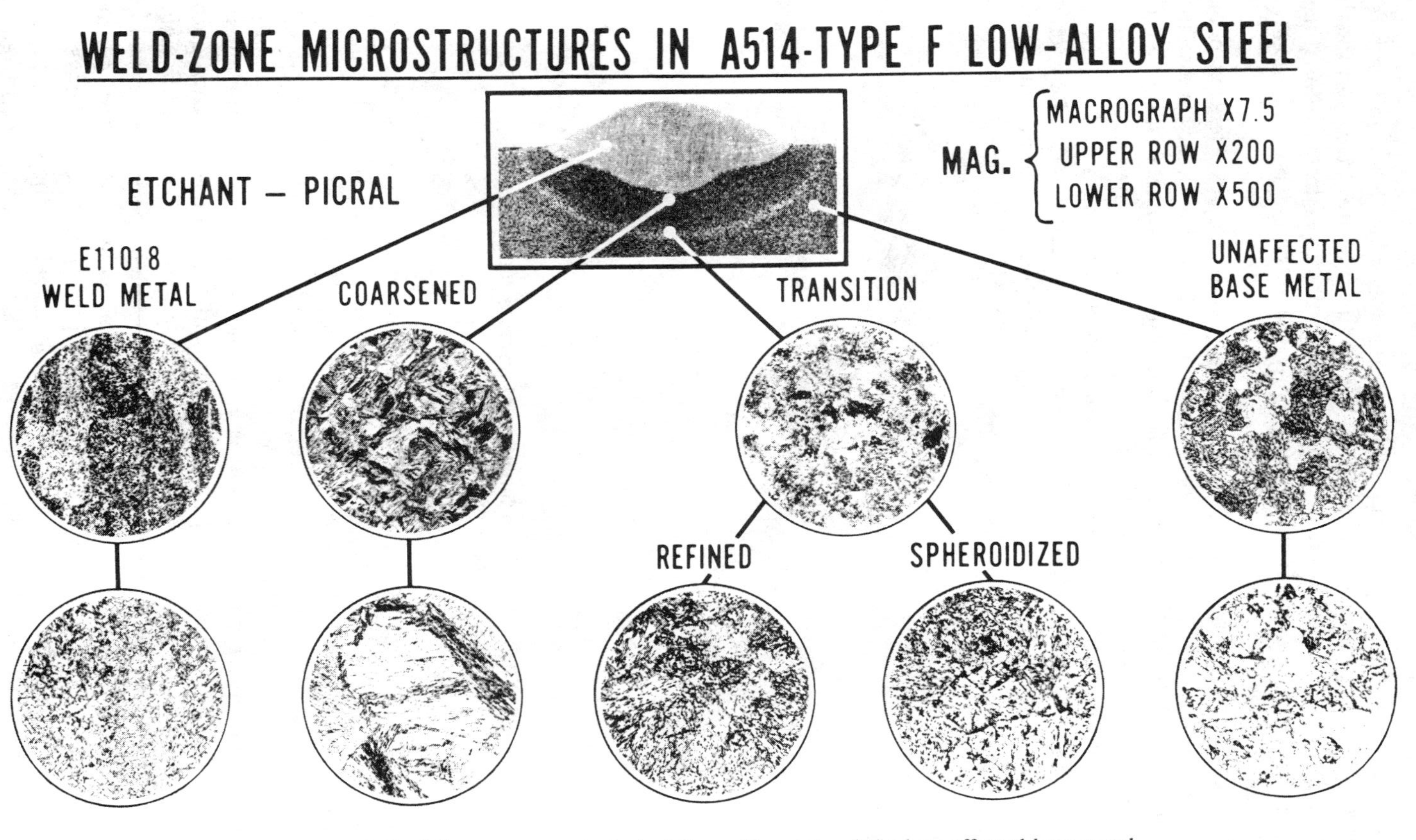

Figure 14.17 Microstructures typical of the weld metal and the heat-affected base metal in a quenched-and-tempered low-alloy steel weld.

of these tests have advantages and disadvantages over other tests, and each test measures some parameter assumed to represent the fracture toughness for the weldment or some zone within it. Although the subject of fracture-toughness testing of weldments has been studied for many years, there is no general agreement on the best test method to use.

Most fracture-toughness tests have a notch or a fatigue crack. Heterogeneous material properties along the front of the notch or the fatigue crack may cause significant variability in the test results. The magnitude of this variability would depend on many factors, including the rate of change in material properties along and in the vicinity of the notch or crack front, the length and volume of the regions with different properties, the location of the regions along the crack front with respect to each other, and the properties of the surrounding materials.

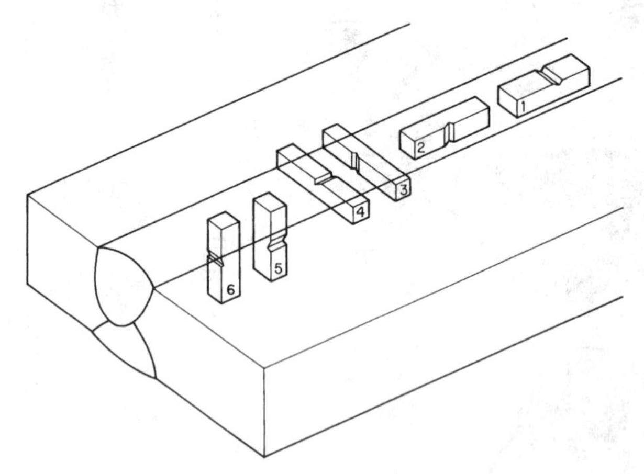

FIG. 10.28 Possible orientation of fracture-toughness specimens in a butt-welded plate.

Weldments exhibit anisotropic heterogeneous material properties. The magnitude and rate of variation in properties can be very large. The volume of a material with essentially uniform properties, especially in the heat-affected zone, can be very small. The low-toughness regions in the heat-affected zone of multipass welds can occur intermittently along the weld interface. These and other factors make the placement of a notch or a fatigue-crack tip in a given microstructural region having uniform properties very difficult. Consequently, significant variability usually is observed in fracture-toughness test results for weldments. Such a variability is observed for Charpy V-notch specimens as well as for K_{Ic} or crack-tip-opening-displacement (CTOD) specimens. Furthermore, the properties for a given region may or may not reflect the properties of the welded joint or its performance in an actual structure. Nevertheless, the criteria for acceptance generally are developed in a manner similar to that for base metal. In this sense, fracture-mechanics concepts have been very helpful.

Fracture-toughness test results obtained by testing Charpy V-notch specimens or fracture-mechanics-type specimens (K_{Ic} or CTOD) are influenced by the orientation of the specimens in a butt-welded plate. Figure 10.28 shows the various orientations for fracture-toughness (CVN or fracture-mechanics type) specimens in a butt-welded plate. Orientation 3 is the most commonly used for Charpy V-notch tests of the weld-metal and heat-affected zone.

Despite the problems in determining the fracture toughness for welded joints, the importance of this property for structural performance makes it necessary to test and attempt to characterize the notch toughness of welded joints. Many codes and standards dictate the specimens and test procedures and establish minimum fracture-toughness values for acceptability. Most of these codes and standards specify the use of the Charpy V-notch impact specimen.

10.10 References

[1] Maddox, S. J., *Fatigue Strength of Welded Structures,* 2nd ed., Abington Publishing, The Welding Institute, Cambridge, England, 1991.

[2] Fisher, J. W., *Fatigue and Fracture in Steel Bridges—Case Studies,* John Wiley & Sons, New York, 1984.

[3] Anonymous, *Fatigue Fracture in Welded Constructions,* Vol. 2, Abington, International Institute of Welding, Paris, 1979.

[4] ASME Section XI Task Group on Fatigue in Operating Plants, "Metal Fatigue in Operating Nuclear Power Plants," WRC Bulletin 376, November 1992.

[5] Bush, S. H., "Failure Mechanisms in Nuclear Power Plant Piping Systems," *Journal of Pressure Vessel Technology,* Vol. 114, November 1992.

[6] Poole, A. B., Battiste, R. L., and Clinard, J. A., "Final Report: Pipe Break Testing of Primary Loop Piping Similar to Department of Energy's New Production Reactor—Heavy Water Reactor," Materials/System Integrity Task of the HWR-NPR Program, Oak Ridge National Laboratory, Martin Marietta, ORNL/NPR-92/64, January 1993.

[7] Barsom, J. M., Review of in-house files, United States Steel, Pittsburgh, PA, 1994.

[8] Vecchio, R. S., Review of in-house files, Lucius Pitkin, Inc., New York, NY, 1994.

[9] ASME Boiler and Pressure Vessel Code, Section III, Division 1, "Rules for Construction of Nuclear Power Plant Components," American Society of Mechancial Engineers, New York NY, 1992.

[10] *Welding Inspection,* American Welding Society, Miami, 1980.

[11] Stout, R. D. and Doty, W. D., *Weldability of Steels,* Welding Research Council, New York, 1978.

[12] Campbell, H. C., *Certification Manual for Welding Inspectors,* American Welding Society, Miami, 1977.

[13] Signes, E. G., et al., "Factors Affecting the Fatigue Strength of Welded High Strength Steels," *British Welding Journal,* Vol. 14, No. 3, 1967.

[14] Fisher, J. W., et al., "Effect of Weldments on the Fatigue Strength of Steel Beams," *NCHRP Report 102,* Transportation Research Board, Washington, DC, 1970.

[15] Watkinson, F., et al., "The Fatigue Strength of Welded Joints in High Strength Steels and Methods for Its Improvement," *Proceedings: Conference on Fatigue of Welded Structures,* The Welding Institute, Brighton, England, July 1970.

[16] Fisher, J. W., et al., "Fatigue Strength of Steel Beams with Transverse Stiffness and Attachments," *NCHRP Report 147,* Transportation Research Board, Washington, DC, 1974.

[17] Roberts, R., Barsom, J. M., Fisher, J. W., and Rolfe, S. T., *Fracture Mechanics for Bridge Design and Student Workbook—Fracture Mechanics for Bridge Design,* FHWA-RD-78-69, Federal Highway Administration, Office of Research and Development, Washington, DC, July 1977.

[18] Bannantine, J. A., Comer, J. J., and Handrock, J. L., *Fundamentals of Metal Fatigue Analysis,* Prentice-Hall, Englewood Cliffs, NJ, 1990.

[19] Stout, R. D., "Postweld Heat Treatment of Pressure Vessels," WRC Bulletin 302, Welding Research Council, New York, February 1985.

[20] Spaeder, C. E. and Doty, W. D., "ASME Postweld Heat Treating Practices: An Interpretive Report," *WRC Bulletin 407,* December 1995.

[21] American Welding Society, *Specifications for Welded Highway and Railway Bridges,* New York, 1941.

[22] Fisher, J. W. and Viest, I. M., "Fatigue Life of Bridge Beams Subjected to Controlled Truck Traffic," Preliminary Publication 7th Congress, IABSE, Zurich, Switzerland, 1964.

[23] American Association of State Highway and Transportation Officials, *Standard Specifications for Highway Bridges,* Washington, DC, 1977.

[24] Fisher, J. W., *Bridge Fatigue Guide—Design and Details,* American Institute of Steel Construction, Chicago, 1977.

[25] Schilling, C. G., Klippstein, K. H., Barsom, J. M., and Blake, G. T., "Fatigue of Welded Steel Bridge Members Under Variable-Amplitude Loadings," *NCHRP Report 188,* Transportation Research Board, Washington, DC, 1978.

[26] Bower, D. G., "Loading History—Span No. 10, Yellow Mill Pond Bridge I-95, Bridgeport, Conn.," Highway Research Record 428, Transporation Research Board, Washington, DC, 1973.

[27] Slockbower, R. E. and Fisher, J. W., "Fatigue Resistance of Full Scale Cover-Plated Beams," *Fritz Engineering Laboratory Report No. 386-9(78),* Lehigh University, Bethlehem, PA, June 1978.

[28] Phillips, A. L., Ed., *Fundamentals of Welding, Welding Handbook,* Section One, American Welding Society, Miami, 1968.

[29] Davidson, J. A., Konkol, P. J., and Sovak, J. F., "Assessing Fracture Toughness and Cracking Susceptibility of Steel Weldments—A Review," Final Report Number FHWA-RD-83, Federal Highway Administration, Washington, DC, December 1983.

[30] Grotke, E., "Indirect Tests for Weldability," in *Weldability of Steels,* by R. D. Stout and W. D. Doty, Welding Research Council, New York, 1978.

[31] Stout, R. D., "Direct Weldability Tests for Fabrication," in *Weldability of Steels,* Welding Research Council, New York, 1978, p. 252.

11

K_{Iscc} and Corrosion Fatigue Crack Initiation and Crack Propagation

11.1 Introduction

FAILURE OF structural components subjected to an aggressive environment may occur under static or fluctuating loads. Failure under static loads is caused by general corrosion or by stress-corrosion cracking. Failure under fluctuating loads in an aggressive environment is caused by the initiation and the propagation of corrosion fatigue cracks. The following sections present the use of fracture mechanics concepts to study stress corrosion cracking, corrosion fatigue crack initiation, and corrosion fatigue crack propagation.

11.2 Stress-Corrosion Cracking

Stress-corrosion cracking has long been recognized as an important failure mechanism. Although many tests have been developed to study this mode of failure (Figure 11.1) [*1*], the underlying mechanisms for stress-corrosion cracking are yet to be resolved, and quantitative design procedures against its occurrence are yet to be established. These difficulties are caused by the complex chemical, mechanical, and metallurgical interactions; the many variables that are known to affect the behavior; the extensive data scatter; and the relatively poor correlation between laboratory test results and service experience.

The traditional approach to study the stress-corrosion susceptibility of a material in a given environment is based on the time required to cause failure of smooth or mildly notched specimens subjected to different stress levels. This time-to-failure approach, like the traditional *S-N* approach to fatigue, combines the time required to initiate a crack and the time required to propagate the crack to critical dimensions. The need to separate stress-corrosion cracking into initiation and propagation stages was emphasized by experimental results for tita-

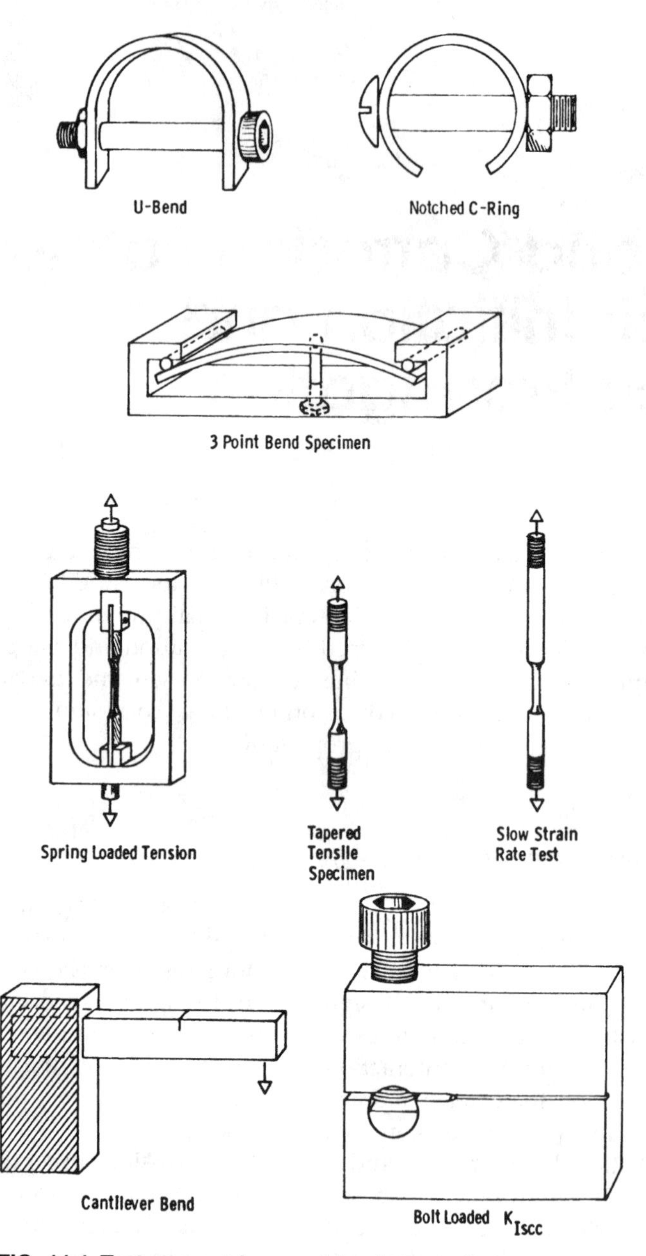

FIG. 11.1 Test geometries used to characterize environmentally assisted cracking behavior.

nium alloys [2]. These results showed that some materials that appear to be immune to stress corrosion in the traditional smooth-specimen tests may be highly susceptible to stress-corrosion cracking when tested under the same condition using precracked specimens. The behavior of such materials was attributed to their immunity to pitting (crack initiation) and to their high intrinsic susceptibility to stress-corrosion cracking (crack propagation). The following discussion presents the use of fracture-mechanics concepts to study the stress-corrosion cracking of environment–material systems by using precracked specimens.

11.2.1 Fracture-Mechanics Approach

The application of linear-elastic fracture-mechanics concepts to study stress-corrosion cracking has been very successful. Because environmentally enhanced crack growth and stress-corrosion attack would be expected to occur in the highly stressed region at the crack tip, it is logical to use the stress-intensity factor, K_I, to characterize the mechanical component of the driving force in stress-corrosion cracking. Sufficient data have been published to support this observation [2–7].

The use of the stress-intensity factor, K_I, to study stress-corrosion cracking is based on assumptions and is subject to limitations similar to those encountered in the study of fracture toughness. The primary assumption that must be satisfied when K_I is used to study the stress-corrosion-cracking behavior of materials is the existence of a plane-strain state of stress at the crack tip. This assumption requires small plastic deformation at the crack tip relative to the geometry of the test specimen and leads to size limitations on the geometry of the test specimen. Because these limitations must be established experimentally, the limitations established for plane-strain fracture-toughness, K_{Ic}, tests (Chapter 3) are usually applied to stress-corrosion-cracking tests.

ASTM has developed a new standard, E-1681, Standard Test Method for Determining a Threshold Stress Intensity Factor for Environment-Assisted Cracking of Metallic Materials Under Constant Load, Volume 03.01. In that specification, the more general subscript, K_{EAC} (Environment-Assisted Cracking) is used instead of K_{Iscc}.

Various investigators [*1–14*] have used fracture-mechanics concepts to study the effects of the environment on precracked specimens. However, the fracture-mechanics approach to environmental testing did not become widely used until Brown [*2,15,16*] introduced the K_{Iscc} threshold concept by using pre-cracked cantilever-beam specimens. The K_{Iscc} value at a given temperature for a particular material–environment system represents the stress-intensity-factor value below which subcritical crack extension does not occur under a static load in the environment. Since that time, the cantilever-beam test specimens have been used widely to study the stress-corrosion-cracking characteristics of steels [*4,5,10,11,13,15,17*], titanium alloys [*14,18–20*], and aluminum alloys [*8,21,22*].

11.2.2 Experimental Procedures

Experimental procedures for stress-corrosion-cracking tests of precracked specimens may be divided into two general categories. They are time-to-failure tests and crack-growth-rate tests. The time-to-failure tests are similar to the conventional stress-corrosion tests for smooth or notched specimens [*23–28*]. This type of test using precracked specimens has been used widely since the early work of Brown and Beachem [2]. The crack-growth-rate tests are more complex and require more sophisticated instrumentation than the time-to-failure tests. However, data obtained by using crack-growth-rate tests should provide information necessary to enhance the understanding of the kinetics of stress-corrosion cracking and to verify the threshold behavior K_{Iscc}.

Various precracked specimens and methods of loading can be used to study the stress-corrosion-cracking behavior of materials in both time-to-failure and crack-growth-rate tests. The cantilever-beam specimens under constant load [*4,13,27,28*], Figure 11.2, and the wedge-opening-loading (WOL) specimen under constant displacement conditions (modified WOL specimen) that was developed by Novak and Rolfe [*9*], Figure 11.3, are described as follows.

The cantilever-beam specimen is usually face notched 5 to 10% of the thickness, and the notch is extended by fatigue-cracking the specimens at low stress-intensity-factor levels. Then the specimens are tested in a stand similar to that shown in Figure 11.2. Usually, two specimens are monotonically loaded to failure in air to establish the critical stress-intensity factor for failure in the absence of environmental effects (K_{Ic} if ASTM requirements [*29*] are satisfied and K_{Ix} if they are not satisfied). Subsequently, specimens are immersed in the environment and dead-weight loaded to various lower initial stress-intensity-factor, K_{Ii}, levels that are lower than K_{Ix}. If the material is susceptible to the test environment, the fatigue crack will propagate. As the crack length increases under constant load,

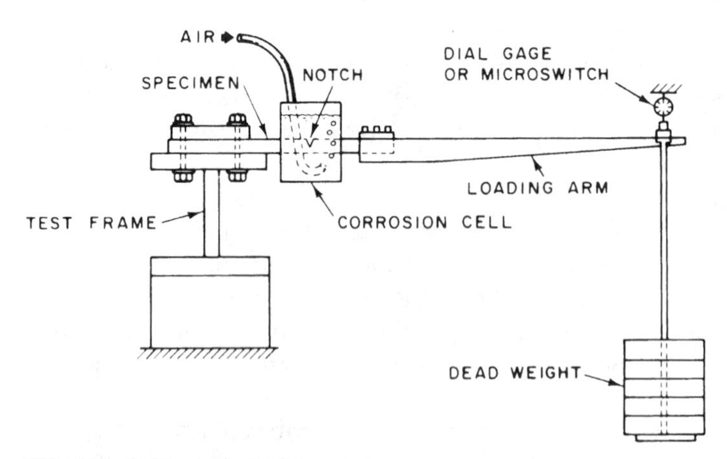

FIG. 11.2 Schematic drawing of fatigue-cracked cantilever-beam test specimen and fixtures.

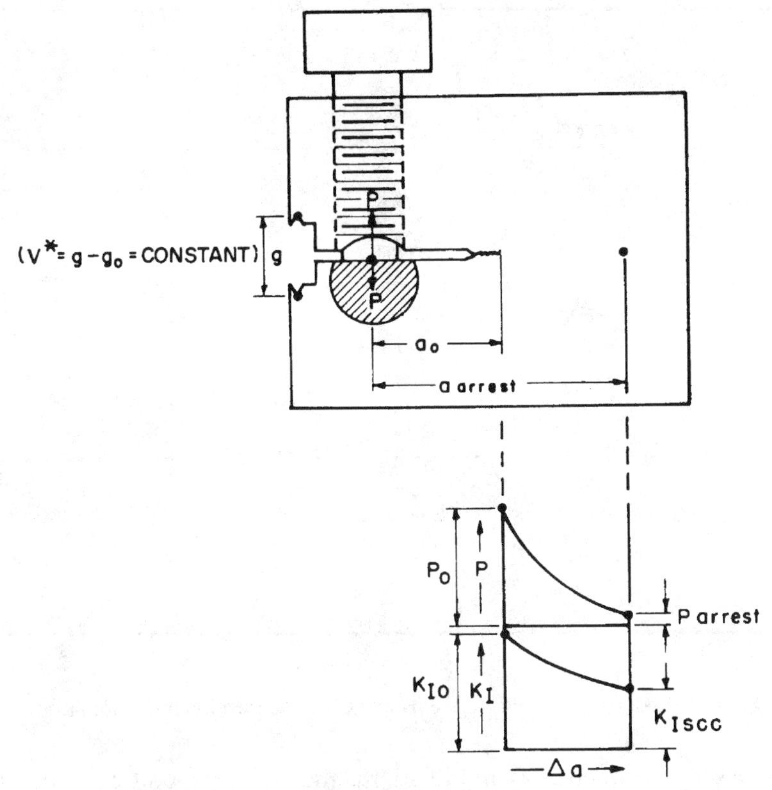

FIG. 11.3 Schematic showing basic principle of modified WOL specimen.

the stress-intensity factor at the crack tip increases to the K_{Ic} (or K_{Ix}) level, and the specimen fractures. The lower the value of the initial K_I, the longer the time to failure. Specimens that do not fail after a long period of test time, usually 1000 h for steels, should be fractured and inspected for possible crack extension. The highest plane-strain K_{Ii} level at which crack extension does not occur after a long test time corresponds to the stress-corrosion-cracking threshold, K_{Iscc}.

Figure 11.4 is a schematic representation of test results obtained by using cantilever-beam test specimens. Approximately ten precracked cantilever specimens are needed to establish K_{Iscc} for a particular material and environment.

The wedge-opening-loading (WOL) specimens [30–32] have been used to study fracture toughness [33], fatigue-crack initiation [34] and propagation [27], stress-corrosion cracking [9], and corrosion-fatigue-crack-growth [27–35] behavior of various materials. The geometry of 1-in.-thick (1-T) WOL specimens is shown in Figure 11.5.

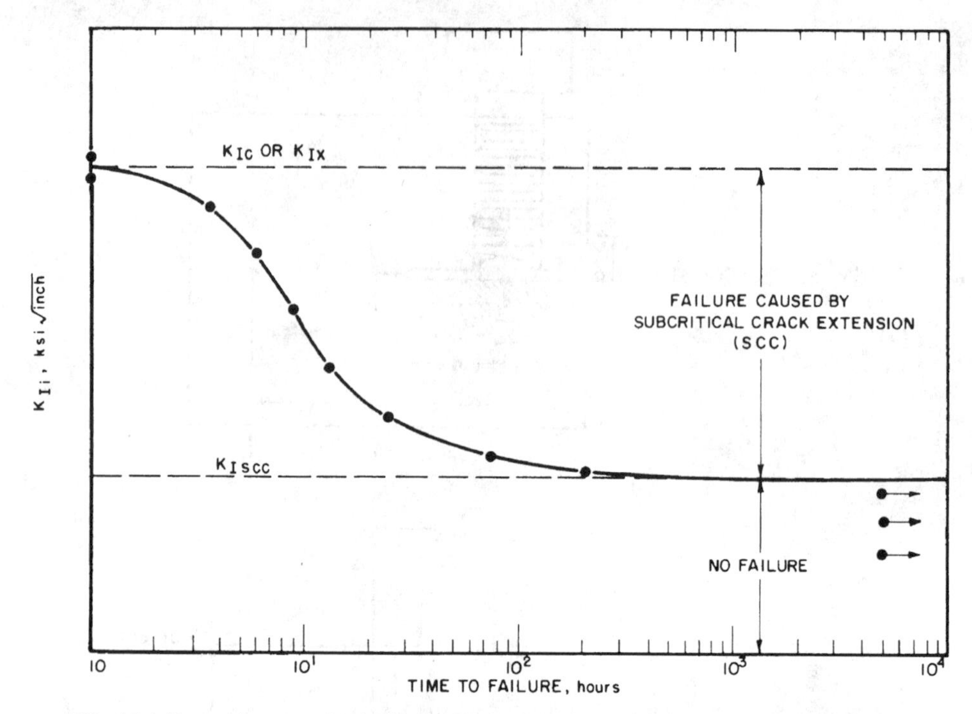

FIG. 11.4 Procedure to obtain K_{Iscc} with precracked cantilever-beam specimens.

The WOL specimen was modified by the use of a bolt and loading tup (Figure 11.3) so that it can be self-stressed without using a tensile machine [9]. The crack opening is fixed by the bolt, and the loading is by constant displacement rather than by constant load as in the cantilever-beam specimen. Because a constant crack-opening displacement is maintained throughout the test, the force, P, decreases as the crack length increases (Figure 11.3). In cantilever-beam testing, the K_I value increases as the crack length increases under constant load, which leads to fracture for each specimen. In contrast, for the modified WOL specimen, the K_I value decreases as the crack length increases under a decreasing load. The decrease in load more than compensates for the increase in crack length and leads to crack arrest at K_{Iscc}. A comparison of these two types of behavior is shown schematically in Figure 11.6. Thus, only a single WOL specimen is required to establish the K_{Iscc} level because K_I approaches K_{Iscc} in the limit. However, duplicate specimens are usually tested to demonstrate reproducibility. Because the bolt-loaded WOL specimen is self-stressed and portable, it can be used to study the stress-corrosion-cracking behavior of materials under actual operating conditions in field environments.

11.2.3 K_{Iscc}—A Material Property

Brown and Beachem [36] investigated the K_{Iscc} for environment–material systems by using various specimen geometries. Their results show that identical

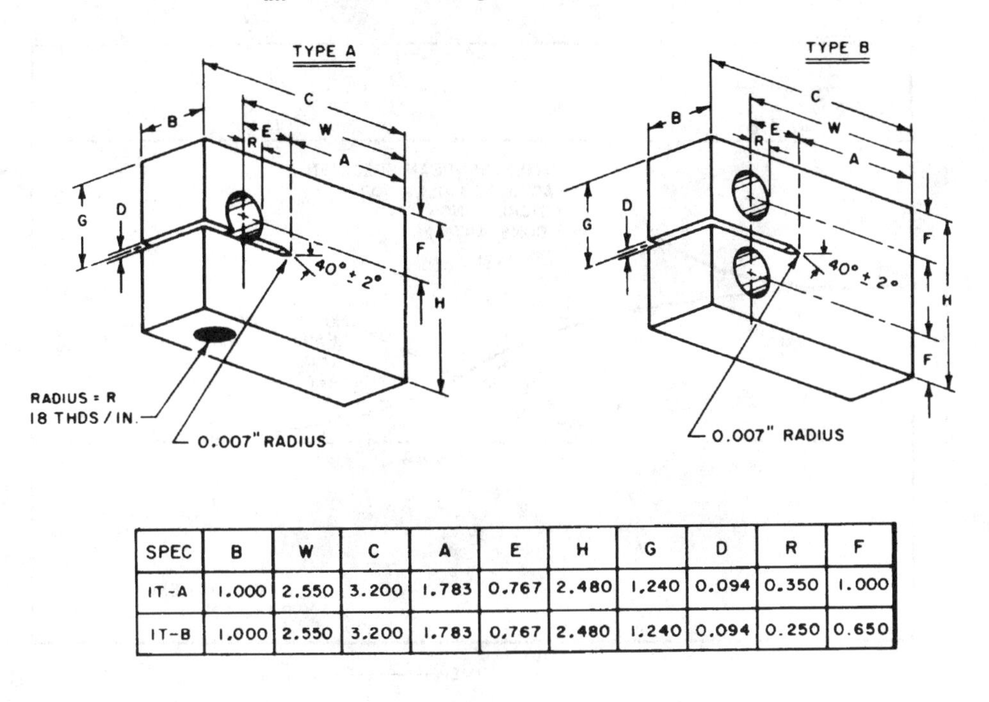

SPEC	B	W	C	A	E	H	G	D	R	F
IT-A	1.000	2.550	3.200	1.783	0.767	2.480	1.240	0.094	0.350	1.000
IT-B	1.000	2.550	3.200	1.783	0.767	2.480	1.240	0.094	0.250	0.650

1 inch = 25.4 mm
1 degree = 0.017 rad

FIG. 11.5 Two types of 1-T WOL specimens.

K_{Iscc} values were obtained for a given environment–material system by using center-cracked specimens, surface-cracked specimens, and cantilever-beam specimens, Figure 11.7. Smith et al. [*18*] measured K_{Iscc} values of 10-25 ksi $\sqrt{\text{in.}}$ and 10-22 ksi $\sqrt{\text{in.}}$ by testing center-crack specimens with end loading and wedge-force loading, respectively, for specimens of Ti-8Al-1Mo-1V alloy in 3.5% solution of sodium chloride.

K_{Iscc} tests using cantilever-beam specimens and bolt-loaded WOL specimens resulted in identical K_{Iscc} values for each of two 12Ni-5Cr-3Mo maraging steels tested in synthetic seawater [*9*]. Further test results showed that K_{Iscc} for a specific environment–material system was independent of specimen size above a prescribed minimum geometry limit [*4*]. On the other hand, the nominal stress corresponding to K_{Iscc}, σ_{Nscc}, was highly dependent on specimen geometry (Figure 11.8), particularly specimen in-plane dimensions such as the height of a cantilever-beam specimen, *W*, and the crack length, *a*. The preceding results indicate that K_{Iscc} for a specific environment–material system is a property of the particular system.

Corrosion products can change the magnitude of the crack-tip driving force by wedging the crack surfaces open. This would be especially applicable for

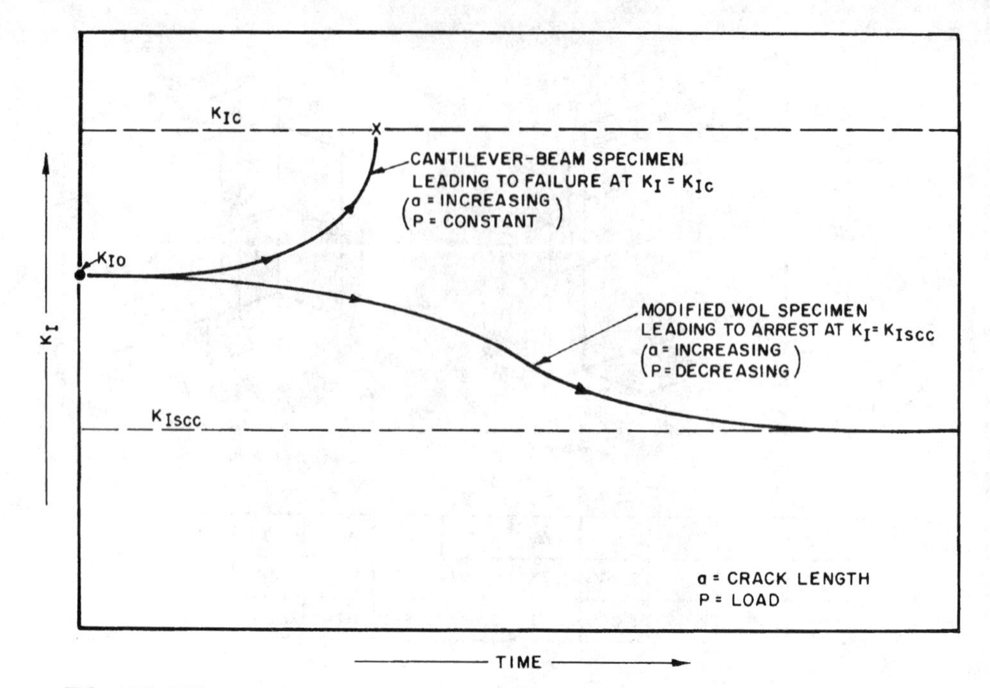

FIG. 11.6 Difference in behavior of modified WOL and cantilever-beam specimens.

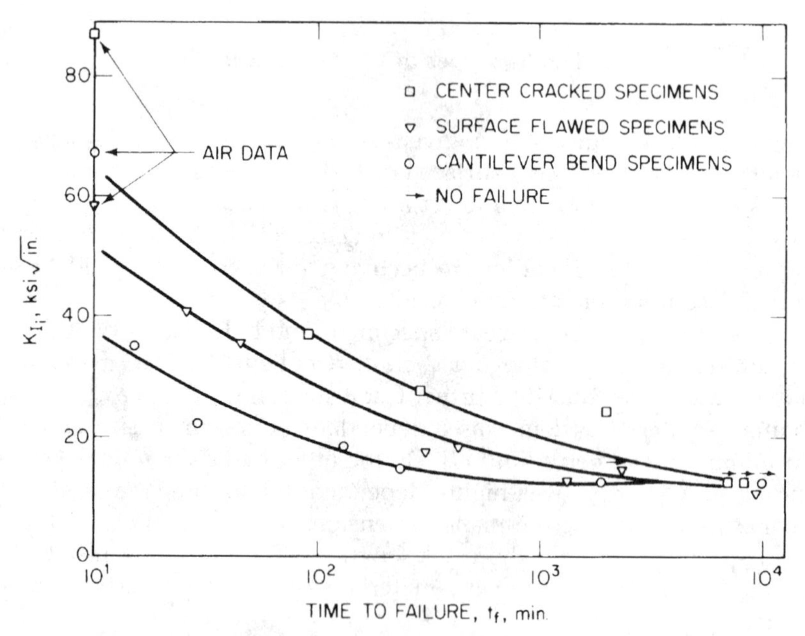

FIG. 11.7 Influence of specimen geometry on the time to failure for an AISI 4340 steel.

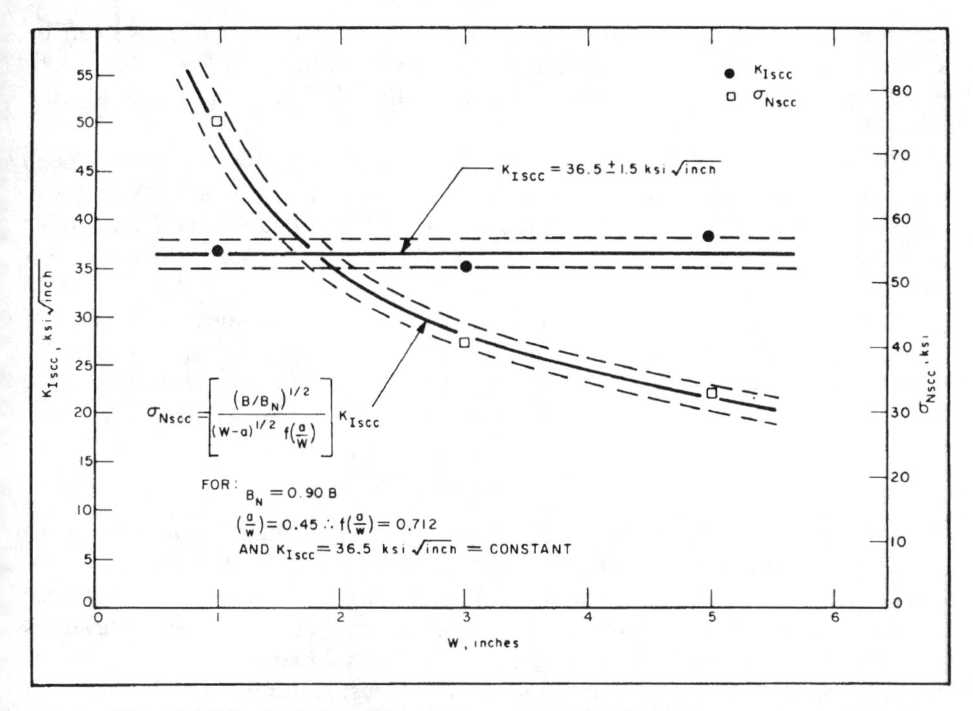

FIG. 11.8 Effect of *W* on K_{Iscc} and σ_{Nscc} for an 18Ni (250) maraging steel.

specimens, like the bolt-loaded WOL specimen, that utilize crack arrest as a measure of K_{Iscc}. Also, because crack-arrest tests appear to be more prone to crack branching, the K_{Iscc} values obtained by increasing K_I tests and decreasing K_I tests can, in some cases, be different.

Imhof and Barsom [27] investigated the effects of thermal treatments on the K_{Iscc} behavior of 4340 steel. Three pieces of 4340 steel were cut from a single plate, and each piece was heat treated to a different strength level. The three pieces were heat treated to a 130-, 180-, and 220-ksi yield strength. The K_{Iscc} for the 130-, 180-, and 220-ksi yield strengths were 111, 26, and 10.5 ksi$\sqrt{\text{in.}}$, respectively. These and other results show that thermomechanical processing may alter the K_{Iscc} for a given material composition in a specific environment. Available data also show that K_{Iscc} for a given material composition, thermomechanical processing, and environment may be different for different orientations of test specimens. An example of this behavior is observed in aluminum alloys where the susceptibility to stress-corrosion cracking in the short-transverse direction (crack plane parallel to plate surfaces) is greater than in the other directions. Consequently, although K_{Iscc} is a unique property of the tested environment–material system, extreme care should be exercised to ensure the use of the correct K_{Iscc} value for a specific application.

Ideally, K_{Iscc} for a particular material–environment system at a given temperature represents the stress-intensity-factor value below which subcritical crack extension does not occur under static load. However, in practice, K_{Iscc} can be defined as the stress-intensity-factor value corresponding to a low rate of subcritical crack extension commensurate with the design service life for the structure.

The mechanisms of stress-corrosion cracking depend on complex chemical, mechanical, and metallurgical interactions that are presently not understood. This lack of understanding contributed to a significant data scatter. Even under the most ideal test conditions, Clark [37] observed a $\pm 20\%$ scatter in data for a single high-quality AISI 4340 steel bar tested in research-grade hydrogen sulfide gas. Thus, small variations in the chemical composition of the environment or of the material, and in the thermomechanical processing and microstructure of the material, may cause significant differences in the stress-corrosion-cracking behavior.

11.2.4 Test Duration

Test duration is the second primary parameter that must be understood to ensure correct test results. The schematic representation for obtaining K_{Iscc} by using precracked cantilever-beam specimens (Figure 11.4) suggests that the true K_{Iscc} level was established with test durations larger than 1000 h. Test durations less than 200 h (Figure 11.4) would have resulted in apparent K_{Iscc} values larger than the true K_{Iscc} value obtained after a 1000-h test duration. The influence of test duration on the apparent K_{Iscc} value obtained by using cantilever-beam test specimens of a 180-ksi yield-strength high-alloy steel in room temperature synthetic seawater is shown in Table 11.1. The data show that an increase of test duration from 100 h to 10,000 h decreased the apparent K_{Iscc} value from 170 ksi$\sqrt{\text{in.}}$ to 25 ksi$\sqrt{\text{in.}}$ Proper test durations depend on specimen configuration, specimen size, and nature of loading, as well as the environment–material system. Test durations for bolt-loaded WOL specimens are longer than for cantilever-beam specimens [9]. In general, test durations for titanium, steel, and aluminum alloys are on the order of 100, 1000, and 10,000 h, respectively. The differences in test duration for different metal alloys are related partly to the incubation-time behavior in stress-corrosion cracking for the particular environment–material system. The incubation-time behavior represents the test time prior to crack extension during which a fatigue crack under sustained load in an aggressive environment appears to be dormant. The existence of incubation pe-

TABLE 11.1. Influence of Cutoff Time on Apparent K_{Iscc}; Constant-Load Cantilever Bend Specimens (Increasing K_I) (Ref *38*).

ELAPSED TIME, h	APPARENT K_{Iscc} (ksi$\sqrt{\text{in.}}$)
100	170
1,000	115
10,000	25

riods for precracked specimens has been demonstrated by various investigators [34,36,38]. Benjamin and Steigerwald [39] demonstrated the dependence of incubation time on prior loading history. Novak demonstrated the dependence of incubation time on the magnitude of the stress-intensity factor (Table 11.2) [38]. It is apparent that as the applied stress-intensity factor, K_{Ii}, approaches K_I at fracture, the incubation time must approach zero, and as the applied K_I approaches K_{Iscc}, the incubation time approaches infinity. Consequently, specimens subjected to K_{Ii} values between K_{Ic} (or K_{Ix}) and K_{Iscc} exhibit different initiation times that may be 1000 h or longer. The preceding observations show that to evaluate correctly the effect of an environment on a statically loaded structure, the test duration and criterion used to obtain the K_{Iscc} value must be known and evaluated.

11.2.5 K_{Iscc} *Data for Some Material–Environment Systems*

In general, the higher the yield strength for a given material, the lower the K_{Iscc} value in a given environment [20–27]. The K_{Iscc} for a single plate of 4340 steel tested in 3.5% solution of sodium chloride decreased from 111 to 10.5 ksi$\sqrt{\text{in.}}$ as the yield strength was increased from 130 to 220 ksi. In general, the K_{Iscc} in room temperature sodium chloride solutions for steels having a yield strength greater than about 200 ksi is less than 20 ksi$\sqrt{\text{in.}}$ Similar generalizations cannot be made for steels of lower yield strengths. Moreover, stress-corrosion-cracking data for steels having yield strengths less than 130 ksi are very sparse. The results of 5000-h (30-week) stress-corrosion tests for five steels having yield strengths less than 130 ksi obtained by testing precracked cantilever-beam specimens in room-temperature 3% sodium chloride solution are presented in Figures 11.9 through 11.13 [40]. The results presented in these figures show apparent K_{Iscc} values that ranged from 80 to 106 ksi$\sqrt{\text{in.}}$ (88 to 117 MN/m$^{3/2}$). The apparent K_{Iscc} values measured for most of these steels corresponded to conditions involving substantial crack-tip plasticity. Consequently, linear-elastic fracture-mechanics concepts cannot be used for quantitative analysis of the respective stress-corrosion-cracking behavior [13]. Furthermore, the apparent K_{Iscc} values are suppressed to various degrees below the intrinsic K_{Iscc} values for these steels in a manner similar to the suppression effect for fracture (K_{Ic}) behavior [13]. Aluminum alloys show high susceptibility to stress corrosion cracking in the short-transverse direction [8,26]. Some aluminum alloys do not exhibit a threshold

TABLE 11.2. Influence of K_I on Incubation Time: Constant-Displacement WOL Specimens (Decreasing K_I) Extent of Crack Growth (in.).

K_I (ksi$\sqrt{\text{in.}}$)	200 h	700 h	1400 h	2200 h	3500 h	5000 h
180	ND*	0.35	0.76	1.00	1.12	—
150	ND	ND	ND	0.28	0.52	0.61
120	ND	ND	ND	ND	0.03	0.045
90	ND	ND	ND	ND	ND	0.045

*ND: no detectable growth.

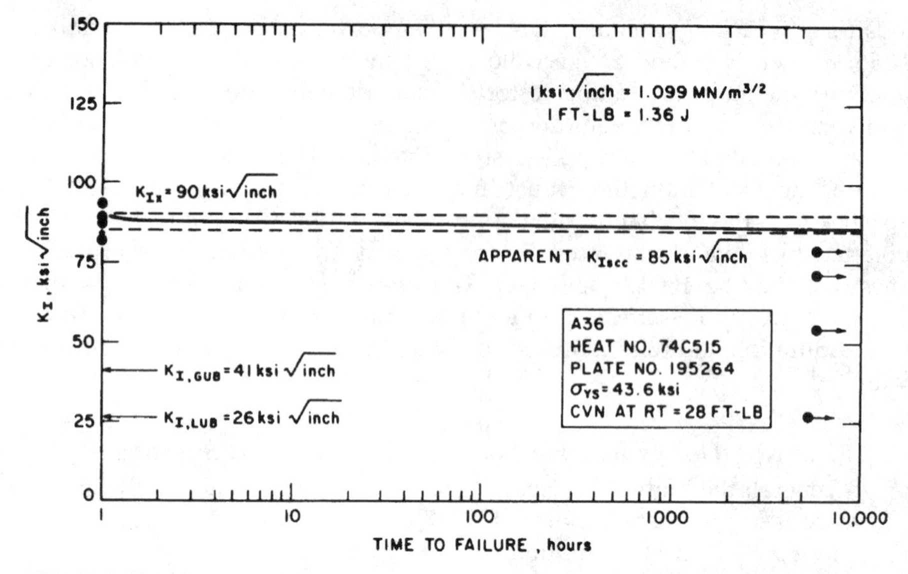

FIG. 11.9 K_I **stress-corrosion results for an A36 steel in aerated 3% NaCl solution of distilled water.**

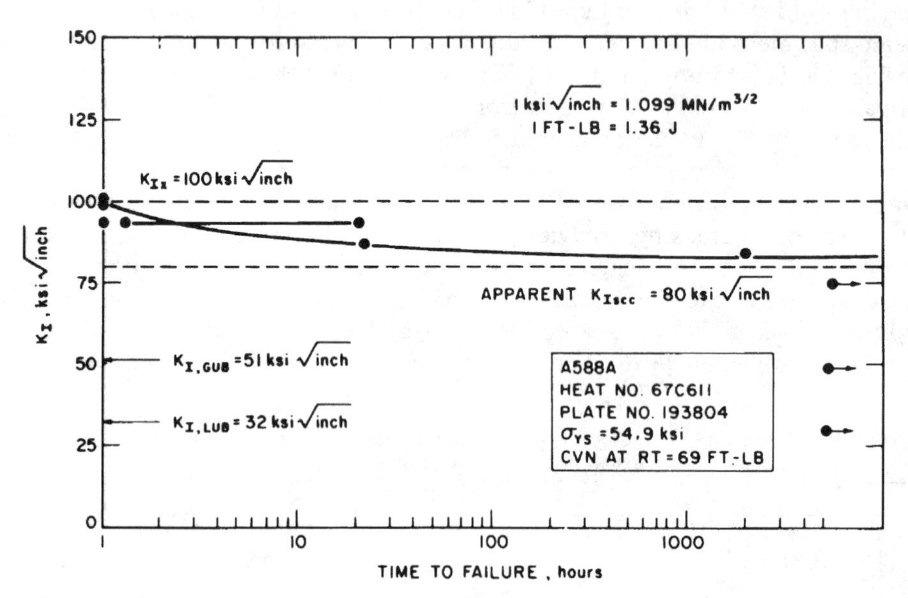

FIG. 11.10 K_I **stress-corrosion results for an A588 Grade A steel in aerated 3% NaCl solution of distilled water.**

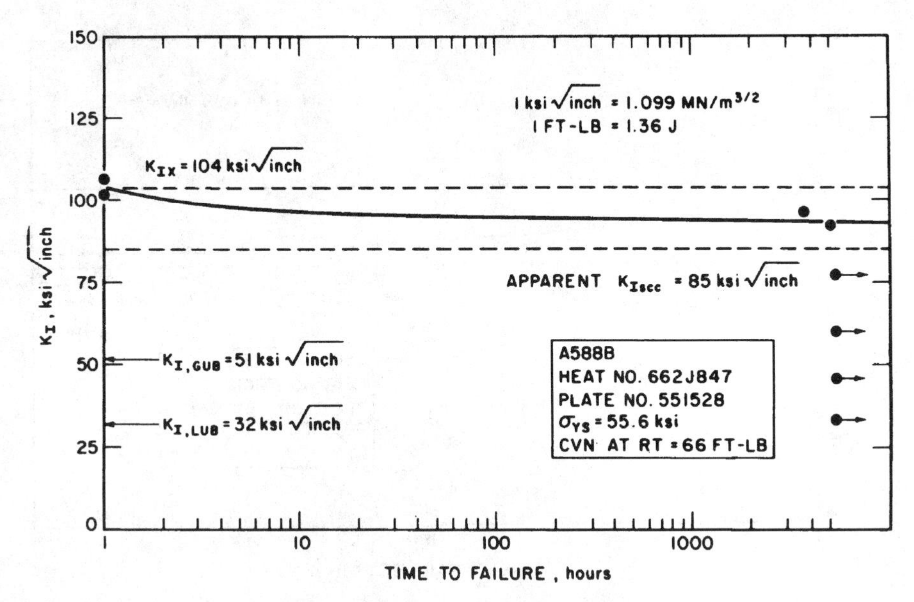

FIG. 11.11 K_I stress-corrosion results for an A588 Grade B steel in aerated 3% NaCl solution of distilled water.

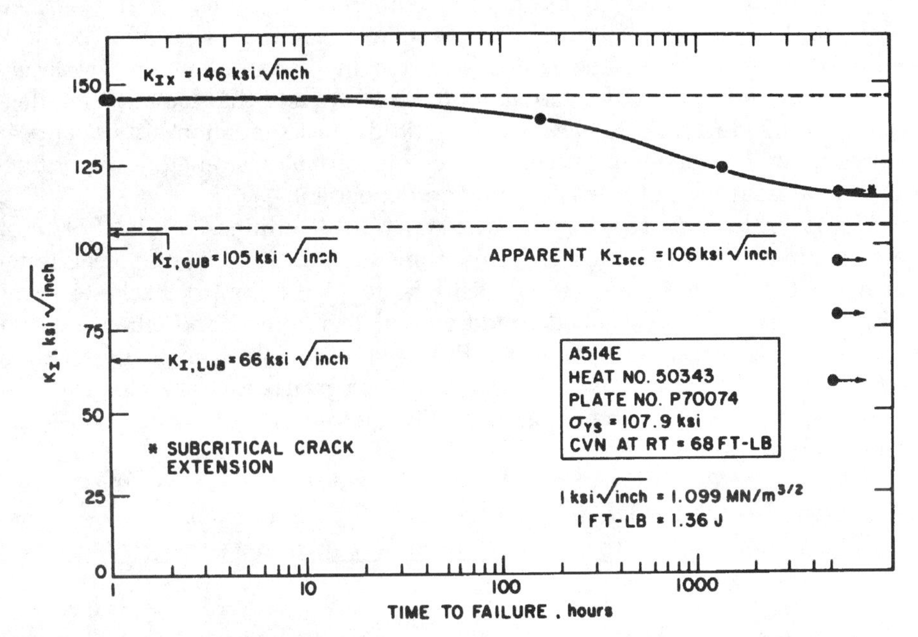

FIG. 11.12 K_I stress-corrosion results for an A514 Grade F steel in aerated 3% NaCl solution of distilled water.

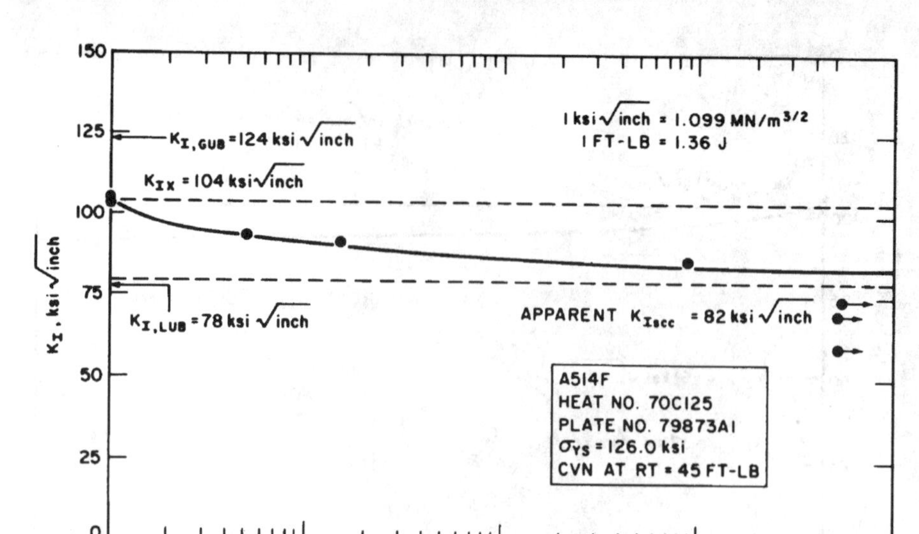

FIG. 11.13 K_I stress-corrosion results for an A514 Grade F steel in aerated 3% NaCl solution of distilled water.

behavior for stress-corrosion cracking [*21,22*]. Some titanium alloys show sustained-load cracking under plane-strain conditions in room temperature air environment. Yoder et al. [*41*] tested plate samples of eight alloys of the Ti-6Al-4V family. They concluded that sustained-load cracking in room-temperature air environment is widespread and serious in these alloys and that the resulting degradations ranged from 11 to 35%. The magnitude of degradation did not appear to correlate with interstitial contents, processing variables, strength level, or fracture toughness level, but it was orientation dependent.

Novak [*13*] conducted a systematic study to determine the effect of prior plastic strain on the mechanical and environmental properties of four steels ranging in yield strength from 40 to 200 ksi (550 to 1400 N/mm^2). Each steel was evaluated first in the unstrained condition and then after 1 and either 3 or 5% plastic strain. The results showed that the value of the stress-intensity factor at fracture decreased with increased magnitude of prior plastic strain. However, the corresponding change in the apparent K_{Iscc} value did not follow any consistent pattern of behavior.

Extensive investigation of the influence of test temperature on the K_{Iscc} behavior of environment–material systems is yet to be conducted.

K_{Iscc} data for various environment–material systems have been gathered and published [*42–46*].

11.2.6 Crack-Growth-Rate Tests

The crack-growth-rate approach to study stress-corrosion-cracking behavior of environment–material systems involves the measurement of the rate of crack growth per unit time, *da/dt*, as a function of the instantaneous stress-intensity factor, K_I. Stress-corrosion crack growth has been investigated in various environment–material systems by using different precracked specimens [3,38]. In general, the results suggest that the stress-corrosion crack-growth-rate behavior as a function of the stress-intensity factor can be divided into three regions (Figure 11.14). In Region I, the rate of stress-corrosion crack growth is strongly dependent on the magnitude of the stress-intensity factor, K_I, such that a small change in the magnitude of K_I results in a large change in the rate of crack growth. The behavior in Region I exhibits a stress-intensity factor value below which cracks do not propagate under sustained loads for a given environment–material system. This threshold stress-intensity factor corresponds to K_{Iscc}. Region II represents the stress-corrosion-crack-growth behavior above K_{Iscc}. In this region the rate of stress-corrosion cracking for many systems is moderately dependent on the magnitude of K_I, Type A behavior in Figure 11.14. Crack-growth rates in Region II for high-strength steels in gaseous hydrogen as well as other material environment systems [43–46] appear to be independent of the magnitude of the stress-intensity factor, Type B behavior (Figure 11.14). In such cases, the primary driving force for crack growth is not mechanical (K_I) in nature but is related to other processes occurring at the crack tip such as chemical, electrochemical, mass-

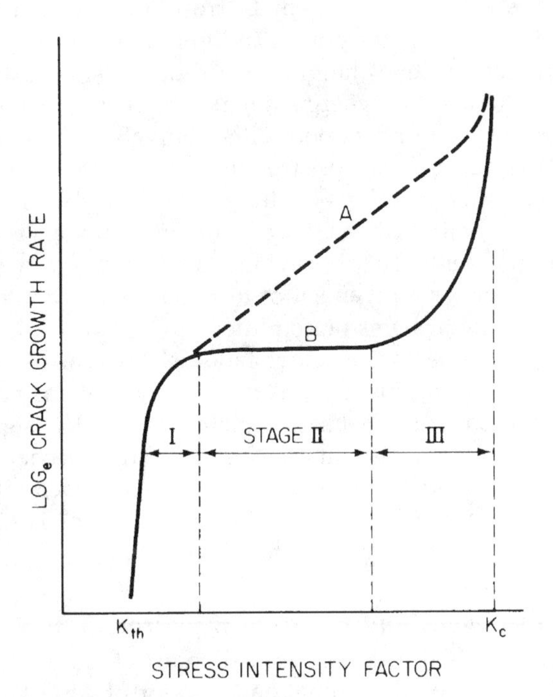

FIG. 11.14 Schematic illustration of the functional relationship between stress-intensity factor, *K*, and subcritical-crack-growth rate (*da/dt*).

transport, diffusion, and adsorption processes. The crack-growth rate in Region III increases rapidly with K_I as the value of K_I approaches K_{Ic} (or K_c) for the material.

The characteristic crack-growth behavior for a given environment–material system is determined by the mechanical properties, chemical properties, or both, for the system. The crack-growth approach to study stress-corrosion cracking is of great value in determining the mechanism of crack extension. It can be used to analyze the safety and reliability of structures subjected to stress-intensity-factor values above K_{Iscc} and for environment–material systems that do not exhibit a threshold behavior. The bolt-loaded WOL specimen can be used best to determine the stress-corrosion-crack-growth-rate behavior and the K_{Iscc}. The rate of growth at K_{Iscc} should be equal to or less than 10^{-5} in./h.

11.3 Corrosion-Fatigue Crack Initiation

Corrosion-fatigue behavior of a given material–environment system refers to the characteristics of the material under fluctuating loads in the presence of a particular environment. The corrosion-fatigue behavior of a given material–environment system depends on the metallurgical, mechanical, and electrochemical components of the particular system. Corrosion-fatigue damage occurs more rapidly than would be expected from the individual effects or from the algebraic sum of the individual effects of fatigue, corrosion, or stress-corrosion cracking. The individual effects with the synergism make the corrosion-fatigue mechanism very complex and not well understood with relatively little or no capabilities to predict, a priori, the performance of structural components. Generally, different environments have different effects on the cyclic behavior of a given material. Similarly, the corrosion-fatigue behavior of different materials is usually different in the same environment. The behavior established for a given material–environment system or for a given set of test conditions should not be applied indiscriminately to other systems or conditions.

Significant developments in understanding the corrosion-fatigue crack-propagation behavior of various metal–environment systems has been achieved by using linear-elastic fracture-mechanics methodology. The significant findings are presented in Section 11.4. Also, linear-elastic fracture-mechanics methodology was used in a systematic study of the corrosion-fatigue crack-initiation behavior for constructional steels in 3.5% sodium chloride solution. The most significant findings of this study are presented in this section.

11.3.1 Test Specimens and Experimental Procedures

Taylor and Barsom [47] used a modified compact-tension (CT) specimen to study the corrosion-fatigue crack-initiation behavior of A517 Grade F steel. The specimens contained notches that were milled to a length (a) of about 17.0 mm (0.67 in.) or 25.4 mm (1.0 in.) and had tip radii (ρ) of 3.3 mm (0.128 in.), Figure 11.15.

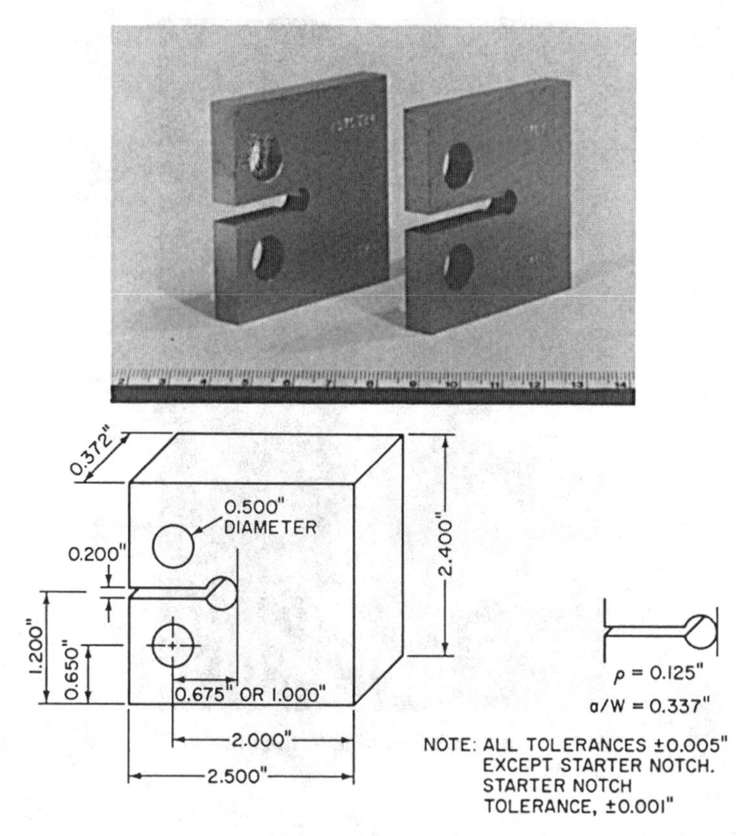

FIG. 11.15 Modified compact-tension specimens.

The specimens were submerged in room-temperature 3.5% solution of sodium chloride in a distilled water environment that was maintained at a pH of 6.5 ± 0.5 and was replaced every 100 h for all tests that exceeded this period of time. The oxygen content of the solution was not controlled. However, some aeration of the solution occurred as a result of the movement of the specimen and exposure of the surface of the solution to the room air. Crack extension was measured on the specimen surface by lowering the environmental tank below the specimen, Figure 11.16.

The test was terminated when a crack of 0.25 mm (0.010 in.) or longer was observed at the surface of the notch tip. For high test frequencies (300 cycles per minute), the frequency was temporarily lowered during inspection to facilitate observation of cracking at the notch tip. The test results were presented as the range of cycles for crack initiation bounded by the number of cycles corresponding to the last inspection of the notch tip where no crack was observed and the following inspection where a crack existed.

Novak [*48,49*] studied the corrosion-fatigue crack-initiation behavior of constructional steels by using single-edge-notched specimens subjected to cyclic

FIG. 11.16 Corrosion tank in the lowered position.

loading under cantilever-bending conditions. All specimens were tested in 3.5% solution of NaCl in distilled water maintained at a 72 ± 4°F temperature and a pH of 6.0 ± 0.8. The solution was changed once a week, and the surface of the solution was exposed to laboratory air.

Crack initiation was determined by a range of cycles bounded by the number of cycles corresponding to the last inspection at which no crack was detected and the following inspection where a crack existed.

11.3.2 Corrosion-Fatigue-Crack-Initiation Behavior of Steels

The corrosion-fatigue behavior for metals subjected to load fluctuation in the presence of an environment to which the metal is immune is identical to the fatigue behavior of the metal in the absence of that environment. Consequently, the effect of an environment on the behavior of a material subjected to load fluctuation can be studied by establishing the deviation of the corrosion-fatigue behavior for the environment–material system from the fatigue behavior of the material in a benign environment.

The difference between fatigue and corrosion-fatigue crack-initiation behavior of steels has been studied by Taylor and Barsom [47] and by Novak [48,49] and is presented in the following sections.

11.3.2.1 Fatigue-Crack-Initiation Behavior The fatigue-crack-initiation behavior for various steels was presented in Chapter 8. The data showed that each steel exhibited a distinct fatigue-crack-initiation threshold at a $\Delta\sigma_{max}$ (or $\Delta K/\sqrt{\rho}$) that can be related to the yield strength of the material.

The fatigue-crack-initiation behavior in air of the ASTM A517 Grade F steel plate whose corrosion-fatigue crack-initiation behavior was investigated by Taylor and Barsom [47] and by Novak [48,49] is presented in Figure 11.17. The test results in this figure include data obtained by Barsom [50] using compact-tension specimens at $R = +0.1$. The data show that fatigue-crack-initiation life is governed by the total (tension plus compression) maximum-stress range, $\Delta\sigma_{max}$, (or $\Delta K/\sqrt{\rho}$) at the notch tip and that different specimen geometries and loadings give similar results. Moreover, the data show a distinct fatigue-crack-initiation threshold that occurred at a $\Delta K/\sqrt{\rho}$ of about 100 ksi (690 MPa), which corresponds to a $\Delta\sigma_{max}$ of 120 ksi (828 MPa). This value is in good agreement with predictions obtained by using the empirical relationships presented in Chapter 8. These relationships indicate that the fatigue-crack-initiation threshold for the A36, A588 Grade A, A517 Grade F, and V150 tested by Novak [48,49] and by Taylor and Barsom [47] occur at a $(\Delta K/\sqrt{\rho})_{th}$ of about 65, 80, 100, and 165 ksi (450, 550, 750, and 1150 MPa), respectively.

11.3.2.2 Corrosion Fatigue Crack-Initiation Behavior The corrosion-fatigue crack-initiation behavior for A36, A588 Grade A, A517 Grade F, and V150 steels under full immersion conditions in 3.5% solution of sodium chloride in distilled water at 12 cycles per minute are presented in Figures 11.18, 11.19, 11.20, and 11.21, respectively. Also shown in these figures is the best-fit equation and the standard deviation for the data and the fatigue-crack-initiation threshold in air for each steel. The data in these figures are presented as a range of cycles bounded by the number of cycles corresponding to the last inspection at which no crack was detected and the following inspection at which a crack existed. The data show that under these test conditions, the environment caused substantial reduction in the crack-initiation life at $\Delta K/\sqrt{\rho}$ values significantly lower than the $\Delta K/\sqrt{\rho}$ value corresponding to the fatigue-crack-initiation threshold in a benign environment. The long-life data points with arrows represent test specimens that were terminated with no indication of corrosion-fatigue crack initiation. These data points may suggest the possible existence of a threshold below which corrosion-fatigue crack-initiation would not occur. Further discussion of corrosion-fatigue crack-initiation threshold is presented in a later section on long-life behavior.

A comparison of the fatigue-crack-initiation behavior in a benign environment and the corrosion-fatigue crack-initiation behavior in 3.5% solution of so-

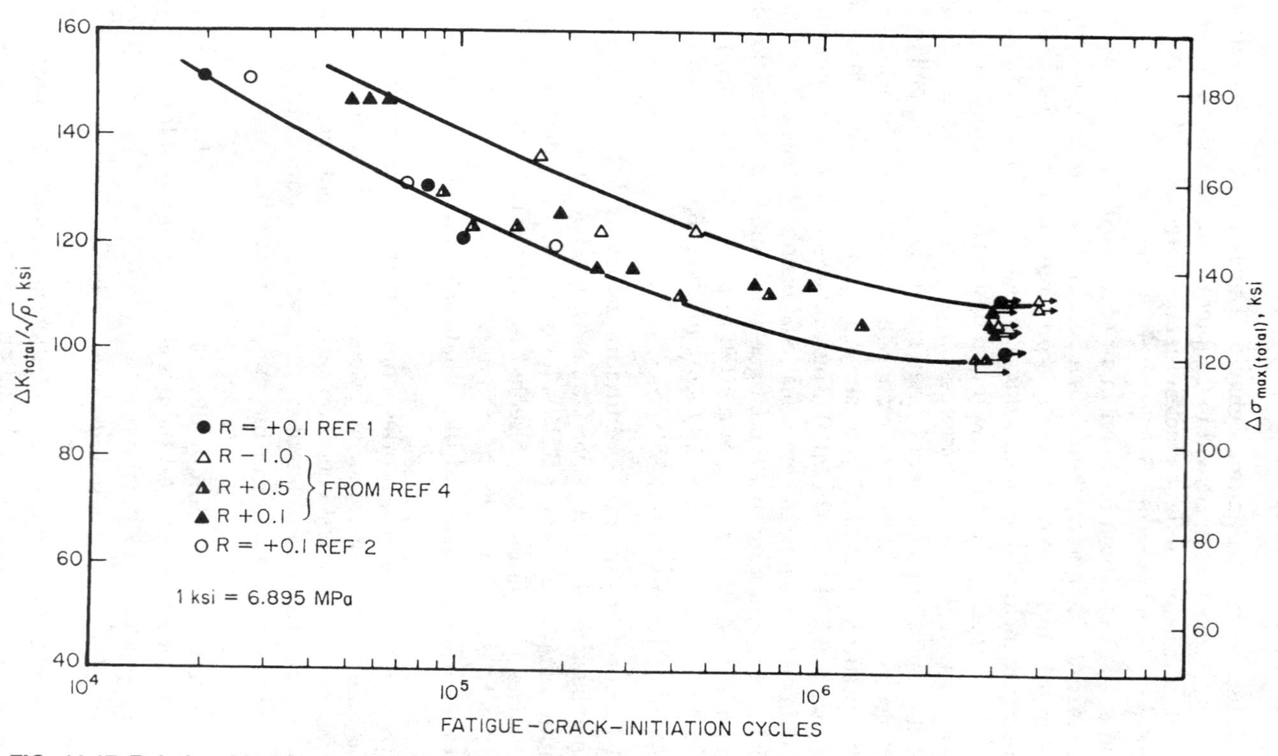

FIG. 11.17 Relationship between fatigue-crack initiation and various stress ratios for an A517 Grade F steel.

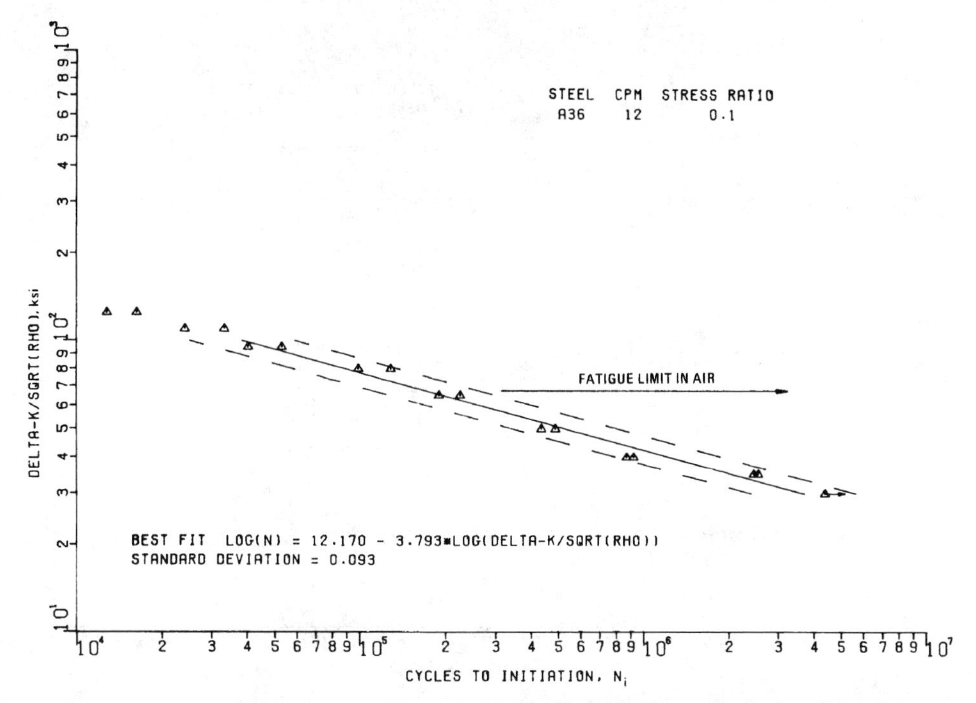

FIG. 11.18 Corrosion-fatigue crack initiation for an A36 steel.

dium chloride shows that the effect of the environment is small in the region of high-stress range and increases as the stress range decreases. Furthermore, the data show that the corrosion-fatigue crack-initiation life can be represented by a linear relationship of log $\Delta\sigma_{max}$ (or log $\Delta K\sqrt{\rho}$) and log N_i, where N_i is the number of cycles for crack initiation. A generalized relationship for predicting the corrosion-fatigue crack-initiation life for various steels in 3.5% solution of sodium chloride is presented in a later section.

The best-fit equations for A36, A588 Grade A, and A517 Grade F suggest the possible dependence of the slope and intercept on the yield strength of the steels such that as the yield strength increases, the slope and intercept value decreases. However, the data for the V150 steel do not support this observation. Further research is needed to establish the relationships, if any, between corrosion-fatigue crack-initiation behavior and properties of the material, the environment, or both.

Figure 11.22 is a superposition of the data presented in Figures 11.18 through 11.21 and shows that, within experimental scatter, the corrosion-fatigue crack-initiation behavior for the four steels was essentially identical. This observation indicates that the corrosion-fatigue crack-initiation behavior for these steels under full immersion conditions in the test environment is independent of chemical composition, microstructure, and mechanical properties (in particular, yield strength) which were significantly different for the steels investigated.

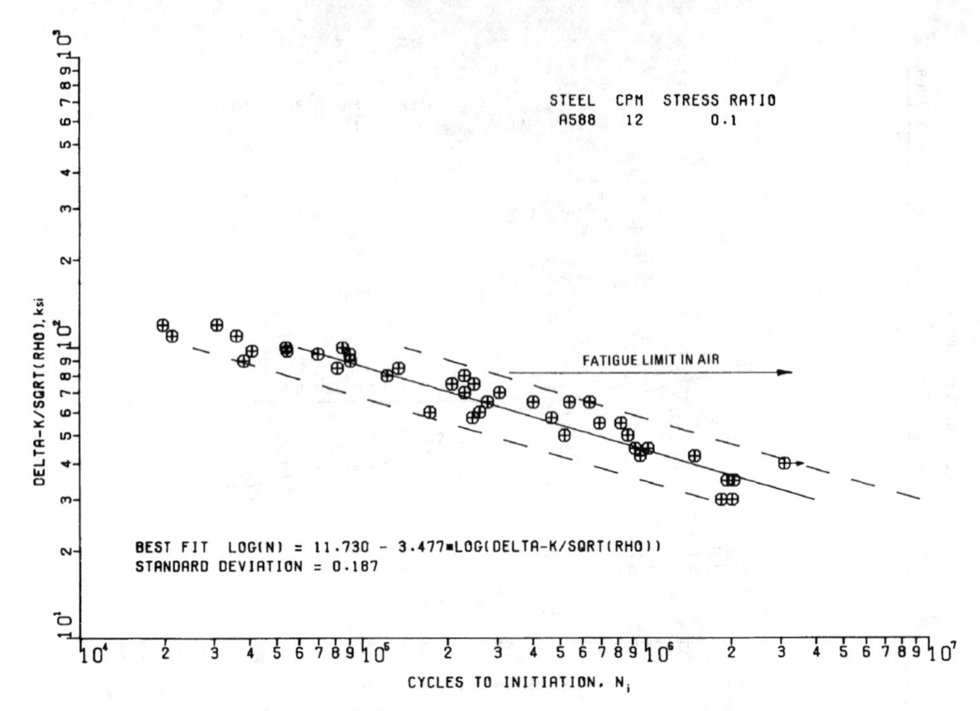

FIG. 11.19 Corrosion-fatigue crack initiation for an A588 Grade A steel.

11.3.2.3 Effect of Cyclic-Load Frequency The effects of cyclic-load frequency on the corrosion-fatigue crack-initiation behavior for steels has been investigated in 3.5% solution of sodium chloride in distilled water at frequencies equal to 1.2 cycles per minute (cpm) and higher [*47,49,51*]. The tests were conducted on A588 Grade A and A517 Grade F steels, and the results are presented in Figures 11.23 and 11.24, respectively. The data show a distinct but small increase in the corrosion-fatigue crack-initiation life with increased cyclic-load frequency from 1.2 to 120 cpm. However, the A517 Grade F corrosion-fatigue crack-initiation life at 300 cpm was less than for the life for 120 cpm.

The data also show that a 100- to 250-fold increase in cyclic-load frequency from 1.2 to 300 cpm resulted in only a 3-fold increase in the mean corrosion-fatigue crack-initiation life. Moreover, the scatter band for the data obtained by testing A588 Grade A and A517 Grade F steels at frequencies of 1.2 to 300 cpm is essentially identical to the scatter band for the various steels tested at 12 cpm (Figure 11.22).

11.3.2.4 Effect of Stress Ratio The effect of stress ratio on corrosion fatigue crack initiation behavior was investigated by testing A588 Grade A and A517 Grade F steels under full-immersion conditions in room temperature 3.5% solution of sodium chloride and at stress ratios, *R* (ratio of the minimum and maximum nominal stresses, $\sigma_{min}/\sigma_{max}$), of −1.0, +1.0, and +0.5. The data show

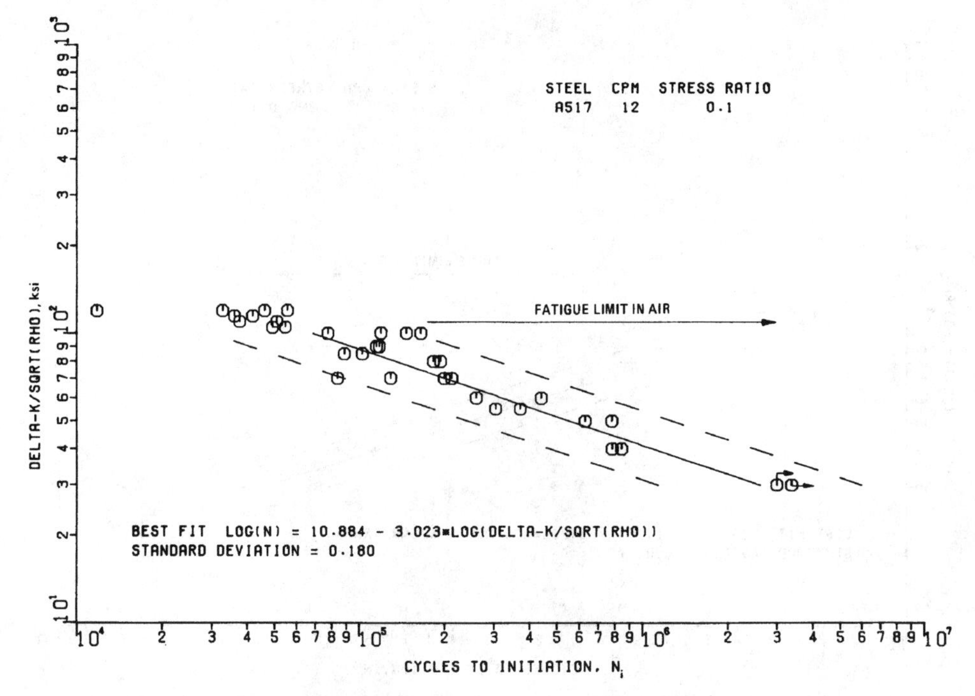

FIG. 11.20 Corrosion-fatigue crack initiation for an A517 Grade F steel.

a slight increase in corrosion fatigue crack initiation life with decreased *R* values. However, the scatter band for these data was essentially identical to the scatter band for the various steels tested at *R* of +0.1.

11.3.2.5 Long-Life Behavior The long-life behavior ($N > 4 \times 10^6$ cycles) and possible existence of a corrosion-fatigue crack-initiation threshold were investigated by testing A588 Grade A and A517 Grade F at various $\Delta K/\sqrt{\rho}$ values between 30 and 14 ksi. The specimens were subjected to 120 cpm at $R = +0.5$ under full-immersion conditions in room temperature 3.5% sodium chloride solution that was changed once a week. The specimens were in test from 4×10^6 to 7×10^7 cycles.

The long-life test results are presented in Figure 11.25 [*51*]. The data fall along the linear extension of the log $\Delta K/\sqrt{\rho}$ and N_i relationship established for the short-life behavior (Figure 11.22) and within the same scatter band.

Also, for the conditions used to conduct these tests, no corrosion-fatigue crack-initiation threshold was observed. However, this observation should not be extended indiscriminately to other material–environment systems or frequencies. For example, preliminary observations suggest the possible existence of thresholds for these steels in this environment with high pH or with cathodic protection.

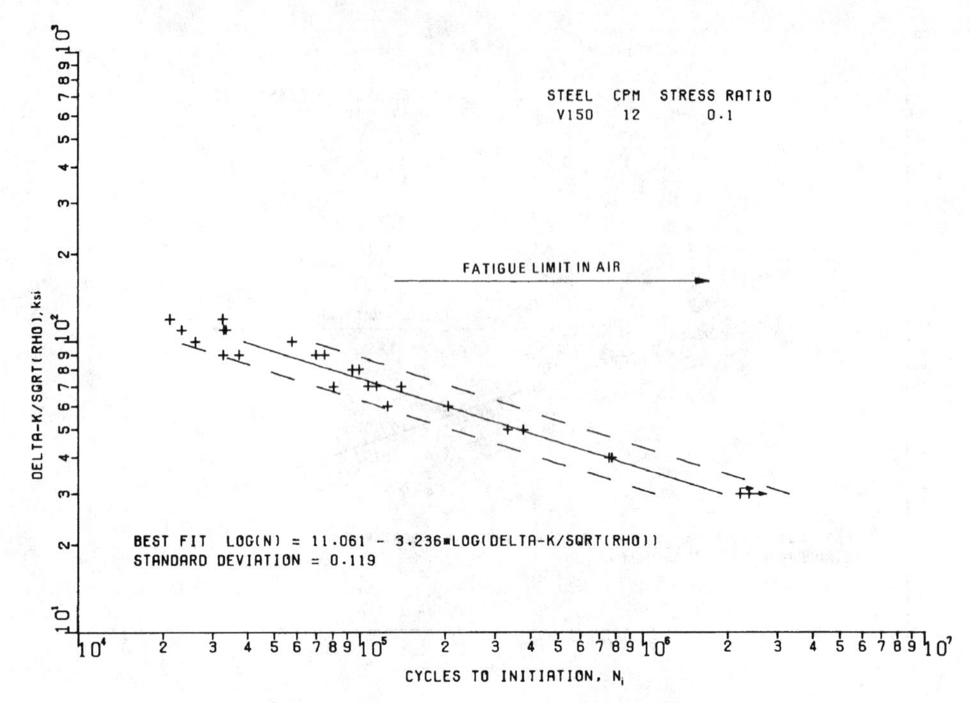

FIG. 11.21 Corrosion-fatigue crack initiation for a V150 steel.

11.3.2.6 Generalized Equation for Predicting the Corrosion-Fatigue Crack-Initiation Behavior for Steels Figure 11.26 presents all the corrosion-fatigue crack-initiation data presented in the preceding sections for A36, A588 Grade A, A517 Grade F, and V150 steels under full-immersion conditions in room temperature 3.5% solution of sodium chloride in distilled water. These steels represent large variations in chemical composition, thermomechanical processing, microstructure, and mechanical properties (tensile strength, yield strength, elongation, strain hardening, fracture toughness, etc.). The combined data encompass frequencies of 1.2, 12, 60, 120, and 300 cpm and stress ratios of -1.0, $+1.0$, and $+0.5$ and span four orders of magnitude in corrosion-fatigue crack-initiation life between about 10^4 and 10^8 cycles.

Considering the large variation in materials and test conditions, the data fall within a surprisingly narrow scatter band and can be represented by a single linear relationship between log $\Delta K/\sqrt{\rho}$ and log N_i. The equation for the best-fit line for all data is

$$N_i = 3.56 \times 10^{11} (\Delta K/\sqrt{\rho})^{-3.36} \tag{11.1}$$

where ΔK is in ksi $\sqrt{\text{in.}}$ and ρ is in inches.

Figure 11.26 includes data that represent the number of cycles corresponding to the last inspection at which no crack was detected and the following inspection at which a crack existed. Thus, Equation (11.1) should represent closely the num-

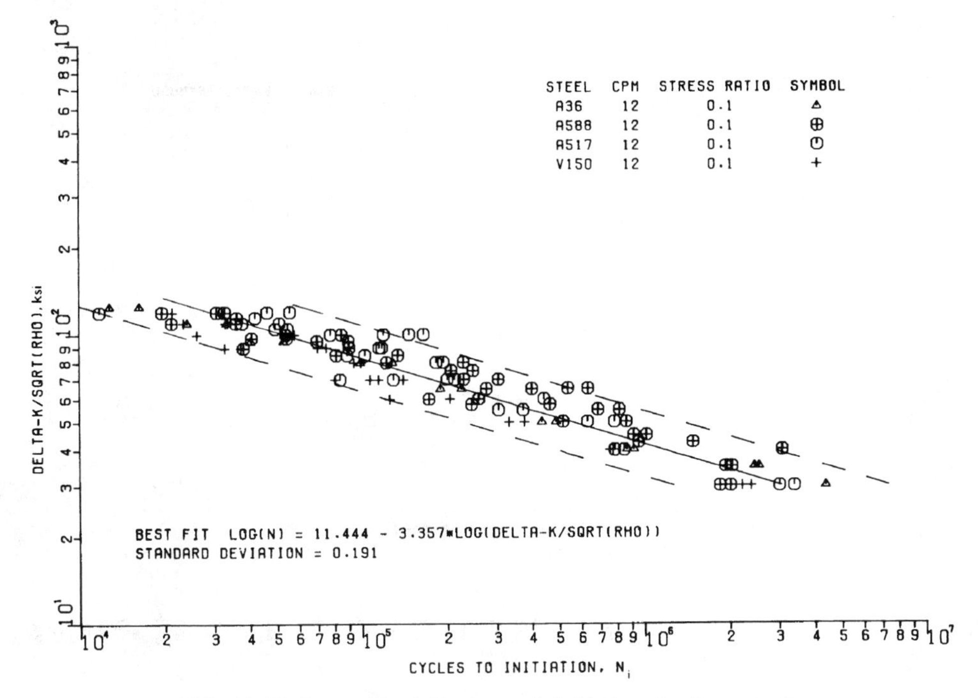

FIG. 11.22 Corrosion-fatigue crack initiation for four steels.

ber of cycles for corrosion-fatigue crack initiation. Prediction of corrosion-fatigue crack-initiation life by using the lower bound for the data may be too conservative. Finally, the corrosion-fatigue crack-initiation life for notches having severe stress raisers or severe imperfections on the notch surface or in the immediate vicinity of the notch tip could be significantly less than predicted by using Equation (11.1).

11.4 Corrosion-Fatigue-Crack Propagation

Several investigators have studied the corrosion-fatigue behavior of various environment–material systems [*3,35,40,46,52–66*]. The results of these investigations have helped greatly in the selection of proper materials for a given application. Despite the significant progress that has been achieved to establish the effects of various mechanical parameters on the corrosion-fatigue behavior of environment–material systems, little has been achieved to establish mechanisms of corrosion fatigue in these systems. The available information shows the high complexity of the corrosion-fatigue behavior and suggests that a significant understanding of this behavior can be achieved only by a synthesis of contributions from various fields.

The generalized fatigue-crack-growth behavior in a benign environment (Figure 11.27) is a special case of the corrosion-fatigue crack propagation behavior

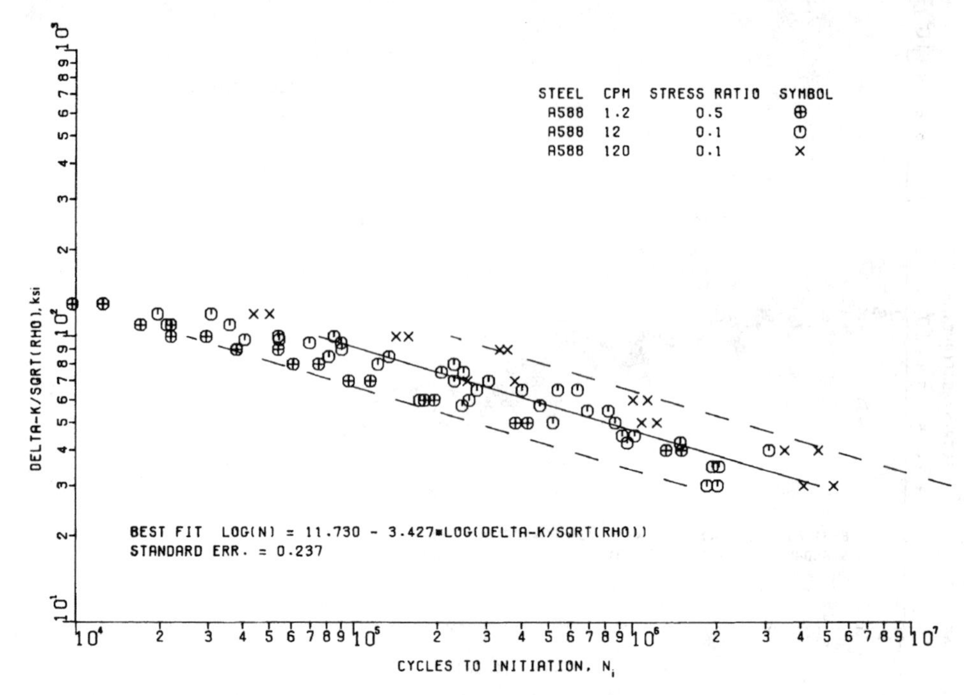

FIG. 11.23 Corrosion-fatigue crack initiation for an A588 Grade A steel.

for metals. It represents the "corrosion-fatigue" behavior of metals subjected to load fluctuations in the presence of any environment that does not affect the fatigue-crack-growth behavior for the metal. Thus, the corrosion-fatigue behavior for a given environment–material system could be investigated by establishing the base-line fatigue behavior and then by determining the effect of the environment on the fatigue behavior in Regions I, II, and III (Figure 11.27). However, because K_{Iscc} for an environment–material system defines the plane-strain K_I value above which stress-corrosion crack growth can occur under static loads, the corrosion-fatigue crack-propagation behavior for the environment–material system could be altered when the maximum value of K_I, K_{Imax}, in a given load cycle becomes greater than K_{Iscc}. Consequently, the corrosion-fatigue crack-propagation behavior should be divided into below-K_{Iscc} and above-K_{Iscc} behaviors.

11.4.1 Corrosion-Fatigue Crack-Propagation Threshold

Several investigations have significantly increased our understanding of environmental effects on the threshold behavior [*52–55*]. The fatigue-crack-propagation threshold behavior for steels in room temperature air environment was presented in Chapter 9. The data showed, among other things, that the ΔK_{th} is strongly dependent on the stress ratio, *R*, and that its value decreases as the value of *R* increases.

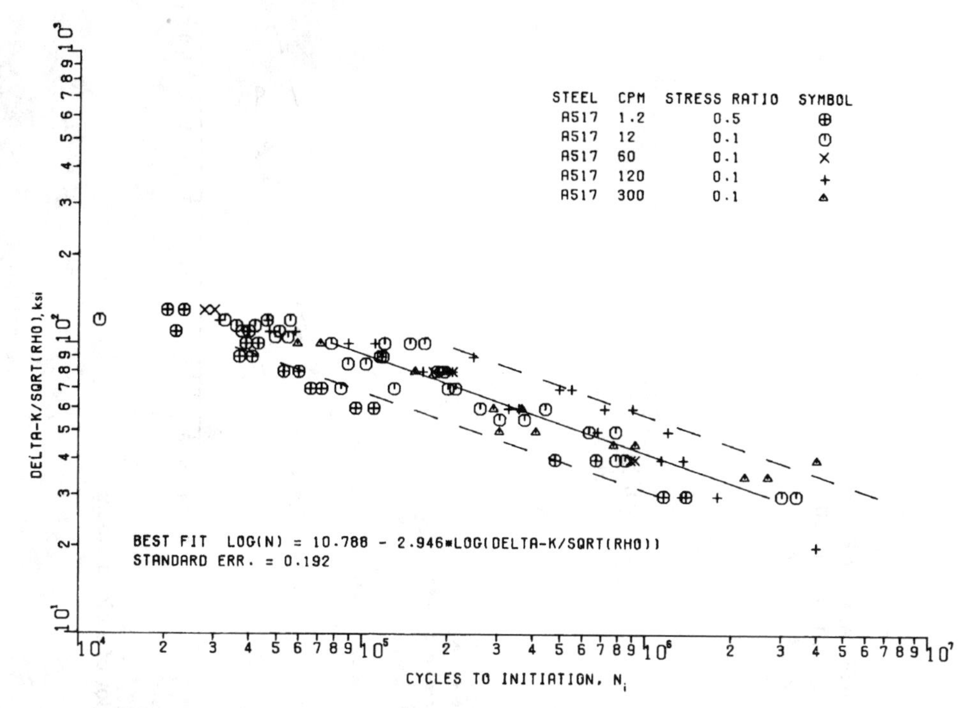

FIG. 11.24 Corrosion-fatigue crack initiation for an A517 Grade F steel.

Bucci and Donald [67] observed that the environmental ΔK_{th} in a 200-grade maraging steel forging was higher than the ΔK_{th} in air and that the "salt water appears to produce an inhibitive effect on fatigue cracking at the very low ΔK levels." Paris et al. [63] also observed that the threshold ΔK of ASTM A533 Grade B, Class 1 steel in distilled water was greater than that established in room-temperature air. "Since the distilled water retardation of very low crack extension

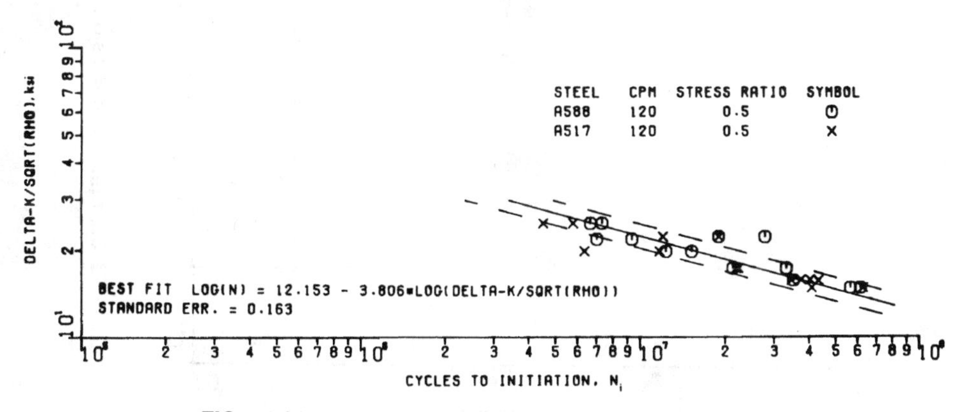

FIG. 11.25 Long-life corrosion-fatigue crack initiation.

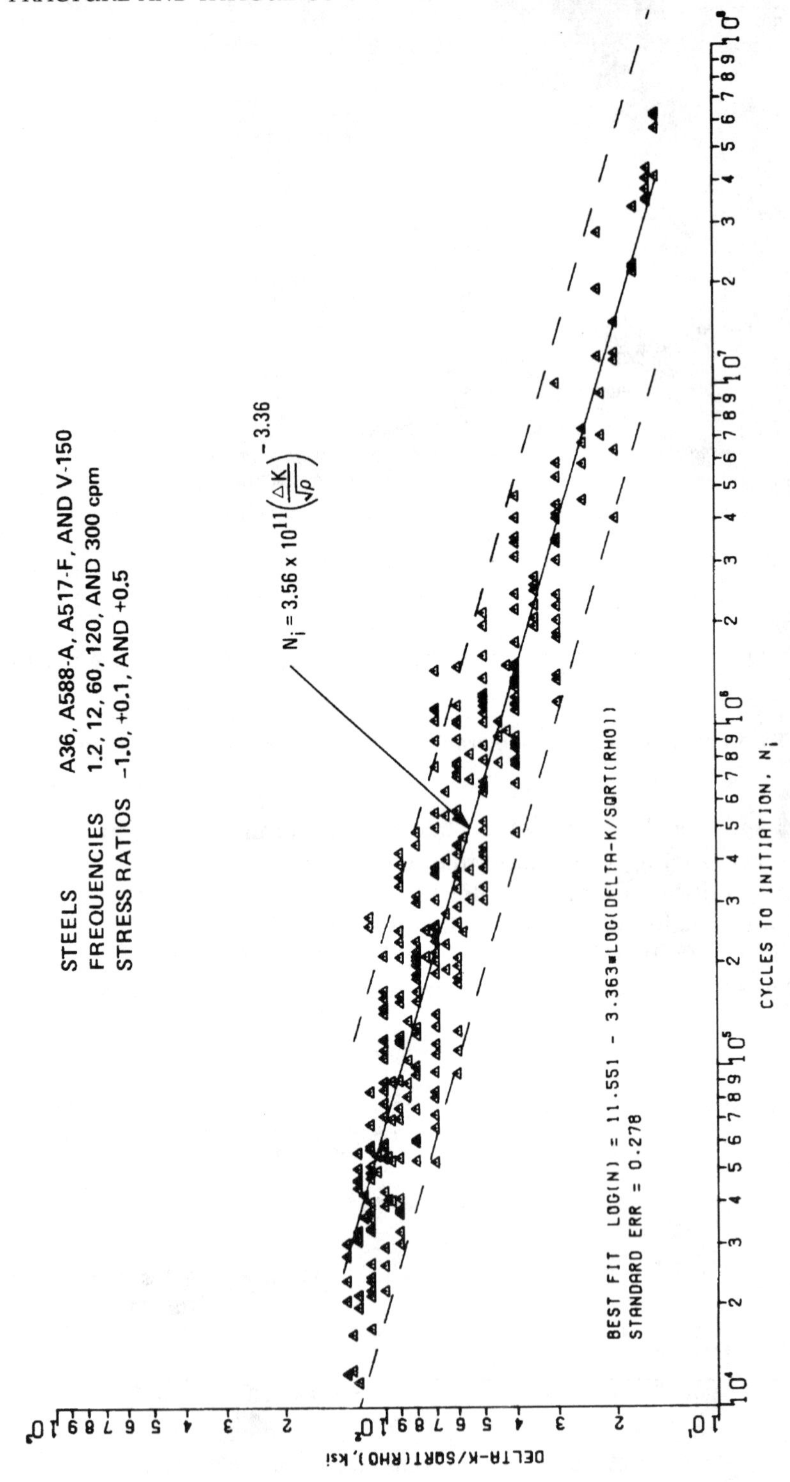

FIG. 11.26 Summary of corrosion-fatigue crack initiation behavior of various steels.

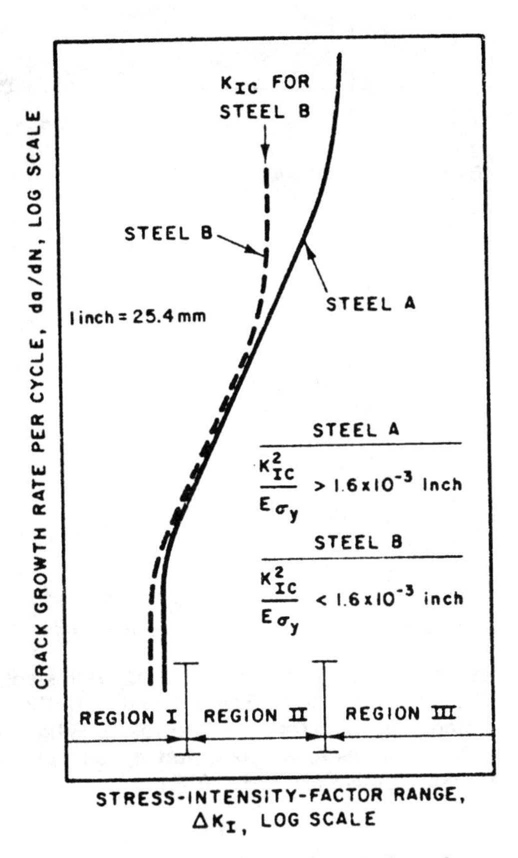

FIG. 11.27 Schematic representation of fatigue-crack growth in steels.

rates was a somewhat surprising result, an additional specimen was tested for which a distilled water environment was provided to the crack tip and its surroundings only after an initial, slow, crack extension rate had been established in room air. Upon application of the distilled water, the rate of crack growth decreased from those initially obtained in air [63].

Figure 11.28 [53] presents the threshold and near-threshold fatigue-crack-propagation behavior for 2 1/4Cr-1Mo steel in hydrogen gas and in humid air at a frequency of 50 Hz and a stress ratio $R = 0.05$. The near-threshold fatigue-crack-propagation rate in dry hydrogen was significantly higher than in air and the ΔK_{th} for dry hydrogen was lower than the value in air. However, unlike the behavior for $R = 0.05$, the near-threshold fatigue-crack-propagation rate and the threshold for $R = 0.75$ in both environments were identical. Similar trends were observed for other gaseous environments and steels [53,54]. The data for inert gaseous environments indicate that the increase in the near-threshold propagation rate and the decrease in ΔK_{th} in dry hydrogen from the values in humid air

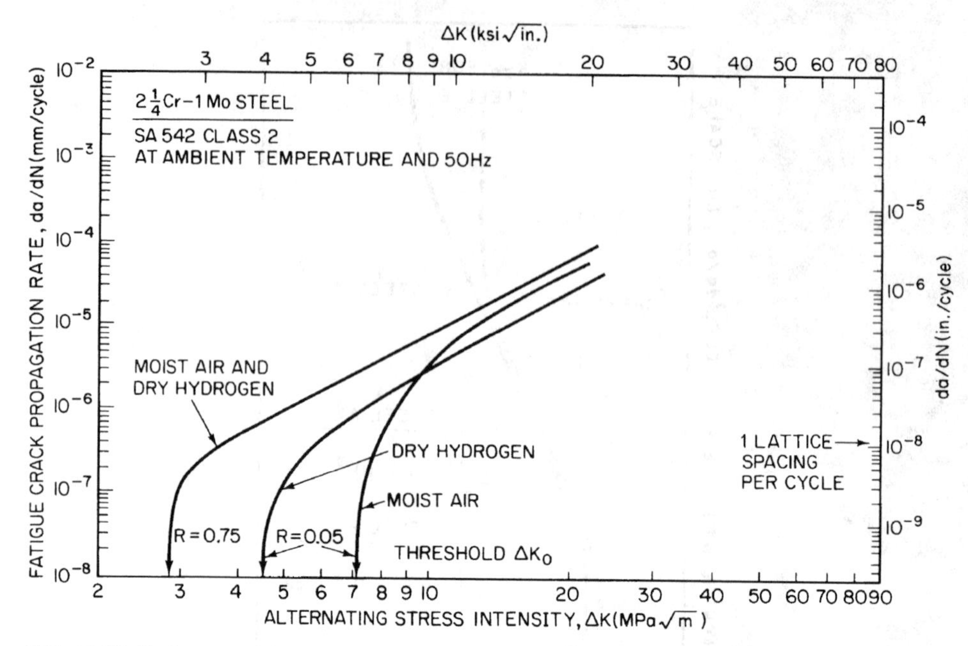

FIG. 11.28 Fatigue-crack propagation in martensitic 2¼Cr-1Mo steel (SA542-2) in moist air and dry hydrogen. (From S. Suresh, G. F. Zamiski, and R. O. Ritchie, "Oxide-Induced Crack Closure: An Explanation of Near-Threshold Corrosion Fatigue Crack Growth Behavior," *Metallurgical Transactions A,* Vol. 12A, Aug. 4, 1981, pp. 1435–1443. American Society for Metals, Metals Park, OH 44073.)

environment are not necessarily related to hydrogen embrittlement. Hydrogen embrittlement and other mechanisms may influence the threshold value or the near-threshold propagation behavior in a given environment-metal system. However, the differences observed between dry gaseous environment and humid air for low-strength steels appear to be related primarily to the formation of corrosion products within the crack in the air environment and their effect on the closure of the crack (see Chapter 9).

The concept of reduced cyclic crack-opening displacement caused by corrosion products within the crack was proposed by Paris et al. [*63*] to explain a higher ΔK_{th} value in distilled water than in air for pressure-vessel steels tested at an R value of 0.1. This concept suggests that a corrosive environment produces corrosion products, which at low stresses wedge the crack open, resulting in a decreased driving force [*52*]. The concept was also used by Skelton and Haigh [*68*] to explain the effect of stress ratio on ΔK_{th} for CrMoV steels tested at 550°C in vacuum and in oxidizing environments. They found that both inert environments and high stress ratios reduced ΔR_{th}, and that compressive stresses resulted in reduced ΔK_{th} by compacting the corrosion products, which increased the effective cyclic crack-opening displacement.

Skelton and Haigh [*68*] reported a systematic decrease of ΔK_{th} as the cyclic frequency was increased from 0.01 to 10 Hz. This finding is very important because many engineering structures are subjected to cyclic frequencies less than 1 Hz. Unfortunately, because of the very large number of cycles necessary to establish the ΔK_{th}, most tests are conducted at cyclic frequencies that are higher than 10 Hz [52]. An increase in cyclic frequency to shorten the test time for determining ΔK_{th} and the use of the results to evaluate the performance of a structure subjected to low cyclic frequencies in the service environment can lead to erroneous conclusions.

Limited data related to the effect of cyclic frequency on ΔK_{th} in aqueous environments has been obtained for various constructional steels [*40*]. The corrosion-fatigue-crack-growth-rate data for A36 steel tested at 12 cpm indicated that the rate of crack growth decreased significantly at ΔK_I values less than 20 ksi$\sqrt{\text{in.}}$ (22.0 MN/m$^{3/2}$) (Figure 11.29). Similar behavior was observed in the A588 Grade A, A588 Grade B, A514 Grade E, and A514 Grade F steels tested, Figure 11.30. The data show that a corrosion-fatigue crack-growth-rate threshold, ΔK_{th}, does exist in A514 steels at a value of the stress-intensity-factor fluctuation below which corrosion-fatigue cracks do not propagate at 12 cycles per minute (cpm) in the environment–steel system tested. The value of ΔK_{th} in the A514 steels tested at 12 cpm in 3% sodium chloride solution was twice as large as the value of 5.5 ksi$\sqrt{\text{in.}}$ (6.0 MN/m$^{3/2}$) for room temperature air [*69*].

11.4.2 Corrosion-Fatigue-Crack-Propagation Behavior Below K_{Iscc}

The first systematic investigation into the effect of environment and loading variables on the rate of fatigue-crack growth below K_{Iscc} was conducted on 12Ni-5Cr-3Mo maraging steel (yield strength = 180 ksi) in 3% solution of sodium chloride [*28,35,70*]. The data showed that environmental acceleration of fatigue-crack growth does occur below K_{Iscc} (Figure 11.31) and that the magnitude of this acceleration is dependent on the frequency of the cyclic-stress-intensity fluctuations. In the sodium chloride solution at high frequencies (cpm > 600), corrosion fatigue crack growth rate was essentially the same as in air; thus, the environment had negligible effects on the fatigue-crack-growth rate. In the sodium chloride solution at 6 cpm, however, corrosion fatigue crack growth rate was three times higher than the value in air, which indicated that the fatigue-crack-growth rate was increased significantly by the environment. The data in Figure 11.31 may be used to predict the corrosion fatigue crack growth rate for any sinusoidal frequency equal to or greater than about 6 cpm in the environment–material system investigated. Because these results were obtained below K_{Iscc}, Barsom [*28,70*] concluded that the corrosion-fatigue-crack-growth rate for 12Ni-5Cr-3Mo maraging steel in a 3% solution of sodium chloride increases to a maximum value and then decreases as the sinusoidal cyclic-stress frequency decreases from 600 cpm to frequencies below 6 cpm. Similar behavior has been established for steels [*28,35,40,61,62,70*], aluminum alloys [*46,59,65*], and titanium alloys [*46,64*]. How-

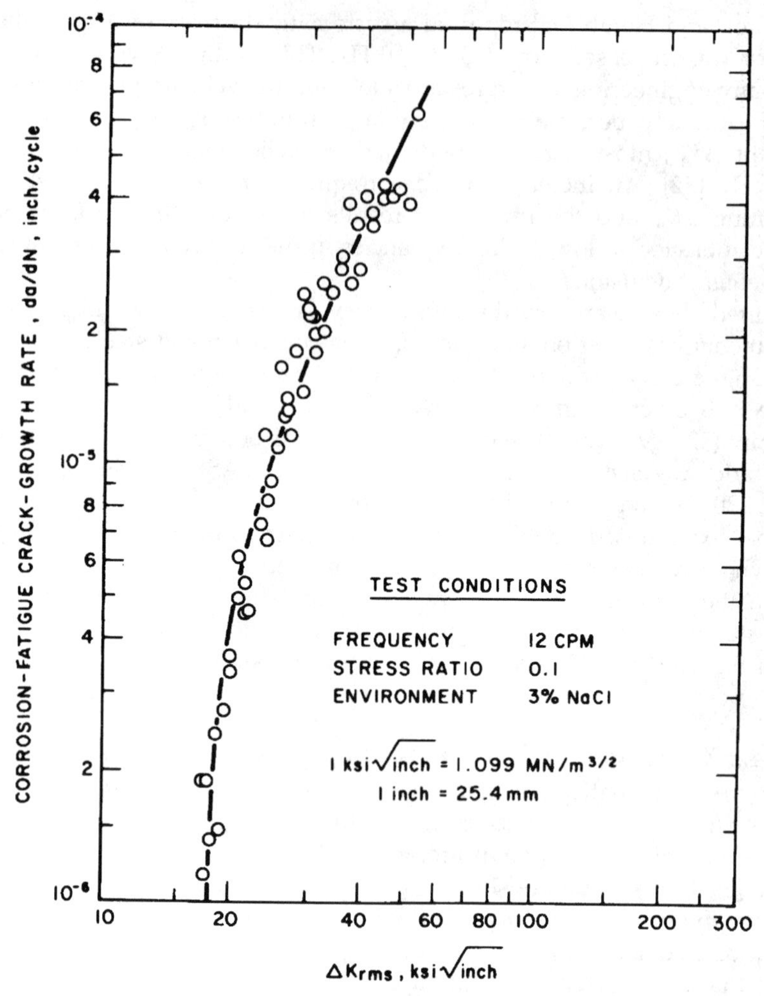

FIG. 11.29 Corrosion-fatigue-crack-growth rate as a function of the root-mean-square stress-intensity factor for an A36 steel.

ever, unlike the data presented in Figure 11.31, curves obtained for different cyclic frequencies in various environment–metal systems do not appear to be parallel to each other.

The magnitude of the effect of cyclic frequency on the rate of corrosion-fatigue-crack growth depends strongly on the environment–material system [27,72]. The data presented in Figure 11.32 [72] show that the 10Ni-Cr-Mo-Co steels tested were highly resistant to the 3% solution of sodium chloride and that, of the four steels tested, the 12Ni-5Cr-3Mo steel was the least resistant to the 3% solution of sodium chloride. Similarly, fatigue-crack-growth rates below K_{Iscc} were accelerated by a factor of 2 when 4340 steel of 130-ksi yield strength was

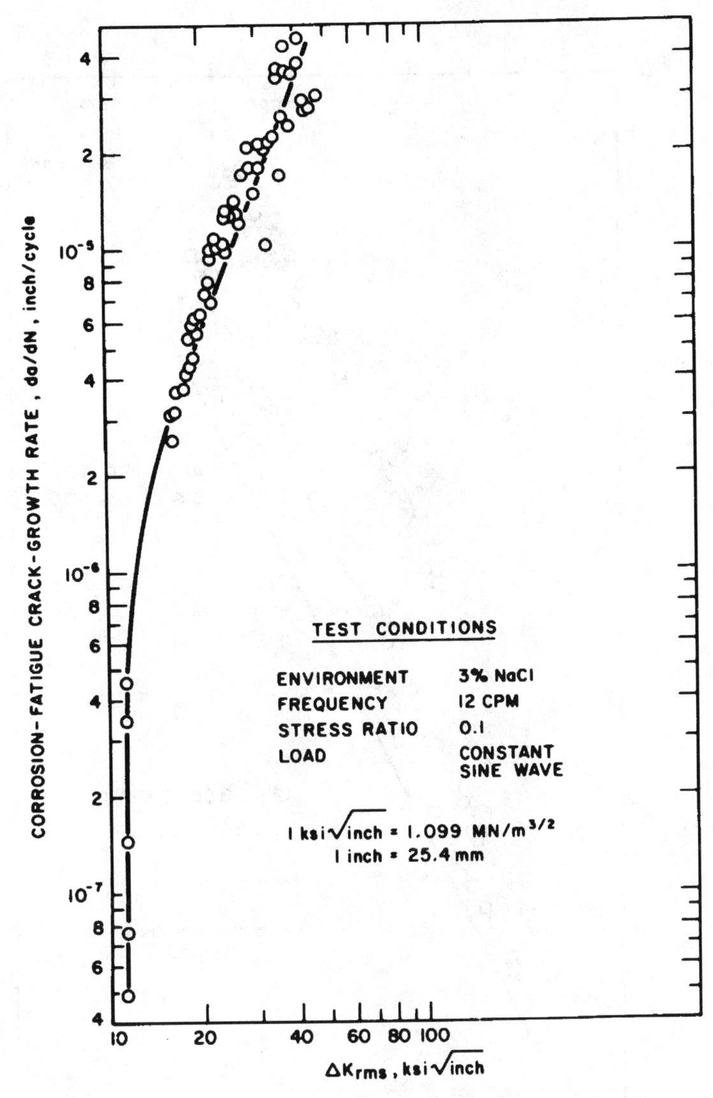

FIG. 11.30 Corrosion-fatigue-crack-growth rate as a function of the root-mean-square stress-intensity factor for an A514 Grade E steel.

tested at 6 cpm in a sodium chloride solution (Figure 11.33) [27]. Under identical test conditions, the corrosion-fatigue-crack-growth rates in the same 4340 steel heat treated to 180-ksi yield strength were five to six times higher than the fatigue-crack-growth rates in room temperature air environments (Figure 11.33).

The effect of various additions to aqueous solutions on the stress-corrosion cracking and on the corrosion-fatigue-crack-growth rates at a given cyclic fre-

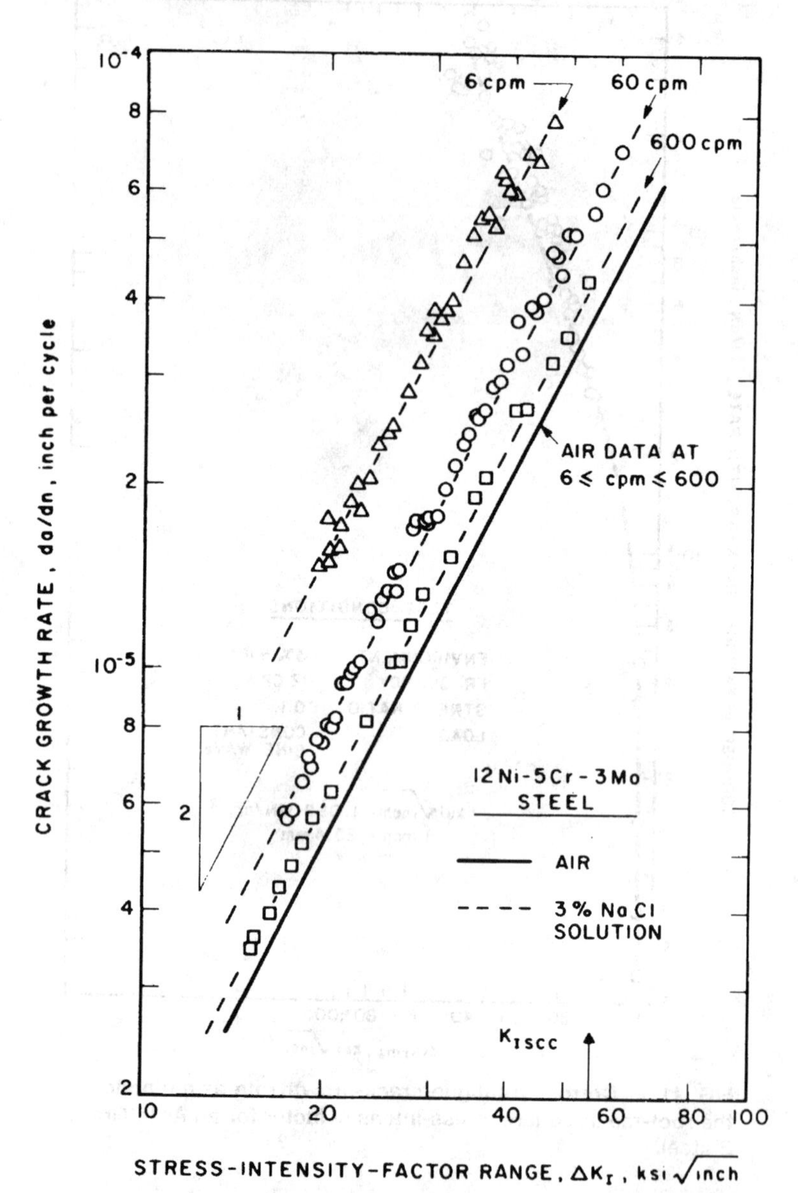

FIG. 11.31 Corrosion-fatigue-crack-growth data as a function of test frequency.

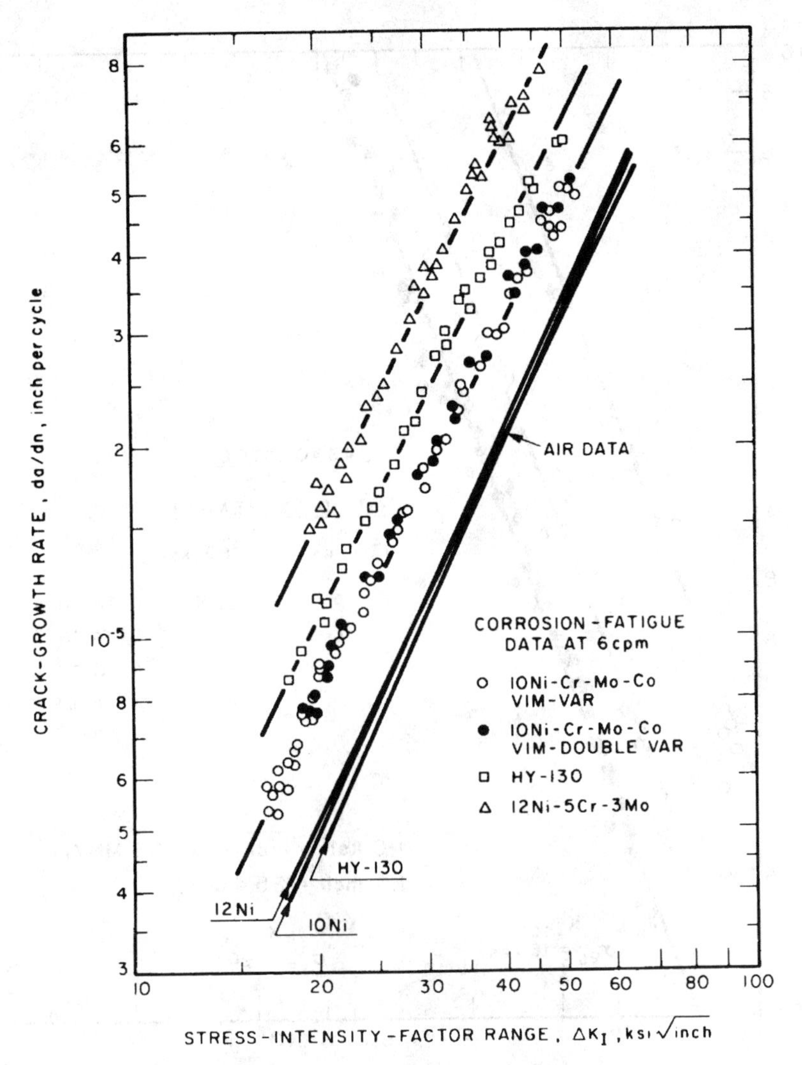

FIG. 11.32 Fatigue-crack-growth rates in air and in 3% solution of sodium chloride below K_{Iscc} for various high-yield-strength steel.

quency have been investigated for high-strength aluminum alloys, magnesium alloys, and titanium alloys [46]. The data suggest that the additions to aqueous solutions that accelerate stress-corrosion crack growth also accelerate the corrosion-fatigue-crack growth and those additions that do not affect stress-corrosion cracking have no effect on corrosion-fatigue-crack growth.

Extensive corrosion-fatigue data have been obtained for various aluminum alloys in water and water vapor environments. The results indicate that these environments have a significant effect on the fatigue-crack-growth rate for alu-

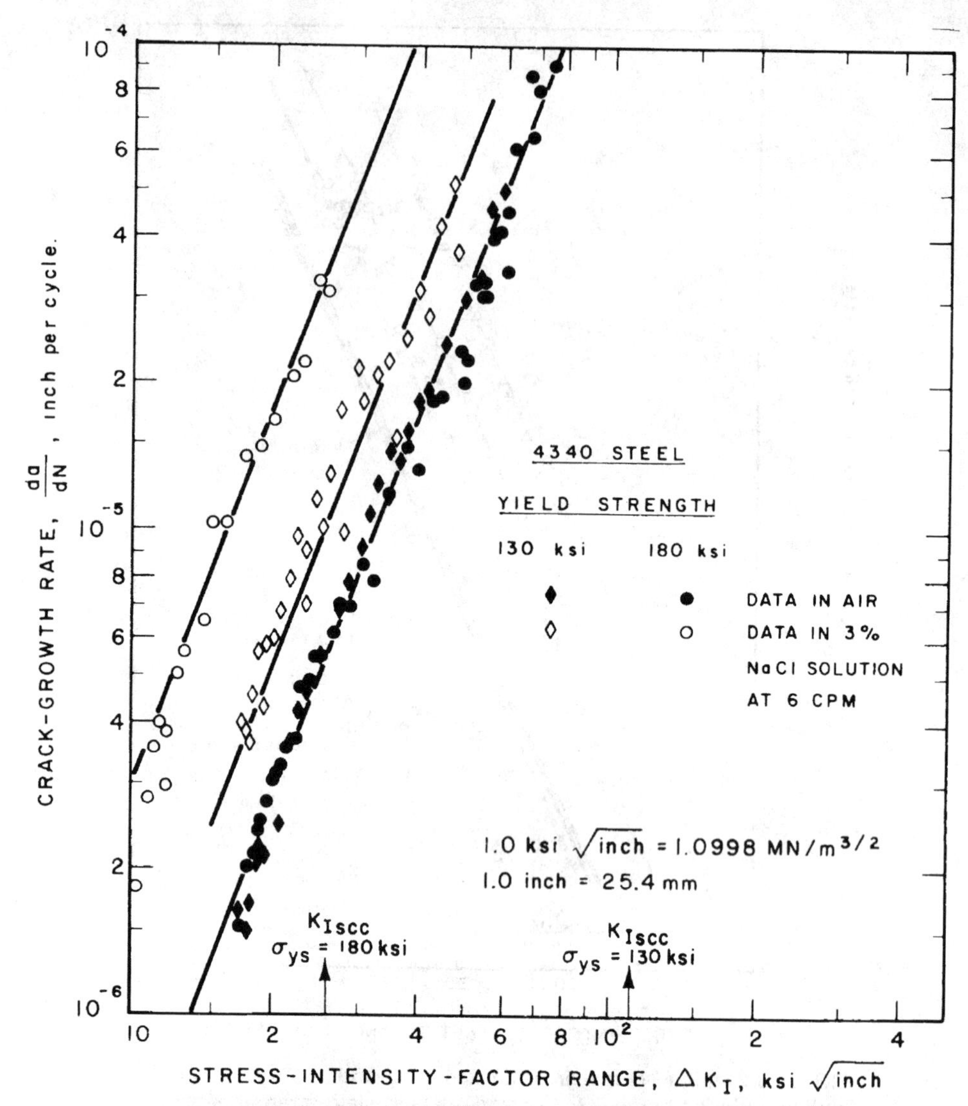

FIG. 11.33 Corrosion-fatigue-crack-growth data for a 4340 steel.

minum alloys [*59,65,73–77*]. Corrosion-fatigue-crack-growth data for various low-yield-strength constructional steels have been investigated below K_{Iscc} to determine the susceptibility of these steels to aqueous environments [*40*]. The steels tested were A36, A588 Grade A, A588 Grade B, A514 Grade E, and A514 Grade F steels in distilled water and in 3% solution of sodium chloride in distilled water. The tests were conducted under constant-amplitude and variable-amplitude random-sequence sinusoidal load fluctuations at frequencies of 60 and 12 cpm. The data presented in Figures 11.34, 11.35, and 11.36 show that the addition of 3% (by weight) sodium chloride to distilled water had no effect on the corrosion-fatigue behavior of these steels. Similar data were obtained for the other steels that were tested [*40*]. The data also showed that the corrosion-fatigue-crack-growth-rate behavior at 60 cpm under sinusoidal loads and under square-wave loads were essentially identical. The increase in scatter was caused by the general corrosion of the specimen surfaces, which decreased the accuracy for determining the exact location of the crack tip. Also, the data show that the corrosion fatigue crack growth rate in distilled water and in 3% solution of sodium chloride in distilled water were essentially identical for the constructional steels investigated. Corrosion-fatigue data obtained by testing these steels in 3% solution of sodium chloride at a stress ratio, *R*, of 0.5 were identical to those obtained at $R = 0.1$, and corrosion-fatigue data obtained for five different heats of A588 steel were also identical [*40*]. Furthermore, corrosion-fatigue crack growth rate for the constructional steels tested at 12 cpm and at stress-intensity-factor fluctuations

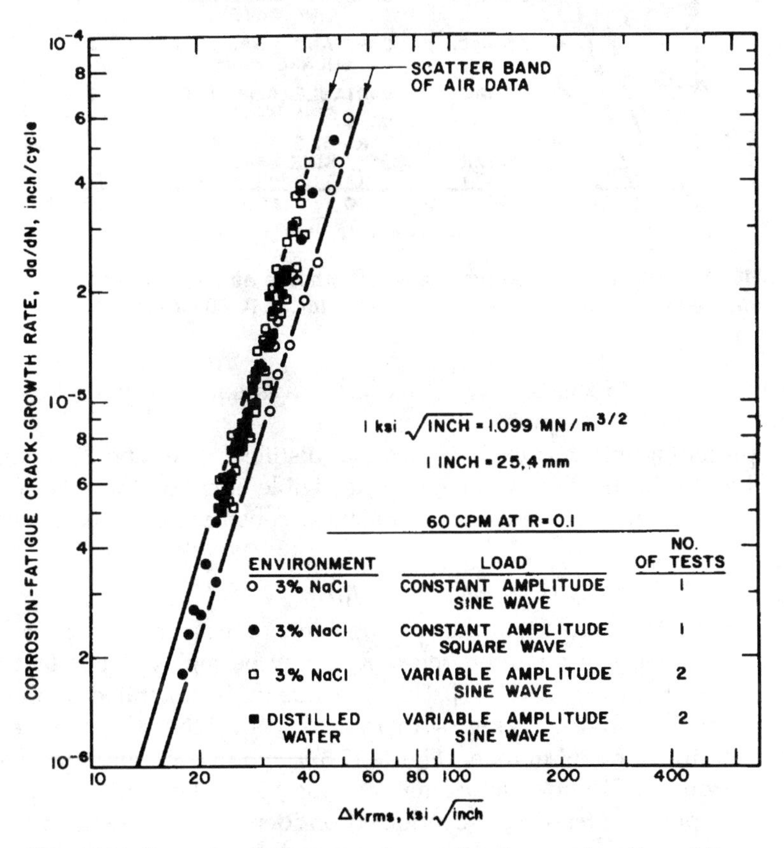

FIG. 11.34 Corrosion-fatigue-crack-growth rate as a function of the root-mean-square stress intensity factor for an A36 steel.

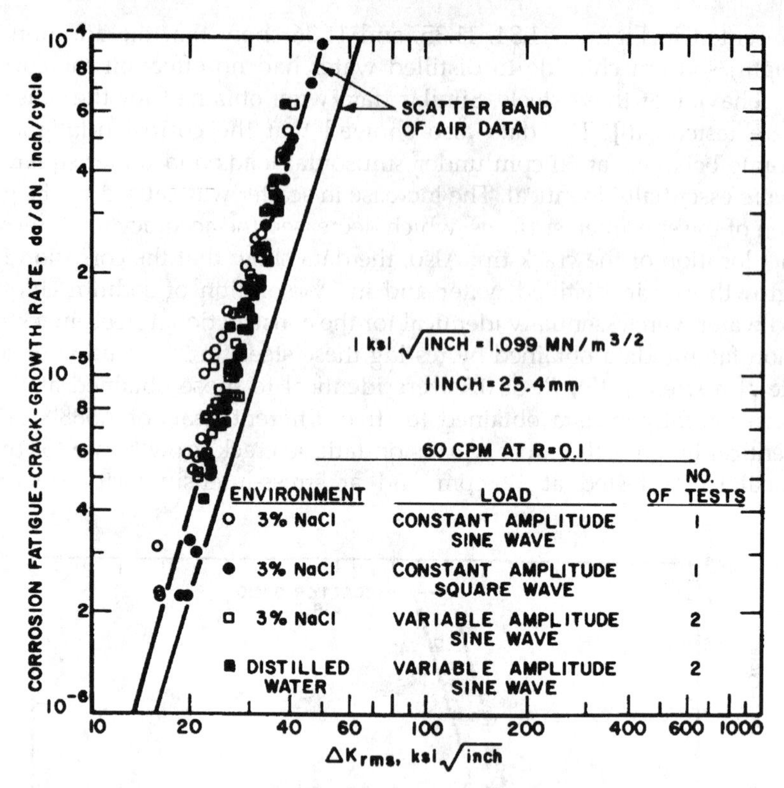

FIG. 11.35 Corrosion-fatigue-crack-growth rate as a function of the root-mean-square stress-intensity factor for an A588 Grade A steel.

greater than about 15 ksi$\sqrt{\text{in.}}$ (16.5 MN/$m^{3/2}$) was equal to or slightly greater than that observed at 60 cpm.

The preceding information indicates that distilled water and 3% solution of sodium chloride in distilled water had a negligible effect on the rate of fatigue crack propagation in the five carbon-manganese constructional steels tested.

11.4.3 Effect of Cyclic-Stress Waveform

Available data indicate that the environmental effects on the rate of fatigue-crack-growth in corrosion fatigue below K_{Iscc} may be highly dependent on the shape of the cyclic-stress wave [35]. This dependence is illustrated by the difference between the fatigue-crack-growth rate data for 12Ni-5Cr-3Mo steel in a room temperature air environment (Figure 11.37), and in a 3% solution of sodium chloride (Figure 11.38) under sinusoidal loading, triangular loading, and square loading at 6 cpm. The tests were conducted on identical specimens in the same bulk environment and at the same maximum and minimum loads. The effect of the cyclic wave shape on the corrosion-fatigue behavior below K_{Iscc} was obtained

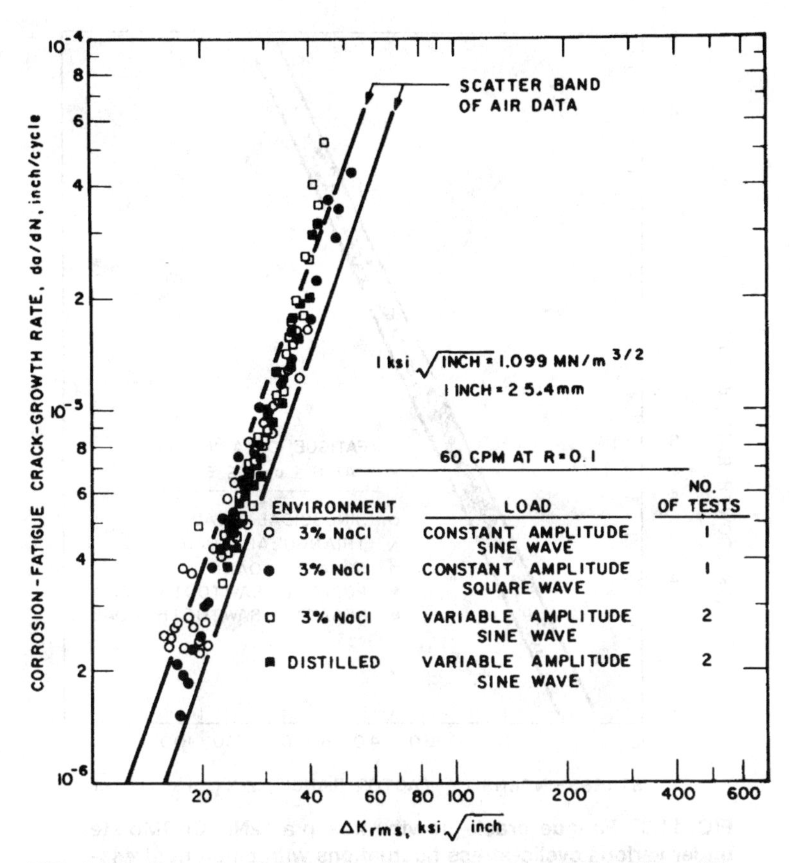

FIG. 11.36 Corrosion-fatigue-crack-growth rate as a function of the root-mean-square stress-intensity factor for an A514 Grade F steel.

from direct comparison between the crack-growth rates per cycle at a constant value of ΔK_I.

The data presented in Figure 11.37 show that the fatigue-crack-growth rates in room temperature air environment are identical under various stress fluctuations and are independent of frequency. The data in Figure 11.38 show that in a sodium chloride solution, the crack-growth rates per cycle under sinusoidal and triangular stress fluctuations are almost identical. At a constant frequency, the environment increased the crack-growth rate by the same amount under sinusoidal stress fluctuations as under triangular stress fluctuations. The data also show that environmental effects are negligible when the steel is subjected to a square-wave stress fluctuation. Corrosion-fatigue-crack-growth rates under square-wave loading at 6 cpm for the 12Ni-5Cr-3Mo steel tested in sodium chloride solution were essentially the same as they were in the absence of environmental effects. By establishing the sinusoidal cyclic frequency that would result

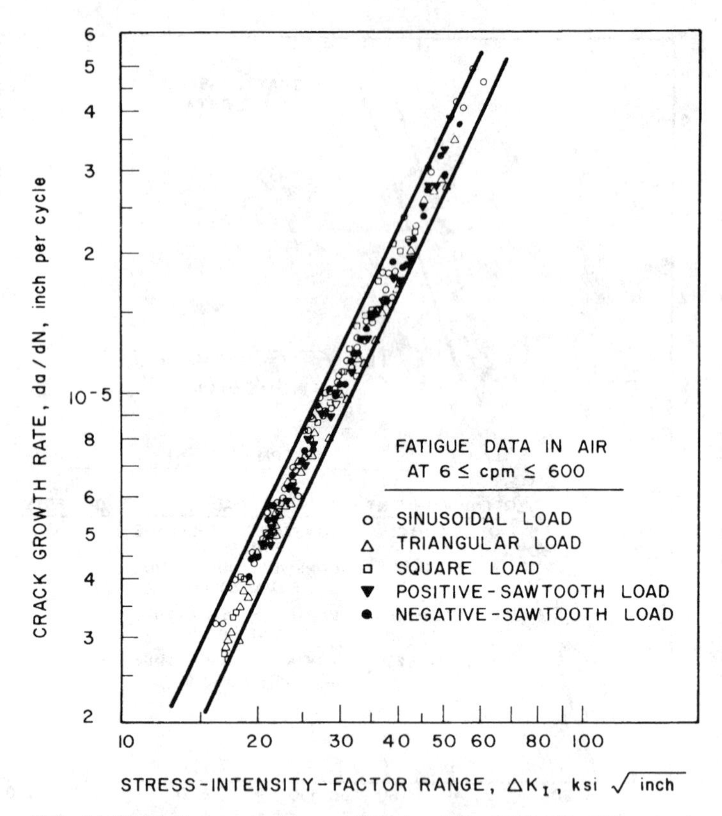

FIG. 11.37 Fatigue-crack-growth rates in a 12Ni-5Cr-3Mo steel under various cyclic-stress fluctuations with different stress-time profiles.

in the same environmental effects on the rate of fatigue crack growth, Barsom [*35*] showed that the environmental damage below K_{Iscc} occurred only during transient loading. Corrosion-fatigue data obtained below K_{Iscc} by using square waves having different dwell times at maximum and minimum loads also showed no environmental effects at constant tensile stresses.

Fatigue-crack-growth data for an aluminum alloy (DTD 5070A) tested in air at 60 cpm showed no difference in the rate of growth under sinusoidal, square, and pulsed waveforms [*78*]. Corrosion-fatigue crack-growth data for 7075-T6 aluminum alloy tested in salt water at 6 cpm under sinusoidal, triangular, and square waveforms showed a behavior very similar to that presented for 12Ni-5Cr-3Mo steel in salt water [*65*].

11.4.4 Environmental Effects During Transient Loading

Corrosion-fatigue-crack-growth test results for 12Ni-5Cr-3Mo maraging steel in 3% sodium chloride solution under sinusoidal, triangular, and square-wave

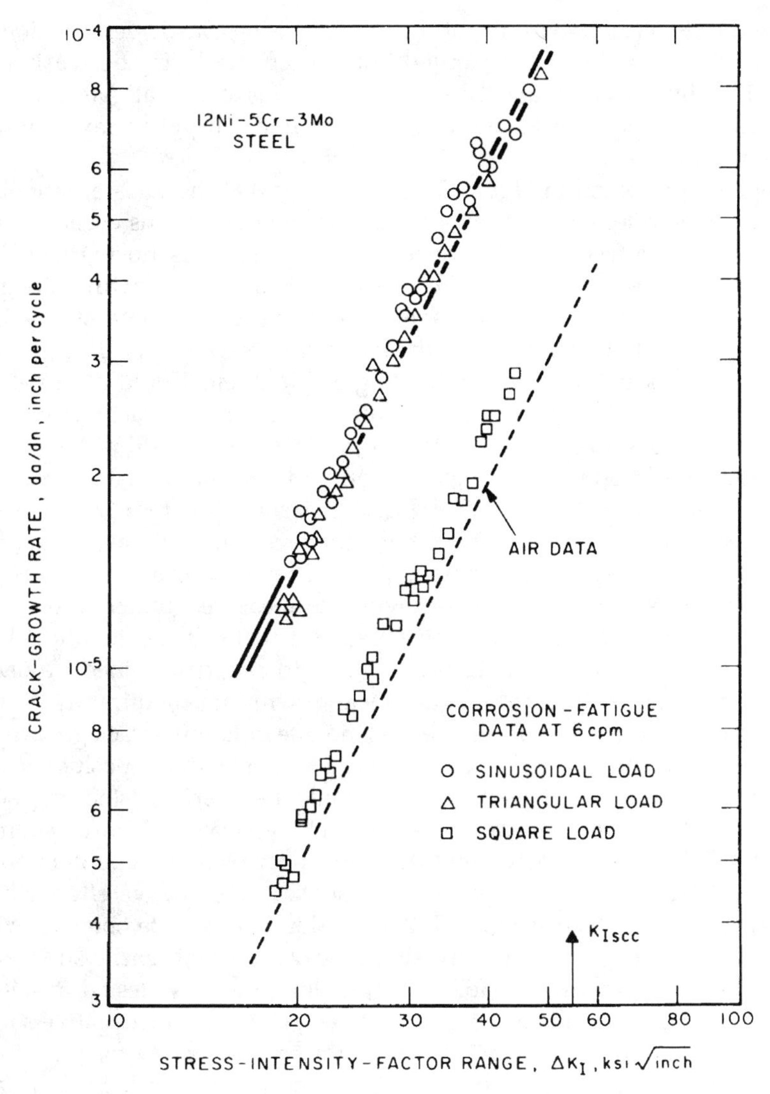

FIG. 11.38 Corrosion-fatigue-crack-growth rates below K_{Iscc} under sinusoidal, triangular, and square loads.

loading showed that environmental effects in the environment–material investigated are significant only during the transient-loading portion of each cyclic-load fluctuation [35]. The difference between the effects of the sodium chloride solution on the rate of fatigue-crack growth during increasing and decreasing plastic deformation in the vicinity of the crack tip was investigated by using test results obtained for specimens subjected to various triangular cyclic-load fluctuations.

The effects of the environment during increasing plastic deformation in the vicinity of the crack tip were separated from the effects during decreasing plastic deformation by studying the differences in the corrosion-fatigue-crack-growth rate obtained under positive-sawtooth (/|) and under negative-sawtooth (|\) cyclic-load fluctuations.

The data presented in Figure 11.37 [*35*] show that the rates of fatigue-crack growth in a room temperature air environment under various cyclic-stress fluctuations are not affected by the form of the cyclic-stress fluctuations. Consequently, differences in the rates of corrosion-fatigue-crack growth among triangular waves, positive-sawtooth waves, and negative-sawtooth waves can be attributed primarily to variations in the interaction between plastically deformed metal at the crack tip and the surrounding environment. These variations result from differences in the pattern of stress fluctuations during each cycle.

The corrosion-fatigue-crack-growth-rate data for 12Ni-5Cr-3Mo maraging steel tested in 3% sodium chloride solution under various cyclic stress fluctuations are presented in Figure 11.39 [*35*]. The data show that the corrosion-fatigue-crack-growth rates determined with the negative-sawtooth wave and with the square wave are essentially the same as the fatigue-crack-growth rate determined in air. The corrosion-fatigue crack-growth rate measured under sinusoidal, triangular, and positive-sawtooth cyclic-stress fluctuations are identical but are three times higher than the fatigue-crack-growth rate determined in air. Thus, the environment increased the fatigue-crack-growth rate significantly.

Because the corrosive effect increased the rate of fatigue-crack growth below K_{Iscc} by the same amount with the triangular wave as with the positive-sawtooth wave, the corrosive processes in the environment–material system investigated were operative *only* while the tensile stresses in the vicinity of the crack tip were increasing. This conclusion is supported by (1) corrosion-fatigue data obtained with the negative-sawtooth wave, which showed no corrosive effect while the tensile stresses were decreasing, and (2) corrosion-fatigue data obtained with the square wave which showed no corrosive effect at constant tensile stresses.

Fatigue-crack-growth data for 7075-T6 aluminum alloys tested in salt water at 6 cpm under sinusoidal, square, positive-sawtooth, and negative-sawtooth loading showed a behavior very similar to the that presented for the 12Ni-5Cr-3Mo maraging steel in salt water [*65*]. However, fatigue-crack-growth data for 7075-T6 aluminum alloy tested at 105 cpm in distilled water showed no significant difference in the rate of growth under positive- and negative-sawtooth loadings [*79*]. Further work is necessary to resolve the apparent discrepancy in the conclusions relating to the effect of waveform on the corrosion-fatigue behavior for high-strength aluminum alloys in water environments.

11.4.5 Generalized Corrosion-Fatigue Behavior

Corrosion-fatigue-crack-propagation behavior is a very complex phenomenon. The preceding discussions show that the behavior is strongly dependent on

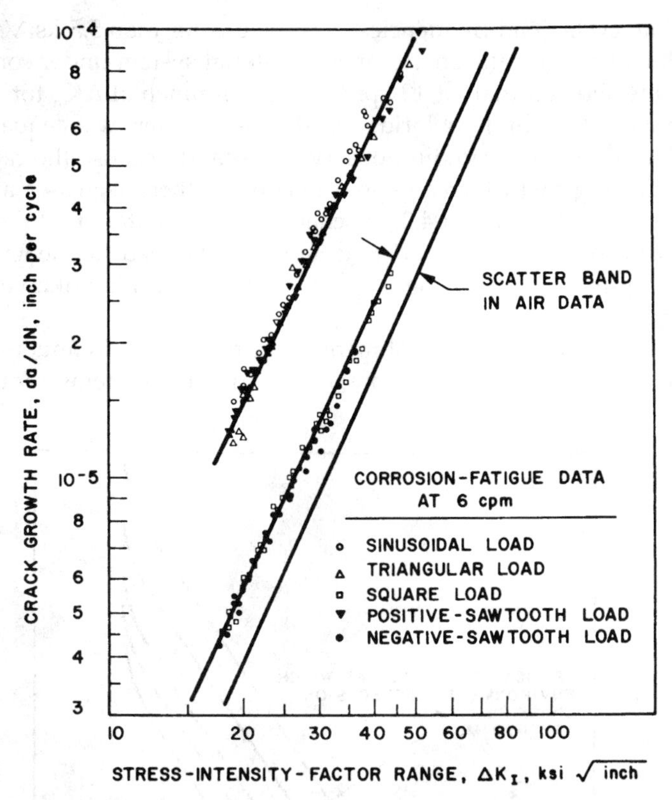

FIG. 11.39 Corrosion-fatigue-crack-growth rates in a 12Ni-5Cr-3Mo steel in 3% solution of sodium chloride under various cyclic-stress fluctuations with different stress-time profiles.

many variables, including frequency, waveform, and stress ratio. Tests at low frequencies are difficult, time consuming, and very costly for Region II behavior and are prohibitive for the threshold behavior. Consequently, a clear understanding of the corrosion-fatigue behavior in the various regions does not exist at the present time. However, based on the available data, a simplified schematic characterization of this behavior may be possible.

In an air environment, the fatigue ΔK_I threshold, ΔK_{th}, in various steels tested at a stress ratio, R, of 0.1 is independent of cyclic-load frequency and is equal to about 5.5 ksi$\sqrt{\text{in}}$. Because hostile environmental effects decrease with increased cyclic-load frequency, the corrosion-fatigue ΔK_{th} at very high cyclic-load frequencies would have a value close to that of fatigue in air. A K_{Iscc} test can be considered a corrosion-fatigue test at extremely low cyclic-load frequency. In such tests, the rate of crack growth at a stress-intensity-factor fluctuation slightly lower than K_{Iscc} is, by definition, equal to K_{Iscc}. Hence, the value of the environmental ΔK_{th}

at intermediate cyclic-load frequencies must be greater than 5.5 ksi$\sqrt{\text{in.}}$ and less than the value of K_{Iscc} for the environment–material system under consideration. The test results showed that, at 12 cpm, the environmental ΔK_{th} for A514 steels in 3% solution of sodium chloride in distilled water was equal to about 11 ksi$\sqrt{\text{in.}}$ Based on the preceding observations, and because the rate of corrosion-fatigue crack growth increases to a maximum then decreases as the cyclic frequency decreases (Section 11.4.2), a schematic representation of the corrosion-fatigue behavior of steels subjected to different cyclic-load frequencies has been constructed (Figure 11.40). This figure is an oversimplification of a very complex phenomenon.

Although significant accomplishments have been made in understanding the corrosion-fatigue behavior for some metal–environment systems, more is needed

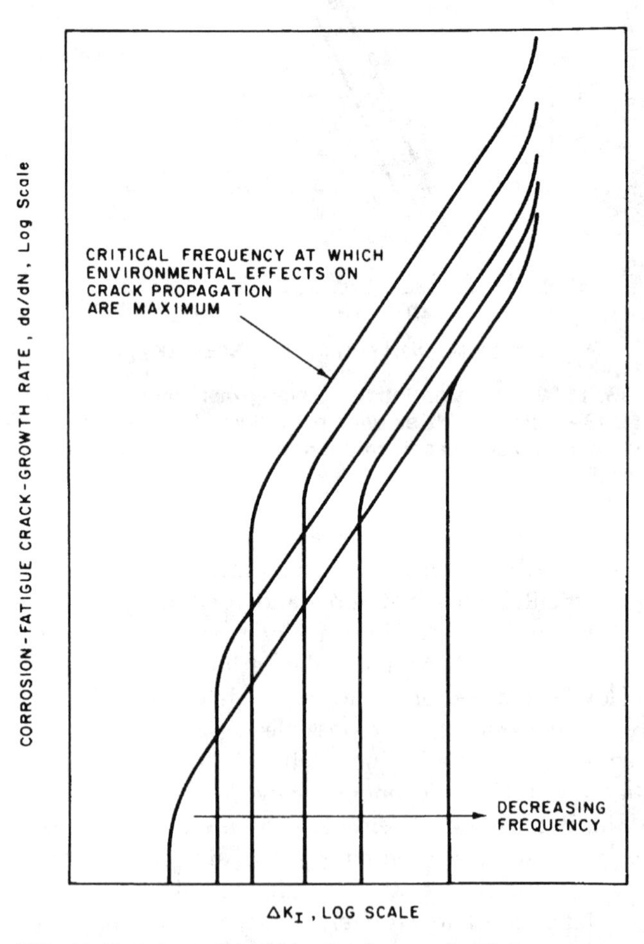

FIG. 11.40 Schematic of idealized corrosion-fatigue behavior as a function of cyclic-load frequency.

to improve the procedures for material selection and for analysis of engineering structures that are subjected to cyclic loads in aggressive environments.

11.5 Prevention of Corrosion-Fatigue Failures

Several methods can be used to prevent or circumvent the detrimental effects of corrosion fatigue on structural performance. These methods may be easy to identify, but their implementation may be difficult or very costly. The effectiveness of a given method or a combination of methods depends on the particular material–environment system under consideration. Although a number of parameters are involved in the selection process, several preventive methods can be eliminated by establishing the basic function of the material and of the environment in the system. For example, changing the characteristics of the environment may not be a viable method if the environment is the desired product of the manufacturing operation.

Some methods that are currently in use to prevent or circumvent the detrimental effects of corrosion fatigue on the performance of structural components are presented in this section. A discussion of the advantages and limitations of these methods is beyond the scope of this document.

1. *Isolate the environment and the material.* This can be accomplished by placing a barrier between the environment and the material. Barriers that have been used include metallic coatings (e.g., zinc, chrome), organic coatings (e.g., paint), inorganic coatings (e.g., glass), ceramic and rubber liners, and clading.
2. *Alter the severity of the environment.* This can be accomplished by chemically removing the aggressive constituents in the environment, by increasing the pH of the environment, or by decreasing the temperature, flow rate, and concentration of the environment.
3. *Apply cathodic protection.* This can be accomplished by externally imposed negative potential or a galvanically generated potential. Sacrificial anodes are frequently used for cathodic protection of structures.
4. *Alter the surface characteristics of the material.* This can be accomplished by inducing favorable compressive stresses on the material surfaces that are exposed to the environment. Compressive surface stress can be induced by using induction hardening (which is used successfully for sucker rods), shot peening, as well as others. These favorable compressive surface stresses may not minimize corrosion-fatigue crack-initiation or general corrosion but could significantly decrease and possibly eliminate crack propagation.
5. *Substitute more resistant material.* In general, materials resistant to a given environment are available. However, care should be exercised in the selection process to ensure: (a) that the substitute material possesses all the other properties that are essential for its use in the particular application; (b) that the substitute material will not fail by a mechanism other than

corrosion fatigue (for example, grain-boundary embrittlement, chloride cracking, etc.) to which the other material was immune; and (c) that the material selection is based on its resistance to corrosion fatigue in the environment rather than its resistance to other damage mechanisms such as corrosion or stress-corrosion cracking in the environment.

6. *Design the components to prevent the initiation or the propagation of cracks to a critical size.* For a given material–environment system, this can be accomplished by using data similar to that presented in this chapter to design the structural components properly against corrosion-fatigue damage or to establish inspection procedures and inspection intervals that would ensure the safe operation of the structure in the environment of interest for the design life of the component (safe life approach).

The preceding presents possible options to prevent or circumvent structural damage caused by corrosion fatigue. The usefulness of any of these methods depends on the particular application. For example, isolating the environment and the material eliminates the detrimental effects of the environment but does not minimize damage caused by fatigue. Also, the effectiveness of these methods to prevent corrosion-fatigue crack initiation may be different than for propagation. For example, cathodic protection can have a significant role in preventing or retarding the initiation of corrosion-fatigue cracks but may have little or no effect on the corrosion-fatigue crack-propagation behavior. Because of these and other considerations, extreme care and sound engineering judgement should be exercised in the selection of the appropriate method to prevent or circumvent corrosion-fatigue damage and to ensure the safety, reliability, and cost effectiveness of the structure.

11.6 References

[1] *Characterization of Environmentally Assisted Cracking for Design—State of the Art*, NMAB-386, National Materials Advisory Board, National Research Council, Washington, DC, January 1982.

[2] Brown, B. F. and Beachem, C. D., "A Study of the Stress Factor in Corrosion Cracking by Use of the Precracked Cantiliver-Beam Specimen," *Corrosion Science*, Vol. 5, 1965.

[3] Johnson, H. H. and Paris, P. C., "Subcritical Flaw Growth," *Engineering Fracture Mechanics*, Vol. 1, No. 1, 1968.

[4] Novak, S. R. and Rolfe, S. T., "Comparison of Fracture Mechanics and Nominal Stress Analysis in Stress Corrosion Cracking," *Corrosion*, Vol. 26, No. 4, April 1970.

[5] Novak, S. R. and Rolfe, S. T., "K_{Ic} Stress-Corrosion Tests of 12Ni-5Cr-3Mo and 18Ni-8Co-2Mo Maraging Steels and Weldments," *AD482761L*, Defense Documentation Center, Arlington, VA, January 1, 1966.

[6] Novak, S. R., "Comprehensive Investigation of the K_{Iscc} Behavior of Candidate HY-180/210 Steel Weldments," *U.S. Steel Applied Research Laboratory Report No. 89.021-024(1)(B-63105)*, Defense Documentation Center, Arlington, VA, December 31, 1970.

[7] Novak, S. R. and Rolfe, S. T., "Fatigue-Cracked Cantilever Beam Stress-Corrosion Tests of HY-80 and 5Ni-Cr-Mo-V Steels," *AD482783L*, Defense Documentation Center, Arlington, VA, January 1, 1966.

[8] Sprowls, D. O., Schumaker, M. B., Walsh, J. D., and Coursen, J. W., "Evaluation of Stress-Corrosion Cracking Using Fracture Mechanics Techniques," *Contract NAS 8-21487, Final Report,* George C. Marshall Space Fight Center, Huntsville, AL., May 31, 1973.

[9] Novak, S. R. and Rolfe, S. T., "Modified WOL Specimen for K_{Iscc} Environmental Testing," *Journal of Materials, JMLSA,* ASTM, Philadelphia, Vol. 4, No. 3, September 1969.

[10] Steigerwald, E. A., "Delayed Failure of High-Strength Steel in Liquid Environments," *Proceedings of the American Society for Testing Materials,* Vol. 60, Philadelphia, 1960.

[11] Johnson, H. H. and Willner, A. M., "Moisture and Stable Crack Growth in a High-Strength Steel," *Applied Materials Research,* January 1965.

[12] Tiffany, C. F. and Masters, J. N., "Applied Fracture Mechanics," *Fracture Toughness Testing and Its Applications, ASTM STP 381,* American Society for Testing and Materials, Philadelphia, April 1965.

[13] Novak, S. R., "Effect of Prior Uniform Plastic Strain on the K_{Iscc} of High-Strength Steels in Sea Water," *Engineering Fracture Mechanics,* Vol. 5, No. 3, 1973.

[14] Judy, Jr., R. W. and Goode, R. J., "Stress-Corrosion Cracking Characteristics of Alloys of Titanium in Salt Water," *NRL Report 6564,* Naval Research Laboratory, Washington, DC, July 21, 1967.

[15] Brown, B. F., "Stress-Corrosion Cracking and Corrosion Fatigue of High-Strength Steels," *Problems in the Load-Carrying Application of High-Strength Steels, DMIC Report 210,* Defense Metals Information Center, Battelle Memorial Institute, Columbus, OH, October 26–28, 1964.

[16] Brown, B. F., "A New Stress-Corrosion Cracking Test for High Strength Alloys," *Materials Research and Standards,* Vol. 6, No. 3, March 1966.

[17] Leckie, H. P., "Effects of Environment on Stress Induced Failure of High-Strength Maraging Steels," *Fundamental Aspects of Stress Corrosion Cracking, NACE-1,* National Association of Corrosion Engineers, Houston, 1969.

[18] Smith, H. R., Piper, D. E., and Downey, F. K., "A Study of Stress-Corrosion Cracking by Wedge-Force Loading," *Engineering Fracture Mechanics,* Vol. 1, No. 1, 1968.

[19] Huber, R. W., Goode, R. J., and Judy, Jr., R. W., "Fracture Toughness and Stress-Corrosion Cracking of Some Titanium Alloy Weldments," *Welding Journal,* Vol. 46, No. 10, October 1967.

[20] Peterson, M. H., Brown, B. F., Newbegin, R. L., and Groover, R. E., "Stress Corrosion Cracking of High Strength Steels and Titanium Alloys in Chloride Solutions at Ambient Temperature," *Corrosion,* Vol. 23, 1967.

[21] Speidel, M. O., "Stress Corrosion Cracking of Aluminum Alloys," *Metallurgical Transactions,* Vol. 6A, No. 4, April 1975.

[22] Speidel, M. O. and Hyatt, M. V., "Stress Corrosion Cracking of High-Strength Aluminum Alloys," *Advances in Corrosion Science and Technology,* Vol. II, M. G. Fontana and R. W. Staehle, Eds., Plenum, New York, 1972.

[23] Uhlig, H. H., *The Corrosion Handbook,* John Wiley, New York, 1972.

[24] Logan, H. L., *The Stress Corrosion of Metals,* John Wiley, New York, 1966.

[25] Loginow, A. W., "Stress Corrosion Testing of Alloys," *Materials Protection,* Vol. 5, No. 5, May 1966.

[26] Sprowls, D. O. and Brown, R. H., "What Every Engineer Should Know About Stress Corrosion of Aluminum," *Metals Progress,* Vol. 81, Nos. 4 and 5, 1962.

[27] Imhof, E. J. and Barsom, J. M., "Fatigue and Corrosion-Fatigue Crack Growth of 4340 Steel at Various Yield Strengths," *ASTM STP 536,* American Society for Testing and Materials, Philadelphia, 1973.

[28] Barsom, J. M., "Corrosion-Fatigue Crack Propagation Below K_{Iscc}," *Engineering Fracture Mechanics,* Vol. 3, No. 1, July 1971.

[29] "Standard Method for Test for Plane-Strain Fracture Toughness of Metallic Materials," ASTM E 399, *Annual Book of ASTM Standards,* American Society for Testing and Materials, Philadelphia, 1974.

[30] Wilson, W. K., "Review of Analysis and Development of WOL Specimen," *67-7D7-BTLPV-R1,* Westinghouse Research Laboratories, Pittsburgh, March 1967.

[31] Wilson, W. K., "Analytical Determination of Stress Intensity Factors for the Manjoine Brittle Fracture Test Specimen," *WERL-0029-3,* Westinghouse Research Laboratories, Pittsburgh, August 1965.
[32] Leven, M. M., "Stress Distribution in the M4 Biaxial Fracture Specimen," *65-1D7-STRSS-S1,* Westinghouse Research Laboratories, Pittsburgh, March 1965.
[33] Wessel, E. T., "State of the Art of the WOL Specimen for K_{Ic} Fracture Toughness Testing," *Engineering Fracture Mechanics,* Vol. 1, No. 1, June 1968.
[34] Clark, Jr., W. G., *Evaluation of the Fatigue Crack Initiation Properties of Type 403 Stainless Steel in Air and Steam Environment, ASTM STP 559,* American Society for Testing and Materials, Philadelphia, 1974.
[35] Barsom, J. M., "Effect of Cyclic-Stress Form on Corrosion-Fatigue Crack Propagation Below K_{Iscc} in a High-Yield-Strength Steel," *Corrosion Fatigue: Chemistry, Mechanics and Microstructure, NACE-2,* National Association of Corrosion Engineers, Houston, 1972.
[36] Brown, B. F. and Beachem, C. D., "Specimens for Evaluating the Susceptibility of High Strengths Steels to Stress Corrosion Cracking," Internal Report, U.S. Naval Research Laboratory, Washington, DC, 1966.
[37] Clark, Jr., W. G., "Applicability of the K_{Iscc} Concept to Very Small Defects," *ASTM STP 601,* American Society for Testing and Materials, Philadelphia, 1976.
[38] Wei, R. P., Novak, S. R., and Williams, D. P., "Some Important Considerations in the Development of Stress Corrosion Cracking Test Methods," *Materials Research and Standards, MTRSA,* Vol. 12, No. 9, 1972.
[39] Benjamin, W. D. and Steigerwald, E. A., "An Incubation Time for the Initiation of Stress-Corrosion Cracking in Precracked 4340 Steel," *Transactions of the American Society for Metals,* Vol. 60, No. 3, September 1967.
[40] Barsom, J. M. and Novak, S. R., "Subcritical Crack Growth and Fracture of Bridge Steels," NCHRP Report 181, Transportation Research Board, Washington, DC, 1977.
[41] Yoder, G. R., Griffis, C. A., and Crooker, T. W., "Sustained-Load Cracking of Titanium—A Survey of 6Al-4V Alloys," *NRL Report 7596,* Naval Research Laboratory, Washington, DC, 1973.
[42] "Stress Corrosion Testing," *ASTM STP 425,* American Society for Testing and Materials, Philadelphia, December 1967.
[43] Staehle, R. W., Forty, A. J., and Van Rooyen, D., Eds., *Fundamental Aspects of Stress-Corrosion Cracking, NACE-1,* National Association of Corrosion Engineers, Houston, 1969.
[44] Brown, B. F., Ed., *Stress-Corrosion Cracking in High-Strength Steels and in Titanium and Aluminum Alloys,* Naval Research Laboratory, Washington, DC, 1972.
[45] Agrawal, A., Brown, B. F., Kruger, J., and Staehle, R. W., Eds., *U.R. Evans Conference on Localized Corrosion,* NACE-3, National Association of Corrosion Engineers, Houston, 1971.
[46] Speidel, M. O., Blackburn, M. J., Beck, T. R., and Feeney, J. A., "Corrosion Fatigue and Stress Corrosion Crack Growth in High Strength Aluminum Alloys, Magnesium Alloys, and Titanium Alloys Exposed to Aqueous Solutions," *Corrosion Fatigue: Chemistry, Mechanics, and Microstructure, NACE-2,* National Association of Corrosion Engineers, Houston, 1972.
[47] Taylor, M. E. and Barsom, J. M., "Effect of Cyclic Frequency on the Corrosion-Fatigue Crack-Initiation Behavior of ASTM A517 Grade F Steel," *Fracture Mechanics: Thirteenth Conference, ASTM STP 743,* Richard Roberts, Ed., American Society for Testing and Materials, Philadelphia, 1981.
[48] Novak, S. R., "Corrosion-Fatigue Crack-Initiation Behavior of Four Structural Steels," *Corrosion-Fatigue: Mechanics, Metallurgy, Electrochemistry, and Engineering, ASTM STP 801,* American Society for Testing and Materials, Philadelphia, 1983.
[49] Novak, S. R., "Influence of Cyclic-Stress Frequency and Stress Ratio on the Corrosion-Fatigue Crack-Initiation Behavior of A588-A and A517-F Steels in Salt Water," presented at the Fifteenth National Symposium on Fracture Mechanics, University of Maryland, College Park, MD, 1982.

[50] Barsom, J. M., "Concepts of Fracture Mechanics—Fatigue and Fracture Control," *Fracture Mechanics for Bridge Design, Federal Highway Administration, Report No. FHWA-RD-78-69,* U.S. Department of Transportation, Washington, DC, July 1977.
[51] Barsom, J. M. "Long-Life Behavior and the Effect of Low Frequency on the Corrosion-Fatigue Crack-Initiation for Steels," unpublished data.
[52] *Fatigue Thresholds—Fundamentals and Engineering Applications,* J. Backlund, A. F. Blom, and C. J. Beevers, Eds., Engineering Materials Advisory Services Ltd., Vols. I and II, The Chameleon Press Ltd., London, 1982.
[53] Suresh, S., Zamiski, G. F., and Ritchie, R. O., "Oxide-Induced Crack Closure: An Explanation of Near-Threshold Corrosion Fatigue Crack Growth Behavior," *Metallurgical Transactions A,* Vol. 12A, August 1981, pp. 1435–1443.
[54] Ritchie, R. O., Suresh, S., and Moss, C. M., "Near-Threshold Fatigue Crack Growth in a 2 1/4Cr-1Mo Pressure Vessel Steel in Air and Hydrogen," *Journal of Engineering Materials and Technology, Transactions of ASME,* Series H, Vol. 102, 1980, p. 293.
[55] Suresh, S. and Ritchie, R. O., "Closure Mechanisms for the Influence of Load Ratio on Fatigue Crack Propagation in Steels," Materials Science Division, U.S. Department of Energy, *Report Contract No. DE-AC03-76SF00098,* Washington, DC, April 1982.
[56] *Corrosion Fatigue: Chemistry, Mechanics, and Microstructure,* International Corrosion Conference Series, Vol. NACE-2, National Association of Corrosion Engineers, Houston, 1972.
[57] Clark, Jr., W. G., "The Fatigue Crack Growth Rate Properties of Type 403 Stainless Steel in Marine Turbine Environments," *Corrosion Problems in Energy Conversion and Generation,* C. S. Tedman, Jr., Ed., Electrochemical Chemistry Society, Princeton, NJ, 1974.
[58] Crooker, T. W. and Lange, E. A., "The Influence of Salt Water on Fatigue Crack Growth in High Strength Structural Steels," *ASTM STP 462,* American Society for Testing and Materials, Philadelphia, 1970.
[59] Bradshaw, F. J. and Wheeler, C., "The Influence of Gaseous Environment and Fatigue Frequency on the Growth of Fatigue Cracks in Some Aluminum Alloys," *International Journal of Fracture Mechanics,* Vol. 5, No. 4, December 1969.
[60] Meyn, D. A., "An Analysis of Frequency and Amplitude Effects on Corrosion Fatigue Crack Propagation in Ti-8Al-1Mo-1V," *Metallurgical Transactions,* Vol. 2, 1971.
[61] Gallagher, J. P., "Corrosion Fatigue Crack Growth Behavior Above and Below K_{Iscc}," *NRL Report 7064,* Naval Research Laboratory, Washington, DC, May 28, 1970.
[62] Miller, G. A., Hudak, S. J., and Wei, R. P., "The Influence of Loading Variables on Environment-Enhanced Fatigue-Crack Growth in High Strength Steels," *Journal of Testing and Evaluation,* Vol. 1, No. 6, 1973.
[63] Paris, P. C., Bucci, R. J., Wessel, E. T., Clark, W. G., and Mager, T. R., "Extensive Study of Low Fatigue-Crack-Growth Rates in A533 and A508 Steels, *ASTM STP 513,* American Society for Testing and Materials, Philadelphia, 1972.
[64] Bucci, R., "Environment Enhanced Fatigue and Stress Corrosion Cracking of a Titanium Alloy Plus a Simple Model for Assessment of Environmental Influence of Fatigue Behavior," Ph.D. dissertation, Lehigh University, Bethlehem, PA, 1970.
[65] Selines, R. J. and Pelloux, R. M., *Effect of Cyclic Stress Wave Form on Corrosion Fatigue Crack Propagation in Al-Zn-Mg Alloys,* Department of Metallurgy and Materials Science Report, M.I.T., Cambridge, MA, 1972.
[66] Wei, R. P. and Landes, J. D., "Correlation Between Sustained-Load and Fatigue-Crack Growth in High-Strength Steels," *Materials Research and Standards, MTRSA,* Vol. 9, No. 7, July 1969.
[67] Bucci, R. J. and Donald, J. K., *Fatigue and Fracture Investigation of a 200 Grade Maraging Steel Forging,* Del Research Corporation Report, Hellertown, PA, October 1972.
[68] Skelton, R. P. and Haigh, J. R., "Fatigue Crack Growth Rates and Thresholds in Steels Under Oxidizing Conditions," *Materials Science and Engineering,* Vol. 36, 1978, pp. 17–25.
[69] Barsom, J. M., "Fatigue Behavior of Pressure-Vessel Steels," *Welding Research Council (WRC) Bulletin,* No. 194, May 1974.

[70] Barsom, J. M., "Investigation of Subcritical Crack Propagation," Ph.D dissertation, University of Pittsburgh, 1969.

[71] Crooker, T. W. and Lange, E. A., "Corrosion Fatigue Crack Propagation of Some New High Strength Structural Steels," *Journal of Basic Engineering, Transactions, ASME,* Vol. 91, 1969.

[72] Barsom, J. M., Sovak, J. F., and Imhof, Jr., E. J., "Corrosion-Fatigue Crack Propagation Below K_{Iscc} in Four High-Yield-Strength Steels," *Applied Research Laboratory Report 89.021-024(3),* U.S. Steel Corporation (available from the Defense Documentation Center), Arlington, VA, December 14, 1970.

[73] Feeney, J. A., McMillan, J. C., and Wei, R. P., "Environmental Fatigue Crack Propagation of Aluminum Alloys at Low Stress Intensity Levels," *Metallurgical Transactions,* Vol. 1, June 1970.

[74] Hartman, A., "On the Effect of Oxygen and Water Vapor on the Propagation of Fatigue Cracks in 2024-TB3 Alclad Sheet," *International Journal of Fracture Mechanics,* Vol. 1, No. 3, September 1965.

[75] Bradshaw, F. J. and Wheeler, C., "The Effect of Environment on Fatigue Crack Growth in Aluminum and Some Aluminum Alloys," *Applied Materials Research,* Vol. 5, No. 2, 1966.

[76] Wei, R. P., "Fatigue-Crack Propagation in a High-Strength Aluminum Alloy," *International Journal of Fracture Mechanics,* Vol. 4, No. 2, June 1968.

[77] Wei, R. P. and Landes, J. D., "The Effect of D_2O on Fatigue Crack Propagation in a High Strength Aluminum Alloy," *International Journal of Fracture Mechanics,* Vol. 5, 1969.

[78] Bradshaw, F. J., Gunn, N. J. F., and Wheeler, C., "An Experiment on the Effect of Fatigue Wave Form on Crack Propagation in an Aluminum Alloy," *Technical Memorandum MAT93,* Royal Aircraft Establishment, Farnborough, Hants, England, July 1970.

[79] Hudak, S. J. and Wei, R. P., "Comments," on paper by J. M. Barsom, "Effect of Cyclic Stress Form on Corrosion Fatigue Crack Propagation Below K_{Iscc} in High Yield Strength Steel," *Corrosion Fatigue: Chemistry, Mechanics, and Microstructure,* International Corrosion Conference Series, Vol. NACE-2, National Association of Corrosion Engineers, Houston, 1972.

Part IV: Fracture and Fatigue Control

12

Fracture and Fatigue Control

12.1 Introduction

THE OBJECTIVE in the structural design of large complex structures, such as bridges, ships, pressure vessels, buildings, etc., is to optimize the desired performance, safety requirements, and cost (i.e., the overall cost of materials, design, fabrication, operation, and maintenance). In other words, the purpose of engineering design is to produce a structure that will perform the operating function efficiently, safely, and economically. To achieve these objectives, engineers make predictions of service loads and conditions, calculate stresses in various structural members resulting from these loads and service conditions, and compare these stresses with the critical stresses for the particular failure modes that may lead to failure of the structure. Members are then proportioned and materials specified so that failure does not occur by any of the possible failure modes. Because the response to loading can be a function of the member geometry, an iterative process may be necessary.

Possible failure modes that usually are considered are:

1. General yielding or excessive plastic deformation.
2. Buckling or general instability, either elastic or plastic.
3. Subcritical crack growth (fatigue, stress-corrosion, or corrosion fatigue) leading to loss of section or unstable crack growth.
4. Unstable crack extension, either ductile or brittle, leading to either partial or complete failure of a member.

Although other failure modes exist, such as corrosion or creep, the abovementioned failure modes are the ones that usually receive the greatest attention by structural engineers. Furthermore, of these four failure modes, engineers usually concentrate on only the first two and assume that proper selection of materials and design stress levels will prevent the other two failure modes from occurring. This reasoning is not always true and has led to several catastrophic structural failures. In good structural design, all possible failure modes should be considered.

In the case of fracture or fatigue, many of the fracture-control guidelines that have been followed to minimize the possibility of fractures in structures are familiar to structural engineers. These guidelines include knowing the service conditions to which the structure will be subjected, the use of structural materials with adequate fracture toughness, elimination or minimization of stress raisers, control of welding procedures, proper inspection, and the like. When these general guidelines are integrated into specific requirements for a particular structure, they become part of a fracture-control plan. A fracture-control plan is therefore a specific set of recommendations developed for a particular structure and should not be indiscriminantly applied to other structures.

The development of a fracture-control plan for large structures, such as bridges, pressure vessels, and ships, is complex. Despite the difficulties, attempts to formulate a fracture-control plan for a given application, even if only partly successful, should result in a better understanding of the possible fracture characteristics of the structure under consideration.

The total useful life of a structural component subjected to repeated loading generally is determined by the time necessary to initiate a crack and to propagate the crack from subcritical dimensions to a critical size. The life of the component can be prolonged by extending the crack-initiation, subcritical-crack propagation, and unstable-crack propagation (fracture) characteristics of structural materials. These factors are primary considerations in the formulation of fracture-control guidelines for structures.

Unstable crack propagation is the final stage in the useful life of a structural component subject to failure by the fracture mode. This stage is governed by the material fracture toughness, the crack size, and the stress level. Consequently, unstable crack propagation cannot be attributed only to material fracture toughness, or only to stress conditions, or only to poor fabrication, but rather to particular combinations of the foregoing factors. However, if any of these factors is significantly different from what is usually obtained for a particular type of structure, the possibility of failure may increase markedly.

Structural materials that have adequate fracture toughness to prevent fractures at service temperatures and loading rates are available. However, when these structural materials are used in conjunction with inadequate design or poor fabrication, or both, the safety and reliability of a structural component cannot be guaranteed. Thus, the useful life of a structural component depends on the magnitude and fluctuation of the applied stress, the magnitude of the stress-concentration factors and constraint, the size, shape, and orientation of any initial discontinuities, the stress-corrosion susceptibility, the fatigue characteristics, and the corrosion-fatigue behavior of the structural material in the environment of interest. Because most of the useful life is expended in initiating and propagating cracks at low values of the stress-intensity factor, an increase in the fracture toughness of the steel may have a small effect on the useful life of a structural component whose primary mode of failure is fatigue. Despite these facts, oversimplification of failure analyses has led some to advocate the philosophy that structures should be fabricated of "forgiving materials." Such materials are characterized as having sufficient fracture toughness to fracture in a ductile manner at operating temperatures and under impact loading, even though the structure may be subjected to slow or intermediate loading rates. The use of these materials is advocated to "forgive" any mistakes that may be committed by the designer, fabricator, inspector, or user. While this approach to ensure safety and reliability of structures may work in some cases, it perpetuates a false sense of security, places an unjustifiable burden on the structural material, and unnecessarily increases the cost of the structure.

Another prevalent yet unfounded philosophy advocates that the primary cause of fractures in welded structures is the inherent inferior characteristics of the weldments. These characteristics include yield-strength residual stresses and weld discontinuities. Advocates of this philosophy often neglect to consider environmental effects, cyclic history, and stress redistribution caused by load fluctuation or proof tests. Furthermore, when an obvious weld discontinuity does not exist in the vicinity of the fracture origin, the cause of failure is attributed to "micro-weld" defects. Although residual stress and weld discontinuities can contribute to failure, the foregoing oversimplification can lead to an incorrect fracture analysis of a structural failure. The use of oversimplification or gross exaggerations in the analysis of failures, exemplified by these philosophies, can lead to erroneous conclusions. Correct diagnosis and preventive action can be established only after a thorough study of all the pertinent parameters related to the specific problem under consideration. An integrated look at these parameters and their synergistic effect on the safety and reliability of a structure is necessary to develop a fracture-control plan.

As pointed out earlier, the recent development of fracture mechanics has been extremely helpful in synthesizing the various elements of fracture-control plans into more unified quantitative plans than was possible previously. Specifically, fracture mechanics has shown that although numerous factors (e.g., service loading, material fracture toughness, design, welding, residual stresses) can contribute to fractures in large welded structures, there are three primary factors that control the susceptibility of a structure to fracture, namely,

1. Fracture toughness of a material at a particular service temperature, loading rate, and plate thickness.
2. Size, shape, and orientation of a crack or a discontinuity at possible locations of fracture initiation.
3. Tensile stress level, including effects of residual stress, stress concentrations, and constraint.

When the particular combination of stress and crack size in a structure reaches the critical stress-intensity factor for a particular specimen thickness and loading rate, fracture can occur. It is the specific intent of any fracture-control plan to establish the possible ranges of K_I that might be present throughout the lifetime of a structure because of the various service loads and to ensure that the critical stress-intensity factor (K_c, K_{Ic}, $K_{Ic}(t)$, K_{Id}, or other critical values such as K_{Iscc}) for the materials used is sufficiently large so that the structure will have a safe life. Figure 12.1 is a schematic showing the relation between stress, flaw size, and fracture toughness, as has been discussed previously in this book.

One of the key questions in developing a fracture-control plan for any particular structure is how large must the degree of safety and reliability be for the particular structure in question. The degree of safety and reliability needed, sometimes referred to as the factor of safety, is often specified by a code. However, the degree of safety depends on many additional factors such as consequences of failure or redundancy and, thus, varies even within a generic class of structures.

Accordingly, a fracture-control plan is developed only for the specific structure under consideration and can vary from one which must, in essence, provide assurance of very low probability of service failures to one which may allow for occasional failures. An example of the former situation would be a nuclear power

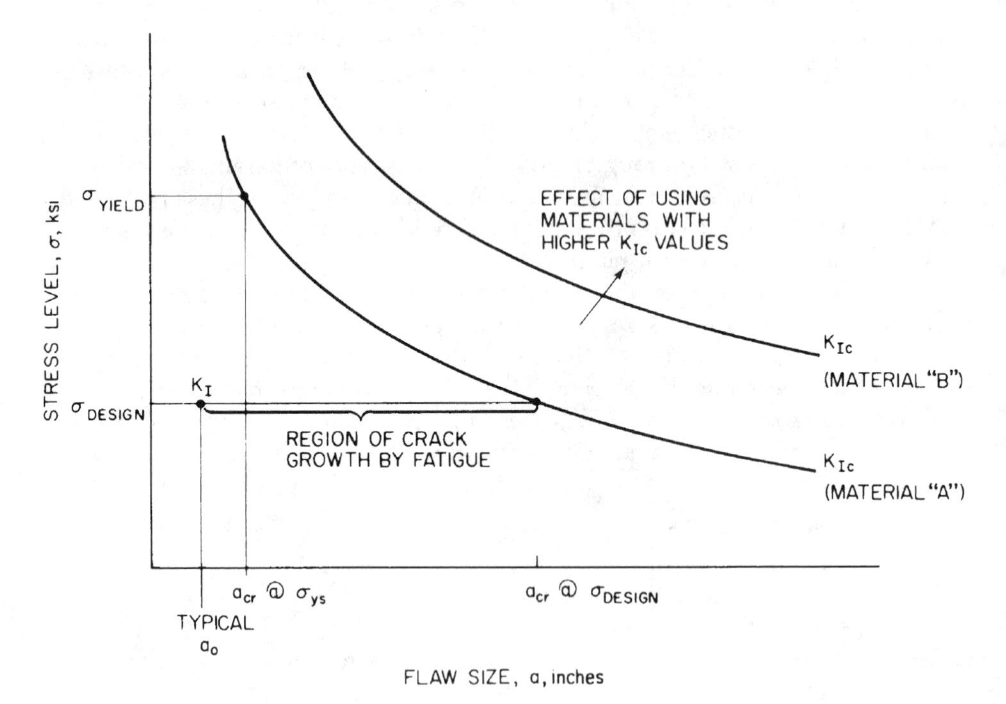

FIG. 12.1 Schematic showing relation among stress, critical flaw sizes, and material fracture toughness.

plant structure where the consequences of a structural failure are such that even a minor failure is not tolerable. In the latter case, where the consequences of failure might be minimal, it would be more efficient and economical to maintain and replace parts periodically rather than design them so that no failures occur. An example of a situation where service failure could be tolerated might be the loading bed of a dump truck where periodic inspection would indicate when a plate would need to be repaired or replaced. In this case, the consequences of failure would be minor.

In commenting on fracture-control plans, George Irwin [*1*] noted that,

> "For certain structures, which are similar in terms of design, fabrication method, and size, a relatively simple fracture control plan may be possible, based upon extensive past experience and a minimum adequate fracture toughness criterion. It is to be noted that the fracture control never depends solely upon maintaining a certain average fracture toughness of the material. With the development of service experience, adjustments are usually made in the design, fabrication, inspection, and operating conditions. These adjustments tend to establish adequate fracture safety with a material quality which can be obtained reliably and without excessive cost."

12.2 Historical Background

Prior to about 1940, fracture-control plans or fracture-safety guidelines did not exist in a formalized sense. Most large structures were built from low-strength materials using thin, riveted plates with the structural members arranged so that in the event of failure of one plate the fracture was usually arrested at the riveted connections. Thus, although there were some exceptions as noted in Chapter 1, most failures were not catastrophic. While it is true that the number of failures, in particular structural situations, was reduced by various design modifications based on experience from previous failures, the *first general overall fracture-control guideline* was merely to use lower design stress levels. One of the early fracture-control applications was in the boiler and pressure-vessel industry where the allowable stress was decreased continually as a certain percentage of the maximum tensile stress, thereby decreasing the number of service failures in succeeding years.

A second general guideline to fracture control was to eliminate stress concentrations as much as possible. In the early 1940s, brittle fracture as a potentially serious problem for large-scale structures was brought to the designer's attention by the large number of World War II ship failures. The majority of fractures in the Liberty ships started at the square hatch corners of square cutouts at the top of the shear strake. The design changes involving rounding and strengthening of the hatch corners, removing square cutouts in the shear strake, and adding rivets and crack arresters in various locations led to an immediate reduction in the incidence of failures in these vessels. Thus, the second general type of fracture-control guideline was that of design improvements to minimize stress concentrations.

The ship-failure problem called attention to a general problem involving welding. The World War II shipbuilding program was the first large-scale use of welding to produce monolithic structures where cracks could propagate continuously from one plate to another, in contrast to arresting in multiple-plate riveted structures. While problems of fracture were apparent prior to the large-scale use of welding, there was evidence that the nature of welded structures provided for continued extension of a fracture, which can result in total failure of a structure. Previous experience with riveted structures indicated that brittle fractures were usually limited to a single plate.

The *third general type of fracture control* was to improve the notch toughness of materials. The first material-control guideline occurred about the late 1940s and early 1950s following extensive research on the cause of the ship failures. The particular fracture-control guideline was the observation that plates having a CVN impact energy value greater than 10 ft-lb at the service temperature did not exhibit fracture but rather arrested any propagating cracks. Studies indicated that generally the ship fractures experienced during World War II did not occur at temperatures above the 10-ft-lb Charpy V-notch impact energy transition temperature. This observation led to the development of the 15-ft-lb transition-temperature criterion as a means of fracture control. Early in the 1950s, development of the nil-ductility transition-temperature test led to another general fracture-control guideline for structural materials, namely, that the NDT temperature of structural steels should be below the service temperature when loaded dynamically.

Since that time, the development of general criteria for prevention of brittle fractures, or at least analyses of methods for designing to prevent brittle fractures, have been numerous. These included the development of the fracture analysis diagram (FAD) by the Naval Research Laboratory, the early use of fracture-mechanics concepts in the Polaris missile motor case failures, and the turbine-rotor generator spin test analyses. In the 1960s and 1970s, rapid development of fracture mechanics as a research tool and eventually as an engineering tool led to the use of fracture-mechanics concepts in the development of fracture criteria for several structures such as nuclear pressure vessels [2], various aircraft structures, and bridges [3]. In the 1980s and 1990s fracture-control plans became more prevalent, for example, the space shuttle and offshore drilling rigs. Two recent documents used in fracture and fatigue control plans are PD6493-Guidance On Methods for Assessing the Acceptability of Flaws in Fusion Welded Structures [4], and ASME Section XI Rules for In Service Inspection of Nuclear Power Plant Components [5]. Recently FEMA issued Interim Guidelines for Welded Steel Moment Frame Structures [6,7].

However, the three basic elements of fracture control, that is, (1) use a lower design stress, (2) minimize stress concentrations, and (3) use materials with improved notch toughness, have long been known to engineers. Fracture mechanics' basic contribution is to make these guidelines *quantitative* and to show the relative importance of each of these elements.

Although the literature has been dominated with the use of fracture mechanics as a research tool, George Irwin [*1*] has stated that:

> "The practical objective of fracture mechanics is a continuing increase in the efficiency with which undesired fractures of structural components are prevented. The fracture control plans most commonly used in the past have not possessed a high degree of efficiency. Generally the visible portions of such plans consist in statements of minimum required toughness and in statements of inspection standards. No quantitative connection between these two fracture control elements is employed in establishing these statements. The unseen parts consist mainly in adjustments of design and fabrication. After enough years of experience so that further adjustments of design and fabrication are unnecessary, the fracture control plan is complete. The plan then consists of certain minimum fracture toughness requirements and adherence to certain inspection requirements plus the state-of-art methods of design and fabrication which fracture failure experience showed to be desirable or necessary. Proof testing has the great advantage, where employed, because much of the fracture failure experience necessary for development of the plan tends to occur in the proof test rather than in service. The use of transition temperature based measurements rather than fracture mechanics based measurements of fracture toughness is not a significant disadvantage to the reliability of such plans, once established. However, the efficiency may be low and use of fracture mechanics methods would be helpful in designing modifications toward improved efficiency."

12.3 Fracture and Fatigue Control Plan

Section 12.1 described general elements of a fracture-control plan. It should be reemphasized that *all* possible failure modes should be considered during a structural design. Textbooks on the design of particular types of structures (bridges, buildings, pressure vessels, aircraft, etc.) should be used to design to prevent failures by typical failure modes such as buckling, yielding, and so on, and appropriate textbooks should be consulted to design to prevent environmental failures such as by corrosion. The purpose of this particular book is to provide technical information and design guidelines that can be used to prevent failure by fracture or subcritical crack growth leading to fracture.

The four general elements of a fracture-control plan are as follows:

1. Identification of factors that may contribute to the fracture of a structural member or to the failure of an entire structure. Include an analysis of service conditions and loadings.

2. Establishment of the relative contribution of each of the factors to a possible fracture or failure of a member or a structure.
3. Determination of the relative efficiency and tradeoffs of the various factors to minimize the possibility of either failure of a member or of the structure.
4. Recommendation of specific design considerations to ensure the safety and reliability of the structure against failure. These would include specific requirements for material properties, design stress levels, fabrication procedures, and inspection requirements. Codes often provide some of this information, but it is the responsibility of the Engineer of Record to ensure that all failure modes have been addressed adequately.

These four elements are described in general in the following sections, realizing that the details of any fracture-control plan depend on the particular structure being analyzed.

12.3.1 *Identification of the Factors*

The first step in all structural design is to established the probable loads and service conditions throughout the design life of a structure. Usually the live loads are specified by codes, for example, American Association of State Highway and Transportation Officials (AASHTO) for bridges, or performance criteria such as operating pressures in pressure vessels or payload requirements in aircraft structures. These types of loadings usually are reasonably well defined, although there are certain types of structures such as ships where the loading is not well defined. Nonetheless, the first step is to establish the probable loads.

From a fracture-toughness viewpoint, a major decision is determining the *rate* at which these loads are applied, since this establishes whether K_c or K_{Ic} for slow loading, or a $K_{Ic}(t)$ value for an intermediate loading rate, or K_{Id} for impact loading should be the controlling fracture toughness parameter.

Wind loads, hurricane loadings, sea-wave loadings, and so on are established by field measurements in the particular location of the structure in the world. Also, field measurements on similar types of structures are used to estimate the effect of these loads on structural response and behavior. Earthquake loadings for particular regions are based on experience plus various code requirements, as well as a judgement factor related to the degree of conservatism desired for a particular structure.

Repeated or fatigue loading is usually established by the particular design function of the structure. That is, bridges are subjected to fatigue by the movement of vehicles, and thus fatigue must be considered in the design of bridges, whereas, except for earthquake loadings, the loads on buildings are primarily dead loads so fatigue is usually neglected. Fatigue loadings can be constant amplitude, such as rotating machinery, or variable amplitude, such as for bridges or aircraft structures. Regardless of the type of loading, fatigue can result in subcritical crack growth by various means, as was described in Chapters 7–11. Thus, even though the initial flaw size may be small (based on quality of fabrication), the possibility of larger cracks is present when the structure is subjected to repeated loading.

Chemical or environmental factors such as corrosion, stress corrosion, cavitation, and so on must be considered for various structures, depending on the particular design function of the structure. Crack growth by stress corrosion and corrosion fatigue were described in Chapter 11.

From a fracture viewpoint, temperature can have a significant effect on the service behavior of structural members. For those structures fabricated from materials that exhibit a brittle to ductile transition in behavior (primarily the low- to medium-strength structural steels), the minimum service temperature must be established.

Quality of fabrication and assurance that quality is actually achieved generally controls the initial crack or defect size. These items are very important factors that should be established so that some estimate of possible initial flaw can be made.

Obviously, the inherent fracture toughness (K_c, K_{Ic}, $K_{Ic}(t)$, or K_{Id}) of a structural material based on the particular chemistry and thermo-mechanical treatment is a primary factor that may contribute to the fracture behavior of the material.

The point is that for each structure for which a fracture-control plan is desired, the designer should consider all possible loadings and service conditions *before* the selection of materials or allowable stresses. In this sense, the basis of structural design in all large complex structures should be an attempt to optimize the *desired performance requirements* in terms of the service loadings relative to *cost considerations* so that the probability of failure is low. Accordingly, if the possible failure mode is brittle fracture, then the $(K_c/\sigma_{ys})^2$ ratio (at the appropriate loading rate, temperature, and plate thickness) should be selected to minimize the probability of fracture. However, if the desired performance requirements are such that the overall weight of the structure must be minimized, then *another* possible "failure mode" would be *nonperformance* because of excessive weight, and therefore a higher allowable design stress must be used. Because the allowable design stress is usually some percentage of the yield strength, a high-yield-strength material usually is specified.

These two requirements of (1) a high $(K_c/\sigma_{ys})^2$ ratio and (2) a high yield strength (σ_{ys}) are sometimes conflicting, and a compromise often must be reached. However, as long as this analysis is made *prior* to material selection and establishment of design or allowable stresses, the designer has a very good chance of achieving a "balanced design" in which the *performance requirements* as well as the *prevention of failure requirements* both are met as economically as possible. A design example for selection of a material for a specific pressure-vessel application based both on performance and minimum weight was presented in Chapter 6 and showed the advantages of the fracture-mechanics approach to design.

Often it is not possible to change materials for a particular structure because of existing codes, past practices, or economics. Also, in the case of a fitness-for-service, or life-extension analysis, the materials and fabricated details already are in existence. In these cases, a limited fracture-control plan still can be effective by reducing the design stress levels, restricting the range of operating temperatures, improving the quality of inspection, and so on. Thus, even though it is desirable to develop complete fracture-control plans during the early design stages, it is not always possible to do so. However, considerable benefits can still be realized by limited fracture-control plans developed at later stages in the life of a structure. This "plan" would be based on a fitness-for-service analysis. In all cases, the designer should identify as closely as possible all service conditions and loadings to which the structure will be subjected or in the case of a fitness-for-service analysis, the loadings to which the structure is being subjected.

12.3.2 *Establishment of the Relative Contribution*

Fracture mechanics technology has shown that the many factors that can contribute to fracture in large welded structures can be incorporated into the three primary factors that control the susceptibility of a structure to fracture, namely (1) tensile stress level (σ) and stress range ($\Delta\sigma$) if loaded in fatigue, (2) material fracture toughness (preferably obtained in terms of K_c, K_{Ic}, $K_{Ic}(t)$ or K_{Id}) and (3) crack size, shape, and orientation (a). As shown schematically in Figure 12.1, these factors can be related quantitatively using the K_I relationships developed in Chapter 2.

The contribution of the various types of loadings to the possibility of a fracture occurs primarily in the calculation of the maximum value of stress which can occur in the vicinity of a crack. The calculation of these stresses ranges from simple calculation of

$$P/A$$

or

$$\frac{Mc}{I}$$

to extremely complex finite element analysis solutions for various structural shapes such as plates, shells, and box girders.

For welded construction, the possibility of residual stresses exists, and the local stress level can be of yield-strength magnitude. In fact, it is often assumed that local yielding exists in the vicinity of welds or even severe stress concentrations. The ductility of structural materials is relied upon to redistribute these stresses so that premature failure does not occur. However, as will be noted in Chapter 16, the inherent ductility of steels can be greatly constrained by the development of triaxial stress in highly constrained welded connections.

If the possibility of fracture exists, the designer may want to assume that localized yield stresses are present in portions of the structure where cracks can be present and to calculate the resistance to fracture accordingly. That is, determine the critical crack size at yield stress loading (Figure 12.1) and compare it with the maximum possible flaw size based on fabrication and inspection capabilities.

Figure 12.2 is a schematic showing the effect of local residual stresses on crack growth as well as the effects of plane-strain or plane-stress conditions on subsequent fatigue-crack growth. Figure 12.3 is a schematic showing the effect of service temperature on critical flaw size, and Figure 12.4 is a schematic showing the design use of K_{Iscc}.

Referring to Figures 12.1–12.4, the following general conclusions can be made with respect to fracture control regarding the relative contribution of stress level, material fracture toughness, and crack size.

1. In regions of high residual stress, where the actual stress can equal the yield stress over a small region, the critical crack size is computed for σ_{yield} instead of the design stress, σ_{design}. If, for the particular structural material being used, both the base metal and the weld metal are sufficiently tough (e.g., Material B in Figure 12.1), the critical crack size for full yield stress loading should be satisfactory. Under fatigue loading, a crack can grow out of the residual stress zone, and the critical crack size becomes the value at

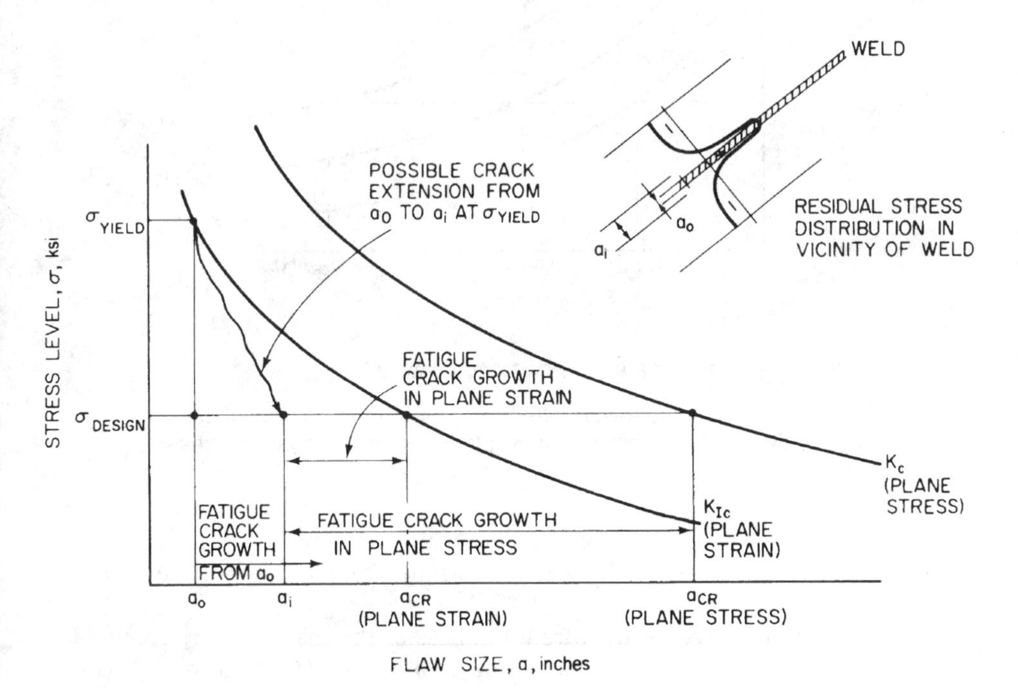

FIG. 12.2 Schematic showing effect of local residual stresses and plane-strain to plane-stress transition (loss of constraint) on fatigue-crack growth.

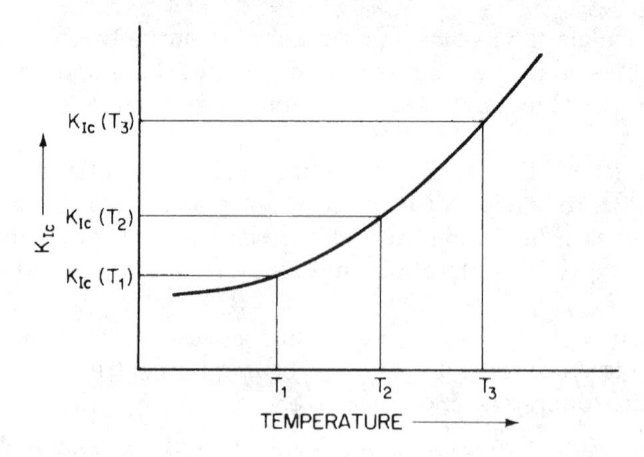

(a) EFFECT OF TEMPERATURE ON K_{Ic}

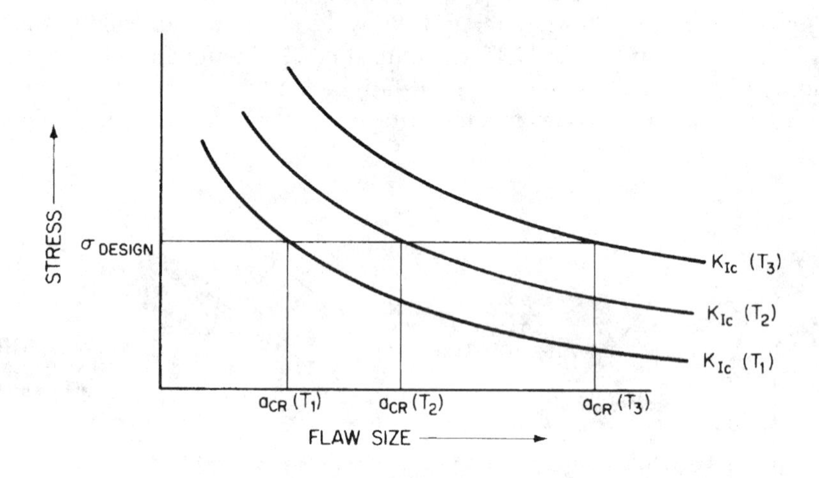

(b) EFFECT OF K_{Ic} ON CRITICAL FLAW SIZE

FIG. 12.3 Schematic showing effect of temperature on critical flaw size: (*a*) Effect of temperature on K_{Ic}; (*b*) effect of K_{Ic} on critical flaw size.

the design stress level. Note that the "critical crack size" in a particular material is dependent on the particular design stress level, and, therefore, is *not* a material property.

2. Depending on the level of fracture toughness of the material, any crack which does initiate in the presence of residual stresses (a_o, Figure 12.2) could arrest quickly as soon as the crack propagates out of the region of high residual stress. However, the initial crack size for any subsequent fatigue-crack growth then may be fairly large (a_i, Figure 12.2).

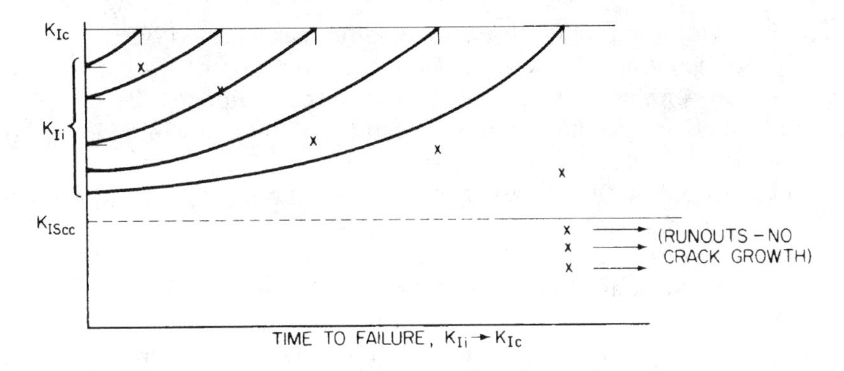

(a) RELATION BETWEEN K_{Ii}, K_{Ic}, K_{Iscc}

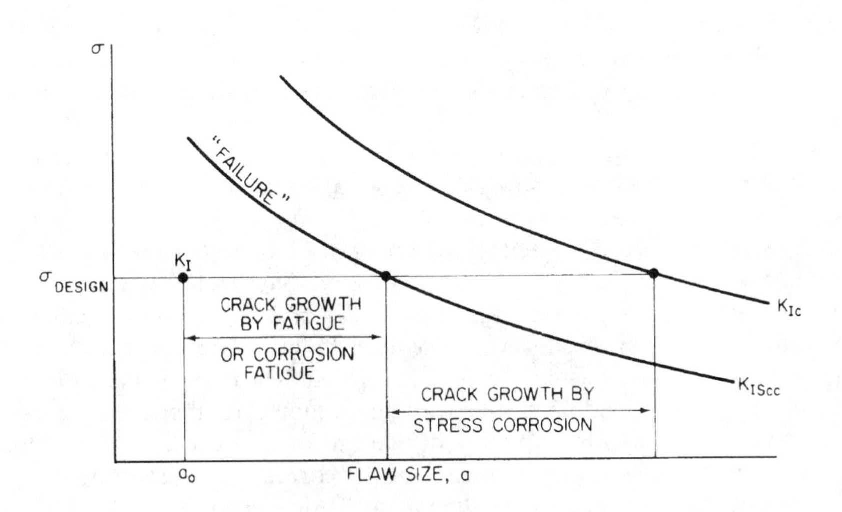

(b) RELATION BETWEEN CRACK GROWTH BY FATIGUE OR CORROSION FATIGUE AND BY STRESS CORROSION

FIG. 12.4 Schematic showing design use of K_{Iscc}: (*a*) relation among K_{Ii}, K_{Ic} and K_{Iscc}; (*b*) relation between crack growth by fatigue or corrosion fatigue and by stress corrosion.

3. For the design stress level, determine the calculated critical crack size. If it is large compared to the plate thickness, subcritical crack growth (by fatigue) should lead to relaxation of the constraint ahead of the crack, resulting in plane-stress or elastic-plastic behavior. For this case the critical stress-intensity factor for plane stress, K_c, will be greater than K_{Ic} or K_{Id}, which is an additional degree of conservatism (Figure 12.2).

4. Materials with low fracture toughness values can be used if
   (a) the applied stresses under tensile loads are reduced,
   (b) residual compressive stresses applied to the component by shotpeening or induction case hardening are used to inhibit crack initiation such as the case for gear teeth or landing gears,
   (c) the distribution of stresses causes the initiation, propagation and arrest of cracks in a decreasing stress field,
   (d) the crack orientation is not in a critical plane to cause unstable crack extension such as the case for "shelling" in rails.
5. The relative contribution of temperature is, of course, to establish the particular level of fracture toughness to be used in calculating the stress crack-size trade-offs. That is, as shown in Figure 12.3*a*, K_{Ic} (or K_{Id}) can vary with temperature for certain structural materials. Thus, the K_{Ic} values at temperatures T_1, T_2, and T_3 are different and lead to different values of critical flaw size even though the design stress remains constant, Figure 12.3*b*.
6. The relative effect of fatigue loading is to grow a crack of initial size a_0 to a_{cr} (Figure 12.1). The number of cycles (or time in the case of stress corrosion) required to grow an existing crack to critical size was discussed in Chapters 7–11.
7. Subcritical crack growth by stress corrosion was discussed in Chapter 11. For materials susceptible to stress corrosion, a reasonable design practice for sustained-load applications is to use K_{Iscc} as a limiting design curve (rather than K_{Ic} so that "failure" is defined as the start of stress-corrosion crack growth, as shown in Figure 12.4*b*. This conservative practice is followed because once stress-corrosion crack growth is started, it is only a matter of time until the K_I level reaches K_{Ic} and complete failure occurs. That is, fatigue-crack growth is usually considered to be deterministic because the number of cycles of loading can be estimated. However, the rate of stress-corrosion crack growth is extremely difficult to predict because of the large number of variables such as chemistry of the corrodent, concentration of corrodent at the crack tip, and temperature. Thus, once the K_{Iscc} level is reached, a realistic design approach for sustained-load applications is to consider failure to be imminent.

Using these guidelines, the relative contribution of material properties, (K_c, K_{Ic}, $K_{Ic}(t)$, K_{Id}, K_{Iscc}, da/dN), stress level (σ), and crack size (a) can be established for a given set of service conditions for a particular structure.

12.3.3 *Determination of Relative Efficiency*

Previously, it has been established that the three primary factors that control the susceptibility of a structure to fracture are:

1. Tensile stress level, including effects of constraint, residual stresses, and stress concentrations.

2. Size, shape, and orientation of the crack or the discontinuities.
3. Material fracture toughness at the particular temperature, loading rate, and plate thickness.

Thus, there are three general design approaches to minimize the possibility of fracture in a structural member, and each of these is directly related to the preceding factors:

1. Decrease the tensile stress level.
2. Minimize initial discontinuities.
3. Use materials with improved fracture toughness.

Each of these design approaches has been used by engineers in various types of structures for many years. The technology of fracture mechanics merely makes the process more quantitative. Using fracture-mechanics terminology, fracture control is simply making sure that $K_I < K_c$ (or K_{Ic}, K_{Id}, etc.) at all times, much like keeping $\sigma < \sigma_{ys}$ to prevent yielding.

For example, Figure 12.5 shows that for the same quality of fabrication and inspection as well as the same critical material fracture toughness, K_{Ic} or K_c, reducing the design stress or stress fluctuation leads to a new margin of safety. This new margin of safety can be either a larger margin of safety against fracture or an increased fatigue life because of the possibility of larger subcritical crack growth before failure.

Figure 12.6 shows the general effect of improving the quality of fabrication and inspection while using the same design stress level and material. Finally,

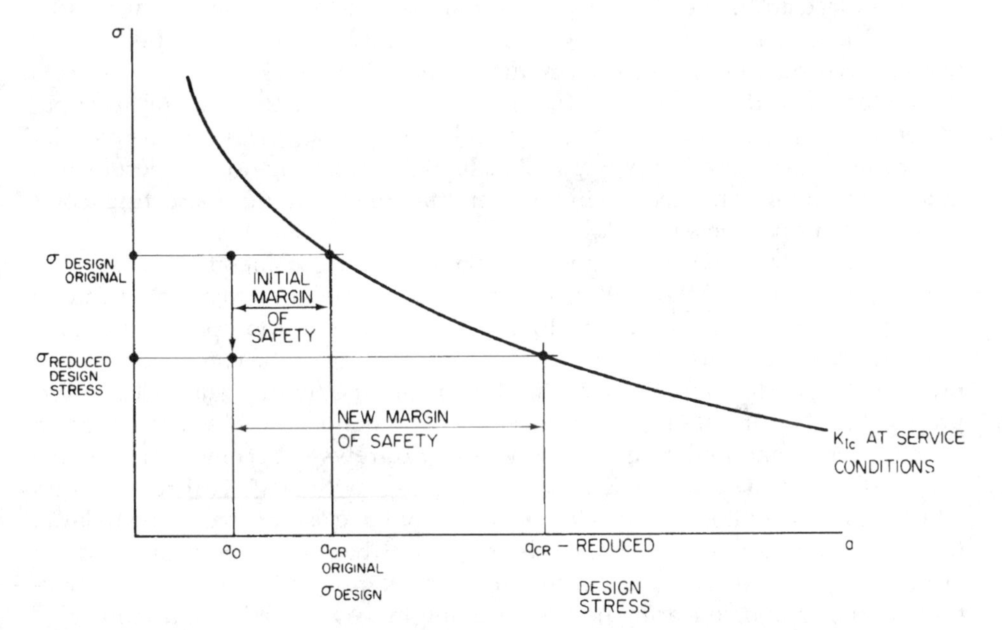

FIG. 12.5 Schematic showing effect of lowering the design stress on fracture control.

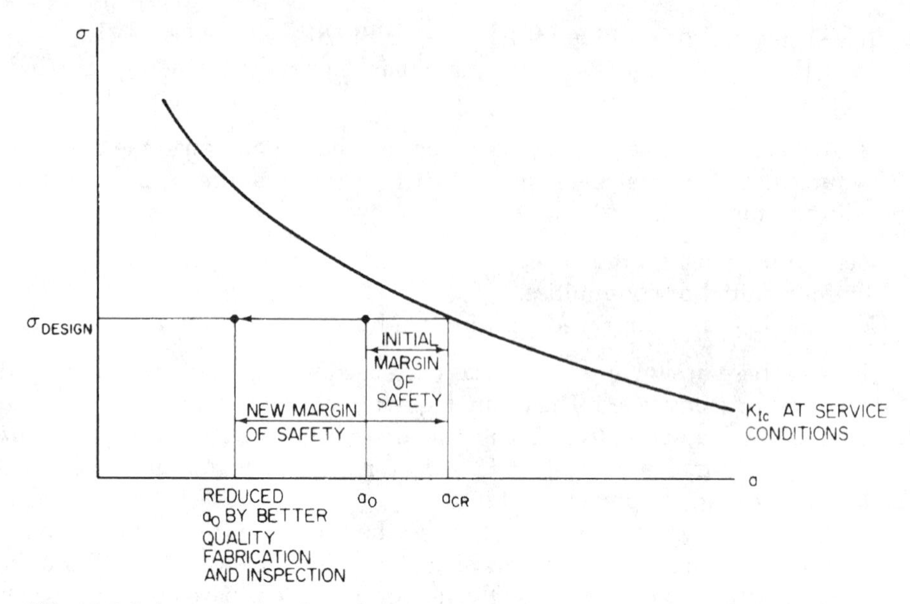

FIG. 12.6 Schematic showing effect of reducing the initial flaw size on fracture control.

Figure 12.7 shows the general effect of using a structural material with improved fracture toughness and the same design stress and quality of fabrication. These examples lead to the general conclusion that because $K_{Ic} \simeq C\sigma\sqrt{a}$, it would be expected that increasing K_{Ic} or decreasing σ would have a greater effect on the resistance to fracture than would reducing the initial crack size, a_o. Also, it is usually easier to determine σ or K_{Ic} than a_o. However, because the primary parameter that governs the rate of subcritical crack growth is the stress-intensity factor range, ΔK_I, raised to power of 2 or higher, decreasing σ or $\Delta\sigma$ results in a much more significant increase in the useful fatigue life of most structural components than does increasing K_{Ic}.

For those cases where fatigue-crack growth is a consideration, such as in bridges or ships, the total useful design life of a structural component can be estimated from the time necessary to initiate a crack and to propagate the crack from subcritical dimensions to the critical size. The life of the component can be prolonged by extending the crack-initiation life and/or the subcritical-crack-propagation life. In an engineering sense, the initiation stage is that region in which a very small initial discontinuity or crack grows to become a measurable propagating fatigue crack. The subcritical-crack-growth stage is that region on which a propagating fatigue crack follows one of the existing crack-growth laws, for example, $da/dN = A \cdot \Delta K^m$, as described in Chapter 9. The unstable crack growth stage is that region in which fatigue-crack growth is very rapid, or fracture occurs, or ductile tearing occurs, resulting in loss of section and failure.

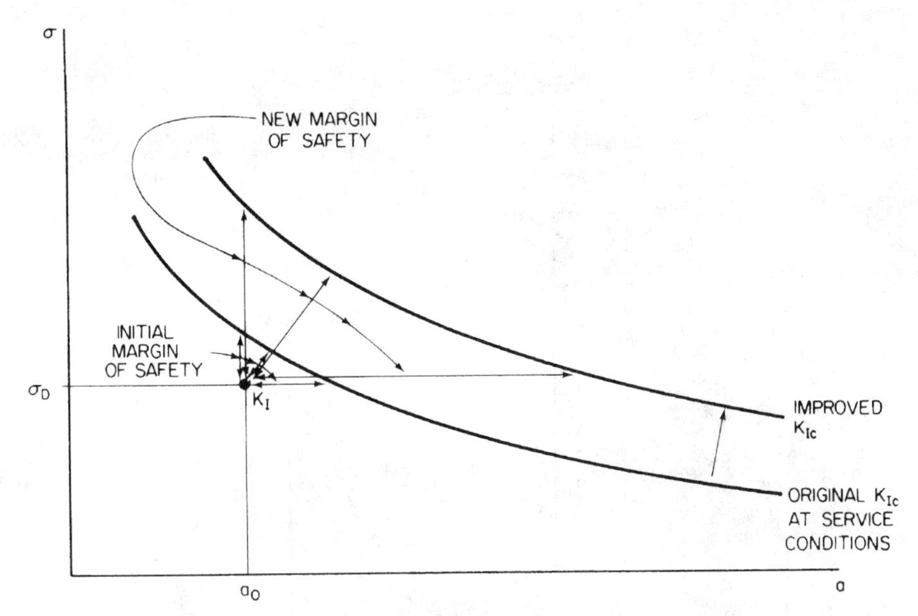

FIG. 12.7 Schematic showing effect of using material with better fracture toughness (Improved K_{Ic}) on fracture control.

The effect of each of three primary factors that control the total life of a structure subjected to *fatigue* loading may be summarized as follows:

Tensile stress — Large effect on life (Region I—Figure 12.8) because the rate of fatigue-crack growth is decreased significantly as the applied stress range is decreased (σ_1 compared with σ_2, Figure 12.8). Design stress range ($\sigma_{max} - \sigma_{min}$) is the primary factor to control.

Flaw size — Large effect on life (Region II—Figure 12.8) because the rate of fatigue crack growth decreases asymptotically as the flaw size decreases. Quality of fabrication and quality assurance by inspection are the primary factors to control.

Material fracture toughness — (A) Large effect on life in moving from plane-strain behavior to elastic-plastic behavior (Region III—Figure 12.8). For example, the AASHTO Material Fracture Toughness Requirements ensure this level of performance under intermediate rates of loading. For most structural applications, some moderate level of elastic-plastic behavior at the service temperature and loading rate constitutes a satisfactory criterion.

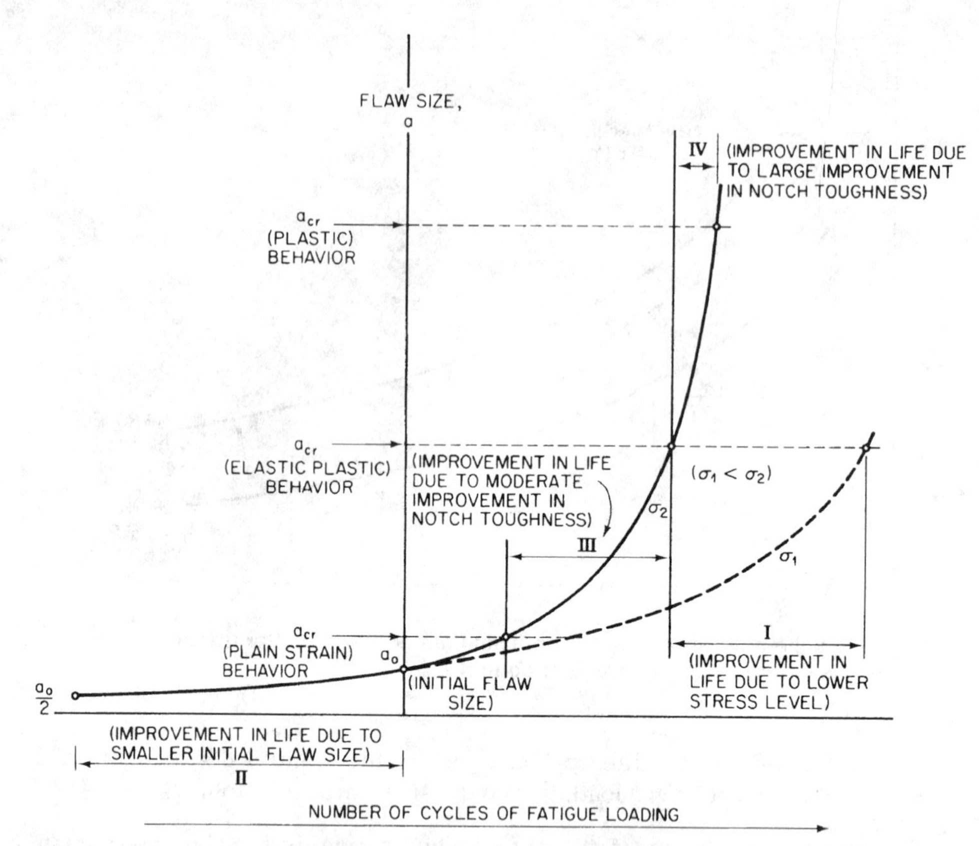

FIG. 12.8 Schematic showing effect of fracture toughness, stress, and flaw size on improvement of life of a structure subjected to fatigue loading.

(B) Small effect on life in moving from elastic-plastic behavior to plastic behavior (Region IV—Figure 12.8) because the rate of fatigue-crack growth becomes so large that even if the critical crack size (a_{cr}) is doubled or even tripled, the effect on the remaining fatigue life is small.

In summary, stress–crack-size–fracture toughness relations such as presented in Figures 12.1–12.8 should be used to determine the general design approach to minimize the possibility of fracture in the particular type of structure considered. Although the preceding three methods of minimizing the possibility of fracture (*use lower design stress, better fabrication, or tougher materials*) are the basic design approaches, other design methods can be used which are also very effective. These methods include the following:

1. Use *structural materials* whose notch toughness is such that the materials would fail by plastic deformation even under the most severe operating

conditions to which the structure may be subjected. The use of HY-80 steel for submarine hull structures is an example of this method. This method is an extreme use of the method just described of using materials with improved notch toughness. However, as shown in Figure 12.8, this method is not that effective for structures subjected to fatigue loading. Also, it is expensive, and the structure still may fail by other modes.

2. Provide *multiple-load* fracture paths so that a single fracture cannot lead to complete failure. If the geometry of a single-span bridge structure is a single box girder such that the failure of the single tension flange leads to collapse of the bridge, then the structure is a *single-load-path* structure. However, if the geometry consists of eight independent structural members supporting the deck, then the structure is a *multiple-load-path* structure and is much more resistant to complete failure than is the single box girder.

   Another example is a truss member composed of one structural shape (e.g., a wide-flange shape in tension) or multiple shapes (e.g., four to ten eye-bar members parallel to each other). The former is a single-load path structure, while the latter is a *multiple-load-path structure.*

   The distinguishing feature is whether or not, in the event of fracture of a primary structural member, the load can be transferred to and carried by other members. If so, the structure is a multiple-load-path one; if not, the structure is a single-load-path one. In this sense, multiple-load-path structures are usually more resistant to failure than single-load-path structure. For example, if a single member fails in a single-load path structures, the entire structure may collapse, as occurred with the Silver Bridge at Point Pleasant, West Virginia [8]. Conversely, if a single member fails in a multiple-load-path structure, the entire structure may not collapse. This type of behavior was demonstrated in the failure of the Kings Bridge in Australia [9]. That is, at the instant of failure, the failed span in the Kings Bridge contained three cracked girders. One member had cracked while the girders were still in the fabrication shop. A second girder failed during the first winter the bridge was opened to traffic, a full 12 months before the failure of the bridge. Failure of the third girder led to final failure, although architectural concrete sidewalls (which added to the multiple-load paths of the overall structure) prevented complete collapse.

   Fatigue-crack propagation in multiple-load-path structures may occur under constant deflection, which corresponds to a decreasing stress field intensity, rather than under constant load. Thus, cracks propagating in multiple-load-path structures may eventually arrest and, although individual structural components will have to be replaced or repaired, complete failure of the structure is not expected to occur so long as redistribution of load can occur.
3. Provide *crack arresters* so that in the event that a crack should initiate, it will be arrested before complete failure occurs. Crack arresters or a fail-safe

philosophy (i.e., in the event of "failure" of a member, the structure is still "safe") have been used extensively in the aircraft industry as well as the shipbuilding industry. A properly designed crack-arrest system must satisfy four basic requirements, namely,

a. Be fabricated from structural material with a high level of fracture toughness.
b. Have an effective local geometry, such as is shown in Figure 12.9.
c. Be located properly within the overall geometry of the structure.
d. Be able to act as an energy-absorbing system or deformation-restricting system in cases such as gas-pressurized line pipes where the primary crack-driving force is the separation of the crack surfaces behind the crack front.

4. Control fatigue-crack growth. If fatigue-crack growth is possible, the same general design methods described for fracture control also apply because of the correspondence between a fracture-control plan based on restriction of fatigue-crack initiation, fatigue-crack propagation, and fracture toughness of materials and a fracture-control plan based on fabrication, inspection, design, and fracture toughness of materials. Fatigue-crack initiation, fatigue-crack propagation, and fracture toughness are functions of the stress-intensity-factor fluctuation, ΔK, and of the critical stress-intensity factor, K_{Ic}, which are in turn related to the applied nominal stress (or stress fluctuation), the crack size, and the structural configuration. Thus, a fracture-control plan for various structural applications in which fatigue is a consideration depends on the same factors in which just fracture is a consideration. This type of fracture control was described in Figure 12.8.

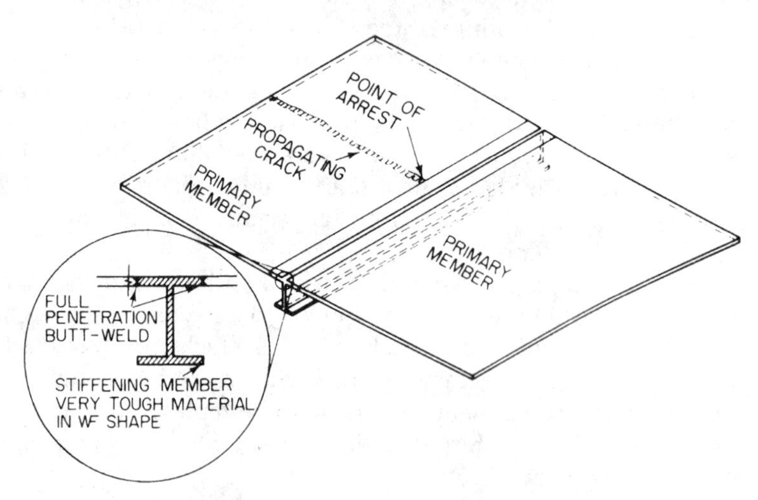

FIG. 12.9 Schematic showing out-of-plane crack arrester where crack must grow through entire wide-flange (WF) shape before it can continue.

5. Reduce the rate of load application. The fact that many structures are loaded at slow to intermediate loading rates leads to an understanding of why there are so few fractures in older structures that have low impact fracture toughness levels but that are loaded slowly. The recent understanding of the loading-rate shift (Chapter 4) leads to this conclusion. Thus, if the structure can be designed such that it is loaded *slowly*, so that the controlling fracture toughness parameter is K_{Ic} rather than $K_{Ic}(t)$ or K_{Id}, the possibility of fracture is reduced considerably. For example, load isolation systems to prevent earthquake damage to buildings and bridges decrease the magnitude of the loads and displacements and improve the fracture characteristics of the material by decreasing the rate of loading. Thus, control of the loading rate is a very effective method of fracture control.

12.3.4 Recommendation of Specific Design Considerations

One would expect that this step in the development of a fracture control plan would be the easiest if the preceding steps have all been followed and all technical and performance factors have been properly considered. However, it is usually the most difficult because it involves making the difficult decision of just how much fracture control can be justified economically.

The goal of good sound engineering design is to optimize structural performance consistent with economic considerations such that the probability of structural failure is minimized. For the yielding and general buckling modes of failure, sufficient experience has been acquired and has been incorporated in the various codes and specifications so that the designer can usually prevent these types of failures quite economically. However, this is not always true for fractures where the interdependence of fracture toughness, design stress (including effects of constraint), and flaw size, which is governed by fabrication and inspection requirements, are to be considered. One element of a fracture control plan is to establish the necessary fracture toughness to ensure safe, reliable, and economical structures. The required fracture toughness for a particular structure, or type of structure (i.e., bridges), must be based on the specific loads (maximum, rate, and fluctuating) and fabrication and inspection requirements for that structure.

All factors related to fracture toughness criterion must be considered, and an economic decision must be made based on technical input obtained regarding the level of performance to be specified. Once this level of performance (plane strain, elastic plastic, fully plastic) has been established for the service loadings and conditions, some material fracture toughness property can be specified. Even if this fracture toughness level can be specified directly in terms of the K_c, J_c, or CTOD fracture toughness values described in Chapter 3, material specifications based on these values are economically and technically prohibitive. These tests are too complex and too expensive to conduct on a routine quality-control basis. Hence, some auxiliary test specimen must be used based on the various correlations described in Chapter 5.

Several fracture criteria for various types of structures are based on concepts of fracture mechanics. However, the actual material requirements (in addition to strength, ductility, etc.) generally are specified in terms of CVN impact or NDT temperature requirements rather than K_c, J_c, or CTOD requirements. Two such examples of the development of fracture-control plans and material specifications are the ASME Code Section III requirements for nuclear pressure vessels [5] and the AASHTO Guide Specification for Steel Bridge Members [3]. In the first case, the material fracture toughness requirements have been specified using NDT and CVN impact specimens. Proposed new procedures are the use of a master curve to establish K_{Jc} values as described in Chapter 3. In the second case the fracture toughness requirements are specified using CVN impact specimens as described in the following section.

12.4 Fracture Control Plan for Steel Bridges

12.4.1 General

The fracture control plan for steel bridges presents special requirements for the design, materials, fabrication, and inspection of member components in steel bridges. The development of these requirements was based on a systematic evaluation of causes of cracking and fracture in bridge members and the consequences of failure. Thus, bridge members are separated into fracture-critical members and nonfracture critical members. Fracture-critical members or member components (FCMs) are defined as tension members or tension components of members whose failure would be expected to result in collapse of the bridge. In addition to being stressed in tension, an FCM must be the only load path available. This is defined by AASHTO as nonredundant. Loads and forces are redistributed to adjoining members and along alternate paths when a non-FCM redundant member fractures. The significant differences in the consequences of fracture dictates that more stringent requirements be imposed on FCMs than on non-FCMs. This section presents a brief discussion of the differences in the design, fabrication, and material requirements of FCMs and non-FCMs.

12.4.2 Design

The majority of bridges do not have fracture-critical members or member components. However, when they do exist, they must be identified by the engineer to ensure that they meet the special requirements of the specification. The location of all FCMs shall be clearly delineated on the contract plans, and the engineer shall ensure that their location and extent are clearly shown on the shop drawings. Also, the engineer shall verify that the fracture control plan is properly implemented in compliance with contract documents at all stages of fabrication and erection.

Fractures in bridges are almost always proceeded by fatigue from cyclic live loads. The fatigue life of a component is governed by the stress range raised to a power of two or greater as described in Chapter 9. Thus, small decrease in stress range results in a significant increase in the fatigue life of a component. Based on this fact, the allowable stress range of FCMs was limited to 80% of that for non-FCMs.

12.4.3 Fabrication

The AASHTO Standard Specifications for Welding Structural Highway Bridges apply to non-FCMs and are supplemented for FCMs by the provisions of Section 12 of the ANSI/AASHTO/AWS D1.5-96 Welding Code [*10*]. This section defines all the provisions regarding fabrication of bridge members designated FCMs. In addition to defining the engineer's responsibilities, and the base metal requirements, it addresses in detail: (1) qualification and certification of the fabricator, welding inspector, and nondestructive testing personnel; (2) welding processes; (3) consumable requirements; (4) welding procedure specification; (5) contractor requirements; (6) thermal cutting; (7) repair of base metal; (8) straightening, curving, and cambering, (9) tack welds and temporary welds; (10) weld inspection; (11) repair welding. The FCM requirements for all these items are more restrictive than for non-FCM.

12.4.4 Material

The fracture toughness requirements for bridge steels were based on several technical considerations that were discussed in Chapters 4 and 5. These include:

1. Effect of constraint and temperature on the fracture-toughness behavior of steels established by testing fracture-mechanics-type specimens.
2. Effect of rate of loading on fracture toughness behavior of steels.
3. Correlation between impact fracture toughness (K_{Id}) and impact Charpy V-notch (CVN) energy absorption.
4. Specification of CVN fracture toughness values that would ensure elastic-plastic or plastic behavior for fracture initiation of fatigue-cracked specimens subjected to minimum operating temperature and maximum in-service rates of loading.

Fracture toughness varies with the degree of localized constraint to plastic flow along the tip of a fatigue crack. Thus, cracks in very thick members are subjected to higher constraint than are cracks in thinner members. The AASHTO fracture toughness requirements were based on the conservative assumption that severe constraint (plane-strain) conditions governed regardless of component thickness or geometry.

Steels undergo a transition from cleavage (low fracture toughness) to ductile (high fracture toughness) crack initiation as the temperature increases. Examples of this behavior were presented in Chapter 4. Also, the fracture transition curve under static loading is shifted to higher temperatures as the rate of loading is increased. The magnitude of the shift decreases as the yield strength, σ_{ys}, increases such that the shift between static and impact plane-strain fracture toughness curves can be calculated from the following relationship:

$$T_{shift}, \text{°F} = 215 - 1.5\sigma_{ys}, \text{ ksi} \tag{12.1}$$

for

$$28 \leq \sigma_{ys} < 140 \text{ ksi}$$

The shift in transition curves between bridge loading (intermediate; 1.0 second loading to fracture), and impact ($\leq$0.001 s to fracture) is about 0.75 the value calculated from Equation (12.1) for the shift between static and impact curves. The shift between a static curve and any intermediate or impact loading curve is given by the relationship:

$$T_{shift}, \text{°F} = (150 - \sigma_{ys}, \text{ ksi})\dot{\varepsilon}^{0.17} \tag{12.2}$$

where $\dot{\varepsilon}$ is strain rate in seconds as described in Chapter 4. A proper use of fracture-mechanics methodology for fracture control of structures requires the determination of fracture toughness for the material at the temperature and loading rate representative of the intended application.

12.4.5 AASHTO Charpy V-Notch Requirements

The information presented in Chapter 4 on the effects of temperature and loading rate on the fracture toughness of steels and the correlation of fracture mechanics data and CVN energy absorption provides the technical foundation that can be used to develop fracture toughness requirements for any steel structure. The developed fracture toughness requirements are dictated by the desired fracture behavior (i.e., fracture criterion) at the minimum operating temperature and maximum loading rate for the particular structure. The selected fracture criterion for the AASHTO fracture toughness was the development of elastic-plastic or plastic fracture initiation at the minimum operating temperature and maximum loading rate for bridges. The conservative bridge loading rate of 1.0 s loading to fracture was used. The minimum operating temperature in the United States varies from location to location. To simplify the specification, the United States was divided into three temperature zones as follows:

MINIMUM SERVICE TEMPERATURE	TEMPERATURE ZONE
0° (−18°C) and above	1
−1°F (−19°C) to −30°F (−34°C)	2
−31°F (−35°C) to −60°F (−51°C)	3

This procedure increased the level of conservatism by requiring a steel bridge subjected to a high temperature within a zone to meet the requirement for the lowest temperature within the zone.

Combining the basic fracture behavior of steels as a function of constraint, temperature, and loading rate with the fracture criterion and the operating temperature zones resulted in the AASHTO CVN requirements for FCMs and non-

FCMs, Table 12.1 and 12.2, respectively [*11*]. The higher fracture toughness requirements for FCMs over non-FCMs reflect the difference in the consequences of failure.

The fatigue provisions of the AASHTO fracture control plan would preclude the presence of a critical fatigue crack at the end of the service life of a steel bridge.

12.4.6 *Verification of the AASHTO Fracture Toughness Requirement*

Extensive testing has been conducted by Schilling et al. 1975 [*12*] and Roberts et al. 1974 [*13*] to ensure the adequacy of the AASHTO fracture toughness requirements. The tests were conducted on full-size beams of A36, A572, A588, and A514 steels. The beams had various types of welded details and were subjected to cyclic loading conditions specified by AASHTO for the particular detail. Subsequently, the fatigue-cracked beams were cooled and loaded to failure. The test results confirmed the methodology used to develop the AASHTO fracture toughness requirements and demonstrated that "fatigue cracked, simulated highway bridge members of various steels that satisfied the minimum 1975 AASHTO fracture toughness requirements exhibited sufficient resistance at the minimum operating temperature and at maximum in-service loading rate to prevent premature fracture even after the members had been subjected to the AASHTO design fatigue loads."

12.4.7 *High-Performance Steels*

Recent technological advances in steelmaking and steel processing have resulted in improved quality and new products. These developments enhanced further the outstanding properties of steels not found in other constructional materials. These properties include a combination of strength, ductility, fracture toughness, fabricability, atmospheric corrosion resistance, repairability, and cyclability.

A cooperative program among the Federal Highway Administration (FHWA), the Navy, the American Iron and Steel Institute (AISI), and the AISI plate producing members (i.e., Bethlehem Steel, Lukens Steel, and U.S. Steel) has led to the development of high-performance weathering steels with 70-ksi minimum yield strength (HPS 70W) and 100-ksi minimum yield strength (HPS 100W) [*14*]. These steels are highly weldable and exhibit a fracture toughness significantly better than the most severe Zone 3 AASHTO requirements. One of the consequences of this development is the elimination of the need to differentiate between fracture-critical and nonfracture-critical members.

12.5 Comprehensive Fracture-Control Plans—George R. Irwin

Dr. George Irwin has prepared the following general comments on fracture-control plans [*1*], which are reprinted in their entirety.

TABLE 12.1. Fracture Critical Member Impact Test Requirements.

GRADE	THICKNESS, in., AND JOINING METHOD	MINIMUM TEST VALUE ENERGY, ft-lbf	MINIMUM AVERAGE ENERGY, ft-lb		
			ZONE 1	ZONE 2	ZONE 3
36F	to 4, mechanically fastened or welded	20	25 at 70°F	25 at 40°F	25 at 10°F
50F	to 2, mechanically fastened or welded	20	25 at 70°F	25 at 40°F	25 at 10°F
	over 2 to 4, mechanically fastened	20	25 at 70°F	25 at 40°F	25 at 10°F
	over 2 to 4, welded	24	30 at 70°F	30 at 40°F	30 at 10°F
70WF	to 2 1/2, mechanically fastened or welded	24	30 at 50°F	30 at 20°F	30 at −10°F
	over 2 1/2 to 4, mechanically fastened	24	30 at 50°F	30 at 20°F	30 at −10°F
	over 2 1/2 to 4, welded	28	35 at 50°F	35 at 20°F	35 at −10°F
100F	to 2 1/2, mechanically fastened or welded	28	35 at 30°F	35 at 0°F	35 at −30°F
	over 2 1/2 to 4, mechanically fastened	28	35 at 30°F	35 at 0°F	35 at −30°F
	over 2 1/2 to 4, welded	36	45 at 30°F	45 at 0°F	not permitted

TABLE 12.2. Nonfracture Critical Member Impact Test Requirements.

		MINIMUM AVERAGE ENERGY, ft-lb		
GRADE	THICKNESS, in., AND JOINING METHOD	ZONE 1	ZONE 2	ZONE 3
36T	to 4, mechanically fastened or welded	15 at 70°F	15 at 40°F	15 at 10°F
50T	to 2, mechanically fastened or welded	15 at 70°F	15 at 40°F	15 at 10°F
	over 2 to 4, mechanically fastened	15 at 70°F	15 at 40°F	15 at 10°F
	over 2 to 4, welded	20 at 70°F	20 at 40°F	20 at 10°F
70WT	to 2 1/2, mechanically fastened or welded	20 at 50°F	20 at 20°F	20 at −10°F
	over 2 1/2 to 4, mechanically fastened	20 at 50°F	20 at 20°F	20 at −10°F
	over 2 1/2 to 4, welded	25 at 50°F	25 at 20°F	25 at −10°F
100T	to 2 1/2, mechanically fastened or welded	25 at 30°F	25 at 0°F	25 at −30°F
	over 2 1/2 to 4, mechanically fastened	25 at 30°F	25 at 0°F	25 at −30°F
	over 2 1/2 to 4, welded	35 at 30°F	35 at 0°F	35 at −30°F

"For certain structures, which are similar in terms of design, fabrication method, and size, a relatively simple fracture control plan may be possible, based upon extensive past experience and a minimum adequate toughness criterion. It is to be noted that fracture control never depends solely upon maintaining a certain average toughness of the material. With the development of service experience, adjustments are usually made in the design, fabrication, inspection, and operating conditions. These adjustments tend to establish adequate fracture safety with a material quality which can be obtained reliably and without excessive cost. A fair statement of the basic philosophy of fracture control for such structures might be as follows. Given that the material possesses strength proeprties within the specific limits, and given that the fracture toughness lies above a certain minimum requirement (nil-ductility temperature, fracture appearance transition temperature, plane-strain, or plane-stress crack toughness), it is assumed that past experience indicates well enough how to manage design, fabrication, and inspection so that fracture failures in service occur only in small tolerable numbers.

With the currently increasing use of new materials, new fabrication techniques, and novel designs of increased efficiency, the preceding simple fracture control philosophy has tended to become increasingly inadequate. The primary reason is the lack of suitable past experience and the increased cost of paying for this experience in terms of service fracture failures. Indeed, modern technology is beginning to exhibit situations with space vehicles, jumbo-jet commercial airplanes, and nuclear power plants for which not even *one* service fracture failure would be regarded as acceptable without consequences of disaster proportions. Consideration must be given, therefore, to comprehensive plans for fracture control such as one might need in order to provide assurance of zero service fracture failures. A review of the fracture control aspects of a comprehensive plan may be advantageous even for applications such that the required degree of fracture control is moderate. One reason for this would be that an understanding of how to minimize manufacturing costs in a rational way is assisted when we assemble all of the elements which contribute to product quality and examine their relative effectiveness and cost. In the present case, the quality aspect of interest is the degree of safety from service fractures.

After these introductory comments, it is necessary to point out that the concept termed comprehensive fracture control plan is quite recent and cannot yet be supported with completely developed illustrations. We know in a general way how to establish plans for fracture control in advance of extensive service trials. However, until a number of comprehensive fracture control plans have been formulated and are available for study, detailed recommendations to guide the development of such plans for selected critical structures cannot be given.

The available illustrative examples of fracture control planning which might be helpful are those for which a large number of the elements contributing to fracture control are known. At least in terms of openly available information, these examples are incomplete in the sense that the fracture control elements require collection, re-examination with regard to relative efficiency, and careful study with regard to adequacy and optimization. Substantial amounts of information relative to fracture control are available in the case of heavy rotating components for large steam turbine generators, components for commercial jet airplanes (fuselage, wings, landing gear, certain control devices), thick-walled containment vessels for BW and PW cooled nuclear reactors, large-diameter underground gas transmission pipelines, and pressure vessel components carried in space vehicles. Certain critical fracture control aspects of these illustrative examples are as follows:

A. Large Steam Turbine Generator Rotors and Turbine Fans
   1. Vacuum de-gassing in the ladle to reduce and scatter inclusions and to eliminate hydrogen.
   2. Careful ultrasonic inspection of regions closest to the center of rotation.
   3. Enhanced plane-strain fracture toughness.

B. Jet Airplanes
   1. Crack arrest design features of the fuselage.
   2. Fracture toughness of metals used for beams and skin surfaces subjected to tension.
   3. Strength tests of models.
   4. Periodic re-inspection.

C. Nuclear Reactor Containment Vessels
   1. Quality uniformity of vacuum degassed steel.
   2. Careful inspection and control of welding.
   3. Uniformity of stainless cladding.
   4. Proof testing.
   5. Investigations of low-cycle fatigue crack growth at nozzle corners and of cracking hazard from thermal shock.

D. Gas Transmission Pipelines
   1. Adequate toughness to prevent long running cracks.
   2. High-stress-level, in-place hydrotesting.
   3. Corrosion protection.

E. Spacecraft Pressure Vessels
   1. Surface finishing of welds so as to enhance visibility of flaws.
   2. Heat treatments to remove residual stress and produce adequate toughness within given limits of strength.
   3. Adjustments of hydrotesting to assure adequate life relative to stable crack growth in service.

In a large manufacturing facility, the inter-group cooperation necessary to achieve successful fracture control on the basis of a comprehensive fracture control plan may require special attention. In general, the comprehensive fracture plan will contain various elements pertaining to design, materials, fabrication, inspection, and service operation. These elements should be directly or indirectly related to fracture testing information. However, the coordination of the entire plan to ensure its effectiveness is not a priori a simple task. The following outline lists certain fracture control tasks under functional headings which might, in some organizations, imply separate divisions or departments.

I. Design
- **A.** Stress distribution information.
- **B.** Flaw tolerance of regions of largest fracture hazard due to stress.
- **C.** Estimates of stable crack growth for typical periods of service.
- **D.** Recommendations of safe operating conditions for specified intervals between inspections.

II. Materials
- **A.** Strength properties and fracture properties.
  σ_{ys}, σ_{UTS}, K_{Ic}, K_c.
  K_{Iscc} for selected environments.
  da/dN for selected levels of ΔK and environments.
- **B.** Recommended heat treatments.
- **C.** Recommended welding methods.

III. Fabrication
- **A.** Inspections prior to final fabrication.
- **B.** Inspections based upon fabrication control.
- **C.** Control of residual stress, grain coarsening, grain direction.
- **D.** Development or production of suitable strength and fracture properties.
- **E.** Maintain fabrication records.

IV. Inspection
- **A.** Inspection prior to final fabrication.
- **B.** Inspections based upon fabrication control.
- **C.** Direct inspection for defects using appropriate non-destructive evaluation (NDE) techniques.
- **D.** Proof testing.
- **E.** Estimates of largest crack-like defect sizes.

V. Operations
- **A.** Control of stress level and stress fluctuations in service.
- **B.** Maintain corrosion protection.
- **C.** Periodic in-service inspections.

From the above outline, one can see that efficient operation of a comprehensive fracture control plan requires a large amount of inter-group coordination. If a complete avoidance of fracture failure is the goal of the plan, this goal cannot be assured if the elements of the fracture control plan are supplied by different division or groups in a voluntary or independent way. It would appear suitable to establish a special fracture control group for coordination purposes. Such a group might be expected to develop and operate checking procedures for the purpose of assuring that all elements of the plan are conducted in a way suitable for their purpose. Other tasks might be to study and improve the fracture control plan and to supply suitable justifications, where necessary, of the adequacy of the plan."

12.6 References

[1] George R. Irwin, private communication.

[2] PVRC Ad Hoc Task Group on Toughness Requirements, "PVRC Recommendations on Toughness Requirements for Ferritic Materials," WRC Bulletin No. 175, August 1972.

[3] AASHTO, *Guide Specification for Fracture Critical Non-Redundant Steel Bridge Members,* September 1978.

[4] British Standards Institution Published Document 64 93 (PD-6493), "Guidance on Method for Assessing the Acceptability of Flaws in Fusion Welded Structures," BSI Standards, 1991.

[5] American Society of Mechanical Engineering, ASME Boiler and Pressure Vessel Code Section XI, Rules for In Service Inspection for Nuclear Power Plant Components, Appendix A, ASME, New York, 1989.

[6] FEMA 267—Interim Guidelines: Evaluation, Repair, Modification and Design of Welded Steel Moment Frame Structures, August 1995.

[7] FEMA 267—Interim Guidelines Advisory No. 1—Supplement to FEMA 267, March 1997.

[8] Point Pleasant Bridge, National Transportation Safety Board, "Collapse of U.S. 35 Highway Bridge, Point Pleasant, West Virginia, December 15, 1967," *Report No. NTSB-HAR-71-1,* Washington, DC, 1971.

[9] Madison, R. B. and Irwin, G. R., "Fracture Analysis of Kings Bridge, Melbourne," *Journal of the Structural Division, ASCE,* Vol. 97, No. ST9, September 1971.

[10] ANSI/AASHTO/AWS D1.5-96 Bridge Welding Code, 1996.

[11] ASTM A709/A709M-97a, Standard Specification for Carbon and High-Strength Low-Alloy Structural Steel Shapes, Plates, and Bars and Quenched-and-Tempered Alloy Structural Steel Plates for Bridges.

[12] Schilling, C. G., Klippstein, K. H., Barsom, J. M., Novak, S. R., and Blake, G. T., "Low-Temperature Tests of Simulated Bridge Members," *Journal of the Structural Division,* Vol. 101, No. ST1, American Society of Civil Engineers, January 1975.

[13] Roberts, R., Irwin, G. R., Krishna, G. V., and Yen, B. T., "Fracture Toughness of Bridge Steels—Phase II Report," FHWA Report No. FHWA-RD-74-59, Federal Highway Administration, Washington, DC, September 1974.

[14] Focht, E. M., "High Performance Steels Development for Highway Bridge Construction: A Cooperative Effort," 1997 Transportation Research Board Meeting, Washington, DC, January 1997.

13

Fracture Criteria

13.1 Introduction

A FRACTURE CRITERION is a standard against which the expected fracture behavior of a structure can be judged. In general terms, fracture criteria are related to the three levels of fracture performance, namely plane strain, elastic plastic, or fully plastic, as shown in Figure 13.1. Although it would appear desirable always to specify fully plastic behavior, this is rarely done because it is almost always unnecessary as well as economically unfeasible in most cases. Furthermore, relying primarily on notch toughness to prevent failure rather than good robust design may lead to problems. Finally, it is unsound engineering because good design is defined as an optimization of satisfactory structural performance, safety, and economic considerations.

For most structural applications, some level of elastic-plastic behavior at the service temperature and loading rate is a satisfactory fracture criterion. While there may be some cases where fully plastic behavior is necessary (e.g., large dynamic loadings such as submarines being subjected to depth charges), or where plane-strain behavior can be tolerated (e.g., certain short-life aerospace applications where the loading and fabrication can be precisely controlled), for the majority of large complex structures (bridges, buildings, ships, pressure vessels, offshore drilling rigs, etc.) some level of elastic-plastic behavior is appropriate. The questions become, "What level of elastic-plastic behavior is required and how can this level of performance be ensured?" The purpose of this chapter is to develop rational engineering approaches to answering these questions.

Unfortunately, the selection of a fracture criterion is often quite arbitrary and is based on service experience for other types of structures that may have no relation to the particular structure an engineer may be designing. An example of the use of a fracture criterion developed for one application yet widely used in many other situations is the 15-ft-lb CVN impact criterion at the minimum

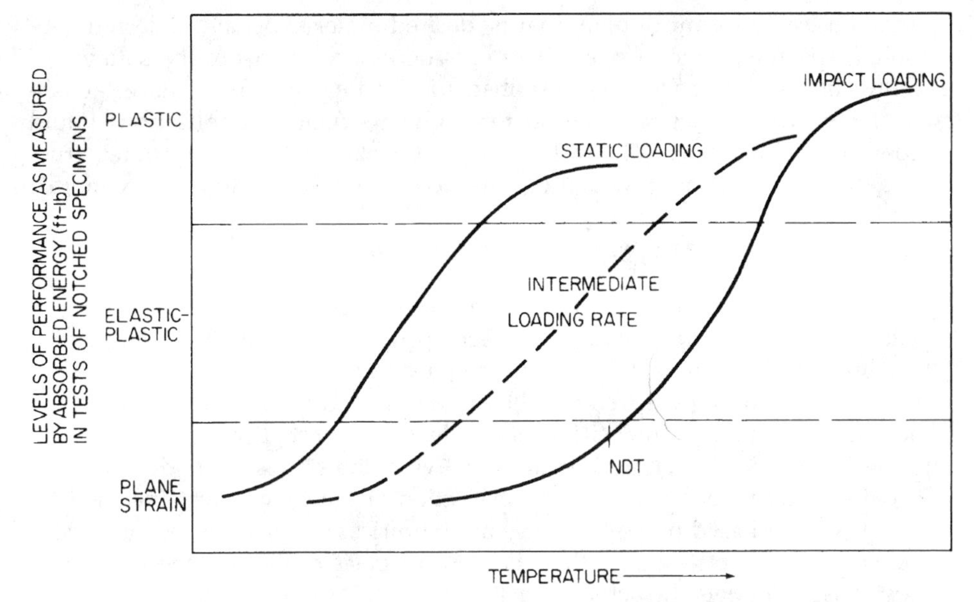

FIG. 13.1 Schematic showing relation between notch-toughness test results and levels of structural performance for various loading rates.

service temperature, which was established based on the World War II ship failures. This criterion has been widely used for various types of structures, even though the material, service conditions, structural redundancy, and so on may be considerably different from that of the World War II ships for which the criterion was established. Also, because a fracture criterion is only a part of a fracture-control plan, selection of a fracture criterion alone, without considering the other factors involved in fracture control (for example, loading or quality of fabrication), will not necessarily result in a safe structure.

Criterion selection should be based on a very careful study of the particular requirements for a particular structure. In this chapter we shall describe the various factors involved in the development of a criterion, which include

1. The service conditions (loadings, temperature, loading rate, etc.) to which the structure will be subjected.
2. The significance of loading rate on the performance of the structural materials to be used.
3. The desired level of performance in the structure.
4. The consequences of failure.

It should be emphasized that there can be no single *best* criterion for all structures because optimum design should involve economic considerations as well as technical ones. That is, selecting a fracture criterion that is far more conservative than necessary is unsound engineering. Furthermore, a study of all aspects of a fracture-control plan, that is, desired material behavior, design, fabrication, inspection, operation, and so on, is necessary to ensure the safety and reliability of structures (as was described in Chapter 12). Thus, the establishment of the proper fracture criterion for a given structural application should be coordinated with service loading requirements, material selection, structural design, fabrication procedures, inspection requirements, and economic considerations.

There are two general parts to a fracture criterion:

1. *The general test specimens to categorize the material behavior.* Throughout the years, various fracture criteria have been specified using notch-toughness tests such as CVN impact, NDT, dynamic tear, and, more recently, the fracture-mechanics test specimens described in Chapter 3 which are used to measure critical stress-intensity factors. Ideally, the test specimen used for a particular application should be that one which most closely models the actual structural behavior. However, selection of the general test specimen to use is often based on past experience, empirical correlations, economics, and convenience of testing rather than on the basis of the test specimen that most closely models the actual structure.
2. *The specific notch-toughness value or values.* The second and more difficult part of establishing a fracture criterion is the selection of the specific level of performance in a particular test specimen in terms of measurable values for material selection and quality control.

The specified values in any criterion should consider both the structural performance and cost. One of the main objectives of this chapter is to provide some rational guidelines for the engineer to follow in establishing fracture toughness criteria for various structural applications.

13.2 General Levels of Performance

The primary design criterion for most large structures such as bridges, buildings, pressure vessels, ships, and so on, is still based on strength and stability requirements such that nominal elastic behavior is obtained under conditions of maximum loading. Usually the strength and stability criteria are achieved by limiting the maximum design stress to some percentage of the yield strength. In many cases, fracture toughness is also an important design criterion, and yet specifying a notch-toughness criterion is much more difficult, primarily because:

1. Establishing the specific level of required fracture toughness (i.e., the required CVN or K_{Ic} value at a particular test temperature and rate of loading) is costly and time consuming and is a subject with which many design engineers are not familiar.

2. There is no well-recognized single "best" approach. Therefore, different experts will have different opinions as to the "best" approach, although the science of fracture mechanics is helping to overcome this difficulty.
3. The cost of structural materials increases with increasing levels of inherent notch toughness. Thus, economic considerations must be included when establishing any toughness criterion.

Materials that have extremely high levels of fracture toughness under even the most severe service conditions are available, and the designer can always specify that these materials be used in critical locations within a structure. However, because the cost of structural materials generally increases with their ability to perform satisfactorily under more severe operating conditions, a designer does not wish to arbitrarily specify more fracture toughness than is required for the specific application. Also, a designer does not specify the use of a material with a very high yield strength for a compression member if the design is such that the critical buckling stress is very low. In the first case, the additional fracture toughness is unnecessary, and in the second case, the excessive yield strength is unnecessary. Both cases are examples of unsound engineering.

The problem of establishing specific fracture-toughness requirements that are not excessive but are still adequate for normal service conditions is a long-standing one for engineers. However, by using concepts of fracture mechanics, rational fracture criteria can be established for fracture control in different types of structures. If a structure is loaded "slowly," i.e., 10^{-5} in./in./s, the K_{Ic}/σ_{ys} ratio is the controlling fracture toughness parameter. If, however, the structure is loaded "dynamically," $\sim 10^1$ in./in./s or impact loading, the K_{Id}/σ_{yd} ratio is the controlling parameter. At intermediate loading rates, the $K_{Ic}(t)/\sigma_{ys}(t)$ ratio determined at the appropriate loading rate is the controlling parameter.

Because high constraint (thick plates and plane-strain conditions) at the tip of a crack can lead to premature fracture, the engineer should strive for the lowest possible degree of constraint (thin plates and plane-stress conditions) at the tip of a crack.

Structural materials whose fracture toughness and plate thickness are such that the critical K-to-yield-strength ratio for service loading rates is less than about $\sqrt{t/2}$ (actually $t = 2.5\,(K_{Ic}/\sigma_{ys})^2$ for plane strain behavior) exhibit elastic plane-strain behavior as described in Chapter 3 and generally fracture in a brittle manner. These materials generally are not used as primary tensile load-carrying members for most structural applications because of the high level of constraint at the tip of a crack and the rather small critical crack sizes at design stress levels. Fortunately, most structural materials have fracture toughness levels such that they do *not* exhibit plane-strain behavior at service temperatures, service loading rates, and common structural sizes normally used. However, very thick plates or plates used to form complex geometries where the constraint can be very high may be susceptible to brittle fractures even though the inherent notch toughness as measured by small-scale laboratory tests appears satisfactory.

Structural materials whose fracture toughness levels are such that they exceed the plane-strain limits described above and in Chapter 3 exhibit *elastic-plastic* fractures with varying amounts of yielding prior to fracture. The tolerable flaw sizes at fracture vary considerably and can be fairly large. Fracture is usually preceded by the formation of large plastic zones ahead of the crack. Most structures are built of materials that exhibit some level of elastic-plastic behavior at service temperatures and loading rates.

Some structural materials exhibit ductile plastic fractures preceded by large deformation at service temperature and loading rates. This type of behavior is very desirable in structures and represents considerable notch toughness. However, this level of toughness is rarely necessary and thus usually is not specified, except for unusual cases such as submarine hulls or nuclear structures.

For structural steels, the levels of performance are usually described in terms of the transition from brittle to ductile behavior as measured by various types of notch-toughness tests. This general transition in fracture behavior was related schematically to the three levels of behavior schematically in Figure 13.1.

For static loading, the transition region occurs at lower temperatures than for impact (or dynamic) loading for those structural materials that exhibit a transition behavior. Thus, for those structures subjected to static loading, a static transition curve (i.e., K_{Ic} or K_c test results) should be used to predict the level of performance at the service temperature. For structures subjected to impact or dynamic loading, the impact transition curve (i.e., K_{Id} test results) should be used to predict the level of performance at the service temperature. For structures subjected to some intermediate loading rate, an intermediate loading-rate transition curve (i.e., $K_{Ic}(t)$ test results) should be used to predict the level of performance at the service temperature. Because the effect of loading rate on fracture toughness of structural steels was not well defined, the impact loading curve (Figure 13.1) often has been used to predict the service performance of structures. However, this may be unduly conservative and often does not properly model the service behavior for many types of structures loaded statically or at intermediate loading rates, such as bridges.

13.3 Consequences of Failure

Although rarely stated as such, the basis of structural design of large complex structures is an attempt to optimize the desired performance requirements relative to cost considerations (including materials, design, and fabrication), so that probability of failure (and its economic consequences) is low. Generally, the primary criterion is the requirement that the structure support its own weight plus any applied loads and have nominal stresses less than either the tensile yield strength (to prevent excessive deformation) or the critical buckling stress (to prevent premature buckling).

Although brittle fractures can occur in riveted or bolted structures, the evolution of welded construction with its emphasis on monolithic structural members has led to the desirability of including some kind of fracture criterion for most structures, in addition to the strength and buckling criteria already in existence. That is, if a fracture does initiate in a welded structure, there usually is a continuous path for crack extension. However, in riveted or bolted structures, which generally consist of many individual plates or shapes, a continuous path for crack extension rarely exists. Thus, any cracks that may extend generally are arrested as soon as they traverse a single plate or shape. Consequently, there can be a large difference in the possible fracture behavior of welded structures compared with either riveted or bolted structures.

Failure of most engineering structures is caused by the initiation and propagation of cracks to critical dimensions. Because crack initiation and propagation for different structures occur under different stress and environment conditions, no single fracture criterion or set of criteria should be used for all types of structures. Most criteria are developed for particular structures based on extensive service experience or empirical correlations and thus are valid only for a particular design, fabrication method, and service use.

However, one of the biggest reasons that no single fracture criterion should be applied uniformly to the design of different types of structures is the fact that the *consequences of structural failure* are vastly different for different types of structures. For example, a fracture criterion for steels used in some seagoing ship hull structures is that the NDT (nil-ductility) temperature be 30°F for a minimum service temperature of 30°F. This criterion was based on the assumption that ships are subjected to full-impact loading. In view of the consequences of failure of a ship, that is, either in terms of loss of life or of cargo, this criterion is conservative for most ships but may be justified for ships loaded dynamically.

In contrast, the fracture criterion for the steels that were proposed to be used in the hull structure of stationary but floating nuclear power plants inside a protective breakwater was that the NDT be −30°F for minimum service temperature of +30°F. Thus, for *less severe loading* (because the platforms are stationary and wave action is controlled, the steels in the floating nuclear power plants were required to have an NDT temperature of 60°F *below* their service temperature, compared with seagoing ship steels whose NDT temperature generally is at their service temperatures. However, for the hull structure of floating nuclear power plants, where service experience is nonexistent and the occurrence of a brittle fracture might have resulted in significant environmental damage and the drastic curtailment of an entire industry, the fracture criterion was extremely conservative because of the consequences of failure. This was true even though the design of the stationary floating hull structure was similar to that of seagoing ship hull structures. It should be noted that even though the floating nuclear power plant concept would have been safe and would have performed satisfactorily, none were ever built, primarily for economic reasons. That is, the safety requirements were so high that the project became uneconomical. As stated before, good design should be an optimization of performance, safety, and cost.

Another example of consequences of failure is the *lack* of the necessity for specifying a fracture criterion for a piece of earthmoving equipment where the consequences of failure of a structural member may be loss of the use of the equipment for a short time until the part can be repaired or replaced. If the consequences of failure are minor, then specification of a fracture toughness requirement that might increase the cost of each piece of equipment significantly may be unwarranted. Thus, the consequences of failure should be a major consideration when determining:

1. The need for some kind of fracture criteria.
2. The level of performance (plane strain, elastic plastic, or fully plastic) to be established by the fracture toughness criteria.

Each class or type of structure must be evaluated carefully and the consequences of failure factored into the selected fracture criterion. In fact, in his 1971 AWS Adams Memorial Lecture [*1,2*], Bill Pellini stated that "one should not use a design criterion in excess of real requirements because this results in specifications of lower NDT and therefore, increased costs." Needless to say, determining "real requirements," that is, balancing safety and reliability against economic considerations, is a difficult task.

13.4 Original 15-ft-lb CVN Impact Criterion for Ship Steels

Although occasional brittle fractures were reported in various types of structures (both welded and riveted) prior to the 1940s [*3,4*] it was not until the rapid expansion in all-welded ship construction during the early 1940s that brittle fracture became a well-recognized structural problem. During the early 1940s over 2500 Liberty ships, 500 T-2 tankers, and 400 Victory ships were constructed as a result of World War II. Because the basic designs of each of these three types of ships (Liberty, Victory, and tankers) were similar, it was possible to analyze the structural difficulties on a statistical basis.

The first of the Liberty-type were placed in service near the end of 1941, and by January 1943 there were ten major fractures in the hull structures of the ships in service at that time. Numerous additional failures throughout the next few years led to the establishment of a board to investigate the design and methods of construction of welded steel merchant ships. In 1946, this board made its final report [*5*] to the Secretary of the Navy.

The role of materials, welding, design, fabrication, and inspection is described in various extensive reviews of this problem [*6–12*]. Of interest in this particular section is the work leading to the development of the 15-ft-lb notch-toughness criterion, which still is widely used for many other types of structures.

In the development of the 15-ft-lb criterion, samples of steel were collected from approximately 100 fractured ships and submitted to the National Bureau of Standards for examination and tests. Their study resulted in the collection of an extremely complete body of data relative to the failure of large welded ship hull structures [*11*]. The plates from fractured ships were divided into three groups:

1. Those plates in which fractures originated, called *source plates.*
2. Those plates through which the crack passed, called *through plates.*
3. Those plates in which a fracture stopped, called *end plates.*

An analysis of the Charpy V-notch impact test results showed that the source plates had a high average 15-ft-lb transition temperature, about 95°F. The through plates had a more normal distribution of transition temperatures, and the average 15-ft-lb transition temperature was lower, that is, about 65°F. The *end* plates had the lowest average 15-ft-lb transition temperature (about 50°F) and a distribution with a long tail at lower transition temperatures.

Parker [*12*] points out that "the character of these distribution curves was not wholly unanticipated; one would suspect that the through plates would be most representative of all ship plates and hence might tend toward a normal distribution, whereas the *source* and *end plates* were *selected* for their role in the fracturing of the ship. A factor in this selection is the notch toughness of the plate as measured in the Charpy test. The overlapping of the *source-* and *end-*plate distributions with the through-plate distribution can be explained by factors involved in the fracturing in addition to the notch toughness of the plate. For example, a plate in the through-fracture category having a transition temperature between about 60 and 90°F might have been a fracture *source* plate under more severe stress conditions such as those in the region of a notch; under less severe conditions of average stress, it might have been an *end* plate. Even though factors other than notch toughness contributed to the selection of a plate for its role in fracturing, it should be pointed out that statistical analyses have indicated that the differences in transition temperature and energy at failure temperature between the *source* and *through* and *through* and *end* plates are not due to chance."

The extremely low probability of the differences in Charpy properties of plates in the three fracture categories being due to chance permitted the development of several criteria very important to engineers. Figure 13.2 shows that only 10% of the *source* plates absorbed more than 10-ft-lb in the V-notch Charpy test at the failure temperature; the highest value encountered in this category was 11.4 ft-lb. At the other extreme, 73% of the *end* plates absorbed more than 10 ft-lb at the failure temperature. Based on these data, it was concluded that brittle fractures were not likely to initiate in plates that absorb more than 10 ft-lb Charpy V-notch impact energy conducted at the anticipated operating temperature in large ship hull structures.

Thus, a slightly higher level of performance, namely, the 15-ft-lb transition temperature as measured with a CVN impact test specimen, was selected as a fracture criterion on the basis of actual service behavior of a large number of similar-type ship hull structures. Since the establishment of this relation between CVN values and service behavior in ship hulls, the 15-ft-lb transition temperature

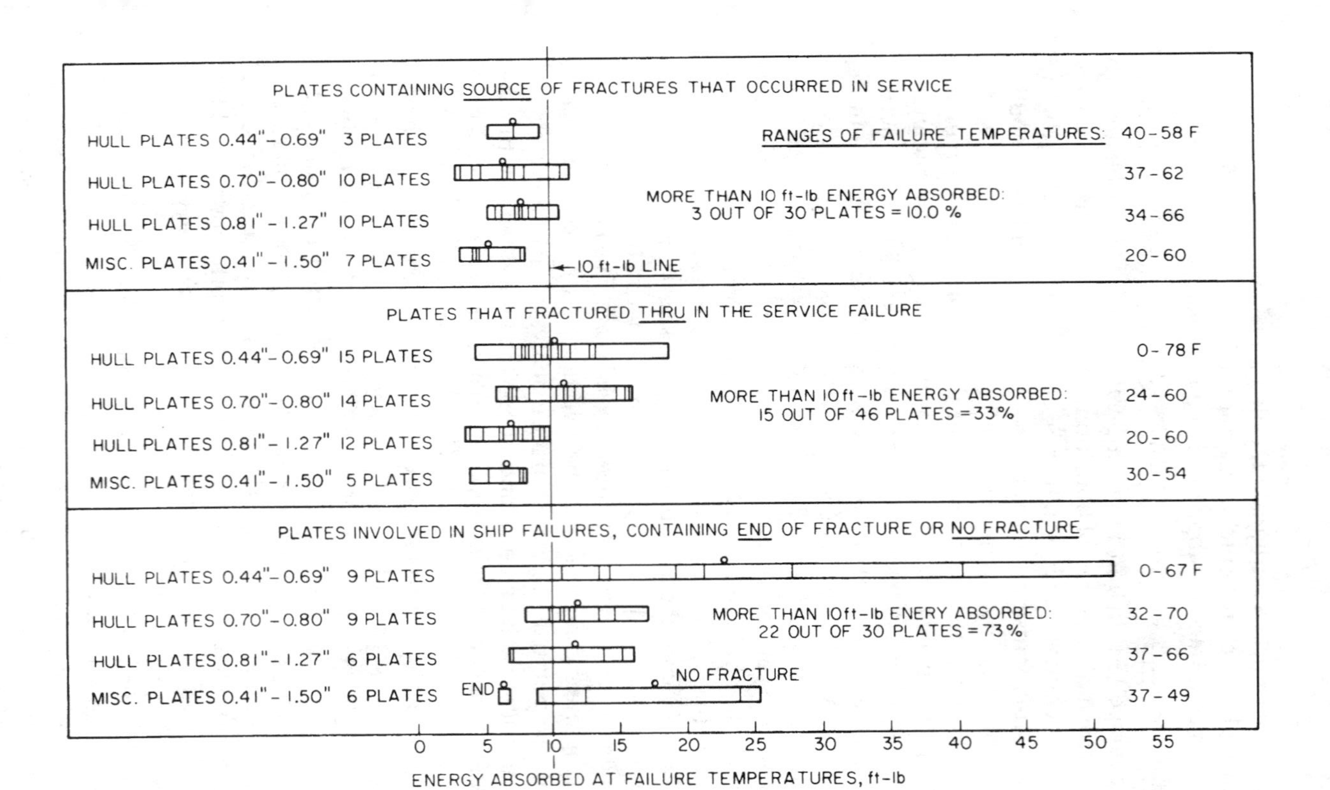

FIG. 13.2 Relation of energy absorbed by V-notch Charpy specimens at the temperature of ship failure to the nature of the fractures in ship plates (Ref. *10*, reproduced with permission of the National Academy of Sciences).

has been a widely used fracture criterion, even though it was developed only for a particular type of steel and a particular class of structures, namely, ship hulls.

Fortunately for the engineering profession and the general public safety, similar statistical correlations between test results and service failures do not exist for any other class of structures because there have not been such a large number of failures in any other type of structure. However, the difficulty of obtaining service experience creates a problem for the design engineer in establishing fracture toughness criteria for new types of structures. Obviously fracture mechanics concepts have become extremely valuable in this process.

13.5 Transition-Temperature Criterion

Since the time of the World War II ship failures, the fracture characteristics of low- and intermediate-strength steels generally have been described in terms of the transition from brittle to ductile behavior as measured by impact tests. This transition in fracture behavior can be related schematically to various fracture states, as shown in Figure 13.1.

For static loading, the transition region occurs at lower temperatures than for impact (dynamic) loading, depending on the yield strength of the steel. Thus, for structures subjected to static loading, the static transition curve should be used to predict the level of performance at the service temperature. For structures subjected to impact or dynamic rates of loading, the impact transition curve should be used to predict the level of performance at the service temperature. For structures subjected to some intermediate loading rate, an intermediate loading-rate transition curve should be used to predict the level of performance at the service temperature. If the loading rate for a particular type of structure is not well defined, and the consequences of failure are such that a fracture will be extremely harmful, a conservative approach is to use the impact loading curve to predict the service performance. As is shown in Figure 13.1, the nil-ductility transition (NDT) temperature is close to the upper limit of plane-strain conditions under conditions of impact loading.

After establishing the loading rate for a particular structure, and thus the corresponding loading rate for the test specimen to be used, the next step in the transition-temperature approach to fracture-resistant design is to establish the *level* of material performance required for satisfactory structural performance. That is, as shown schematically in Figure 13.3 for impact loading of three arbitrary steels 1, 2, and 3, one of the following three general levels of material performance should be established at the service temperature for primary load-carrying members in a structure.

a. Plane-strain behavior.
b. Elastic-plastic (mixed-mode) behavior.
c. Fully plastic behavior.

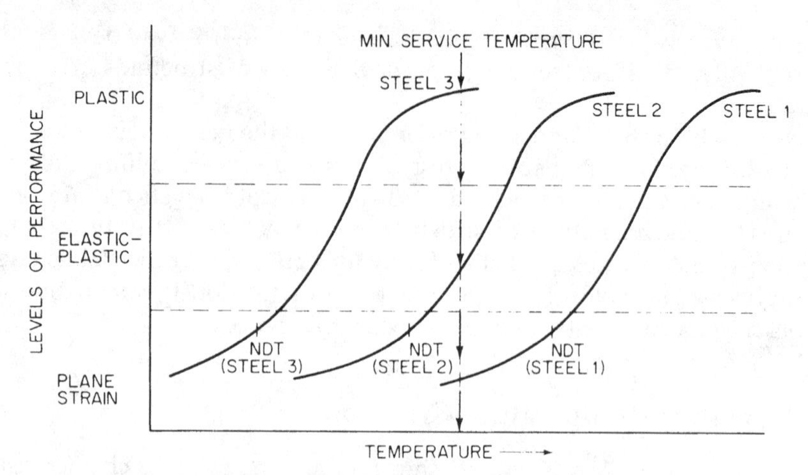

FIG. 13.3 Schematic showing relation between level of performance as measured by impact tests and NDT for three arbitrary steels.

Using the schematic results shown in Figure 13.3 and an arbitrary minimum service temperature as shown, Steel 1 would exhibit plane-strain behavior at the minimum service temperature, whereas Steels 2 and 3 would exhibit elastic-plastic and fully plastic behavior, respectively.

As an example of the transition-temperature approach, assume that a 30-ft-lb Charpy V-notch impact test value is required for ship hull steels. Figure 13.4 compares the average notch toughness levels of several grades of ABS ship hull steels and shows that, according to a 30-ft-lb CVN criterion, the CS-grade steel can be used at service temperatures as low as −90°F, whereas the CN-grade steel can be used only to about −30°F, and the B-grade steel meets this requirement only down to service temperatures of about +10°F.

One limitation to the transition-temperature approach sometimes occurs with the use of materials that do not undergo a distinct transition-temperature behavior or with materials that exhibit a low-energy shear behavior. Figure 13.5 shows the relationship of low-energy performance compared with normal behavior of a steel with a high-level of fracture toughness.

Low-energy shear behavior usually does not occur in low- to intermediate-strength structural steels ($\sigma_{ys} < 100$ ksi), but sometimes is found in high-strength materials. For example, if the desired level of performance is as shown in Figure 13.5, a material exhibiting low-energy shear behavior may never achieve this level of performance at *any* temperature. For these high-strength materials, the through-thickness yielding and the leak-before-burst methods described in the following sections are very useful to establish fracture criteria.

13.6 Through-Thickness Yielding Criterion

The through-thickness yielding criterion for structural steel is based qualitatively on two observations [*13*]. First, increasing the design stress in a particular ap-

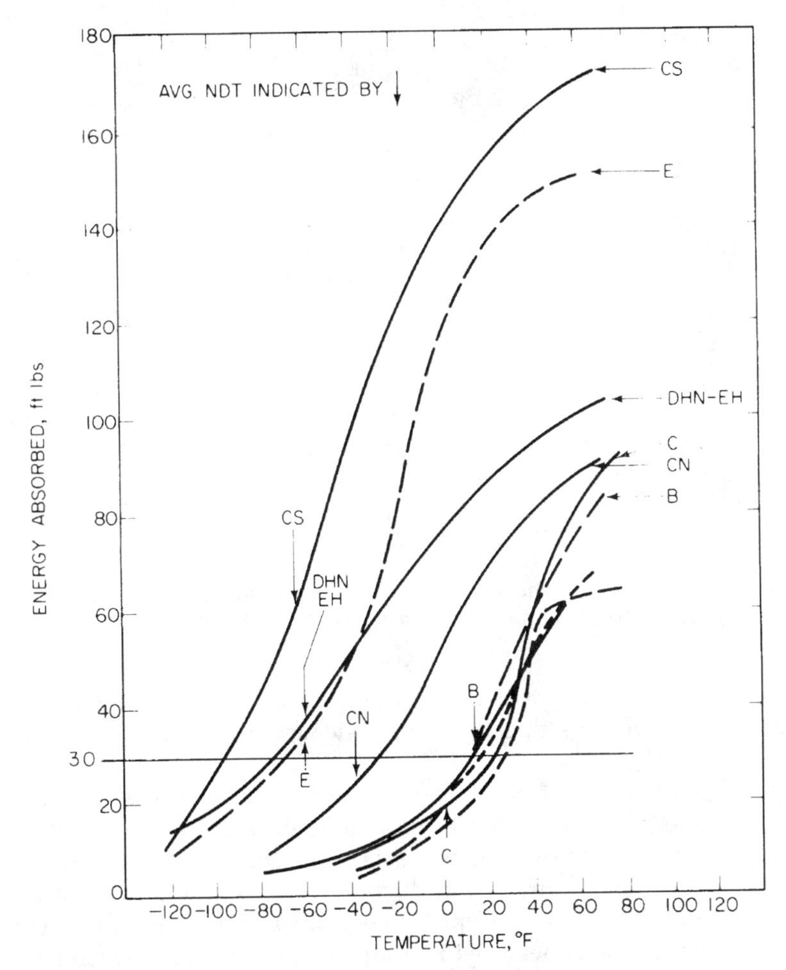

FIG. 13.4 Comparison of minimum service temperature of ship steels using an arbitrary criterion of 30 ft-lb.

plication (which generally requires a higher-yield-strength material) results in more stored elastic energy in a structure. This higher amount of stored elastic energy means that the fracture toughness of the steel also should be increased to have the same degree of safety against fracture as a structure with a lower working stress and a lower stored energy. Note that it is the stored energy available that propagates a crack. Second, increasing plate thickness promotes a more severe state of stress, namely, plane strain. Thus, a higher level of fracture toughness is required to obtain the same level of performance in thick plates as would be obtained in thin plates.

By using concepts of linear-elastic fracture mechanics, a quantitative approach to the development of fracture toughness requirements is based on the requirement that in the presence of a large sharp crack in a large plate, through-

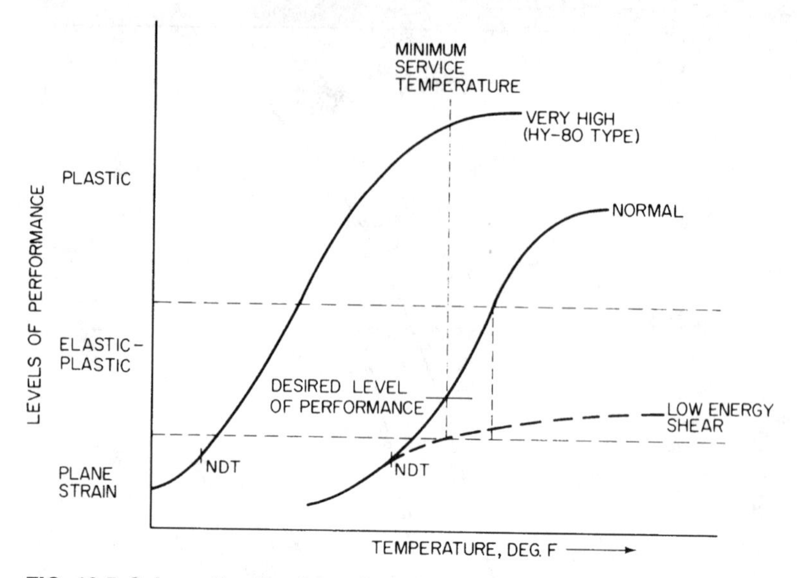

FIG. 13.5 Schematic showing relation among normal-, high-, and low-energy shear levels of performance as measured by impact tests.

thickness yielding should occur before fracture. Specifically, the requirement is based on the ratio of plate thickness to plastic-zone size ahead of a large sharp flaw.

From a qualitative viewpoint, the effect of plate thickness on the fracture toughness of steel plates tested at room temperature has been generally established in Chapter 4. As the plate thickness is decreased, the state of stress changes from plane strain to plane stress, and ductile fractures generally occur along 45° planes through the thickness. Except for very brittle materials, failure is usually preceded by through-thickness yielding and is not catastrophic. Conversely, in thick plates the state of stress is generally plane strain, and fractures usually occur normal to the direction of loading. Except for very ductile materials, through-thickness yielding does not occur prior to fracture, and failure may be unstable. The transition in state of stress from plane strain to plane stress is responsible for a large increase in fracture toughness and is quite desirable. Thus, if plane-stress behavior can be assured, structural members should fail when plastically loaded or when the critical size is very large.

The behavior of most structural members is somewhere between the two limiting conditions of plane-stress and plane-strain. Hahn and Rosenfield [*14–16*] have shown that in terms of either through-thickness strain or crack-opening displacement, there is a significant increase in the rate at which through-thickness deformation occurs when the following relation exists.

$$\left(\frac{K_c}{\sigma_{ys}}\right)^2 \frac{1}{t} \geq 1 \tag{13.1}$$

where K_c = critical stress-intensity factor, ksi$\sqrt{\text{in.}}$, and
σ_{ys} = yield strength, ksi.

As shown in Chapter 3, the following relation must be satisfied to insure plane-strain behavior:

$$B = t \geq 2.5(K_{Ic}/\sigma_{ys})^2 \text{ or } \left(\frac{K_c}{\sigma_{ys}}\right)^2 \frac{1}{t} \leq 0.40 \tag{13.2}$$

Thus, it seems reasonable that Equation (13.1) defines the plane stress condition at which considerable through-thickness yielding begins to occur. This type of behavior is desirable in structural applications and can be used as a criterion to obtain satisfactory performance in structures where through-thickness yielding can occur, such as in large thin plates that contain through-thickness cracks and in which prevention of fracture is an important consideration.

Equation (13.1), which is based on Hahn and Rosenfield's experimental observation of the plastic-zone size in silicon steel, is similar to Irwin's "leak-before-break" criterion discussed in the next section. Thus, for through-thickness yielding to occur in the presence of a large sharp crack in a large plate, a fracture-toughness criterion based on Equation (13.1) appears reasonable on the basis of both experimental and theoretical considerations. This condition is conservative and there are many design applications for which this type of performance is not required.

Rearranging Equation (13.1), a fracture toughness criterion for steels to obtain through-thickness yielding before fracture can be developed in terms of yield strength and plate thickness as follows:

$$K_c \geq \sigma_{ys}\sqrt{t} \text{ for } t \leq 2 \text{ in.}$$

From this relation, the K_c values required for through-thickness yielding before fracture at any temperature were developed for steels with various yield strength levels and plate thickness. These values, shown in Table 13.1, demon-

TABLE 13.1. K_c Values Required for Through-Thickness Yielding Before Fracture.

YIELD STRENGTH, σ_{ys} (ksi)	PLATE THICKNESS, t (in.)	K_c, ksi$\sqrt{\text{in.}}$
40	½	28
	1	40
	2	57
60	½	42
	1	60
	2	85
80	½	57
	1	80
	2	113
100	½	71
	1	100
	2	141

strate the marked dependence of fracture toughness on yield strength and plate thickness if through-thickness yielding is to precede fracture. That is, the desired fracture toughness increases linearly with yield strength and with the square root of plate thickness. Note that because plane-strain conditions do not exist (by definition) the fracture toughness is presented in terms of K_c.

13.7 Leak-Before-Break Criterion

The leak-before-break criterion was proposed by Irwin et al. [*17,18*] as a means of estimating the necessary fracture toughness of pressure-vessel steels so that a surface crack could grow through the wall and the vessel "leaks" before fracturing. That is, the critical crack size at the design stress level of a material meeting this criterion would be greater than the wall thickness of the vessels so that the mode of failure would be leaking (which would be relatively easy to detect and repair) rather than fracture.

Figure 13.6 shows schematically how such a surface crack might grow through the wall into a through-thickness crack having a length approximately equal to $2B$ or $2t$. Thus, the leak-before-break criterion assumes that a crack of twice the wall thickness in length should be stable at a stress equal to the nominal design stress. The value of the general stress intensity, K_I, for a through crack in a large plate where the applied stress approaches yield stress (Figure 13.7) is

$$K_I^2 = \frac{\pi\sigma^2 a}{1 - \frac{1}{2}(\sigma/\sigma_{ys})^2} \tag{13.3}$$

where $2a$ = effective crack length,
σ = tensile stress normal to the crack, and
σ_{ys} = yield strength.

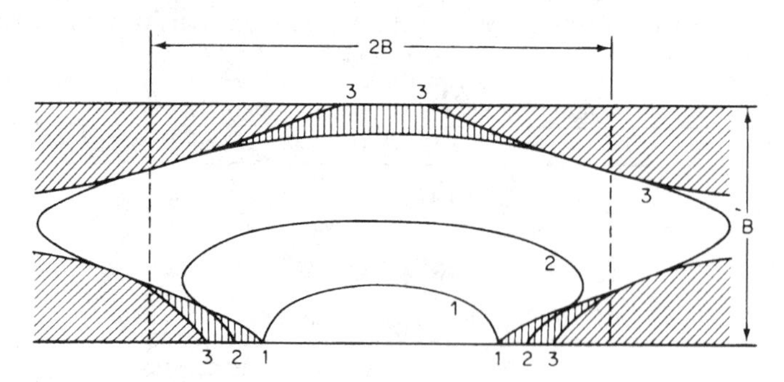

FIG. 13.6 Spreading of a part-through crack to critical size (Line 3) for the short-crack failure models. Lines 1, 2, and 3 are assumed to represent crack edge positions during increase of crack size. Shaded regions are shear lip (vertical shading) or potential shear lip (slant shading).

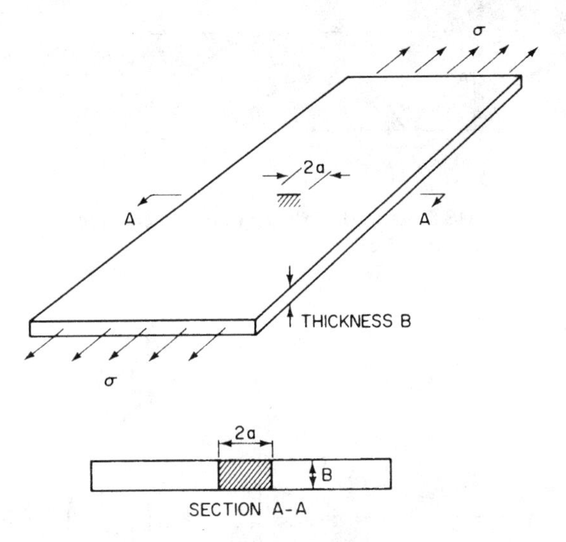

FIG. 13.7 Final dimensions of through-thickness crack.

(Note that for low values of design stress, σ, this expression reduces to $K_I = \sigma\sqrt{\pi a}$.)

At fracture, $K_I = K_c$ (assuming plane-stress behavior) and

$$K_c^2 = \frac{\pi\sigma^2 a}{1 - \frac{1}{2}(\sigma/\sigma_{ys})^2} \tag{13.4}$$

Because standard material properties were usually obtained in terms of K_{Ic}, the following relation between K_c and K_{Ic} was used by Irwin to establish the leak-before-break criterion in terms of K_{Ic},

$$K_c^2 = K_{Ic}^2(1 + 1.4\beta_{Ic}^2) \tag{13.5}$$

where $\beta_{Ic} = \frac{1}{B}\left(\frac{K_{Ic}}{\sigma_{ys}}\right)^2$, a dimensionless parameter

Thus, substituting for K_c and β_{Ic}, the following general relation is obtained:

$$\frac{\pi\sigma^2 a}{1 - \frac{1}{2}(\sigma/\sigma_{ys})^2} = K_{Ic}^2\left[1 + 1.4\left(\frac{K_{Ic}^2}{B\sigma_{ys}^2}\right)^2\right] \tag{13.6}$$

In the leak-before-break criterion, the depth of the surface crack, *a*, is set equal to the plate thickness, *B*, (Figure 13.8), and we obtain

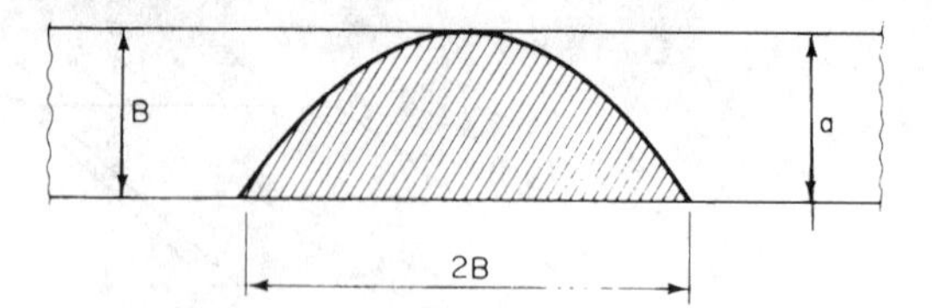

FIG. 13.8 Assumed flaw geometry for leak-before-burst criterion.

$$\frac{\pi\sigma^2 B}{1 - \frac{1}{2}(\sigma/\sigma_{ys})^2} = K_{Ic}^2\left[1 + 1.4\left(\frac{K_{Ic}^4}{B^2\sigma_{ys}^4}\right)\right]$$

or (13.7)

$$\frac{\pi\sigma^2}{1 - \frac{1}{2}(\sigma/\sigma_{ys})^2} = \frac{K_{Ic}^2}{B}\left[1 + 1.4\left(\frac{K_{Ic}^4}{B^2\sigma_{ys}^4}\right)\right]$$

where σ = nominal design stress, ksi,
σ_{ys} = yield strength, ksi,
B = vessel wall thickness, in., and
K_{Ic} = plane-strain fracture toughness (ksi$\sqrt{\text{in.}}$) required to satisfy the leak-before-break criterion for a material with a particular σ_{ys}, a vessel with wall thickness B and design stress σ.

In this expression, σ, σ_{ys}, and K_{Ic} are the design stress, yield strength, and material toughness at the particular service temperature and loading rate.

For the critical situation of $\sigma = \sigma_{ys}$, the criterion reduces to

$$\frac{\pi\sigma_{ys}^2}{1 - \frac{1}{2}(\sigma_{ys}/\sigma_{ys})^2} = \frac{K_{Ic}^2}{B}\left[1 + 1.4\left(\frac{K_{Ic}^4}{B^2\sigma_{ys}^4}\right)\right] \quad (13.8)$$

$$2\pi\sigma_{ys}^2 = \frac{K_{Ic}^2}{B} + (1.4)\frac{K_{Ic}^6}{B^3\sigma_{ys}^4}$$

or

$$\frac{1.4K_{Ic}^6}{B^3\sigma_{ys}^4} + \frac{K_{Ic}^2}{B} = 2\pi\sigma_{ys}^2 \quad (13.9)$$

As an example of the use of this criterion, the engineer must first select the nominal yield-strength steel that he or she wishes to use, then determine the wall thicknesses (these two factors might be established on the basis of a general strength criterion to withstand a given internal pressure) and, finally, select the required minimum fracture toughness level necessary to meet the criterion. *Then,* from the steels available, the engineer must select the one or ones that meet the criterion. Final material selection would be on the basis of the foregoing conditions, plus other criteria, such as cost, fabrication, and so on.

The required K_c values that will satisfy the leak-before-break criterion at yield strength levels ranging from 40 to 120 ksi and for wall thickness ranging from ½ to 2 in. are presented in Table 13.2. Note that these are listed as K_c rather than K_{Ic} as developed by Irwin because all values are well above the plane strain minimum size of $B \geq 2.5\ (K_{Ic}/\sigma_{ys})$. These results illustrate the significant effect of an increase in thickness or yield strength when selecting materials to satisfy the leak-before-break criterion. Note the similarity of the through-thickness yielding and the leak-before-break criterion by comparing the results presented in Tables 13.1 and 13.2. Thus, either criterion can be used to ensure a reasonably high level of notch toughness (elastic-plastic behavior).

The leak-before-break criterion, like the through-thickness yielding criterion, becomes too conservative as the plate thicknesses increase above about 2 in. Both can be used as conservative elastic-plastic fracture toughness criteria that account for the effects of yield strength and thickness. Both criterion ensure non-plane-strain behavior.

13.8 Fracture Criterion for Steel Bridges

A fracture criterion for steel bridges was developed as part of a comprehensive fracture control plan. This fracture criterion was discussed in Chapter 12. It is based, in part, on the fracture-toughness transition behavior of bridge steels as a function of temperature and on the effects of loading rate on fracture-toughness transition. It requires an elastic-plastic fracture initiation at the minimum operating temperature and at the maximum loading rate for bridges and

TABLE 13.2. K_c Values Required to Satisfy Leak-Before-Break Criterion for Yield-Strength Loading.

MATERIAL YIELD STRENGTH AND ASSUMED APPLIED STRESS, ksi ($\sigma = \sigma_{ys}$)	VESSEL THICKNESS, B (OR t) (in.)	REQUIRED K_c (ksi$\sqrt{\text{in.}}$)
40	½	35
	1	50
	2	70
60	½	50
	1	75
	2	105
80	½	70
	1	100
	2	140
100	½	85
	1	120
	2	175

a critical crack size larger than could exist at the end of the design fatigue life. Also, higher fracture toughness values are required for fracture critical than for non-fracture critical members.

13.9 Summary

The preceding sections describe a few notch-toughness criteria that have been proposed for use on structural applications. Obviously, the most desirable criterion is one which is based directly on the appropriate critical *K* values that are used to calculate stress–crack-size trade-offs, as discussed in Chapter 6. However, the *direct* use of fracture mechanics K_I calculations and K_c measurements assumes that material fracture toughness values can be determined using the test methods described in Chapter 3. As has been discussed before, it is difficult to obtain K_c values routinely for most structural materials. Hence many criteria are based on concepts of fracture mechanics but are described in terms of simple tests, such as the CVN impact test.

A key aspect of criteria selection is the proper consideration of the consequences of failure. Also, the engineer should remember that good design is an optimization of performance, safety, and cost. Thus, arbitrarily specifying a fracture toughness criterion more than is necessary may increase the cost of the structure such that it may be prohibitive to build. This was the case for the floating nuclear power plants designed in the early 1970s that were not built for several reasons, including the excessive safety factor that increased the cost considerably.

13.10 References

[1] Pellini, W. S., 1971 AWS Adams Lecture, *Principles of Fracture-Safe-Design*, Part I, *Welding Journal Research Supplement*, March 1971, pp. 91-S–109-S.
[2] Pellini, W. S., 1971 AWS Adams Lecture, *Principles of Fracture-Safe-Design*, Part II, *Welding Journal Research Supplement*, April 1971, pp. 147-S–162-S.
[3] Acker, H. G., "Review of Welded Ship Failure," *Ship Structure Committee Report, Serial No. SSC-63*, U.S. Coast Guard, Washington, DC, Dec. 15, 1953.
[4] Brown, D. P., "Observations on Experience with Welded Ships," *Welding Journal*, September 1952, pp. 765–782.
[5] *Final Report of a Board of Investigations to Inquire into the Design and Methods of Construction of Welded Steel Merchant Vessels*, GAO, Washington, DC, 1947.
[6] Jonassen, F., "A Resume of the Ship Fracture Problem," *Welding Journal, Research Supplement*, June 1952, pp. 316-S–318-S.
[7] Williams, M. L. and Ellinger, A., "Investigation of Fractured Steel Plates Removed from Welded Ships," *Ship Structure Committee Report, Serial No. NBS-1*, U.S. Coast Guard, Washington, DC, Feb. 25, 1949.
[8] Williams, M. L., Meyerson, M. R., Kluge, G. L., and Dale, L. R., "Investigation of Fractured Steel Plates Removed from Welded Ships," *Ship Structure Committee Report Serial No. NBS-3*, U.S. Coast Guard, Washington, DC, June 1, 1951.

[9] Williams, M. L., "Examination and Tests of Fractured Steel Plates Removed from Welded Ships," *Ship Structure Committee Report, Serial No. NBS-4,* U.S. Coast Guard, Washington, DC, April 2, 1953.

[10] Williams, M. L. and Ellinger, G. A., "Investigation of Structural Failures of Welded Ships," *Welding Journal, Research Supplement,* October 1953, pp. 498-S–527-S.

[11] Williams, M. L., "Analysis of Brittle Behavior in Ship Plates," *Ship Structure Committee Report, Serial No. NBS-5,* U.S. Coast Guard, Washington, DC, Feb. 7, 1955. (Also presented in *ASTM STP 158,* American Society for Testing and Materials, Philadelphia, 1954.)

[12] Parker, E. R., *Brittle Behavior of Engineering Structures,* prepared for the Ship Structure Committee, John Wiley, New York, 1957.

[13] Rolfe, S. T., Barsom, J. M., and Gensamer, M., "Fracture-Toughness Requirements for Steels" presented at the Offshore Technology Conference, Houston, May 18–21, 1969.

[14] Hahn, G. T. and Rosenfield, A. R., "Source of Fracture Toughness: The Relation Between K_{Ic} and the Ordinary Tensile Properties of Metals," *ASTM STP 432,* American Society for Testing and Materials, Philadelphia, 1968, pp. 5–32.

[15] Hahn, G. T. and Rosenfield, A. R., "Plastic Flow in the Locale of Notches and Cracks in Fe-3Si Steel Under Conditions Approaching Plane Strain," *Ship Structure Committee Report SSC-191,* U.S. Coast Guard, Washington, DC, Nov. 1968.

[16] Rosenfield, A. R., Dai, P. K., and Hahn, G. T., *Proceedings of the International Conference on Fracture,* Sendai, Japan, 1965.

[17] Irwin, G. R., "Relation of Crack-Toughness Measurements to Practical Applications," *ASME Paper No. 62-MET-15,* presented at Metals Engineering Conference, Cleveland, April 9–13, 1962.

[18] Irwin, G. R., Krafft, J. M., Paris, P. C., and Wells, A. A., "Basic Aspects of Crack Growth and Fracture," *NRL Report 6598,* Naval Research Laboratory, Washington, DC, November 21, 1967.

14

Fitness for Service

14.1 Introduction

FITNESS FOR SERVICE is a term used by Wells [1] in the early 1960s to describe the actual safety or reliability of a structure at a given time. It takes account of the actual stresses, flaw sizes, material toughness, service conditions, etc. in a realistic manner so that the "fitness" for *continued* service of an existing structure can be established on a rational basis. It has also been referred to as common sense engineering [2] and serves as the basis for a decision as to the "life extension" of an existing structure.

Fitness for service (purpose) was defined by Alan Wells [1] in the early 1960s as:

> "Fitness for purpose is deemed to be that which is consciously chosen to be the right level of material [that is, having the appropriate fracture toughness, e.g., K_c, CTOD, *J*-integral, CVN, etc.] and fabrication quality [that is, tolerable flaw size for the given application] for each application [that is, the appropriate loading or stress level for the given application], having regard to the risks and consequences of failure; it may be contrasted with the best quality that can be achieved within a given set of circumstances, which may be inadequate for some exacting requirements, and needlessly uneconomic for others which are less demanding. A characteristic of the fitness-for-purpose approach is that it is required to be defined beforehand according to known facts, and by agreement with purchasers which will subsequently seek to be national and eventually international.
>
> The need for such an approach has already been seen with the development and application of fracture mechanics, but the paper [by Wells] draws attention to a wider scope which also embraces the evolution of the process,

risk analysis and reliability engineering, nondestructive examination, codes and standards, and quality assurance.

It is considered that the assessment of fitness for purpose should relate well to the quality assurance approach, since the latter aims to be comprehensive, and makes provision for updating its own procedures."

It is proposed that fitness for service is indeed common-sense engineering and that both concepts rely heavily on the field of fracture mechanics. The fitness-for-service approach can be used during the design process or to analyze the remaining life or life extension of an existing structure in which a crack has developed. Although several standards [3–5] are available to analyze the fitness for service of existing structures, there is no single all-encompassing methodology. Fitness for service also encompasses risk analysis or probability, nondestructive examination, quality control, and quality assurance as well as other related technologies, most of which incorporate concepts of fracture mechanics. In short, one of the major new directions in fracture mechanics deals with fitness for service and the application of fracture mechanics concepts to extending the life of existing structures.

It should be emphasized that the principles of fracture mechanics used in new design, Parts I and II, and those of fatigue or stress corrosion, Part III, also apply to fitness-for-service analyses for structures in service. The unique aspect of a fitness-for-service analysis for structures in service is the use of the service conditions, flaw sizes, and, when possible (see Section 14.3.3, Step C, for example), material properties actually existing at the time of the analysis rather than extreme loadings, possible flaw sizes, or material properties assumed at the time of new design.

Guidelines on the specific inputs to a fitness-for-service analysis are given in the following sections.

14.2 Use of Fracture Mechanics in Fitness-for-Service Analysis

14.2.1 General

The driving force, K_I, can be calculated as described in Chapter 2. Most of the crack geometries encountered in actual structures can be approximated by the various K_I relations presented in Chapter 2. For unusual crack geometries that cannot be approximated by one of the K_I relations presented in Chapter 2, the reader is referred to several handbooks of the K_I factors [6,7].

Chapter 3 described the various test methods available to measure K_c for a given material. Recall that for structural steels, there were four general regions of fracture behavior, as is shown in Figure 14.1. In many cases, material is not available to test, and estimates of K_c must be made from CVN impact test results (which require much less material) as discussed in Chapter 5.

Chapter 4 discussed the effect of temperature, loading rate, and constraint on fracture behavior. Obviously, the test temperature is the easiest parameter to

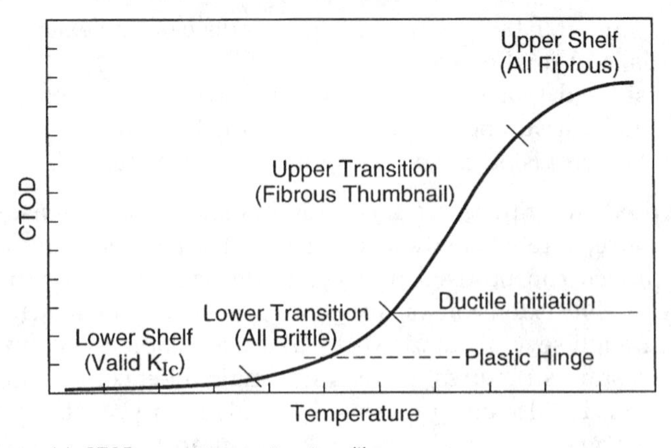

(a) CTOD versus temperature transition

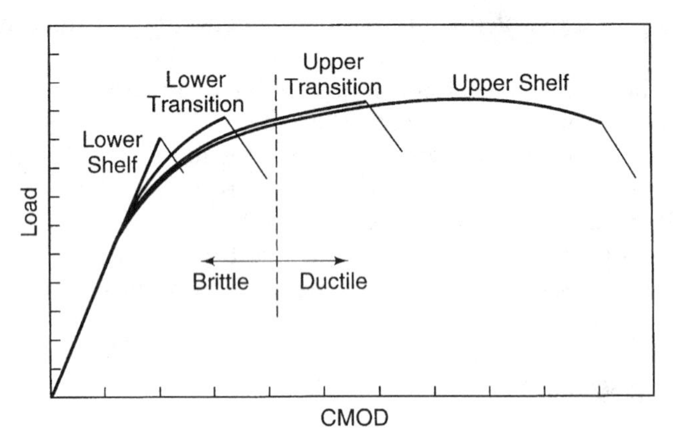

(b) Load versus crack-mouth opening displacement

FIG. 14.1 Schematic toughness behavior and test records of low-strength steels.

establish for the particular structure. The issue of loading rate and whether or not to design for initiation (slow loading) or arrest (dynamic loading) is not as straightforward, however.

14.2.2 Effect of Loading Rate

For materials that exhibit loading-rate or strain-rate effects, such as structural steels with yield strengths less than about 140 ksi, the loading rate at a given temperature can affect the resistance force (fracture toughness) significantly. This behavior is shown in Figure 14.2 for an A572 Grade structural steel. In an ideal fitness-for-service evaluation, the loading rate in the particular fracture toughness

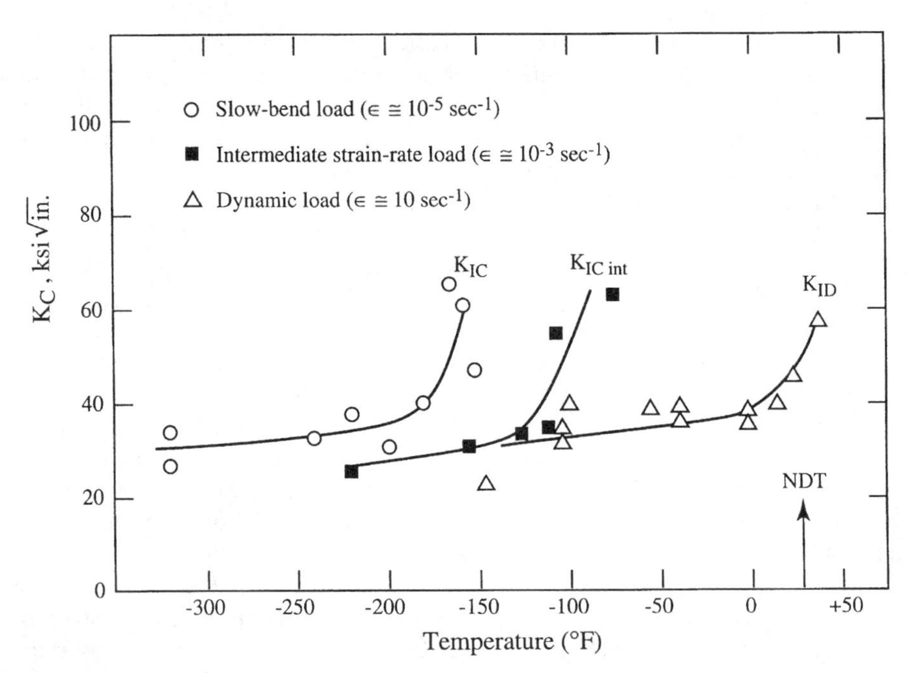

FIG. 14.2 Effect of temperature and strain rate on fracture toughness of 50-ksi yield-strength structural steel.

test (*K*, CTOD, *J*) used to evaluate the behavior of a structural material should be essentially the same as the loading rate in the structure being analyzed. Fracture toughness criteria can then be established by using either static, intermediate, or dynamic fracture toughness test results for a particular structure.

The importance of using test results that model the loading rate in the actual structure should be emphasized. If the engineer can be certain that a structure is loaded statically or that the structural material is not strain-rate sensitive, then slow-loading tests should be used to evaluate the fitness for service. Conversely, if impact loading is the actual loading rate, then an impact test should be used to predict the fitness for service.

Because of the large differences in behavior of structural materials that are strain-rate sensitive, Figure 14.2, the likelihood of dynamic loading vs. slow loading of structures should be considered carefully during a fitness-for-service analysis. There are two different schools of thought regarding the influence of loading rate on the fracture toughness behavior of structures.

The first school of thought assumes that there can be highly localized regions where the fracture toughness is low in all large structures. Examples of these regions might be local brittle zones in weldments, grain-coarsened regions in the heat-affected zone of a weldment, arc strikes on the base plate, nonmetallic inclusions, etc. These highly localized regions of low fracture toughness are assumed to initiate microcracks under statically applied loads so that the surrounding material is suddenly presented with a moving dynamic (but small) crack. Thus, even though the overall applied structural loading rate may be slow, this school of thought believes that the crucial loading rate is that of the crack-tip pop-in. This sudden separation of a few metal grains (<~0.01 in. in size) is referred to as "pop-in" and is assumed to create dynamic loading rates. This line of reasoning implies that the dynamic fracture toughness (K_{Id}/σ_{yd}) always controls fracture extension irrespective of initiation conditions. Accordingly, this assumed behavior leads to the conclusion that the K_{Id}/σ_{yd} parameter always should be used to evaluate the behavior for all structures regardless of the measured structural loading rates. This is referred to as designing for crack-arrest behavior.

The second school of thought assumes that different loading rates can apply for different structures. For those structures where the measured loading rates are slow, the static K_c/σ_{ys} ratio is the controlling fracture-toughness parameter. For those structures loaded at some intermediate loading rate (which applies to many structures such as bridges, ships, offshore rigs, and buildings subjected to earthquake loadings), an "intermediate" $K_{I(t)}/\sigma_{ys}$ ratio is believed to be the controlling fracture-toughness parameter. If indeed a structure is loaded dynamically, then K_{Id}/σ_{yd} is the controlling parameter. Interestingly, many structures whose dynamic toughness (K_{Id}/σ_{yd}) is relatively low have been performing quite satisfactorily for many years because the actual structural loading rate is slow or, at most, intermediate.

A realistic appraisal of these two schools of thought leads to the following observation for fitness-for-service evaluations. While it is true that there may exist highly localized regions of low fracture toughness in complex structures, it seems hard to visualize that a dynamic stress field can be created by the localized extension of a small microcrack, or "pop-in." If this were indeed the case, many hundreds of thousands of structures that continue to perform satisfactorily should have failed. A more realistic mechanism of creating a localized dynamic stress field under nominally static loading would appear to result from the sudden separation of secondary structural members such as a stiffener or a gusset plate. This sudden separation might change the local stress distribution over a reasonably large area rather suddenly in nonredundant structures. Thus, a large dynamic stress field surrounding a pre-existing crack is created such that the dynamic properties (K_{Id}/σ_{yd}) control the resistance to failure.

Accordingly, whereas it may be conservative to analyze existing structures assuming dynamic loading (use of K_{Id}/σ_{yd} toughness values at the service temperature), it does not seem to be appropriate to automatically use impact results in a fitness-for-service analysis. Rather, the actual loading rate should be used to evaluate the remaining life of actual structures.

As an example of the fitness-for-service approach, the AASHTO material-toughness requirements [*8*] for fracture critical tension members of A36 bridge steels having service temperatures down to 0°F require that the steel plates exhibit 25-ft-lb CVN impact energy at +70°F. Because bridges have been shown to be loaded at an intermediate rate of loading, this requirement is designed to ensure non-plane-strain behavior beginning at about −50°F, well below the minimum service temperature of 0°F, as shown in Figure 14.3. The use of the temperature-rate "shift" to establish fracture toughness requirements for steel highway bridges has been demonstrated to be quite satisfactory because of:

1. The intermediate loading rate to which bridges are subjected.
2. The loadings are reasonably well known.
3. The conservative AASHTO fatigue requirements.

Laboratory tests conducted on nonredundant, welded bridge details indicated that the AASHTO fracture toughness requirements were adequate even when the details were subjected to the total fatigue design life, the maximum design stress, the minimum operating temperature, and the maximum expected loading rate [9]. Although not stated directly, the AASHTO fracture toughness requirements for bridge steels follow a fitness-for-service approach and demonstrate the use of actual loading rates in setting fracture toughness requirements.

14.2.3 *Effect of Constraint*

Although the effects of temperature and loading rate on the fracture behavior of structures are reasonably well understood and documented, the relationship between constraint and fracture toughness is less clear.

The out-of-plane constraint (Figure 14.4) generally is handled by either testing very thick specimens or, at least in a fitness-for-service application, testing the thickness of actual interest. However, the in-plane constraint is more difficult to analyze. The in-plane constraint is controlled by the effect of crack depth (*a*)

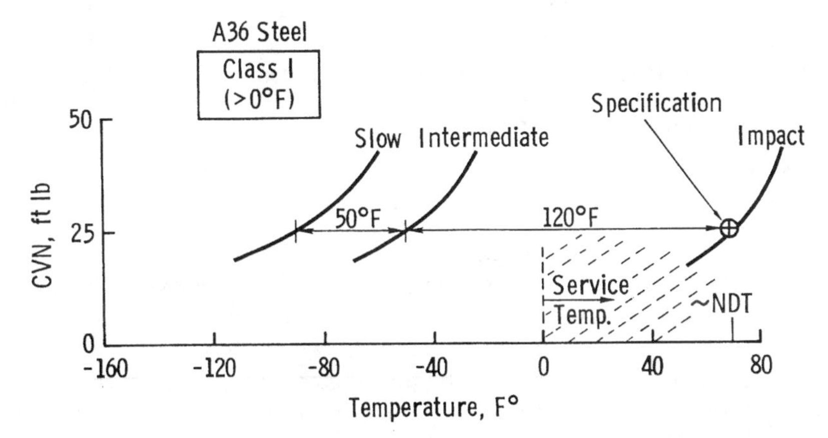

FIG. 14.3 Schematic showing elastic-plastic fracture toughness behavior at a service temperature of 0°F for an AASHTO material toughness specification of 25 ft-lb at +70°F.

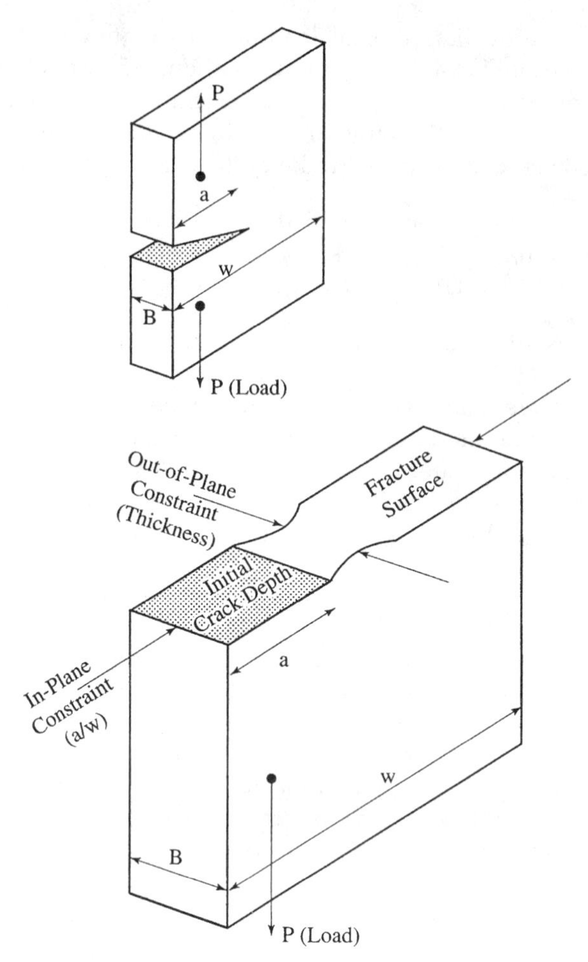

FIG. 14.4 Schematic showing out-of-plane constraint as a function of thickness and in-plane constraint as a function of crack length and *a/W* ratio.

and crack-depth to specimen-width ratio (a/W) in test specimens. Comparison of finite element results of various crack-depth to specimen-width ratios (a/W) shows a fundamental change in the nonlinear stress field at an a/W ratio of approximately 0.15. Specimens with shorter cracks ($a/W = 0.10$ and 0.05) show yielding to the free (tension) surface behind the crack well before the formation of a plastic hinge. Specimens with deeper cracks ($a/W \geq 0.20$) develop a plastic hinge before the plastic zone extends from the crack tip to the back surface.

In the lower shelf region, where valid K_{Ic} results can be obtained, experimental fracture toughness (CTOD and J-integral) results of short-crack specimens are similar to the results of the deep-crack specimens. Correspondingly, the three-dimensional finite element analyses for two specimen a/W ratios (0.50 and 0.15) reveal that the opening mode stresses near the crack tip are essentially identical at the same fracture toughness (K_{Ic}, CTOD, and J_{Ic}) levels. This confirms that the fracture toughness can be expressed as a single parameter characterization of the stress field, which is independent of specimen size and crack depth in the lower shelf or linear-elastic region.

However, in the lower-transition region, where considerable plastic deformation and crack tip blunting occurs prior to brittle fracture, the experimental fracture toughness results (CTOD and J) of the short-crack specimens are approximately two to three times larger than results of the deep-crack specimens at identical temperatures, as shown in Figure 14.5 [*10*]. Thus, at equivalent fracture toughness levels in the elastic-plastic regime, the finite element analyses reveal significant differences in crack tip stresses between the shallow and deep-cracked specimens. The deep-crack specimens exhibit significantly higher opening mode stresses near the crack tip compared to the short-crack specimens, as shown in Figure 14.6. Correspondingly, at equivalent levels of opening-mode stress, the short-crack specimens have CTOD and J values approximately 2.5 times larger than the deep-crack specimens, as was shown in Figure 14.5.

Both crack depth (a) and a/W ratio affect the fracture toughness of structural and pressure vessel steels. For deep-crack geometries ($a/W = 0.5$), crack depth has a limited effect on fracture toughness. That is, even for a large difference in laboratory specimen size ($W = 0.8$ to 4.0 in.), there is little difference in fracture toughness as long as the a/W ratio is 0.5, as shown in Figure 14.7 for an A533B steel. As the in-plane constraint is decreased by decreasing the a/W ratio to 0.1, specimens with a smaller crack length exhibit a higher fracture toughness, Figure 14.8. For specimens with the same crack length ($a = 0.4$), decreasing the a/W ratio increases the fracture toughness, Figure 14.9. Finally, decreasing both crack length, a and a/W ratio results in a significant increase in fracture toughness, Figure 14.10.

The significance of this increase in fracture toughness with respect to structural performance will have a large effect on fitness-for-service evaluations. However, presently it is difficult to quantify this effect, which is the subject of current research. For example, the Nuclear Regulatory Commission is analyzing the positive effect that this observation may have on the prediction of the remaining life of pressure vessels used in the nuclear industry [*11*].

Constraint is the result of a triaxial state of stress at the crack tip. If stresses are relaxed in any direction, either out-of-plane or in-plane, then the constraint is decreased and the fracture toughness increases as was just shown. For structural situations as well as laboratory specimens, constraint can be lost in either the out-of-plane or the in-plane direction.

A current theory [*12*] suggests that at low a/W ratios, the single fracture toughness parameters, such as K, J or CTOD, are not adequate to characterize the fracture toughness. Specifically these studies show that a second parameter,

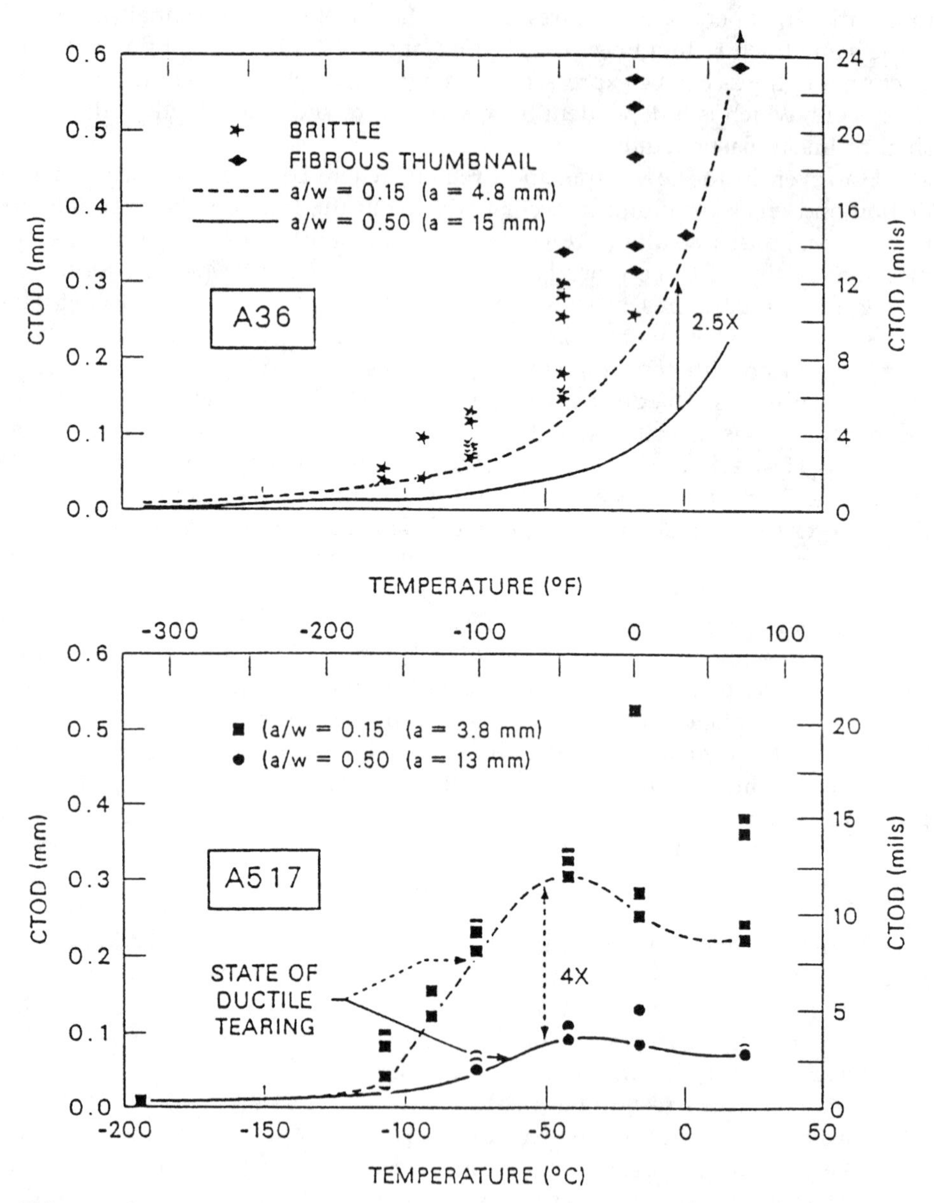

FIG. 14.5 Shallow crack results showing elevated fracture toughness in A36 and A517 steels as a function of *a/W* ratio.

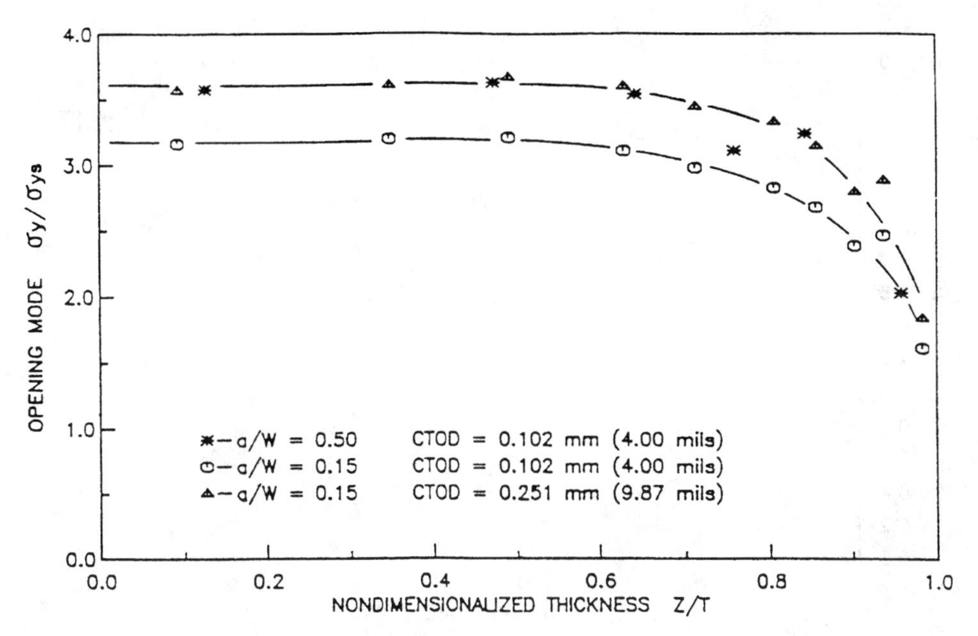

FIG. 14.6 Crack opening stress distributions in 31.8 by 31.8-mm A36 specimens.

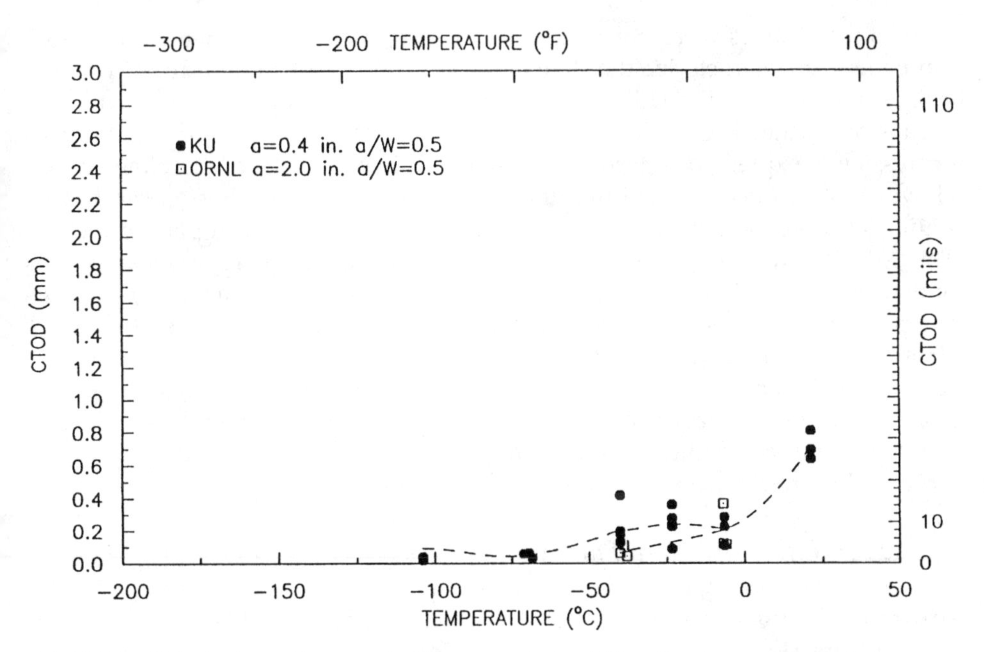

FIG. 14.7 CTOD test results for A533B steel showing little effect of crack depth for *a/W* = 0.5.

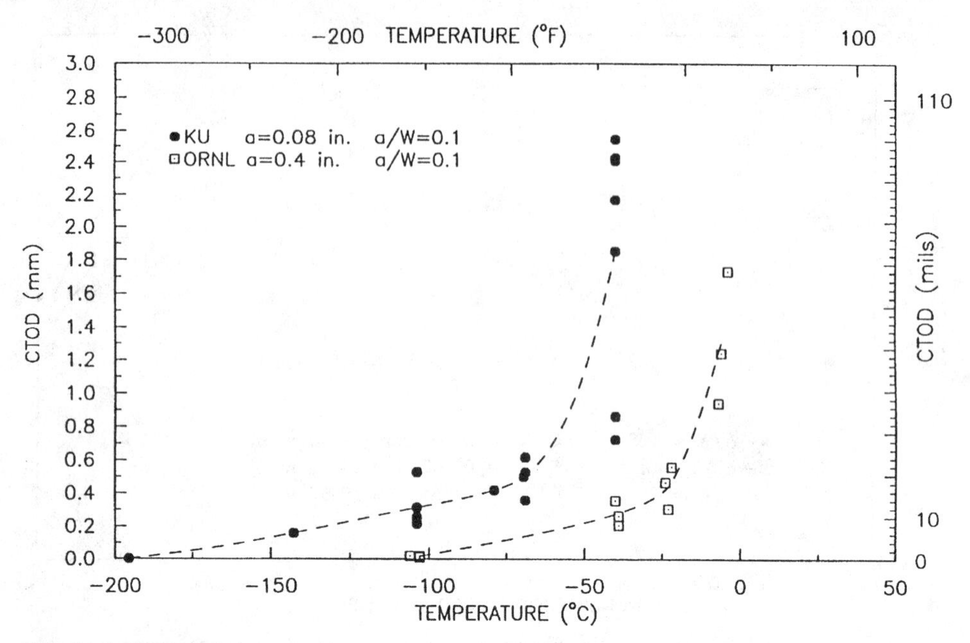

FIG. 14.8 CTOD test results for A533B steel showing large effect of crack depth for *a/W* = 0.1.

such as the *T* stress when referred to K_{Ic} or the *Q* parameter when referred to *J*, can be used to characterize the state of stress at the tip of a crack. In the original Williams stress function [13], the stresses near the crack tip are described by an expansion of the stress field about the crack tip. The first term of the Williams expression is expressed in terms of K_I, which is linear-elastic fracture mechanics. However, it has been shown that the second term of this expansion can have a significant effect on the stress levels near the crack plane in the in-plane direction. This *K-T*, two-term expansion is limited to the linear-elastic region just as K_I is limited to the linear-elastic region. In the elastic-plastic area, a similar analysis exists for a *Q* parameter which is compatible with the *J* integral. Just as the *T* stress affects the fracture toughness for low values of a/W in the linear-elastic region, the *Q* parameter will affect the fracture toughness in the elastic-plastic regime. The *K-T* stress or the *J-Q* analyses may enhance our understanding of the conditions controlling fracture and lead to a better understanding of the structural performance of structures with shallow cracks.

14.2.4 Effect of Many Factors

There are cases where many factors may affect the fracture behavior of a structure. The fracture of welded steel moment frame (WSMF) connections during the Northridge earthquake in January, 1994 has been assumed to be such an example [14]. That earthquake demonstrated the susceptibility to damage of the

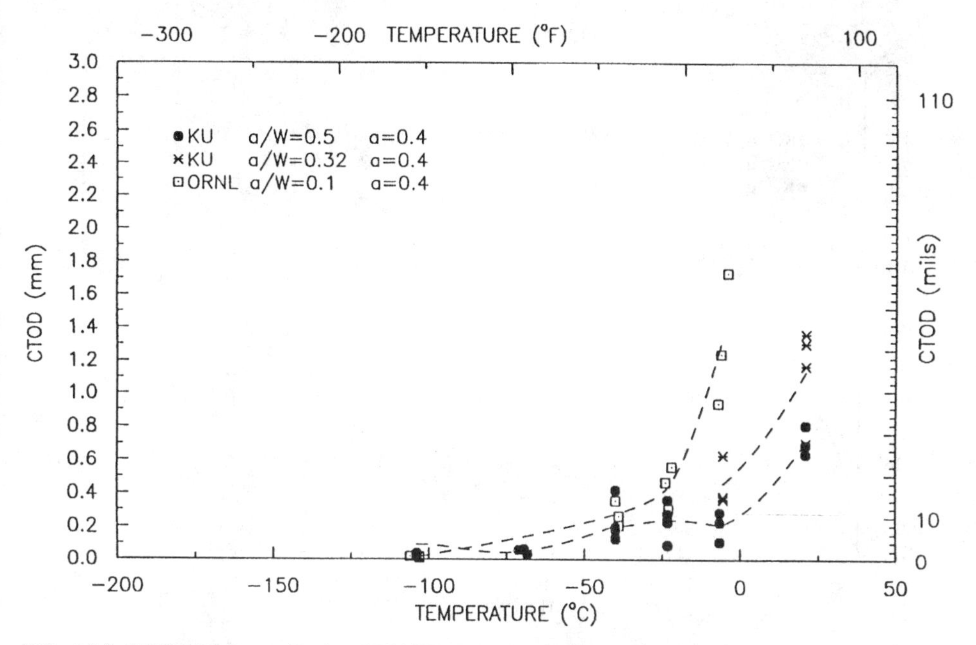

FIG. 14.9 CTOD test results for A533B steel showing effect of *a/W* ratio for constant crack depth.

welded beam-to-column moment connections commonly used in the construction of welded steel moment resisting frames [*15*].

The Preface to SAC Report 95-09, Background Reports [*14*] notes that:

> "It is now known that a large number of factors contributed to the damage sustained by steel frame buildings in the Northridge earthquake. These included [numbered for clarity]:

1. design practice that favored the use of relatively few frame bays to resist lateral seismic demands, resulting in much larger member and connection geometries than had previously been tested
2. standard detailing practice that resulted in the development of large inelastic demands at the beam to column connections
3. detailing practice that often resulted in large stress concentrations in the beam-column connection
4. the common use of welding procedures that resulted in deposition of low toughness weld metal in the critical beam flange to column flange joints
5. relatively low levels of quality control and assurance in the construction process, resulting in welded joints that did not conform to the applicable quality standards
6. detailing practice for welded joints that resulted in inherent stress risers and notches in zones of high stress

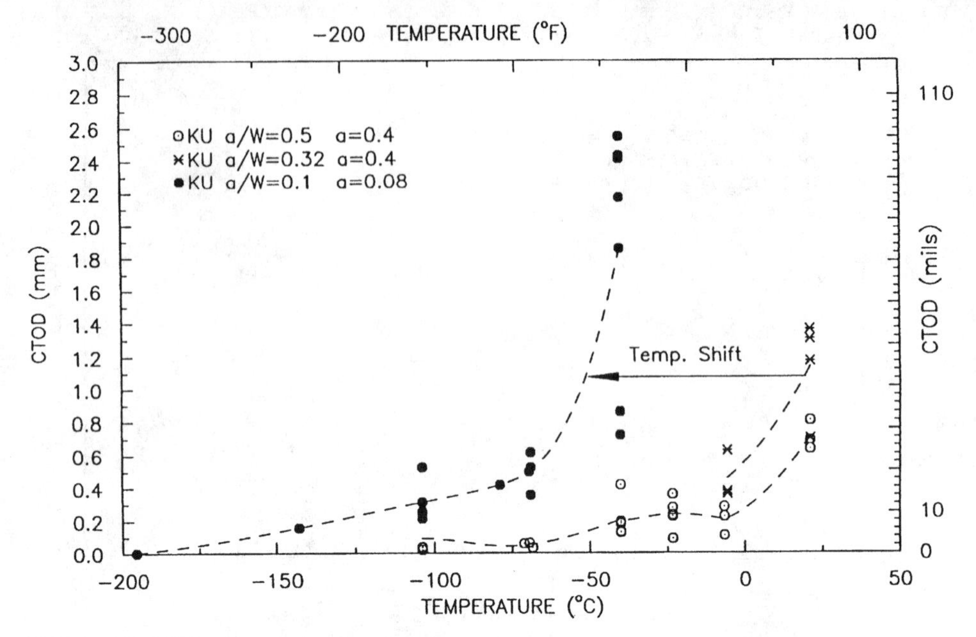

FIG. 14.10 CTOD test results for A533B steel showing large effect of both crack depth, *a*, and *a/W* ratio.

7. excessively weak and flexible column panel zones that resulted in large secondary stresses in the beam flange to column flange joints
8. large variations in material strengths relative to specified values
9. the inherent inability of the material to yield under conditions of high triaxial restraint such as exist at the center of the beam flange to column flange joints."

The relative importance of these factors currently is being studied in numerous investigations [*16*]. Various recommendations and guidelines for reliable connections are being developed and should be available in the year 2000.

14.3 Existing Fitness-for-Service Procedures

14.3.1 General

Two existing fitness-for-service procedures have been used widely by engineers for years. These two methodologies are:

1. PD 6493—Guidance on Methods for Assessing the Acceptability of Flaws in Fusion Welded Structures [3]. This procedure was developed in 1980 and has been updated recently. It is used "to examine critically the integrity of fusion welded joints in new or existing constructions."
2. ASME Section XI—Rules for In Service Inspection of Nuclear Power Plant Components [5]. This procedure is used to determine the acceptability of

flaws detected during in-service inspections of fabricated nuclear vessels and components.

Both of these methodologies are based on concepts of fracture mechanics and generally follow the procedures described in Parts I, II, and III of this book. An introduction to specifics of these procedures is given in the following sections. A third procedure has just been developed by API and also is described.

14.3.2 PD 6493

The British Standards Institute document PD 6493:1991, Guidance on Some Methods for the Derivation of Acceptance Levels for Defects in Fusion Welded Joints, is based on concepts of fracture mechanics as stated earlier. This document establishes specific guidelines to the engineer who is assessing the safety and reliability of existing structures in which defects have been found. The document establishes a failure assessment diagram that is actually an interaction curve relating the limits of both fracture and yielding. The schematic diagram in Figure 14.11 shows that as long as both

$$K_r = \frac{K_I}{K_c} \tag{14.1}$$

and

$$S_r = \frac{\sigma_{nom}}{\sigma_{flow}} \tag{14.2}$$

are below the assessment line, the structure should be safe. In essence, the failure

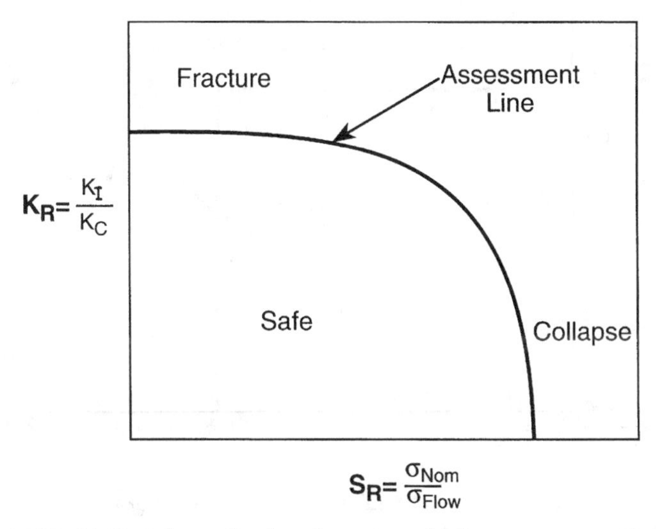

FIG. 14.11 Schematic showing general failure assessment diagram used in PD 6493.

assessment diagram is designed to keep the driving forces (K_I and σ_{nom}) less than the resistance forces (K_c and σ_{flow}) as has been discussed earlier. In the PD 6493 procedure, the flow stress is used rather than the yield stress because it is a more accurate predictor of the limit load in tension. PD 6493 actually establishes a formalized procedure that can be used as a code. Three levels of sophistication are specified.

Level 1 is shown in Figure 14.12 and includes a limit of 0.7 on K_r and 0.8 on S_r. These two limits result in a factor of safety of 2 on flaw size as follows:

$$\left(\frac{\sqrt{1}}{\sqrt{2}} = 0.7\right)$$

Because $\sigma_{flow} \simeq 1.2\sigma_{yield}$, restricting S_r to 0.8 limits σ_{nom} to be less than or equal to the yield stress. Because the Level 1 analysis includes a factor of safety, it is used as a conservative screening procedure for rapid assessment of the safety of an existing structure.

The fracture behavior in Level 1 also can be established in terms of CTOD, as follows:

$$\sqrt{\delta_r} = \sqrt{\frac{\delta_I}{\delta_c}} \tag{14.3}$$

where

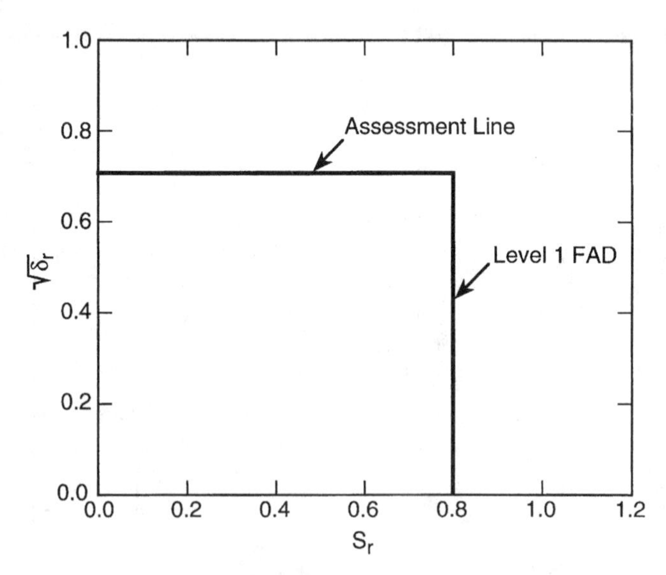

FIG. 14.12 Level 1 failure assessment diagram.

$$\delta_I = \text{the driving force} = \frac{K_I^2}{\sigma_{ys}\, E} \tag{14.4}$$

$$\delta_c = \text{the resistance force as described in Chapter 3}$$

Level 2 is a preferred assessment level for the majority of structural applications. It is based on the strip yield model described in Chapter 2 as follows:

$$K_r, \sqrt{\delta_r} = S_r \left[\frac{8}{\pi^2} \ln \sec \left(\frac{\pi}{2} Sr \right) \right]^{-1/2} \tag{14.5}$$

This relationship is shown in Figure 14.13. Note that the limits of K_r or $\sqrt{Sr}$ and S_r are 1.0, so there is no pre-established factor of safety.

Either K_r or $\sqrt{\delta_r}$ can be used, depending on the availability of material fracture toughness data. Although J_c data can be converted to K_c values, the procedure specifically prohibits using CTOD test results to estimate K_c values. The reasons for this restriction are that, at the time of writing PD 6493, the relation between K_c and δ_c was not well established. The authors of this text believe that the relation presented earlier, namely:

$$K_c = \sqrt{1.7\delta_c \sigma_{flow} E} \tag{14.6}$$

is sufficiently well established that it would give reasonable results. However, any conversion between CTOD test results and K_c result is prohibited in the 1991 version of PD 6493.

The document still allows conversion of the driving force, K_I, to δ_I as follows:

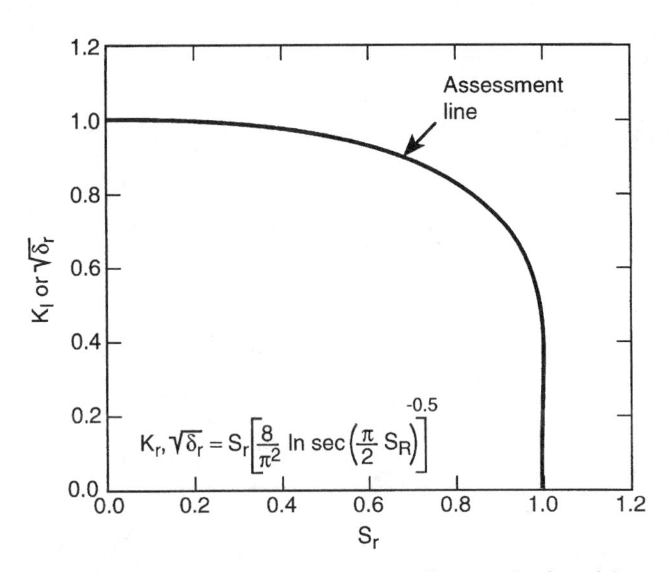

FIG. 14.13 The failure assessment diagram for Level 2.

$$\delta_I = \frac{K_I^2}{\sigma_{ys}E} \tag{14.7}$$

Although there is not inherent safety factor on a Level 2 assessment, welded wide plate test data have been analyzed and compared with Level 2 assessment [*17,19*]. These results are shown in Figure 14.14 and show that the actual factors of safety vary considerably.

Level 3 is the advanced level and normally is used only for the assessment of materials with high work-hardening exponents. However, recent studies at Lehigh University show that there does not appear to be any significant advantages of using the Level 3 assessment, particularly in view of the additional testing and analysis required. Readers interested in the Level 3 approach are referred to PD 6493.

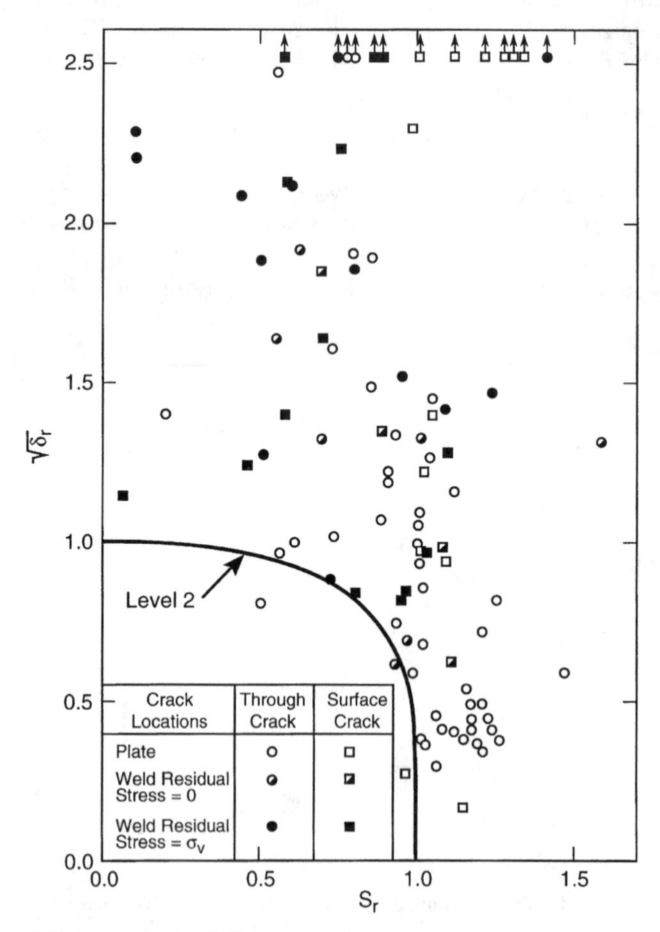

FIG. 14.14 Welded plate data showing variability in factor of safety compared with Level 2 assessment.

The preceding discussion is a brief overview of the PD 6493 fitness-for-service fracture assessment of existing structures. PD 6493 also describes fatigue crack growth behavior, which is similar to that presented in Part III of this book. In addition, there are brief sections on yielding, buckling, and creep in PD 6493.

14.3.3 ASME Section XI

Appendix A of the ASME Code—Section XI [5] provides a procedure for determining the acceptability of flaws (fitness for service) that have been detected during in service inspection which exceed a specified allowable value. The procedure is based on principles of fracture mechanics and applies to ferritic materials 4-in. thick or with $\sigma_{ys} = 50$ ksi or less.

Briefly, the procedure consists of the following steps:

Step A Determine the actual flaw configuration or configurations (if multiple flaws are present) and resolve the flaw(s) into a simple shape representative of the actual severity of the flaw(s). The Code gives procedures for establishing the possible interaction of multiple flaws.

Step B Determine the stresses and stress intensity factors at the location of the observed flaw for normal, emergency, and faulted conditions. The stress intensity factors are calculated using the relations described in Chapter 2.

Step C Determine the necessary material properties, e.g., K_{Ic} or K_{Ia}, including the effects of irradiation if applicable. K_{Ia} is based on the lower bound of critical crack arrest K_I values measured as a function of temperature. K_{Ic} is based on the lower bound of critical static initiation K_I values measured as a function of temperature.

Lower bound K_{Ia} and K_{Ic} versus temperature curves from tests of SA-533 Grades B Class 1, SA-508 Class 2, and SA-508 Class 3 steel are provided in Figure 14.15 for use if data from the actual product form are not available. The temperature scale of these data should be related to the reference nil-ductility temperature RT_{NDT}, as determined for the material prior to irradiation. The curves in Figure 14.15 are intended to be very conservative since the recommended procedure is to determine the material fracture toughness from specimens of the actual material and product form in question.

If fatigue crack growth is a consideration, the Code describes the relations to be used, which are in essence, the ones presented in Part III of this book.

Step D Finally, using the Code-prescribed procedures for different loading conditions, compare K_I to K_{Ia} or K_{Ic} for various loading conditions. The minimum critical flaw size for emergency and faulted conditions, a_i, should be established using K_{Ic} data for flaw initiation considerations and K_{Ia} data for flaw arrest considerations.

In summary, Section XI of the ASME Code provides a fitness-for-service

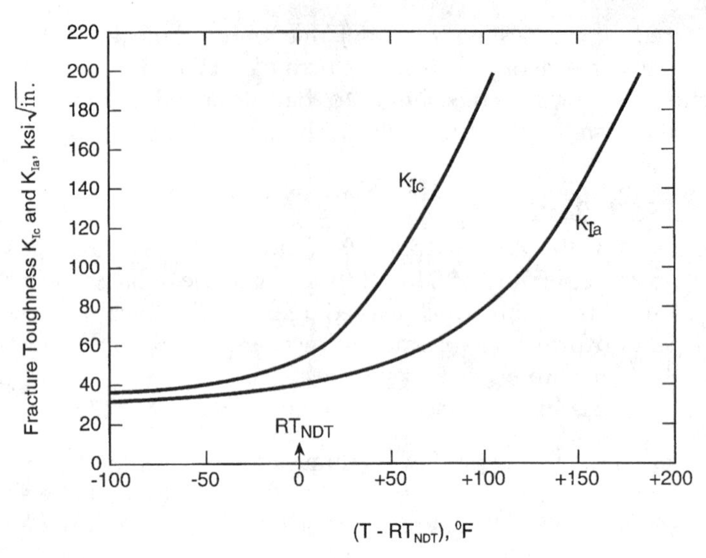

FIG. 14.15 Lower bound K_{Ia} and K_{Ic} test data for SA-533 Grade B Class 1, SA-508 Class 2, and SA-508 Class 3 steels.

analysis of vessels in which flaws have been detected and for which fracture toughness levels may have decreased because of irradiation effects.

14.3.4 API 579

API 579—Recommended Practices for Fitness-For-Service [*20*] is a "result of a need for standardization of fitness-for-service assessment techniques for pressurized equipment in the refinery and chemical industry."

The document is a comprehensive one intended to: (1) ensure that older equipment can be used safely, (2) provide technically sound fitness-for-service assessment procedures to ensure that different service providers furnish consistent life predictions, and (3) help optimize maintenance and operation of existing facilities to maintain availability of older plants and enhance long-term economic viability. API 579 describes fitness-for-service evaluations using the results from a finite element analysis, in particular for inelastic behavior involving the *J*-integral or CTOD procedures as discussed in the appendix to Chapter 2.

Although the section on assessment of crack-like flaws should be of primary interest to readers of this text, the document also includes assessment of general metal loss, corrosion, blisters, weld misalignment, creep and fire damage.

14.4 Benefits of a Proof or Hydro-Test to Establish Fitness for Continued Service

Of the various methodologies to justify continued fitness for service, or life extension of an existing structure (e.g., re-analysis, inspection, probabilistic analysis, etc.), an actual load test, or in the case of a pressure vessel, a hydro-test, is by far the best methodology on which to base a decision to extend the design life of a structure. In fact, periodic load or hydro-testing can be used to extend the life of a structure indefinitely.

Engineers have known this for years. In a proof test, or hydro-test there are no assumptions or modeling involved. The real structure is loaded and the actual material condition is evaluated in the presence of whatever imperfections actually may be in the structure. The engineer does not need to assume crack size or material properties or to model loading conditions, etc. Hydro-testing is a well-established viable procedure for determining fitness for future service, i.e., life extension.

Although a proof or hydro-test is the best fitness-for-service or life-extension methodology that can be used for almost any type of structure, unfortunately it is not practical for some large structures such as ships. However, in the case of many structures, the actual material in its current state, e.g., possibly irradiated in the case of a nuclear pressure vessel, with actual (if any) flaws, can be subjected to the same types of loading that the vessel will see in continued service. Test temperature and pressure can be controlled so that the desired degree of conservatism during a proof or hydro-test can be achieved. Thus, a proof or hydro-test is an ideal methodology to evaluate the continued service of many structures. Boeing used this concept in the 1960s to evaluate the performance of pressure vessels, and NASA used this approach during the Apollo space program. Industry uses this procedure regularly. The methodology is very applicable to many structures, particularly where the loading is slow and can be well controlled at temperatures near ambient.

Fracture mechanics methodology can be used to demonstrate the merits of proof or hydro-testing as shown schematically in Figure 14.16. For a given flaw depth, a, with a given shape factor, Q, the general K_I relation for a surface flaw is:

$$K_I = C \cdot \sigma \sqrt{\sigma \frac{a}{Q}} \tag{14.8}$$

Letting $K_I = K_c$, the critical material fracture toughness at a given level of embrittlement and service temperature ($K_c = K_{Ic}$ if plane strain conditions exist), the locus of failure points can be calculated as shown in Fig. 14.16. As long as K_I, the combination of operating stress and flaw size, is kept below K_c, the vessel will not fracture. During hydro-test, an overpressure is applied such that $\sigma_{\text{Hydro-test}}$ will cause failure if a_{cr} is equal to or greater than the $a_{\text{cr(present time and future time T2)}}$ value shown in Figure 14.16. If failure does not occur during the hydro-test, the vessel is safe to operate at $\sigma_{\text{operating}}$ until continued material degradation (or perhaps reduced temperature) reduces K_c in the future to K_{c2} such that failure would be predicted. Note that at Time T1, the toughness of the material has been reduced only to K_{c1} and a larger flaw size, $a_{\text{cr(future time T1)}}$, would be required for failure.

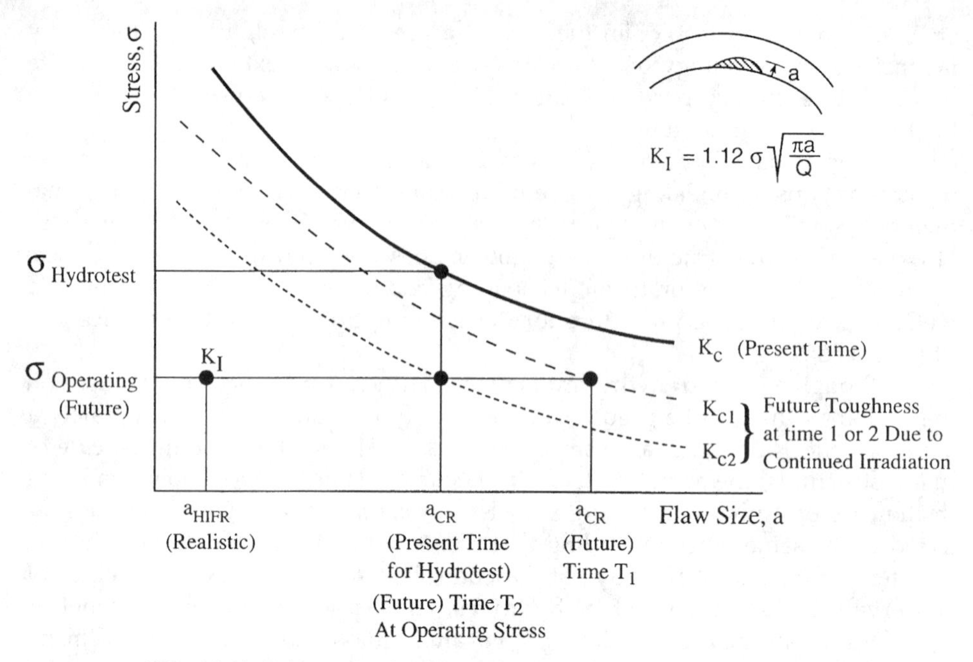

FIG. 14.16 Schematic showing fracture control by hydro-testing.

If fatigue crack growth does not occur, the actual flaw size will never be greater than it is at the present time. However, subsequent proof tests would account for fatigue crack growth as the hydro-test is conducted on the actual vessel with any actual cracks. Also, note that the more likely flaw size in many structures is $a_{HFIR(realistic)}$, not $a_{cr(present\ time)}$. This observation introduces some conservatism into the hydro-test procedure. In summary, fracture mechanics theory can be used to demonstrate that hydro-testing is a viable engineering methodology to insure the safety of a structure. The clear advantage of periodic hydro-testing, coupled with continued surveillance testing, is that if periodic hydro-testing is successful, the process shown in Figure 14.16 can be repeated indefinitely.

14.5 Difference Between Initiation and Arrest (Propagation) Fracture Toughness Behavior

In a fitness-for-service analysis of an existing structure, it is important to understand the difference between crack initiation and crack arrest (propagation) behavior.

The general difference in initiation and propagation behavior of low-to-medium-strength structural steels as related to fracture-toughness test results is shown schematically in Figure 14.17. The curve labeled "static" refers to

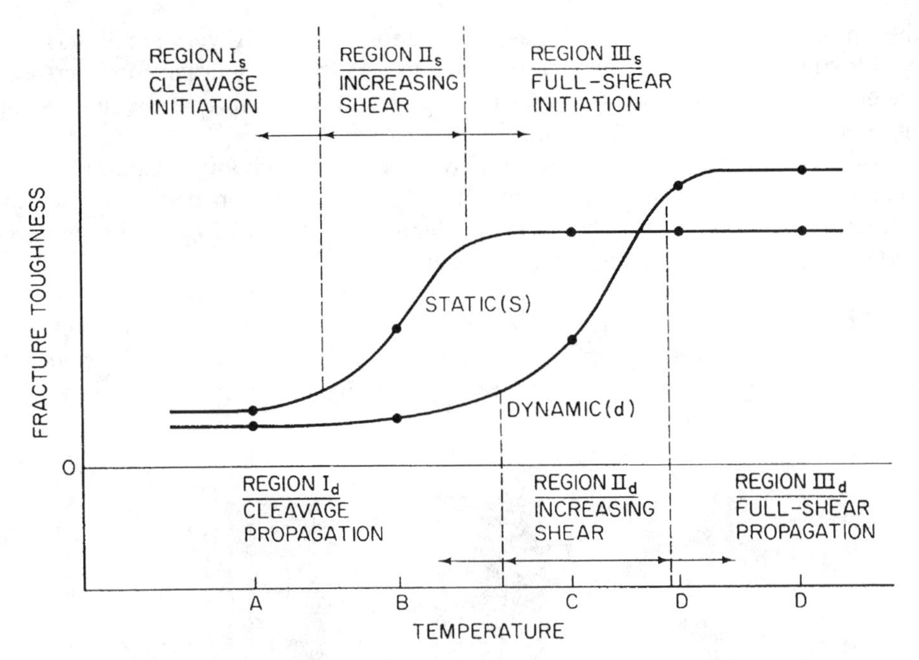

FIG. 14.17 Schematic showing relation between static and dynamic fracture toughness.

the fracture toughness obtained in a standard ASTM K_{Ic} fracture test under conditions of slow loading. [The curve for rapid or intermediate loading-rate tests, K_{Ic} (t), would be slightly to the right of the static curve.] The impact curve is from a K_{Id} or other dynamic test under conditions of impact loading. The difference in the location of these curves is the temperature shift, which is a function of yield strength for structural steels as discussed in Chapter 4.

In Region I$_s$ for the static curve (Figure 14.17), the crack initiates in a cleavage mode from the tip of the fatigue crack. In Region II$_s$ the fracture toughness that will result in initiation of unstable crack propagation increases with increasing temperature. This increase in the crack-initiation fracture toughness corresponds to an increase in the size of the plastic zone and in the zone of ductile tear (shear) at the tip of the crack prior to unstable crack extension. In Region III$_s$ the static fracture toughness is quite large and is more difficult to define (that is, elastic-plastic fracture-mechanics tests are required—Chapter 3), but the fracture initiates by ductile tear (shear).

The ductile tear zone at the tip of a statically loaded crack is confined to the zone of plastic deformation along the crack front. The maximum size for this plastic zone is restricted by the ASTM E-399 K_{Ic} test requirement to ensure valid plane-strain test results. Deviations from this requirement for elastic plane-strain conditions toward elastic-plastic conditions usually result in high initiation fracture-toughness values. As was described in Chapter 3, the ductile tear zone at the tip of a crack under elastic-plastic conditions is usually very small and may be difficult to delineate by visual examination. As the test temperature increases, the elastic-plastic behavior becomes more plastic than elastic and the ductile tear zone at the tip of the crack becomes very evident.

The dynamic curve in Figure 14.17 represents the dynamic (impact) fracture toughness behavior for the steel and the crack propagation behavior once the crack initiates. The fracture behavior in Regions I_d, II_d, and III_d for the dynamic curve are similar to those for the static curve.

In an actual steel structure loaded at Temperature *A*, initiation may be static and propagation dynamic. However, there is no apparent difference between the two because both initiation and propagation are by cleavage. If a similar structure is loaded slowly to failure at Temperature *B*, there will be some localized shear

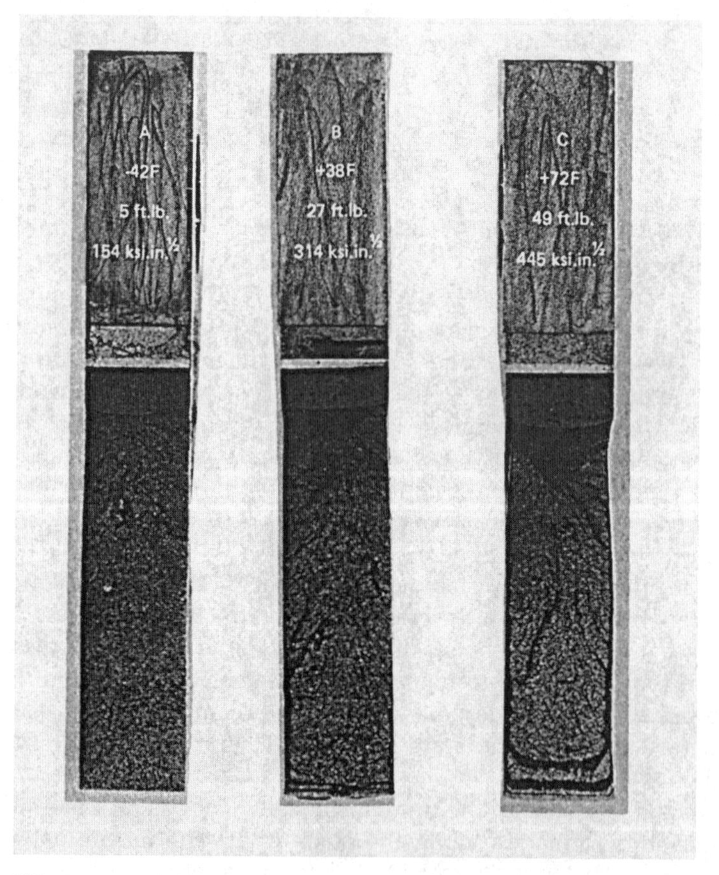

FIG. 14.18 Fracture surfaces of full-thicknesses (*B* = 1.5 in.) 4-T compact tension specimens of A572 Grade 50 steel tested under load-control conditions using a total-unload/reload loading sequence.

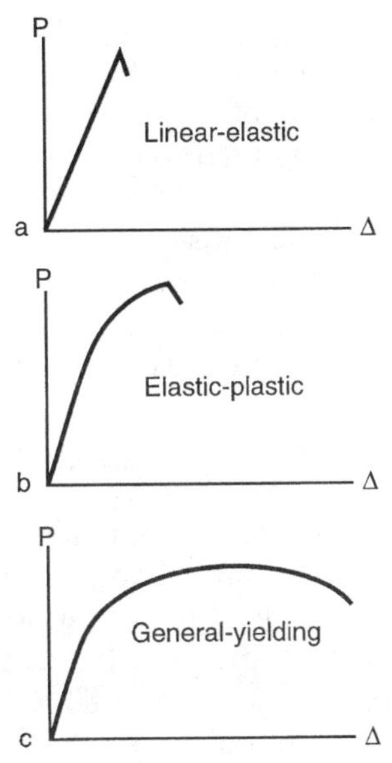

FIG. 14.19 Schematic *P*-Δ curves, (*a*), (*b*), and (*c*), correspnding to Fracture Surfaces A, B, and C, Figure 14.18.

and a reasonable level of static fracture toughness at the initiation of failure. However, for rate-sensitive structural materials, such as structural steels used in bridges, offshore rigs, or ships, once the crack initiates, the fracture toughness is characterized by the dynamic fracture toughness level on the impact curve. Thus, the fracture appearance for the majority of the fracture surface is cleavage. If the structure is loaded slowly to fracture initiation at Temperature C, the initiation characteristics will be full-shear initiation with a high level of plane-stress fracture toughness K_c. However, the fracture surface of the running crack may still be predominately cleavage but with some amount of shear as shown in the lower impact curve at Temperature C in Figure 14.17.

Figure 14.18 shows the fracture surfaces and representative fracture toughness levels (CVN and K_c) for an A572 Grade 50 steel. The specimen tested at −42°F exhibits some small amount of shear initiation, that is, at a temperature slightly below B, Figure 14.17. The specimen tested at +38°F exhibits full shear initiation (between B and C), but primarily cleavage propagation. The specimen tested at +72°F exhibits full shear initiation (Temperature C) but still exhibits a large region of cleavage propagation, Figure 14.18. Thus, ductile crack propagation (dynamic) would only occur at Temperature D, which is essentially dynamic upper-shelf CVN impact behavior (i.e., 80% or above shear fracture appearance).

Schematic *P*-Δ records as would be obtained from a slow-bend fracture test are presented in Figure 14.19 for each of the three fracture surfaces shown in Figure 14.18. Note that although all of them are shown to exhibit some nonlinear or elastic-plastic behavior, there is a considerable difference in the overall test records.

In summary, steels that exhibit considerable initiation fracture toughness at slow loading rates may have fracture propagation surfaces that are primarily cleavage. If the engineer looks only at the propagation surface, he or she may conclude that the cause of the fracture was low fracture toughness. This conclusion is only partially correct. Whereas the arrest (i.e., propagation) or dynamic fracture toughness may be low, e.g., Region I_d, Figure 14.17, the initiation fracture toughness, e.g., Region II_s, Figure 14.17, may be fairly high. Thus, if initiation can be prevented, e.g., by slow loading, the material may be quite satisfactory.

This behavior will be illustrated in Chapter 16 in the case study of the Ingram Barge. A slow overload *2.5 times* the design load in a region of very high constraint led to crack initiation. After the fracture started, it propagated by cleavage around the entire ship, even though the CVN impact notch toughness was 55 ft-lb. Analysis of this failure illustrated the importance of understanding the difference between initiation and arrest (propagation) fracture toughness, as well as the influence of constraint in brittle fracture initiation.

14.6 References

[1] Wells, A. A., "The Meaning of Fitness-for-Purpose and Concept of Defect Tolerance," International Conference, *Fitness-for-Purpose Validation of Welded Constructions*, The Welding Institute, London, 1981.

[2] Rolfe, S. T., "Fitness for Service—Common Sense Engineering," *Proceedings of the Symposium The Art and Science of Structural Engineering*, honoring William J. Hall, University of Illinois, Prentice Hall, April 1993.

[3] British Standards Institution Published Document 6493 (PD-6493): Guidance on Methods for Assessing the Acceptability of Flaws in Fusion Welded Structures, 1991.

[4] API-1104, Standard for Welding Pipelines and Related Facilities, American Petroleum Institute, 1988.

[5] American Society of Mechanical Engineers, ASME Boiler and Pressure Vessel Code, Section XI, Rules for In Service Inspection for Nuclear Power Plant Components, Appendix A, ASME, New York, 1989.

[6] Tada, H., Paris, P., and Irwin, G., *The Stress Analysis of Cracks, Handbook*, Paris Productions Incorporated, St. Louis, MO, 1985.

[7] Rooke, D. P. and Cartwright, D. J., *Compendium of Stress Intensity Factors*, London, Her Majesty's Stationery Office, ISBN 011 771 3368, 1974.

[8] AASHTO, Guide Specifications for Fracture Critical Non-Redundant Steel Bridge Members, September 1978.

[9] Schilling, C. G., Klippstein, K. H., Barsom, J. M., Novak, S. R., and Blake, G. T., "Low-Temperature Tests of Simulated Bridge Members," *Journal of the Structural Division, ASCE*, Vol. 101, No. ST1, January 1975.

[10] Smith, J. A. and Rolfe, S. T., "The Effect of Crack Depth (a) and Crack-Depth to Width Ratio (a/W) on the Fracture Toughness of A533B Steel," WRC Bulletin 418.

[*11*] Theiss, T. J., Shum, D. K. M., and Rolfe, S. T., "Interim Results from the HSST Shallow Crack Fracture Toughness Program," *ASTM 24th National Symposium on Fracture Mechanics,* June 1992, Gatlinburg, TN, *ASTM STP 1207,* 1994.

[*12*] Dodds, Jr., R. H., Shih, C. F., and Anderson, T. L., "Continuum and Micromechanics Treatment of Constraint in Fracture," Department of Civil Engineering, University of Illinois at Urbana-Champaign, Report No. UILU-ENG-92-2014, November 1992.

[*13*] Williams, M. L., "On the Stress Distribution at the Base of a Stationary Crack," *Journal of Applied Mechanics,* American Society of Mechanical Engineers, Vol. 24, March 1957, pp. 104–114.

[*14*] Background Reports: "Metallurgy, Fracture Mechanics, Welding, Moment Connections and Frame Systems Behavior" FEMA-288, March 1997.

[*15*] FEMA, Interim Guidelines Advisory No. 1—"Supplement to FEMA-267 Interim Guidelines: Evaluation, Repairs, Modification and Design of Welded Steel Moment Frame Structures," Report No. SAC-96-03, SAC Joint Venture, Sacramento, CA, January 1997.

[*16*] SAC Steel Project, Technical Office, 1301 S. 46th Street, Richmond, CA 94804-4698.

[*17*] Garwood, S. J., "A Crack Tip Opening Displacement (CTOD) Method for the Analysis of Ductile Materials," *18th ASTM National Symposium on Fracture Mechanics,* Boulder (*STP 945*), June 1985.

[*18*] Kamath, M. S., "The COD Design Curve—An Assessment of Validity Using Wide Plate Tests," *International Journal Pressure Vessel and Piping,* Vol. 9, 1981, pp. 79–105.

[*19*] Towers, O. L., Williams, S., Harrison, J. D., and Jutla, T., "Elastic Plastic Fracture Toughness Testing and Assessment Methods Based on an ECSC Collaborative Programme," Welding Institute Report 7916.01/86/5 36.2, 1986.

[*20*] API 579—Recommended Practice for Fitness-For-Service, American Petroleum Institute Document in Preparation, 1999.

Part V: Applications of Fracture Mechanics—Case Studies

15

Importance of Fracture Toughness and Proper Fabrication Procedures—The Bryte Bend Bridge

15.1 Introduction

FRACTURE toughness is a very important consideration in preventing brittle fractures of welded structures. However, as research and service experience have shown, brittle fracture is not, and never has been, just a material problem. Design, fabrication, materials, inspection, and service conditions all are factors that affect the susceptibility of a structure to brittle fracture. In this chapter the contribution of these factors to the Bryte Bend Bridge failure [*1,2*] is discussed. In addition, the AASHTO Fracture Control Plan for steel bridges is discussed to show how, had the ASSHTO Fracture Control Plan [*3*] been in existence, this failure would not have occurred.

In the Bryte Bend Bridge, the combination of low material toughness and poor fabrication practices led to the failure of this structure. Specifically, an out-of-specification steel in the presence of a severe Category E fatigue detail resulting from welding a lateral brace into the flange combined to cause a brittle fracture.

This case study points out that brittle fractures are caused by complex interrelations between material fracture toughness, design, welded details, fabrication practices, and service conditions. For steel bridges, the AASHTO Fracture Control Plan provides a reasonable set of balanced controls on all of these factors and thus is a satisfactory plan to prevent brittle fractures in steel bridges.

However, it should be emphasized that a particular plan developed for the prevention of fractures in one type of structure such as the AASHTO Fracture Control Plan for steel bridges cannot be used indiscriminately to prevent fracture in other types of structures. As discussed previously, differences in loading rate, design details, fatigue loading, etc. are such that fracture control should be made as specific to a particular type of structure as possible.

15.2 AASHTO Fracture Control Plan For Steel Bridges

Fracture-toughness requirements often are developed to be used in conjunction with good design, fabrication, and inspection procedures, without being specific as to how "good" procedures are defined. In the 1970s AASHTO recognized the need for a more specific fracture control plan for nonredundant steel bridges and developed the "Guide Specifications for Fracture Critical Non-Redundant Steel Bridge Members" [3]. Specifically, this Guide requires additional controls on the material notch toughness, welding, and inspection of fracture-critical bridge members compared with redundant bridge members. Fracture-critical members (FCMs) or member components are tension members or tension components of members whose failure would be expected to result in collapse of the bridge. The Guide specifications are to be used in conjunction with all existing AASHTO requirements and are based on numerous research studies designed to translate research into engineering practice.

The American Association of State Highway and Transportation Officials (AASHTO) approach to fracture control has been to specify the materials, design, fabrication (welding), construction, inspection, and maintenance in four separate specifications as follows:

1. Standard Specifications for Highway Bridges [4].
2. Standard Specifications for Transportation Materials and Methods of Sampling and Testing [5].
3. Standard Specifications for Welding of Structural Steel Highway Bridges [6].
4. Manual for Maintenance Inspection of Bridges [7].

The controls described in these specifications have worked very well, and the incidence of bridge failures has been extremely low after the development of the AASHTO Fracture Control Plan [3]. When failures did occur, they usually were attributed to the fact that the various AASHTO specifications were violated.

15.3 Bryte Bend Bridge Brittle Fracture

The Bryte Bend Bridge, near Sacramento, California, Figure 15.1, consists of twin parallel structures with an overall length of 4050-ft (1234-m) and 55-ft (16.8-m) vertical clearance between low steel and mean high water, Figure 15.2. The main river section consists of four continuous spans of 281, 370, 370, and 281 ft (85.6, 112.8, 112.8, and 85.6 m) with a hinge in each of the 370-ft (112.8-m) spans, as shown in Figure 15.3.

The superstructure is a trapezoidal steel box supported on reinforced concrete piers. The exterior webs of the box were sloped to reduce the width of the

FIG. 15.1 Overall view of Bryte Bend Bridge.

compression flanges in the continuous spans and to improve overall appearance, Figure 15.4. Conventional girder flanges were welded to the tops of the box sides and to a single longitudinal web plate stiffening the center of the box. The bottom plate of the box is longitudinally stiffened with a series of vertical plates.

The bridge was fabricated from plates of A36, A441, and an A517 type steel. The A36 steel was used in all areas of low stress. The A441 steel was used for the web and compression members at the river piers. In this region, the tension flanges were fabricated from 2¼-in. (5.7-mm)-thick A517 type steel plates (σ_{ys} = 100 ksi, 690 MPa) to reduce the size of the members. The A517 type steel was furnished out of specification and had lower notch toughness than was expected, although there were no notch toughness requirements. All fabrication was by welding.

While the concrete deck was being placed in June 1970, a brittle fracture occurred across one of the outer flanges in the negative moment region at Pier 12, Figure 15.3. It initiated at the intersection of a ½-in (12.7-mm)-thick lateral attachment welded to the 2¼-in. (57.1-mm)-thick flange. The fracture propagated across the entire 30-in. (762-mm)-wide flange and about 4 in. (101.6 mm) down into the web where it was arrested. The fracture surface was a classic herring-bone-type brittle fracture with very small shear lips, Figure 15.5.

The nominal yield strength of this material was 100 ksi (690 MPa), and the maximum design stress was 45 ksi (310 MPa). At the time the crack propagated, the dead load stress was about 28 ksi (193 MPa) and the ambient temperature was about 60°F (15.5°C).

FIG. 15.2 View of twin structures over river.

Analysis of the fracture surface indicated that a weld crack about 0.2 in. (5.1 mm) deep, Figure 15.6, initially was present in a residual stress field such that sometime during fabrication or erection, the weld crack initiated. After the initial weld crack propagated out of the residual stress field, it arrested at a distance of about 1.3 in. (33 mm) as determined from the rusted area shown in Figure 15.6. This would be expected because of the residual compressive stress field adjacent to the residual tensile stress field and the fact that there was no applied load. During pouring of the concrete deck, when the dead load stress increased to about 28 ksi (193 MPa), complete fracture of the top flange occurred.

At the service temperature of +60°F (15.5°C) and for slow loading rates, the K_{Ic} value of steel from the flange plate was 55 ksi$\sqrt{\text{in.}}$ (60.5 MPa $\cdot$ m$^{1/2}$). This was a valid K_{Ic} test result that met all the requirements of the ASTM Fracture Mechanics Test Method E 399, Chapter 3. The fracture surface was flat with very small shear lips, similar to that of the actual fracture surface. The K_{Ic} value of 55 ksi$\sqrt{\text{in.}}$ (60.5 MPa $\cdot$ m$^{1/2}$) at +60°F (15.5°C) was considerably lower than would be expected for this material and is not representative of A514-517 steels

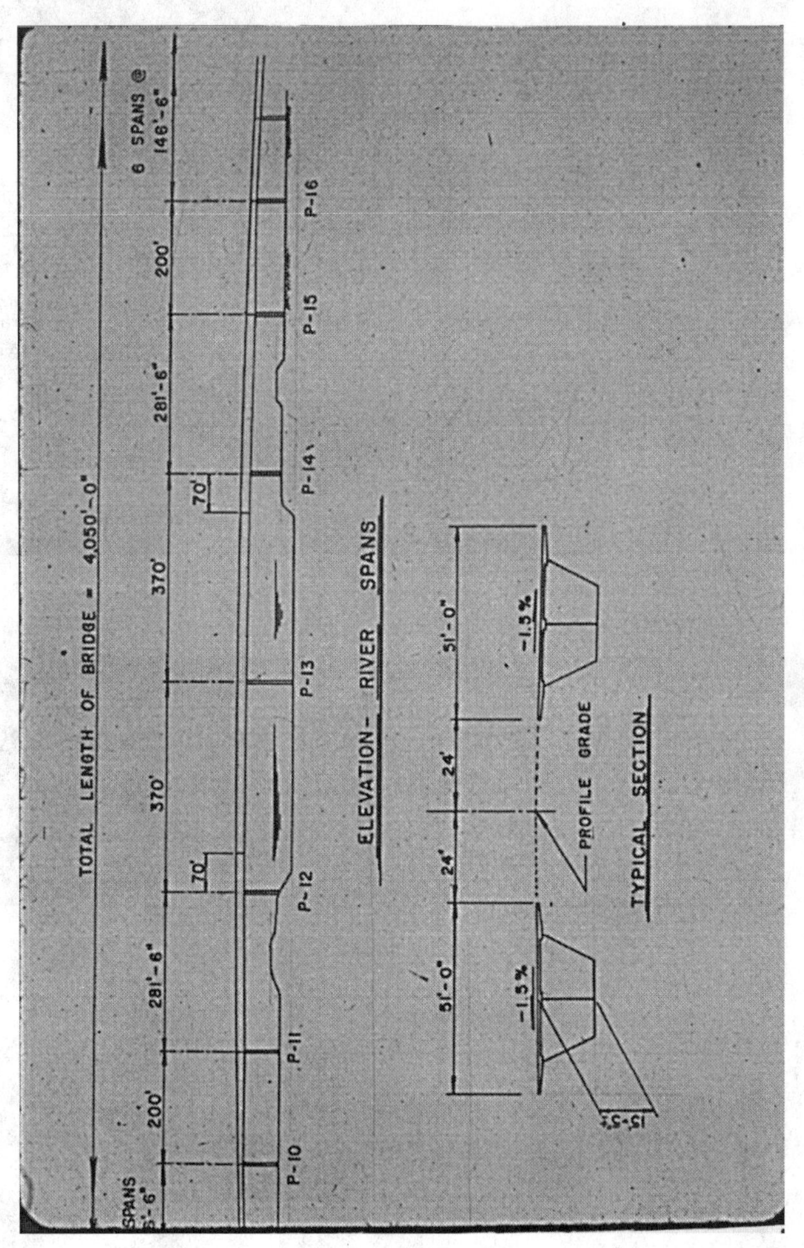

FIG. 15.3 Layout of Bryte Bend Bridge.

FIG. 15.4 Cross section of box girder.

FIG. 15.5 Brittle fracture surface showing classic herringbone pattern and small shear lips.

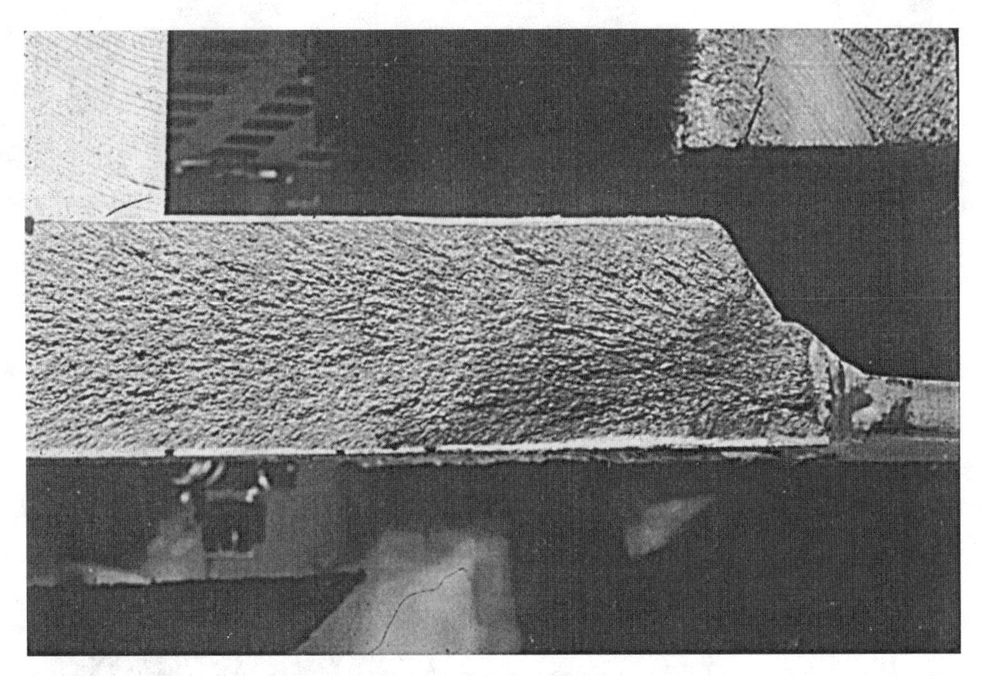

FIG. 15.6 Origin of fracture showing 0.2-inch-deep weld crack and 1.3-inch-deep crack.

because the steel was furnished out of specification. The stress–flaw size relation for $K_{Ic} = 55\ \text{ksi}\sqrt{\text{in.}}$ (60.5 MPa·m$^{1/2}$) is shown in Figure 15.7 using a simple edge-crack relation. It was assumed that sometime during fabrication, transportation, or erection, the 0.2-in. (0.5-mm) crack initiated. For a K_{Ic} of 55 ksi$\sqrt{\text{in.}}$ (60.5 MPa·m$^{1/2}$) and an initial flaw size of 0.2 in. (0.5 mm), crack propagation

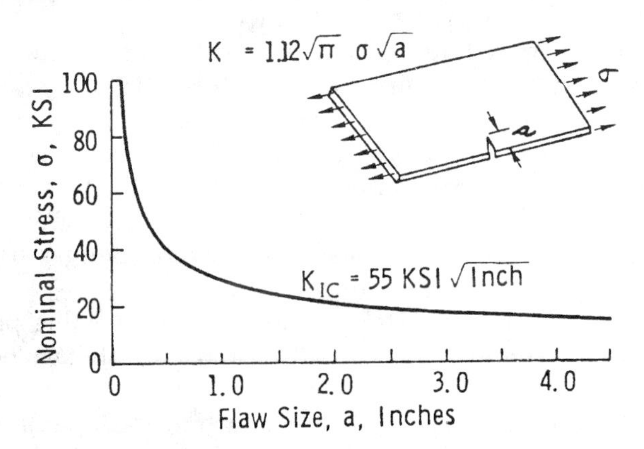

FIG. 15.7 Stress-flaw-size relation for edge crack in steel with K_{Ic} = 55 ksi$\sqrt{\text{in.}}$.

would be expected in a high residual stress region of 60–80 ksi. As the crack propagated out of the residual stress region, it arrested at a distance of about 1.3 in. (33 mm) as was shown by the rusted area in Figure 15.6. As the dead load stress was increased, the combination of an applied stress of 28 ksi (193 MPa) and the 1.3-in. (33-mm)-deep crack caused the stress intensity K_I to reach the critical stress intensity for this material (55 ksi$\sqrt{\text{in.}}$ (60.5 MPa $\cdot$ m$^{1/2}$)), Figure 15.7. Complete failure of the flange resulted under static-loading conditions. As the crack propagated into the thinner web plate that had a higher level of notch toughness and the load was transferred to other members in the structures, the crack was arrested.

15.4 Design Aspects of the Bryte Bend Bridge as Related to the AASHTO Fracture Control Plan (FCP)

Since the Bryte Bend Bridge fracture in 1970, there have been a number of similar fractures in other bridges in the United States. Collectively, these fractures brought about certain changes in the AASHTO Bridge design process, especially for fracture-critical nonredundant members. The following comparison of the AASHTO design criteria used during the design of the Bryte Bend Bridge and the current AASHTO FCP evaluates these changes as they relate to the events which took place at the Bryte Bend Bridge.

The original design for the structure was based on the provisions of the Ninth Edition (1965) of AASHTO's Standard Specifications for Highway Bridges. The significant areas of change in the Standard Specifications since then are the subsequent adoption of provisions for the use of high-yield strength quenched and tempered alloy steels, the modifications of the fatigue design procedure, the addition of specific notch toughness requirements, and more stringent requirements on fabrication and inspection. Also, the Guide Specifications for Fracture Critical Non-Redundant Members was adopted.

The 1965 design procedure for fatigue considered maximum stress, type of loading (lane load or truck), number of cycles of maximum stress for a given use, the ratio of minimum stress to maximum stress, and consideration of the method of joining the pieces making up a structural element. There was no direct relationship drawn between stress range and structural details.

Prior to the design of the Bryte Bend Bridge, the California Department of Transportation had designed and constructed several bridges using quenched and tempered alloy steel. Design stresses were conservative, 45 ksi (310 MPa) maximum for a minimum yield stress of 100 ksi, and welding procedures and quality control were carefully monitored. The structures all have performed well to date.

The Bryte Bend Bridge design called for attaching the top transverse member of the intermediate crossbracing to the upper flanges of the box, Figure 15.8. These members were designed to take horizontal forces that resulted from the

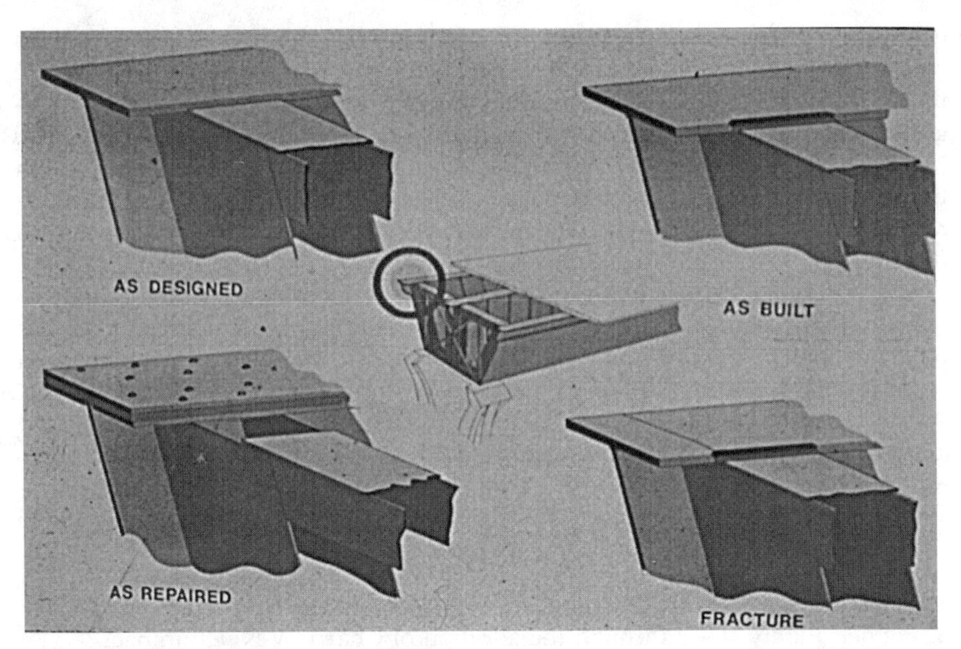

FIG. 15.8 Critical detail at brittle fracture—As Designed—As Built—As Repaired, and Fracture.

slope of the exterior webs. With the exception of the cross frames at the piers (where the fracture occurred), the intermediate braces were attached to flanges under compressive stress. The end frames at the piers of the continuous span were designed to transfer this force through the web stiffeners and were not to be attached to the upper flanges, Figure 15.8 (As Designed).

An A517 pressure vessel quality steel was specified rather than an A514 steel because of the general two-direction rolling used in the manufacture of A517. It was believed that the additional rolling would ensure a superior overall product compared to A514 steel because the A517 steel is "pressure vessel quality." Thus, no impact testing was specified in the contract provisions; also, the material was supplied out of specification according to the ASTM standard.

The contract provisions specified comprehensive requirements for welding procedures, materials, and testing. The basic document used, by reference in the specifications, was the Specifications for Welded Highway and Railway Bridges of the American Welding Society. This was augmented by additional State of California requirements for electrodes, welding processes and procedures, preheat and interpass temperatures, prequalification testing of welds and welders, and nondestructive testing to be done during fabrication.

During the fabrication of the box units, the contractor misinterpreted the design detail for connecting the top transverse bracing at the piers for the continuous spans of the bridge. The member, an inverted "U" made up of a 24-in.

(61-mm) × ½-in. (1.3-mm) plate with two 18-in. (45.7-mm) × 1-in. (2.54-mm) plates welded to it as shown in Figure 15.8 (As Built), was welded to the girder flanges for the entire width of the 24-in. (61-mm) plate (present AASHTO Category E Fatigue Detail). The contractor had then ground the edge of the 2¼-in. (5.7-mm) flange to a slope of approximately 1:3 in an attempt to have a smooth transition to the ½-in. (1.3-mm) plate. The transition was in the transverse direction only with no attempt to feather the sides of the taper in the longitudinal direction.

A design review was made of the incorrectly fabricated detail to evaluate its effect on the flange stresses and fatigue life. While the general stress state was only nominally changed from the original design, the attachment introduced high residual weld stresses along the edge of the flange and re-entrant corners. A design review concluded that the fatigue life would be satisfactory for a sound joint. It was assumed that these welds would be inspected, yet they were not.

A comparison between the 1965 and 1981 AASHTO Specifications, with respect to the Bryte Bend Bridge failure, points out three significant areas where changes have been made:

1. Vastly improved fatigue criteria.
2. Higher quality steels through the addition of Charpy V-Notch impact requirements.
3. A requirement for a design evaluation of structural redundancy.

Had Zone I requirements (25 ft lb @ 0°F) been specified, a representative value for K_{Ic} would have been about 150 ksi$\sqrt{\text{in.}}$ (165 MPa·$m^{1/2}$), Figure 15.9, showing that even at the maximum design stress of 45 ksi (310 MPa), the critical crack size is about 3 in. (7.6 mm). Actually, the critical crack size would be even larger than the 3 in. because elastic-plastic plane-stress behavior (K_c) would govern rather than K_{Ic}. However, fatigue crack growth would still have occurred

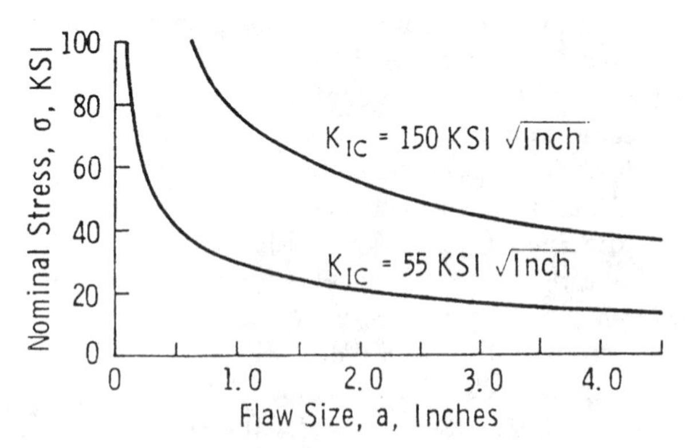

FIG. 15.9 Stress–flaw-size relation for edge crack in steel with K_{Ic} = 55 and 150 ksi$\sqrt{\text{in.}}$

from the 0.2-in. (0.5-mm)-deep weld defect and may have caused problems eventually. However, the critical crack size would have been much larger.

An added dimension of low-notch toughness steels, although difficult to establish quantitatively, is that it may cause problems during welding. In addition to the minimum notch toughness requirements in the current AASHTO Fracture Control Plan, the controls on preheat and interpass temperatures (if followed) help to prevent the formation of cracks during welding.

The repair was accomplished by jacking the entire structure into a zero-stress condition, cutting out the material in the vicinity of the failure, and replacing it with material with a higher level of notch toughness. The notch toughness of the replacement plate was insured from various correlations between fracture toughness tests, including the upper-shelf CVN impact K_{Ic} correlation. Material requirements used met the required AASHTO notch toughness values for fracture-critical members.

Because the notch toughness of other A517 type plates in similar negative moment regions in the structure was questioned, and because it did not appear feasible to remove these plates completely and replace them using field welding, additional plates were bolted to the original plates. As stated earlier, the entire structure was jacked into a zero-stress condition before the additional plates were added. Thus, the design stress in the original plates was reduced considerably, thereby increasing the critical crack size at the lower design stress level markedly. Also, and more importantly, multiple-load paths were established to carry the entire load in the event that additional fractures should ever occur in these A517 type plates throughout the life of the structure.

A closeup of the replacement plates as bolted to the original plates in the negative moment regions over the piers is presented in Figure 15.10. The bridge was opened to traffic in October 1971 and has been in continuous use since that time.

15.5 Adequacy of the Current AASHTO Fracture Control Plan

15.5.1 Implied vs. Guaranteed Notch Toughness

The question is often asked whether it is necessary to specify certain minimum toughness requirements or can the engineer rely on generic properties of various classes of steels. Although the engineer would like to be able to expect certain minimum material properties for a particular grade of steel, notch toughness values should be specified when needed because of their greater sensitivity to thermo-mechanical history than other material properties such as yield strength. Essentially this is the position AASHTO has taken by the development of the Fracture Control Plan material notch toughness requirements for bridge steels. Other code-writing bodies have taken a similar position.

A lower level of notch toughness also may cause problems during welding. This is another reason it is desirable to have some minimum level of notch tough-

FIG. 15.10 Closeup of additional plates attached to original flange plates.

ness, such as that specified by AASHTO. The question of how much notch toughness is necessary for a particular structural application depends on many factors such as service history, design, fabrication, consequences of failure, etc. Obviously fracture control plans consist of a specific set of recommendations developed for a particular type of structure and should not be applied indiscriminately to other types of structures.

15.5.2 Effect of Details on Fatigue Life

The original plans for the Bryte Bend Bridge called for the lateral attachment at the fracture origin to be connected only to the web, not to the flange, Figure 15.8 (As Designed). After discovery of the weld in the fabrication yard, the fatigue stress range was checked and found to be acceptable by the existing design standards. Because the stress range was within the design allowable and because it was thought that cutting the lateral attachments free from the flange might have produced additional lateral loading, it was decided prior to the occurrence of the fracture to leave the attachments welded to the flange.

For steels with low levels of notch toughness (e.g., $K_{Ic} = 55\ \text{ksi}\sqrt{\text{in.}}$ (60.5 MPa $\cdot$ m$^{1/2}$), for a steel with a yield strength of 100 ksi (690 MPa), or $K_{Ic}/\sigma_{ys} = 0.5$, the critical crack size at the design stress loading is small, Figure 15.9.

Figure 15.11 is a schematic plot of flaw size vs. number of cycles of fatigue loading. Given an initial flaw size, a_o, the number of cycles necessary to reach a_{cr} is small for plane strain behavior such as would be expected for a steel with

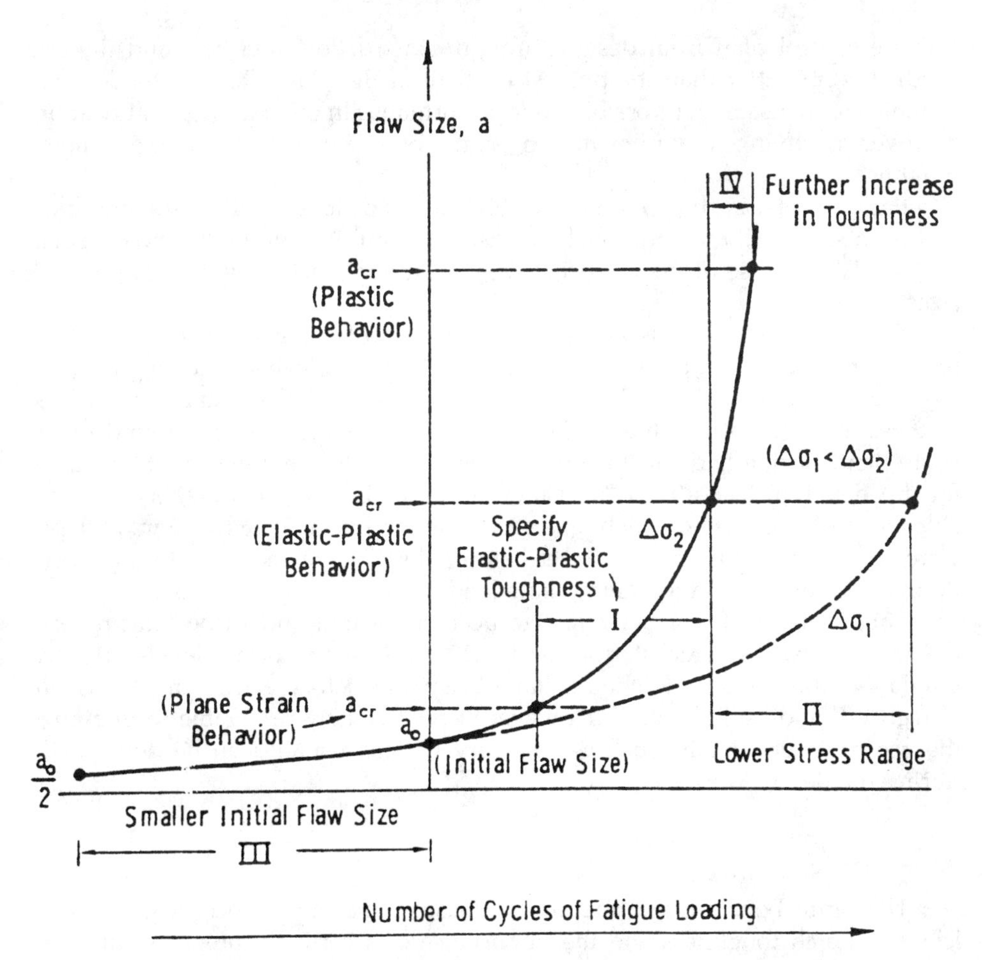

FIG. 15.11 Schematic showing flaw size–fatigue life relations for various critical crack sizes.

low notch toughness. In the case of the Bryte Bend Bridge, for a_o = 1.3 in. (3.3 mm), the propagation fatigue life was essentially zero. Thus, had failure not occurred during erection of the bridge, fatigue crack propagation would have led to failure early in the life of the bridge.

If the Bryte Bend Bridge had been fabricated from a steel with more typical levels of notch toughness (e.g., 25 ft-lb minimum CVN impact value as now required by AASHTO), a_{cr} would have been higher, i.e., elastic-plastic behavior, Figure 15.11. Thus, there would have been considerable improvement in the fatigue life due to a moderate improvement in notch toughness, Region I. However, this particular detail still could have led to a fatigue problem because of the poor fabrication (e.g., the initial 0.2-in. weld crack) and complex geometry that compounded the severity of the detail. Thus, it is obvious that each element of a fracture control plan from design through construction is critical and the end product is no better than the poorest element of the plan. That is, the designer cannot "compensate" for possible poor performance in one area (e.g., fabrication) by over-specifying requirements in some other area (e.g., material notch toughness).

For example, as shown in Figure 15.11, if plastic levels of fracture toughness had been specified, the additional increase in fatigue life would have been small (Region IV) because of the rapid increase in fatigue crack growth at large crack sizes.

FHWA-sponsored tests at Lehigh University of beams with lateral attachments properly fabricated from A514 steel with normal levels of notch toughness have verified that the AASHTO Fracture Control Plan is adequate for bridges. In these tests the fatigue life of Category E details similar to that found in the Bryte Bend Bridge had satisfactory fatigue lives at the low design stress range for this detail. In fact, after 2,000,000 cycles of loading, at the Category E allowable stress range, during which time cracks ranging from 7/16 to 1¼ in. in depth were produced, testing temperatures below −140°F were necessary to cause brittle fractures under bridge intermediate loading rates.

Thus, had the Bryte Bend Bridge been fabricated and inspected properly (i.e., no pre-existing cracks) from A514-517 steels with normal levels of notch toughness, the service life should have been satisfactory even with the severe Category E allowable loading. However, it is obviously preferable to minimize the use of severe details such as Category E. This usually can be done easily during the design stage.

15.5.3 Summary

The Bryte Bend Bridge fracture emphasizes the importance of guaranteed levels of notch toughness and the importance of a carefully prepared and executed quality control and assurance plan. Had appropriate levels of notch toughness been specified, the improperly fabricated joint and resulting weld crack would not have been as critical. Furthermore, after the fabricating error had been discovered, had a quality control and assurance plan been in operation, it would have been apparent that insufficient weld inspection was done. All of these aspects are covered by the current AASHTO Fracture Control Plan, as well as the consequences of using poor details. Thus, the current AASHTO Fracture Control Plan appears to be adequate for steel bridges.

Any rational Fracture Control Plan must recognize the fact that engineering design is an optimization of performance, safety, and cost. The AASHTO Fracture Control Plan recognizes this fact in that the potential contributors to a brittle fracture (materials, design details, fabrication, and severe conditions including maintenance) are controlled so that the bridge would perform satisfactorily throughout its lifetime.

When brittle fractures have occurred, e.g., the Lafayette Street Bridge, the I-75 Bridge in Pittsburgh, or the Dan Ryan Bridge in Chicago [*8*], the cause has been failure to adhere either to the AASHTO Fracture Control Plan or to the principles of good engineering judgement, sometimes referred to as common sense. Specifically, lack of penetration weld defects and the absence of adequate cope holes in the Lafayette Street Bridge were major contributors to that brittle fracture. The brittle fracture in the I-75 Bridge in Pittsburgh originated in a repair weld that contained a large pre-existing crack. The brittle fracture in the Dan Ryan Bridge in Chicago originated at a square cutout with fatigue behavior much more severe than the Category E fatigue details. In all of these cases, had the AASHTO Fracture Control Plan as well as good engineering principles been followed, the failure should not have occurred.

Other examples of bridge and structural failures result in a similar conclusion, i.e., brittle fracture is as much a problem of poor design or poor fabrication practices as it is a problem of low material fracture toughness. In fact, the Bryte Bend Bridge fracture is one of the few failures in which the level of material fracture toughness played a significant role.

It should be emphasized that the AASHTO fracture toughness requirements, as well as most fracture toughness requirements, are not sufficient to prevent brittle fracture propagation under certain possible combinations of poor design, fabrication, or loading conditions.

15.6 References

[1] "State Cites Defective Steel in Bryte Bend Failure," *Engineering News Record,* Vol. 185, No. 8, Aug. 20, 1970.
[2] Barsom, J. M. and Rolfe, S. T., "Fracture Mechanics in Failure Analysis," *ASTM STP 945,* 1988.
[3] Guide Specifications for Fracture Critical Non-Redundant Steel Bridge Members, American Association of State Highway and Transportation Officials (AASHTO), September 1978.
[4] Standard Specifications for Highway Bridges, AASHTO.
[5] Standard Specifications for Transportation Materials and Methods of Sampling and Testing, AASHTO.
[6] Standard Specifications for Welding of Structural Steel Highway Bridges.
[7] Manual for Maintenance Inspection of Bridges.
[8] Fisher, J. W., *Fatigue and Fracture in Steel Bridges,* John Wiley & Sons, New York, 1984.

16

Importance of Constraint and Loading—The Ingram Barge

16.1 Introduction

THIS CHAPTER DISCUSSES the brittle fracture of the Ingram Barge which occurred in 1972. This ship had a notch toughness level of about 55 ft-lb at the service temperature, well above what would generally be considered adequate to prevent a brittle fracture. However, in this case, the three-dimensional constraint based on the design was such that the principle stresses, σ_1, σ_2, and σ_3, were relatively high in tension. This triaxial state of tensile stresses prevented the development of significant shear stresses at the fracture origin. Thus, as would be concluded from strength of materials principles, brittle fracture could occur when the three principal stress components are large.

It should be emphasized that materials with good ductility and notch toughness may fracture in a brittle manner under conditions of triaxial tension. This state of stress can be produced either by direct loading in the three orthogonal directions, which is rare, or in complex welded details that are constrained such that they do not allow for any relaxation of stress in the two directions perpendicular to the primary stress.

16.2 Effect of Constraint on Structural Behavior

The stress field for an element within a structure can be described by three principal stresses that are normal to each other, Figure 16.1 [Barsom (1996)] [*1*]. Shear stresses can be calculated from the principal stress components. Assuming that σ_1 in Figure 16.1 is the largest principal stress and σ_3 is the smallest principal stress, the maximum shear stress component along the two shaded planes is:

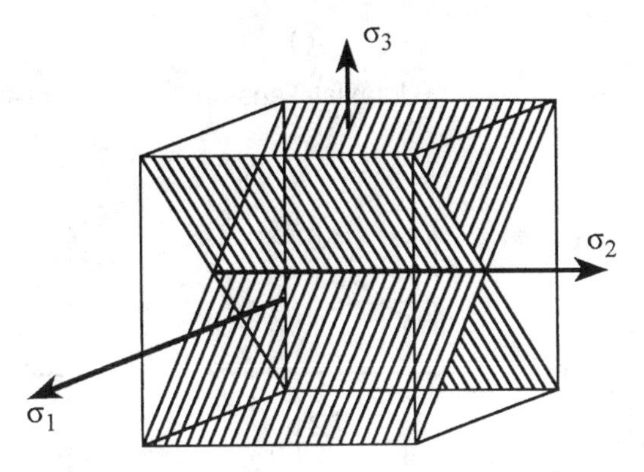

FIG. 16.1 Principal stresses and planes of maximum shear stress.

$$\tau_{max} = \frac{1}{2}(\sigma_1 - \sigma_3) \tag{16.1}$$

In a uniaxial tension test used to specify material properties, $\sigma = \sigma_{max}$ and $\sigma_2 = \sigma_3 = 0$. Therefore

$$\tau_{max} = \frac{\sigma_1}{2} = \frac{\sigma_{max}}{2} \tag{16.2}$$

Since the plastic deformation, i.e., yielding, begins when τ_{max} reaches a critical value, a change in the relationship between τ_{max} and σ_{max} represents a change in the plastic deformation behavior of the material. Note that yielding occurs when the shear stress, τ_{max}, reaches a critical value, not when σ_{max} reaches a critical value.

The relationship between the shear stress and the normal stresses, σ_1, σ_2, and σ_3, can result in either yielding and relaxation of constraint or not yielding and increased constraint. This behavior is illustrated in Figure 16.2 using Mohr's circle of stress. Figure 16.2*a* shows the principle stress directions, with the largest being σ_1. For uniaxial loading, such as the case of a standard tension test, σ_1 = applied stress and $\sigma_2 = \sigma_3 = 0$. At $\sigma_1 = \sigma_{max}$, $\tau_{max} = \sigma_{max}/2$, as shown in Figure 16.2*b*, and yielding occurs when $\tau_{max} = \sigma_{ys}/2$.

In contrast to the simple tension test, Figure 16.2*c* represents a triaxial tensile state of stress such as would be expected in highly constrained connections such as the Ingram Barge. Because of the triaxial stress loading, the stresses approach the ultimate stress and yielding (which is prevented because τ_{max} is low) may never occur.

The effect of increased severity of a structural detail on yield strength and plastic deformation may be illustrated further by considering the inelastic behavior of a material in a smooth tension test and a tension test with a circular notch, Figure 16.3. The reduced section in the notched tension test bar deforms

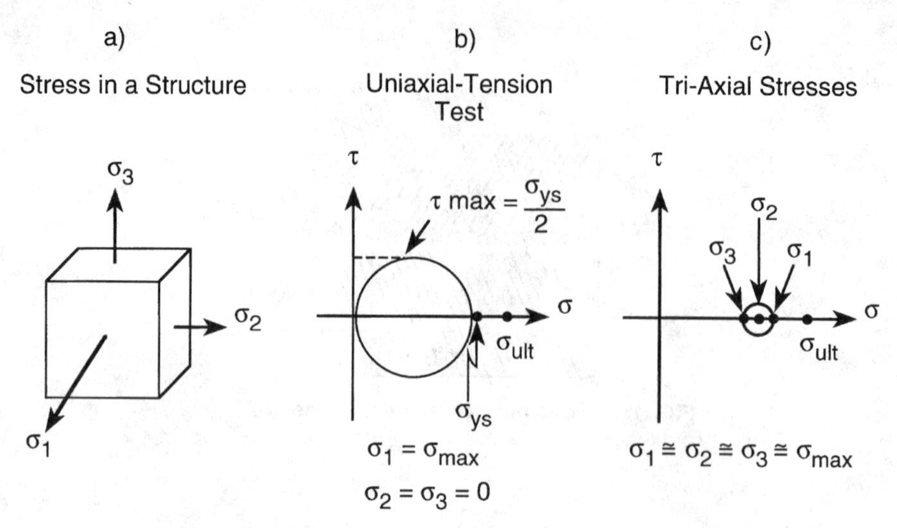

FIG. 16.2 Mohr's circle of stress analysis for stresses in a structure.

inelastically while the surrounding material is still elastic. Since the amount of elastic contraction due to Poissons ratio is small compared with the inelastic contraction of the reduced section, a restriction to plastic flow develops. This restriction corresponds to a reaction-stress system such that the σ_2 and σ_3 stresses restrict or constrain the flow in the σ_1 (σ_y or primary load) direction, Figure 16.3.

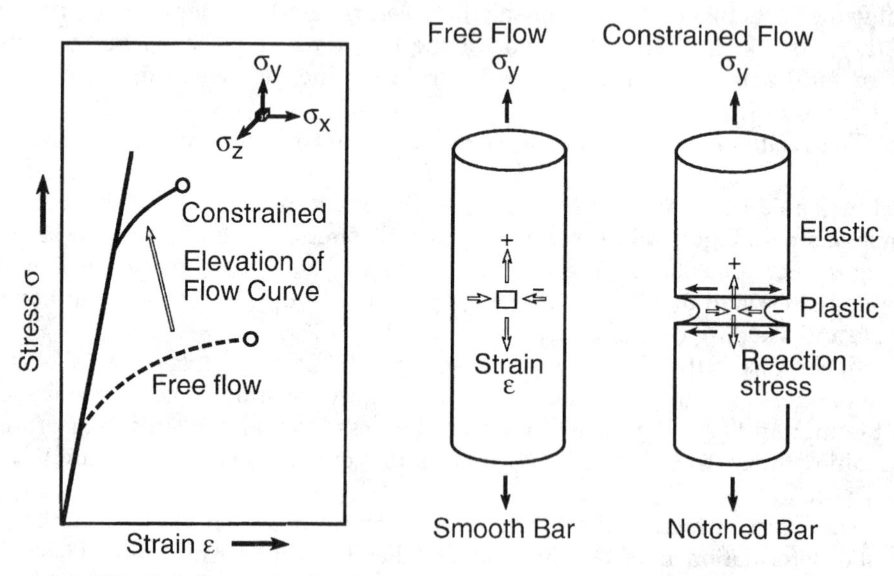

FIG. 16.3 Constraint to plastic flow caused by notched geometries.

Thus, the uniaxial stress state in the smooth bar is changed to a triaxial tensile stress system in the notched bar. Because plastic flow is restricted, the yield strength exhibited by the notched bar is higher than in the smooth bar. In other words, the notched bar behaves elastically at a higher stress than does the smooth bar. Thus, the strength, ductility, and fracture behavior in a smooth tension test referred to in design specifications do not characterize the behavior of highly constrained and notched details like the Ingram Barge.

16.3 Constraint Experiences in the Ship Industry

The published literature contains many examples that demonstrate the effects of various factors on the fracture behavior of steel structures. A brief description of the fractures of ships in the mid 1940s is presented prior to discussing the Ingram Barge fracture in 1972 to illustrate the significant effects of severe stress raisers such as sharp corners and notches and highly constrained connection details on the fracture behavior of steel structures.

More than 20% of the 4694 merchant ships built during World War II developed cracks of considerable size by 1946. Between 1942 and 1952, more than 200 ships had sustained fractures classified as serious, and at least 9 T-2 tankers and 7 Liberty ships had broken completely in two as a result of brittle fractures [2–9]. The first procedure to solve this problem was to contour sharp corners at various openings in the ship deck to decrease their stress concentration. These changes, plus improvements in other fabrication practices, improved the performance of ships significantly. Later, studies of the notch toughness of source, through, and arrest plates showed that plates exhibiting ft-lb levels greater than 10 ft-lb tended to be plates in which cracks had arrested. This study led to the 15 ft-lb CVN fracture criterion.

16.4 Ingram Barge Failure

On January 10, 1972, the 178-m (584-ft)-long Ingram Barge fractured in a brittle manner, Figure 16.4. At the time the air temperature was 7°C (45°F) and the ship was turning slowly at the harbor in calm waters. Failure originated at an unusually high stress level for this type of structure (2.5 times design load or 2.5 times a nominal stress level of about 165 MPa (24 ksi)). The unusually high stress level occurred because there was no approved loading manual and an unusual ballasting of the barge created the large stress. The ship had been in service for only 9 months [*10,11*].

The fracture initiated in a region of very high constraint at a doubler ring welded to the deck plate with a king post welded to the doubler, Figure 16.5. Also, four gusset plates were welded to the king post and deck as stiffeners, Figure 16.6. The fracture origin at the port king post is shown in Figure 16.7, as

FIG. 16.4 Overall fracture of Ingram Barge.

well as the fracture path across the structure. As illustrated in Figure 16.7, a second fracture initiated from the same general location at the starboard king post.

No pre-existing flaw was observed, even though chevron markings on both sides of the king post pointed toward the fracture origin at the king post, as shown in Figure 16.8. The local geometry was such that a high triaxial state of stress existed, and thus failure occurred at a very high load even in the absence of a flaw. A finite element stress analysis verified that the local stress level was well above the yield stress level. Because there was no pre-existing crack, the notch toughness level did not have as significant an effect on the initiation as did the local geometrical constraint and loading. Once a brittle fracture had initiated in the deck plate, the loading conditions (essentially constant load) were such that the cracks would be expected to propagate in both directions until the potential energy was dissipated by a complete fracture.

The steel had very good notch toughness as measured by the Charpy V-Notch impact test specimen, well within the expected range for ABS-B steel, Figure 16.9. The notch toughness as measured by a more severe fracture test, the dynamic tear (DT) test, indicated a lower notch toughness as shown in Figure 16.10. The dynamic notch toughness of the steel, as measured with one of the

FIG. 16.5 Origin of fracture at Port King on Ingram Barge.

most severe notch toughness tests available (i.e., the DT test), was about 10°F, approximately 30°F below the service temperature, near the lower end of the transition range. Although a ductile, rather than a brittle, fracture might have occurred had the notch toughness been greater, full upper shelf dynamic behavior is needed to eliminate completely the risk of failure by brittle fracture.

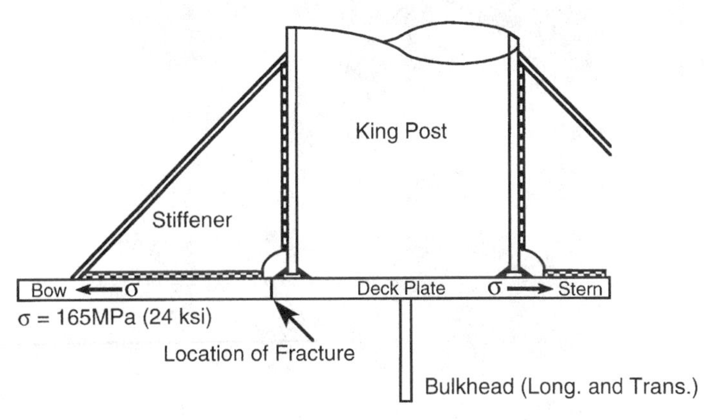

FIG. 16.6 Longitudinal section at king post showing location of fracture origin.

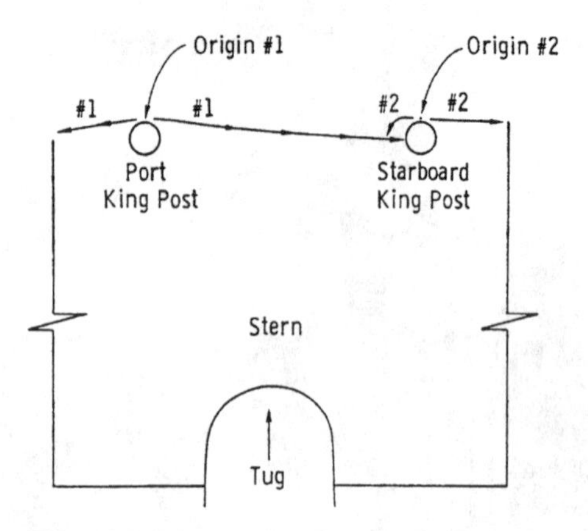

FIG. 16.7 Fracture path across barge showing a primary and a secondary origin.

FIG. 16.8 Primary origin of fracture (ring stiffener above deck plate and longitudinal bulkhead below).

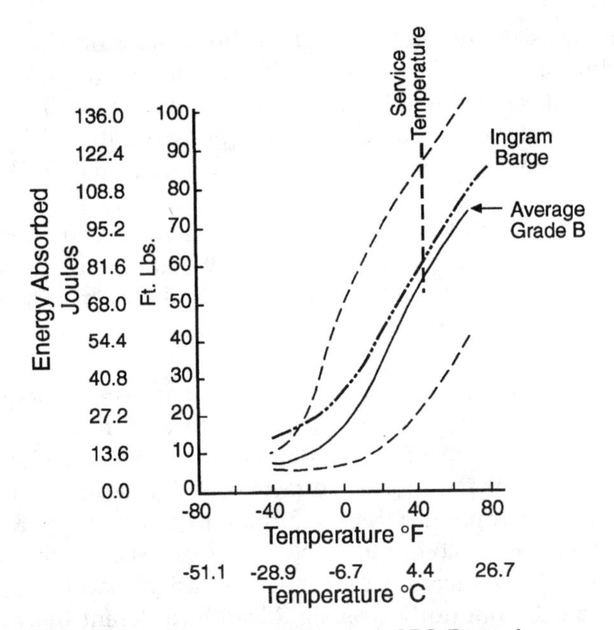

FIG. 16.9 Comparison of average ABS-B steel toughness level with range of toughness values and ABS-B steel from the Ingram Barge.

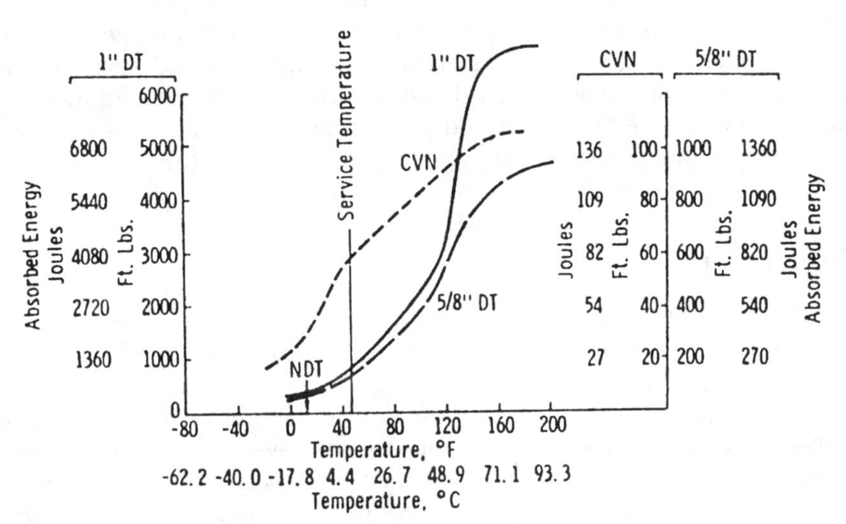

FIG. 16.10 Relations between NDT, CVN, and DT test results for ABS-B steel from the Ingram Barge.

Overall analysis of this failure lead to the conclusion that the very large sustained loading at a highly restrained detail caused by improper ballasting was the primary cause of the failure. It is also possible that the unusually large load caused a premature separation of a stiffener from the deck plate. Such an occurrence would have superimposed a dynamic load onto the large sustained load.

The notch toughness of the steel at the service temperature as measured by the CVN impact test was very good and certainly better than that found in many other types of structures. On the basis of the general service experience of ships, this level of notch toughness should be satisfactory for normal design, fabrication, and loads.

This particular fracture emphasizes the fact that brittle fracture is not, and never has been, just a material problem. Design, fabrication, materials, inspection, and operation (i.e., loads) are all factors that affect the susceptibility of a structure to fracture. When one or more of these factors is significantly more severe in a particular structure, compared with conditions in other structures of the same generic class, then the possibility of fracture is increased. In the case of the Ingram Barge, the sustained load of 2.5 times the design load in the presence of a severe geometrical discontinuity was significantly different than usually found in this type of structure. Therefore, this loading and the king post detail were the dominant factors leading to the brittle fracture.

In contrast, steel surge tanks in North Dakota, built from welded plates with only 2 to 3 ft-lb operating at a minimum service temperature of −40°F, have operated successfully for over 45 years [*12*]. The surge tank design is a simple cylinder, the loading is slow, the stresses are about one half the yield strength, and inspection and usage are such that any flaws would be small.

Thus, heavily constrained structures, such as the Ingram Barge, can fail under severe loads even though the inherent notch toughness and ductility may be very good. In contrast, well-designed simple structures can operate successfully at temperatures where their notch toughness may be very low. Thus, constraint and loading are key factors in the prevention of brittle fracture.

16.5 Summary

In the marine industry, structural integrity is established by adherence to a set of rules and specifications developed and modified over many years of service experience. Thus, only those materials, designs, and fabrication procedures that have had satisfactory performance in previous structures or that have been extensively evaluated are used in new ship designs. Over the years this approach has been used in a conservative manner to ensure a very high level of structural integrity in the marine industry. Accordingly, the state-of-the-art modern day merchant ship design is such that the overall structural integrity is quite good, and brittle fractures occur only when unusual circumstances arise such as in the case of the Ingram Barge.

In the case of the Ingram Barge, the design detail of the king post was highly constrained and is no longer allowed. Furthermore, the loading was excessive and in violation of approved loading conditions.

16.6 References

[1] Barsom, J. M., "Steel Properites—Effects of Constraint, Temperature, and Rate of Loading, *Seismic Design, Evaluation and Retrofit of Steel Bridges—Proceedings of the Second U.S. Seminar,* A. Astaneh-Asl, and J. Roberts, Eds., Report No. UCB / CTE-STEEL-96/09, University of California, Berkeley, November 1996.

[2] Bannerman, D. B. and Young R. T., "Some Improvements Resulting from Studies of Welded Ship Fracture, *Welding Journal,* Vol. 25, No. 3, March 1946.

[3] Acker, H. G., "Review of Welded Ship Failures," Ship Structure Committee Report Serial No. SSC-63, National Academy of Sciences-National Research Council, Washington, December 15, 1953.

[4] *Final Report of a Board of Investigation—The Design and Methods of Construction of Welded Steel Merchant Vessels,* July 15, 1946, Government Printing Press, Washington, DC 1947.

[5] Boyd, G. M. and Bushell, T. W., "Hull Structural Steel—The Unification of the Requirements of Seven Classification Societies, *Quarterly Transactions: The Royal Institution of Naval Architect,* London, Vol. 103, No. 3, March 1961.

[6] Parker, E. G., *Brittle Behavior of Engineering Structures,* Wiley, New York, 1957.

[7] Turnbull, J., "Hull Structures," *The Institution of Engineers and Shipbuilders of Scotland Transactions,* Vol. 100, Part 4, Dec. 1956–1957, pp. 301–316.

[8] Boyd, G. M. "Fracture Design Practices for Ship Structures, *Fracture,* H. Liebowitz, Ed., Vol. 5, *Fracture Design of Structures,* Academic Press, New York and London, 1969, pp. 383–470.

[9] Heller, S. R., Nielsen, R., Lyttle, A. R., and Vasta, Jr., *Twenty Years of Research Under the Ship Structure Committee,* Ship Structure Committee Report Serial No. SSC-182, U.S. Coast Guard Headquarters, Washington, DC, December 1967.

[10] Marine Casualty Report, "Structure Failure of Tank Barge I.O.S. 3301 Involving the Motor Vessel Martha R. Ingram on 10 January 1972 Without Loss of Life," Report No. SDCG/NTSB, March 1974.

[11] Rolfe, S. T., "Structural Integrity in Merchant Ships," *ASME Journal of Engineering Materials and Technology,* January 1980, Vol. 102, pp. 15–19.

[12] The Art and Science of Structural Engineering, Symposium Honoring William J. Hall, University of Illinois, Urbana, Illinois, April 1993, Prentice Hall, 1996.

17

Importance of Loading and Inspection—Trans Alaska Pipeline Service Oil Tankers

17.1 Introduction

OIL TANKERS, such as those in the Trans Alaska Pipeline Service (TAPS), can be subjected to fairly severe wave loadings. The severe wave loadings result in high cyclic stresses, and undetected cracks may grow by fatigue to lengths approaching the critical crack size of the hull steel.

This type of structure is very prone to fatigue cracking because of the multitude of Category D and E fatigue details located throughout the ships and the fact that they are continuously being loaded in fatigue. Inspection is very difficult and yet an extremely important criteria to ensure safe service.

Previously, we discussed the importance of maximum stress in controlling fracture and stress range in controlling fatigue. Ships, as well as airplanes, are subjected to loadings that are heavily influenced by the weather. Accordingly, a fracture control element that can be used for ships or airplanes is referred to as voyage planning, that is, scheduling the passage of a ship to avoid the most severe storms. If this can be done, the extremely large fatigue loading, in which most of the fatigue damage occurs, can be avoided and the life of a ship can be increased significantly.

This chapter presents a general fracture mechanics methodology that can be used to assess the structural reliability of critical area details in structures such as oil tankers that experience fatigue cracking. The methodology is based on principles discussed previously in this text and is primarily deterministic.

Inspection recommendations are made based on the potential fatigue loading of the ships. However, the fatigue loading can be reduced significantly by voyage planning wherein storms are avoided. Although this may increase the travel time of the structure, it will significantly increase the fatigue life of the structure. By reducing the fatigue crack propagation and increasing the fatigue life, the safety and reliability of TAPS vessels have been increased markedly, as described in this case study.

17.2 Background

The United States Coast Guard has conducted an extensive review of cracking reported between 1984 and 1988 on the 69 vessels over 10,000 gross tons in the TAPS trade during that time frame. These studies revealed that while the TAPS fleet comprised only 13% of the U.S. flag fleet, these tankers accounted for 59% of all of the reported fractures. Additionally, 73% of the reported TAPS fractures occurred in only 24 of the 69 ships [*1*].

The Coast Guard review of the vessels in the TAPS trade noted that hulls fabricated from high tensile strength (HTS) steel experienced a disproportionately higher number of structural cracks than did hulls fabricated from mild steel plates. Although the design rules allow the allowable stress to rise as the HTS yield strength increases, the fatigue strength of HTS steel weldments remains about equal to that of mild steel and offers no advantage in this area. As the operating stress range increases, the number of cycles to fatigue failure generally decreases (reduced fatigue life), and the subsequent fatigue damage may end up being greater than would be the case in a similar mild steel detail. This fact, combined with thinner scantlings from the use of HTS steel, as well as possible further reduction in scantlings by corrosion, may lead to early fatigue cracking in tankers fabricated from HTS steel.

The fairly severe wave loads can lead to fatigue crack initiation and propagation at certain fatigue-sensitive details in oil tankers. During inspection of critical details in some oil tankers, fatigue cracks have been discovered [*1*]. Since the fatigue initiation life of these details already is exhausted, the prediction of the remaining propagation fatigue life must be made using a fracture mechanics crack-propagation methodology. Determination of the remaining fatigue crack propagation life is essential in establishing inspection intervals to insure the safety and reliability of these oil tankers for continued safe service in the TAPS trade.

17.3 Fracture Mechanics Methodology

This chapter describes the methodology used to establish the remaining fatigue crack propagation life and representative inspection intervals for a specific ship detail, namely bottom shell plates near longitudinal drainage and master butt weld cutouts. Figure 17.1 is a schematic drawing of this detail showing the location of fatigue cracks at the "rat hole" at the end of the master butt weld cutout. The methodology presented in this chapter can be applied to other types of ship details, provided the steps below are performed for each particular detail. Note

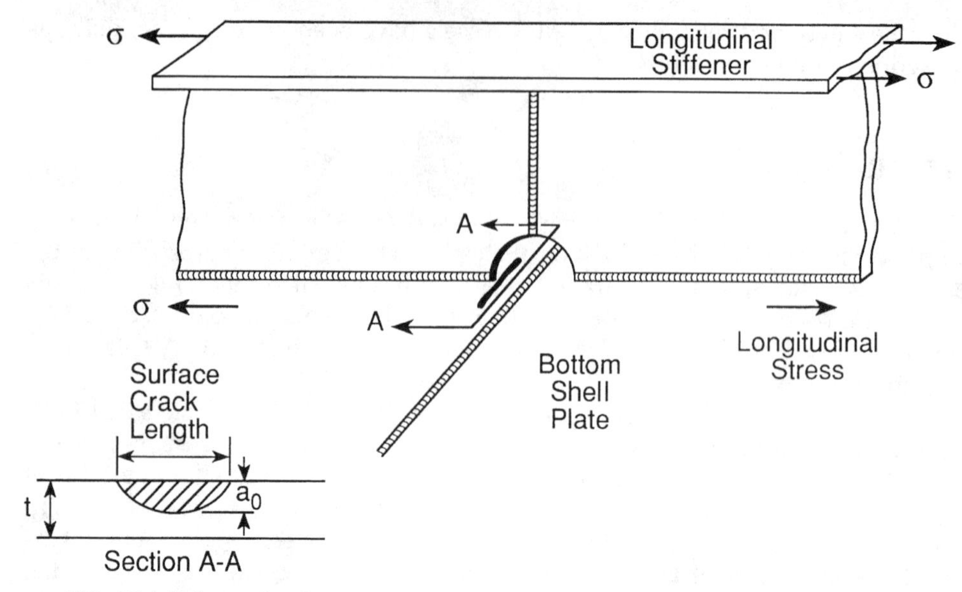

FIG. 17.1 Schematic of master butt weld cutout showing fatigue crack location.

that each of these individual steps are discussed previously in this book. Briefly, the methodology consists of the following steps:

1. Identification of specific details where cracks occur and selection of a stress intensity factor, K_I, that describes the stress field at that detail.
2. Inspection of these details to establish a representative initial flaw size, a_o, to be used in a fatigue crack propagation analysis.
3. Determination of a representative fracture toughness value of the steel plates used in the details under study. By knowing the maximum stress to which these details will be loaded, the critical crack size can be estimated. The critical crack length is the length that a fatigue crack must reach before the crack will propagate unstably. This length depends on material toughness and applied stress level so it is not a material property.
4. Use of histograms to estimate the equivalent root-mean square stress range, $\Delta\sigma_{RMS}$, to which the ship is subjected for a specific loading season. This $\Delta\sigma_{RMS}$ value can be used in existing crack propagation equations to estimate the number of cycles of loading (N_P) it takes a crack to grow from the initial crack size, a_o, to the final or critical crack size, a_{cr}.
5. On the basis of this estimate of the crack propagation life (N_P), establish reasonable inspection intervals for safe and reliable service.

Determination of the fatigue crack propagation life for a particular critical crack size in a specific structural detail is a complex process and cannot be generalized for different details or structures. Each type of detail and loading must be analyzed individually. Accordingly, this case study presents a generalized methodology that can be used with specific details and then presents one single example for what is considered to be a representative loading of one detail in one class of tank ships in the TAPS trade.

17.4 Application of Methodology to a Detail in an Oil Tanker

17.4.1 Identification of Critical Details

Fatigue cracks have been observed in some classes of tankers engaged in the TAPS trade [*1*]. These tankers are subjected to fairly severe service loads on a routine basis, and this loading, plus the use of high-strength steel in fatigue sensitive details, has led to cracking. On one class of tankers in particular, the details where cracking is most severe are:

1. Side shell longitudinal bracket connections to transverse bulkheads and to web frames.
2. Webs of bottom shell longitudinal stiffeners.
3. Bottom shell plates near longitudinal drainage and master butt weld cutouts.

Analysis of these details on this class of tankers indicates that while all cracking in ships potentially can be serious, the first two types of cracks appear to be less severe and are being addressed by inspection and repair, improvement of details, grinding of poor weld contours, hammer peening, and the use of drilled holes as crack arrestors.

Cracking in the third category of details, however, is more difficult to detect and has the potential of leading not only to a through thickness penetration of the bottom shell plating, but possibly to rapid fracture in the tankers. Accordingly, this study has focused on the significance of bottom shell cracks, Figure 17.1, with respect to the overall structural integrity of these tankers. Finally, recommendations are made regarding hull grinder inspection criteria.

17.4.2 Fracture Toughness

Crack-tip opening displacement (CTOD) fracture tests were conducted on 3/4-in.-thick AH-36 steel plates taken from tankers in this one class of TAPS vessels, using ASTM Standard E 1290. Each specimen used the full plate thickness after surface grinding to a uniform thickness. The specimen sizes were approximately 3/4-in. by 1.5 in. Analysis of these results indicated that, as expected, there was considerable variation in the CTOD results for various plates and weldments. Results presented in Table 17.1 show CTOD test results for two typical bottom shell plates plus one weld and one heat-affected zone (HAZ). At 32°F, a representative minimum bottom shell plate temperature, the CTOD values for

TABLE 17.1. CTOD Test Results from Bottom Shell Plates, mils (0.001 in.).

TEMP. °F	PLATE PC7	PLATE PC8-A	PC8-HAZ	PC8-WELD
32	1.9	29.7	7.0	5.4
32	2.8	28.4	29.3	9.9
32	2.4	28.9	9.5	10.4
Avg	2.4	29.0	15.3	8.6

base metal can average as low as 2.4 mils. This value is consistent with unpublished test results obtained from other tankers. Test results for weld metal and HAZ specimens were higher (8.6 and 15.3 mils, respectively).

The base metal toughness is of greatest interest since most fatigue crack growth probably occurs in base metal. Accordingly, a conservative value of 2.4 mils was selected as a representative minimum value to analyze the behavior of the bottom shell plates in this one class of vessels subjected to TAPS service.

A value of 2.4 mils for base metal can be related to an equivalent K_c by:

$$K_c = \sqrt{m\delta_c\sigma_{FL}E} \tag{17.1}$$

where K_c = critical stress intensity factor, ksi$\sqrt{\text{in.}}$,
$m \cong 1.7$ based on research studies of structural grade steels,
δ_c = CTOD value, 2.4 mils,
σ_{FL} = flow stress (average of yield and tensile strength)
$\simeq \dfrac{\sigma_{ys} + \sigma_{ULT}}{2} = 70$ ksi, and
E = modulus of elasticity.

Therefore $K_c = \sqrt{(1.7)(.0024)(70{,}000)(30{,}000{,}000)}$

$$K_c \simeq 92.5 \text{ ksi}\sqrt{\text{in.}}$$

For a CTOD value of 29 mils (see Table 17.1), the estimated K_c is approximately 322 ksi$\sqrt{\text{in.}}$ Thus, there is considerable scatter in the fracture toughness of these steels based on a limited sample analysis.

Charpy V-notch (CVN) test results of these same two plates and results of other tests presented in Table 17.1 show that the toughness of about 2.4 mils is at the lower range of values for this particular steel. Therefore, as a representative value, a fracture toughness level of about 100 ksi$\sqrt{\text{in.}}$ was selected as a reasonable lower bound value to use for subsequent critical crack size calculations. It should be noted that at the time of construction of this particular class of vessels, there were no CVN specifications for AH-36 steel in the ABS Rules for Steel Vessels. The specification in the ABS Rules for Steel Vessels now is 25 ft-lb. Of the five typical 3/4-in.-thick bottom plate samples tested, three had CVN values below this 25 ft-lb minimum. As noted later, the fatigue life of these tankers is not that dependent on notch toughness as long as the critical crack size is reasonably large, as it appears to be for these tankers.

17.4.3 *Stress Intensity Factors and Critical Crack Size for Critical Details*

To predict critical crack lengths, estimates of the material fracture toughness, K_c, and the maximum likely stress level, σ_{max}, that occurs during maximum sea states are required. The fracture toughness and maximum stress level are used in the expression for K_I, the stress intensity factor that best represents the actual structural geometry in the bottom shell plates to calculate critical crack lengths. Different geometries require different K_I relations, as described in Chapter 2.

For an unstiffened bottom shell plate, the relatively simple expression for a through-thickness crack in a semi-infinite wide plate would be appropriate. This expression is:

$$K_I = \sigma\sqrt{\pi a} \tag{17.2}$$

Values of critical fracture toughness ($K_I = K_c$) and maximum stress ($\sigma = \sigma_{max}$) are used to calculate the critical crack size, a_{cr}. Actually the critical crack length is *twice* this value or $2a_{cr}$ because of the nature of the stress-intensity factor equation. Because the bottom shell plate actually is stiffened, the above expression should be modified to account for the effect of the presence of a single stiffener perpendicular to the crack [2], Figure 17.1. A review of the effect of stiffeners on K_I values leads to the conclusion that the K_I value in a stiffened plate is about 0.7 of the K_I value for an unstiffened plate. This value of 0.7 is used to correct the value of stress range in the analysis of fatigue crack growth and is referred to as the single stiffener reduction factor (RF_{SS}) in subsequent fatigue analyses.

For very long cracks that have crossed several stiffeners, the effect of these stiffeners on the stress intensity factor is greater. This observation may help to explain why cracks of several feet in length crossing one or more stiffeners may not lead to a rapid fracture. Thus, in addition to preventing plate buckling during compressive loading, longitudinal stiffeners may act as crack growth retarders (possibly even arrestors) for severe stresses during tensile loading. The fact that stiffeners have this effect emphasizes the need to repair all cracks in the webs (and flanges) of longitudinal stiffeners near drainage and weld cutouts at each inspection.

The K_I expression for an unstiffened plate is modified by reducing the maximum stress by a reduction factor of about 0.6 (RF_{MS}) to account for the beneficial effect of multiple stiffeners. As noted in Reference 2, the actual effect of several stiffeners may be to reduce the K_I by a factor greater than 0.6. Using a RF_{MS} factor of 0.6, the relation for critical crack size, $2a_{cr}$, therefore becomes:

$$K_c = (RF_{MS})\,\sigma_{max}\sqrt{\pi a_{cr}} \tag{17.3}$$

$$\therefore\ 2a_{cr} = \frac{2}{\pi}\left(\frac{K_c}{(0.6)\,\sigma_{max}}\right)^2$$

Previously, it was shown that a reasonable lower bound fracture toughness, K_c, is about 100 ksi$\sqrt{\text{in.}}$ Members of the industry working group [1] estimated the maximum stress to be about 30 ksi, although discussions with ABS personnel have indicated that the actual maximum stress might be slightly higher. Therefore, assuming that the maximum stress, σ_{max}, can be as high as about $2/3\sigma_{ys}$, or about 34 ksi, $2a_{cr}$ is estimated to be:

$$2a_{cr} \cong \frac{2}{\pi}\left(\frac{100}{(0.6)(34)}\right)^2$$

$$2a_{cr} \simeq 15 \text{ inches}$$

It is important to note that the stress RF for a single stiffener, RF_{SS}, is to be applied only when a crack is small as it is during the early stages of fatigue crack propagation. The stress reduction factor for multiple stiffeners, RF_{MS}, is to be used to estimate critical crack length, when the crack may be fairly large.

It should be noted that 15 in. is a fairly conservative value for the critical crack length because the lowest measured fracture toughness level and a fairly high stress level were used to estimate the critical crack length. Also the effect of several stiffeners may result in a reduction factor less than 0.6 and thus increase the critical crack size even further. However, even if the critical crack size were larger, the calculated fatigue crack growth rate is fairly high (because of the large crack length), resulting in only a slight increase in fatigue life. In other words, even if the critical crack length were larger than 15 in., the fatigue life would not be significantly longer. This is why it was stated earlier that the fatigue life is not strongly dependent on fracture toughness as long as the level of fracture toughness is reasonably high. Even if the material had a higher K_c, the crack growth rate is fairly large at this point, and a material with higher fracture toughness would not increase the fatigue crack propagation life significantly. Thus, a critical crack size of about 15 in. is assumed for the bottom shell plates in this example, realizing that in most cases it probably is larger.

17.4.4 Inspection Capability for Initial Crack Size, a_o

Determining a realistic value of the crack size that can be detected reliably in an oil tanker is likely the most difficult aspect of a fracture control methodology. The probability of detection (POD) of a crack varies from inspection to inspection and is dependent on a variety of factors. These include degree of surface cleanliness, lighting, inspection techniques used, inspector experience level and familiarity with the vessel class, vessel loading condition, condition of the coating system, and the location of the critical structural detail in the ship. No POD curves were available for ship structures.

Lacking such POD information, a conservative estimate for each critical area, taking into account the factors listed above, was made about what size cracks could be found with reasonable certainty. This value should be used in fatigue crack propagation studies as the initial flaw size, a_o, assumed to exist in the structure after an inspection has been completed. In an article about their new fatigue guide for tankers, the American Bureau of Shipping [3] recently noted that ship operators constantly detect and repair cracks of 3 to 4 in. It is interesting to note that these values are similar to what was estimated in this case study. For the particular class of TAPS tankers evaluated for this study, U.S. Coast Guard inspectors estimated that surface cracks could be detected in the areas identified as critical with a high degree of confidence. These detectable cracks were estimated by the inspectors to be 3 in. in length using visual means, and 2 in. in length using either ultrasonic or magnetic particle inspection techniques.

17.4.5 Determination of Histogram for Fatigue Loading

In developing the stress histogram, the most accurate estimate of actual stresses experienced by a member in the critical area (both fatigue stress ranges and extreme stress values) was made. The calculations included using seasonal-based wave scatter data to account for the effect of loading history. A hydrodynamic model was used to develop global wave-induced hull girder vertical and horizontal bending moments, external and internal hydrodynamic pressures, and internal and inertial induced pressures; then a finite element analysis was used to develop local critical area stresses. Consideration should be made for the effects of vessel speed, loading conditions, wave direction, and wave spreading, or termed differently, "short" and "long-crestedness," as it varies during each voyage. Statistical analyses of the wave scatter data and the subsequent lifetime fatigue and the extreme stresses in this study were based on the formulation by Ochi [4,5]. The fatigue stress range histogram was then used to calculate the root mean square stress range value for each season, $\Delta\sigma_{RMS}$.

Using these procedures, a dynamic stress range histogram was developed for the subject tankers by the American Bureau of Shipping representatives [6]. National Oceanic and Atmospheric Administration buoy wave data measured at 5 points along the actual vessel's route was used to specify the characteristic seasonal wave environments. Statistical information was developed on the premise of seasonal fatigue loading and 20-year lifetime extreme values. The extreme values were obtained by adding the maximum dynamic stresses to the still water bending and hydrostatic pressure (internal and external, where applicable). The extreme stress calculated in the bottom shell was in the loaded condition and was 207 N/mm^2 (30 ksi). Table 17.2 shows the dynamic stress range histogram developed using this approach for the bottom shell on the subject vessels oper-

TABLE 17.2. Wave Loadings and Number of Cycles and Values of $\Delta\sigma_{RMS}$ for Center of Center Tank, Bottom Shell Plate.

σ_{min} N/mm²	σ_{max} N/mm²	AVERAGE $\Delta\sigma$, N/mm²	FULL LOAD				NORMAL BALLAST				ANNUAL TOTAL
			SPRING M, A, M	SUMMER J, J, A	FALL S, O, N	WINTER D, J, F	SPRING M, A, M	SUMMER J, J, A	FALL S, O, N	WINTER D, J, F	
0	10	5	28886	84717	29781	24563	31225	87494	32243	27171	346080
10	20	15	51032	78971	41011	39814	53548	81520	43165	43309	432370
20	30	25	57664	51971	49465	48155	59495	53013	51162	51100	422025
30	40	35	49488	35064	46511	45123	50117	34751	47424	46539	355017
40	50	45	34963	20102	36928	35396	34933	19971	37079	35491	254863
50	60	55	21342	9579	26038	24452	21170	9738	25807	23910	162036
60	70	65	11654	3892	16769	15373	11496	4084	16497	14676	94441
70	80	75	5824	1367	10002	8974	5681	1479	9823	8332	51482
80	90	85	2699	417	5560	4922	2570	462	5468	4401	26499
90	100	95	1169	111	2887	2553	1066	125	2842	2169	12922
100	100	105	475	25	1403	1257	407	29	1377	999	5972
110	120	115	182	5	639	589	143	5	1377	432	2617
120	130	125	66		273	263	46	1	261	176	1086
130	140	135	23		110	112	14		102	67	428
140	150	145	7		42	45	4		37	24	159
150	160	155	2		15	17	1		13	8	56
160	170	165			5	6			4	3	18
170	180	175			1	2			1	1	5
180	190	185									0
		SUM	265476	286221	267440	251616	271916	292672	273927	258808	2168076
		$\Delta\sigma_{RMS}$	37.11	26.00	42.20	42.01	36.51	25.92	41.62	40.54	36.81

ating in the TAPS service as part of this study. Use of the stress ranges is described in the next section on fatigue crack propagation.

17.4.6 Fatigue Crack Propagation in Bottom Shell Plates

As discussed in the section on inspection capability, there is a strong likelihood of either 2 or 3-in.-long surface cracks being present after any given structural inspection. That is, because there are fatigue-sensitive details that have been subjected to fairly severe fatigue loading throughout the life of these vessels, cracks continue to initiate from these details. These cracks are difficult to detect when they are small, but as they grow they can be found and repaired. However, cracks smaller than either 2 or 3 in. in length, depending on type of inspection, may not be detected. Thus, it is prudent, on the basis of information provided by Coast Guard inspectors, to assume that either 2 or 3-in.-long surface cracks (depending on type of inspection) may be present after a structural inspection.

An unknown factor is the relative shape of cracks with a surface length of either 2 or 3 in. Although the bottom shell is loaded primarily in tension, there are pressure stresses as well as differences in weld contours that may affect the shape of an unknown crack. Analysis of actual cracks found in the plate samples shows that the relative crack depth (a) to surface length ($2c$) ratio, $a/2c$, the crack aspect ratio, can vary from about 0.15 to about 0.35. Figure 17.2 shows the two initial surface crack lengths of 2 and 3 in. for an assumed $a/2c$ ratio of 0.25, which was chosen to model typical crack growth behavior. This assumption ap-

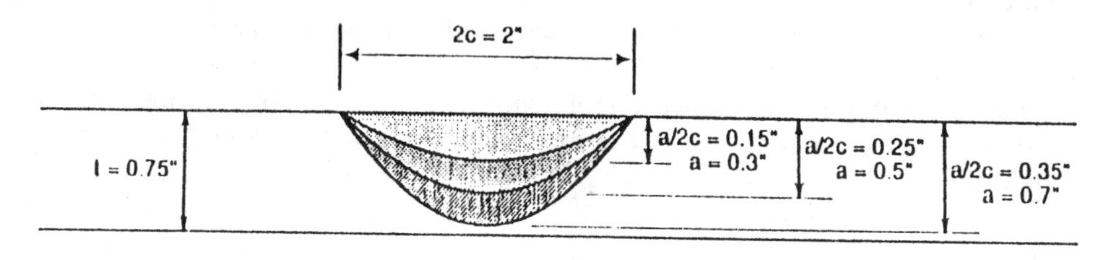

a) 2-Inch Long Surface Crack.

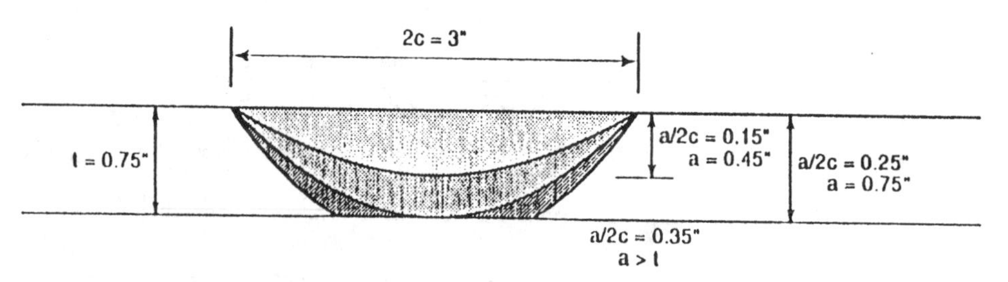

b) 3-Inch Long Surface Crack

FIG. 17.2 Surface crack model for $a/2c$ = 0.15 in. or 0.35 in., and $2c$ = 2 in. or 3 in.

pears to be reasonable on the basis of observations of actual fracture surfaces. Studies of ratios ranging from 0.15 to 0.35 indicate that the shape of a 2 or 3-in.-long surface crack does not have a significant effect on the fatigue propagation life for the 0.75-in.-thick bottom shell plates in these tankers. After the crack grows through the 0.75-in.-thick wall, it becomes a through-thickness crack and grows to the critical crack size, $2a_{cr}$, as shown in Figure 17.3. Note that for surface cracks, Figure 17.2, "*a*" is the dimension through the plate. For through-thickness cracks, Figure 17.3, "*a*" is one half the total crack length. This is common fracture mechanics terminology as described in Chapter 2.

To estimate the time it would take either a 2 or 3-in. surface crack to grow to critical size, the crack shown in Figure 17.2 was subjected to the $\Delta\sigma_{RMS}$ values presented in Table 17.2 and reduced by the reduction factor (RF_{SS}). The stress range histograms shown in Table 17.2 were computed using the formulation by Ochi [4,5] and the buoy measured wave data available from NOAA. These histograms show representative stress ranges and numbers of four seasons in both the fully loaded and normal ballast condition. $\Delta\sigma_{RMS}$ values for each condition were calculated as follows:

$$\Delta\sigma_{RMS} = \sqrt{\frac{\sum (\Delta\sigma_i)^2}{n}} \tag{17.4}$$

These $\Delta\sigma_{RMS}$ values were used to represent the variable loading as described in Chapter 9. Individual $\Delta\sigma_{RMS}$ values are shown at the bottom of each of the eight conditions in Table 17.2. Because the differences in fully loaded and normal ballast conditions were so small, these two conditions were averaged, and thus only the four seasonal loading conditions were used in the fatigue analysis.

Based on the information presented in Table 17.2, it was assumed that a representative oil tanker experiences the following four fatigue loading conditions during a typical year:

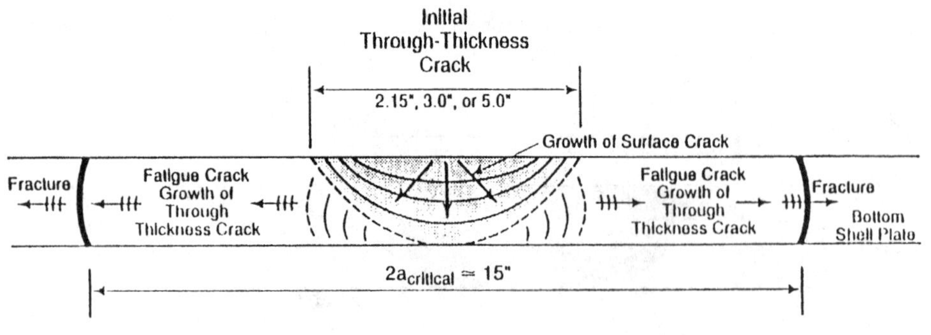

FIG. 17.3 Through thickness crack growth model: crack grows from initial size to $2a_{cr}$ = 15 in. At $2a_{cr}$, rapid fracture occurs.

Winter: $\Delta\sigma_{RMS} = \dfrac{42.01 + 40.54}{2} = \underline{41.28}$ MPa—5.98 ksi

for N = 251,616 (Table 17.2)
+258,808
510,424 cycles — Reduced Loading 0.7 (5.98) = $\underline{4.19}$ ksi

Spring: $\Delta\sigma_{RMS} = \dfrac{37.11 + 36.51}{2} = \underline{36.81}$ MPa—5.34 ksi

for N = 265,476 (Table 17.2)
+271,916
537,392 cycles — Reduced Loading 0.7 (5.34) = $\underline{3.74}$ ksi

Summer: $\Delta\sigma_{RMS} = \dfrac{26.00 + 25.92}{2} = \underline{25.96}$ MPa—3.76 ksi

for N = 286,221 (Table 17.2)
+292,672
578,893 cycles — Reduced Loading 0.7 (3.76) = $\underline{2.64}$ ksi

Fall: $\Delta\sigma_{RMS} = \dfrac{42.20 + 41.62}{2} = 41.91$ MPa—6.08 ksi

for N = 267,440 (Table 17.2)
+273,927
541,367 cycles — Reduced Loading 0.7 (6.08) = $\underline{4.25}$ ksi

The crack growth behavior of ship steels can be represented by the following expression, Chapter 9:

$$\frac{da}{dN} = 3.6 \times 10^{-10}(\Delta K_{RMS})^{3.0} \tag{17.5}$$

Accordingly, the number of cycles, ΔN, that it takes to grow a crack an amount, Δa, is:

$$\Delta N = \frac{\Delta a}{3.6 \times 10^{-10}\,(\Delta K_{RMS})^{3.0}}$$

For a surface crack of Length $2c$ and Depth a:

$$\Delta K_{RMS} = 1.12\,\Delta\sigma_{RMS}\sqrt{\pi a/Q}\cdot M_K$$

where $Q = f(a/2c)$, and
M_K = back-surface magnification factor.

For the through-thickness crack:

$$\Delta K_{RMS} = \Delta \sigma_{RMS} \sqrt{\pi a} \tag{17.6}$$

Figure 17.4 shows the calculated size of either a 2-in.-long or 3-in.-long surface crack versus loading time in months. As a crack grows, it changes from a surface crack (Figure 17.2) to a through-thickness crack (Figure 17.3). Figure 17.4 shows that for the assumptions made earlier (stress ranges, fracture toughness levels, maximum stress levels), the critical crack size, $2a_{cr}$, is about 15 in. and takes about 60 months to grow a surface crack of 2-in. length to a through-thickness crack of 15 in., depending on the $a/2c$ ratio.

Figure 17.4 also shows the calculated size of the 3-in.-long surface crack as a function of loading time. The behavior is similar to that of the 2-in. surface crack but that, as expected, the time to reach a crack size of 15 in. is less, namely about 48 months. Also, for a surface crack length of 3 in., any $a/2c$ ratio larger than 0.25 is already through the 0.75-in. bottom shell plate, and thus any effect of $a/2c$ ratio is smaller than for the 2-in. surface crack.

The calculated lives shown in Figure 17.4 are fairly short and indicate the need for periodic inspection. These results also demonstrate that improved quality of inspection, i.e., an inspection procedure that will find 2-in. surface cracks reliably rather than 3-in-long surface cracks, can lead to increased fatigue lives.

17.5 Effect of Reduced Fatigue Loading

Previously, it was shown that by improving the quality of inspection so that 2-in.-long surface cracks can be found, rather than 3-in.-long surface cracks, the

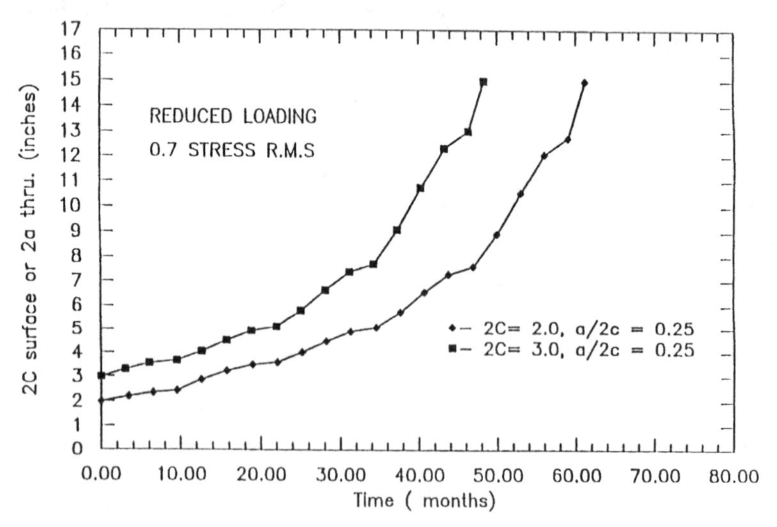

FIG. 17.4 The effect of 2C value on the crack growth time using the reduced stress (0.7 stress R.M.S.).

fatigue life can be extended. If the inspection capability could be improved even more, the increase in fatigue life would be even greater. This improvement in life occurs by reducing the initial crack size, a_o, as was discussed in Chapter 12, Fracture and Fatigue Control.

For structures where the loading can be affected by the weather, such as ships or airplanes, the concept of voyage planning can be very useful in extending the life of a structure. As described in Reference 7, Winter and Lewis made actual strain range measurements for 26 consecutive voyages of a TAPS tanker. By using actual loading (not maximum theoretical as was presented in Table 17.2), and by voyage planning, the actual stress ranges were reduced below those presented in Table 17.2. A comparison of the actual Winter and Lewis (measured) data with that predicted in the current study is presented in Table 17.3. Note that while the measured results were about 3.6 million compared with the predicted results of about 2.2 million cycles of loading, the $\Delta\sigma_{RMS}$ range was only 1.4 ksi compared with 5.2 ksi. Because the crack growth varies as $\Delta\sigma_{RMS}$ to the power of 3.0, this reduction in actual stress range by voyage planning has a *very significant* effect on the fatigue crack propagation life of a ship.

The authors [7] state that: "The comparison of the two datasets is quite striking because the measured stress range/count data is considerably less than the study predictions. The table [Table 17.3] shows:

1) the maximum measured stress-range is less than the predictions for all seasons and for both loading conditions;
2) the measured counts in equal stress-range are less than the predicted counts, except for the two lowest stress-range bins; and
3) the predicted stress-range/count study for the *summer* season is higher than the measured data for the *winter.*

The larger number of counts in the lowest stress-range interval is caused by the hull vibrating nearly constantly at its first natural frequency." These low stress-range results due to hull vibration could not be included in the predicted NOAA results.

"The predicted stress-range count data was used in the Rolfe study to propagate 2-inch and 3-inch cracks using the root mean square stress-range. The table shows the root mean square stress-range for the measured data is considerably less than the study predictions. The Rolfe study calculated it would take 50 months to grow a 3-inch surface crack to 15 inches using the study stress-range/count data shown in Table 17.3. Using the measured data from this table and the same fracture mechanics equations used in the Rolfe study, a 3-inch crack would only grow to 6 inches in 240 months (20 years) [as shown in Figure 17.5]. Since the rate at which cracks grow is roughly proportional to the cube of the root mean square stress range, the lower measured root mean square value has a dramatic effect on fatigue crack propagation. The fact that fatigue damage to the bottom shell plating has not been a problem on these

TABLE 17.3. Comparison of Rolfe et al. Predicted Stress-Range/Counts with Full-Scale Measurements.

STRESS RANGE INTERVAL, ksi	SPRING MAR, APR, MAY		SUMMER JUN, JUL, AUG		FALL SEP, OCT, NOV		WINTER DEC, JAN, FEB		ANNUAL TOTAL	
	MEASURED	ROLFE	MEASURED	ROLFE	MEASURED	ROLFE	MEASURED	ROLFE	MEASURED	ROLFE
0.0–1.5	164,412	28,226	413,964	84,717	278,774	29,781	228,220	24,563	3,023,861	346,080
1.5–3.0	22,108	51,032	40,371	78,971	40,254	41,011	49,071	39,814	416,848	432,370
3.0–4.5	4,282	57,664	3,177	51,971	8,882	49,465	19,997	48,155	94,129	422,025
4.5–6.0	994	49,488	228	35,064	2,744	46,511	7,775	45,123	28,858	355,017
6.0–7.5	227	34,963	34	20,102	910	36,928	2,799	35,396	10,776	254,863
7.5–9.0	73	21,342	5	9,579	341	26,038	973	24,452	3,771	162,036
9.0–10.5	25	11,654		3,892	110	16,769	295	15,373	1,285	94,441
10.5–12.0	10	5,824		1,367	26	10,002	123	8,974	524	51,482
12.0–13.5	4	2,699		417	4	5,560	47	4,922	205	26,499
13.5–15.0	2	1,169		111	1	2,887	23	2,553	91	12,922
15.0–16.5		475		25		1,403	14	1,257	34	5,972
16.5–18.0		182		5		639	5	589	9	2,617
18.0–19.5		66				273	3	263	6	1,086
19.5–21.0		23				110	2	112	2	428
21.0–22.5		7				42		45		159
22.5–24.0		2				15		17		56
24.0–25.5						5		6		18
25.5–27.0						1		2		5
27.0–28.5										
28.5–30.0										
Sum	192,135	265,476	457,779	286,221	332,046	267,440	309,348	251,616	3,580,399	2,168,076
$\Delta\sigma_{rms}$	1.27	5.57	1.03	3.90	1.39	6.33	1.91	6.30	1.39	5.52

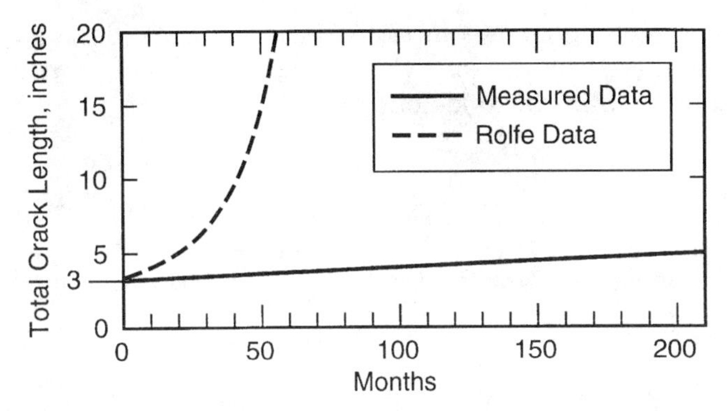

FIG. 17.5 Comparison of growth of a 3-in. bottom shell crack.

tankers tends to indicate that the measured stress-range/count data is more in line with reality."

17.6 Summary

The objective of this study was to present a general fracture mechanics methodology that can be used to assess the structural reliability of critical area details in oil tankers experiencing cracking. A methodology that is primarily deterministic was developed based on concepts described earlier in this book and was used to estimate the behavior of cracks in bottom shell plates. Inspection recommendations were based on conservative but reasonable assumptions. For the example presented, the predicted fatigue lives as well as a reasonable minimum critical crack length are consistent with service experience to date. That is, relatively large bottom-shell fatigue cracks have been observed in service but no complete failures have occurred. Using this methodology, similar analysis would be made on other classes of tankers, or other types of vessels.

Voyage planning was shown to be a very effective method of improving the life of those structures where weather loading can be significant. Because the fatigue crack propagation lives are proportional to ($\Delta\sigma_{RMS}$), even a slight reduction in $\Delta\sigma_{RMS}$ will have a significant effect on fatigue life, as was shown dramatically by Winter and Lewis [7].

Estimation of critical crack lengths and fatigue propagation lives of cracks in ships depends on many factors. Thus, each class of ships as well as each type of detail must be evaluated individually. This case study describes a fracture mechanics methodology that can be used to estimate the critical crack length and fatigue life of bottom shell cracks in tankers. The example deals specifically with the case of one structural detail in one class of tankers subjected to TAPS service, and the results cannot be generalized to other details, ships, or loadings. However, the methodology can be used in other cases provided that the specific loadings, material fracture toughness levels, inspection capabilities, and initial crack sizes are established. Comparison with another ship where actual loads were measured shows that this methodology can be very useful.

17.7 References

[1] Rolfe, S. T., Henn, A. E., and Hays, K. T., "Fracture Mechanics Methodology for Fracture Control in Oil Tankers," *Ship Structures Symposium 1993,* November 16–17, 1993, Arlington, VA.
[2] Rook, D. P. and Cartwright, D. J., *Compendium of Stress Intensity Factors,* Crown, 1976, printed in England for her Majesty's Stationery Office by the Hillingdon Press, Uxbridge, Middex.
[3] "A Guide for Combatting Fatigue," *Surveyor,* A quarterly publication of the American Bureau of Shipping, December 1992, pp. 24–31.
[4] Ochi, M. K., "On Prediction of Extreme Values," *Journal of Ship Research,* March 1973, pp. 29–37.
[5] Ochi, M. K., "Wave Statistics for the Design of Ships and Ocean Structures," *Transactions SNAME,* Vol. 86, 1978, pp. 47–76.
[6] Private communication, American Bureau of Shipping.
[7] Winter, D. J. and Lewis, J. W., "Operational and Scientific Hull Structure Monitoring on TAPS Trade Tankers," *Transactions SNAME,* Vol. 102, 1994, pp. 501–533.

18

Importance of Proper Analysis, Fracture Toughness, Fabrication, and Loading on Structural Behavior—Failure Analysis of a Lock-and-Dam Sheet Piling

18.1 Introduction

THIS CHAPTER DESCRIBES the fracture of a sheet pile in a cofferdam cell and presents a failure analysis to determine the cause of failure. The analysis examines the steel properties, fabrication and construction procedures, and the design of the cell.

A cofferdam is a temporary enclosure placed in a wet area that, when dewatered, permits construction to proceed within it under dry conditions. Lock and Dam 26 replacement project in the Mississippi River at Alton, Illinois, consisted of a dam having nine gate bays and tainter gates and two navigation locks 1200 and 600 ft long. Construction of this lock and dam was performed in three stages to ensure uninterrupted river navigation during its construction. Each stage consisted of construction of a temporary cellular cofferdam, construction of a portion of the permanent lock and dam inside the cofferdam, followed by removal of the cofferdam. Figure 18.1 shows the cofferdam construction for the second stage of this project, which consisted of the construction of a cofferdam, a 1200-ft navigational lock and two one half gate bays of the concrete dam.

The cofferdam for the second phase of construction was oval in shape and comprised of circular cells connected by connection arcs. The cells and arcs were constructed from interlocking PS-32 steel sheet piles. The geometry and dimen-

FIG. 18.1 Construction of Stage II of the Lock and Dam 26 replacement project at Alton, Illinois.

sions of the sheets are presented in Figure 18.2. The sheets were driven around the perimeter of a circular template into the river bed. Variations in the composition of the river bed prevented driving the sheets to the same elevation. Thus, the top of the sheet piles had to be trimmed by acetylene torches to the required elevation. Once the sheets were driven, the cells were filled with river sand. Work would proceed to the adjoining arc and cell until the cofferdam was fully enclosed. Finally, the interior of the cofferdam was dewatered to permit construction within its confines.

18.2 Description of the Failure

In the days proceeding the failure at Lock and Dam 26, river sand was piled daily on top of the cells, Figure 18.3. The next day, a front-end loader, Figure 18.3, would push the sand over the side to create a stability berm. On December 9, 1995, a Caterpillar 973 front-end loader was operating on top of Cell 68. At that time, the sand was being pushed over the side of Cell 68 before all the sheets had been trimmed to elevation along the perimeter of the cell. During this operation, Sheet 55 of Cell 68 split from top to bottom of the cell. The sand in the cell spilled out of the slit, and the front-end loader sank to the bottom of the cell, Figure 18.4. The operator escaped from the loader unharmed. The temperature of the steel at the time of the failure was estimated to be 32°F.

Cell 68 was 63 ft in diameter and contained a 35-ft-diameter cell that became part of the downstream guide wall. The geometry of the cells and adjoining arcs is presented in Figure 18.5.

18.3 Steel Properties

Subsequent to the failure, Sheet 55 was removed from the cell wall and was subjected to chemical and mechanical tests. The tests confirmed that the steel satisfied the specified chemical and physical requirements of ASTM A328 specifications for which it was purchased. The yield strength from two longitudinal tests were 43.4 and 51.1 ksi, and the tensile strengths were 92.1 and 92.0 ksi. The transverse test resulted in a 51.4-ksi yield strength and a 79.3-ksi tensile strength. The percent elongations in 2 in. were 27.5 and 26.0% for the longitudinal tests and 27.5% for the transverse test. These physical properties exceeded the ASTM A328 requirements of 70-ksi minimum tensile strength, 39-ksi minimum yield point, and 17% elongation in 8 in. Charpy V-notch (CVN) fracture toughness was not required for the purchase of the steel. However, CVN tests were conducted on the sheet that fractured and the results are presented in Figure 18.6.

CVN ft-lb data may be converted to impact stress-intensity factors K_{Id} by using the relationship:

$$K_{Id} = \sqrt{5E(\text{CVN, ft-lb})} \tag{18.1}$$

where E is the modulus of elasticity, psi, and K_{Id} is psi$\sqrt{\text{in.}}$

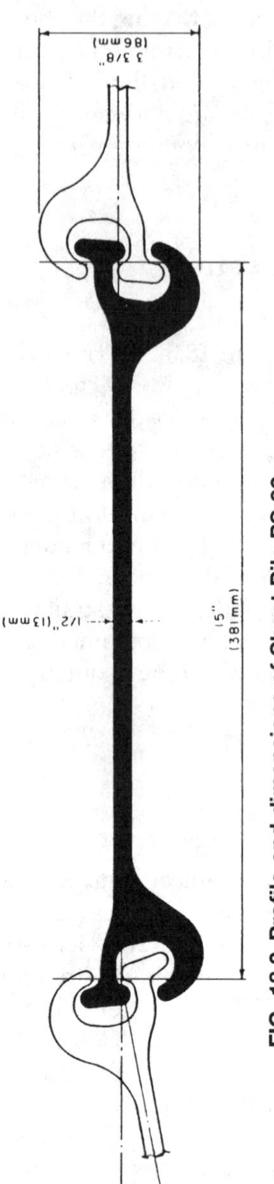

FIG. 18.2 Profile and dimensions of Sheet Pile PS 32.

FIG. 18.3 Sand piled on cofferdam cells and the front-end loader used to push it overboard.

FIG. 18.4 Fracture of a sheet pile in Cell 68.

The calculated K_{Id} values from Equation 1 may be used to estimate the static stress-intensity factors, K_{Ic}, by using the temperature shift relationship between K_{Id} and K_{Ic}:

$$T_{shift} = 215 - 1.5\sigma_{ys} \tag{18.2}$$

where T_{shift} = temperature shift in F° between the impact and the static fracture toughness.

σ_{ys} = room temperature yield strength ksi.

The conversion of CVN ft-lb data into K_{Id} data and the prediction of K_{Ic} behavior are presented in Figure 18.7.

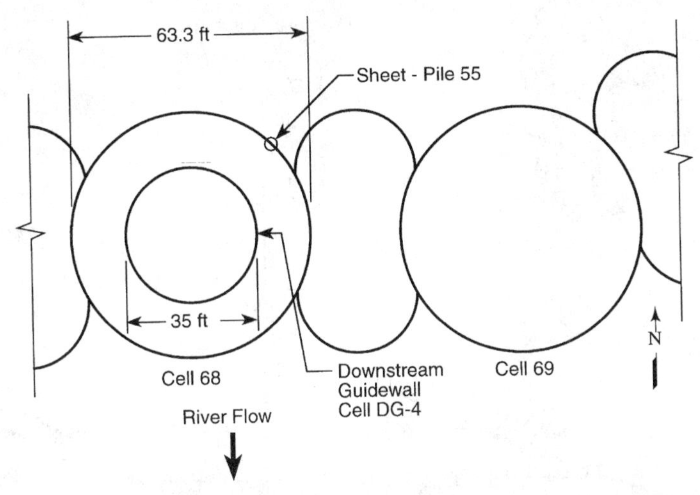

FIG. 18.5 Geometry of Cells 68 and 69 and connecting arcs.

Several crack-tip-opening-displacement (CTOD) fracture toughness tests were conducted on Sheet 55 steel at 0, 32, and 75°F. The CTOD test results were converted to equivalent K_c values by using the relationship

$$K_c = \sqrt{1.7\sigma_{\text{flow}}(\text{CTOD})E} \tag{18.3}$$

where K_c = static critical stress intensity factor, psi,
σ_{flow} = algebric average of the yield and tensile strengths of the steel, psi,
CTOD = crack tip opening displacement at fracture, inch, and
E = modules of elasticity, psi.

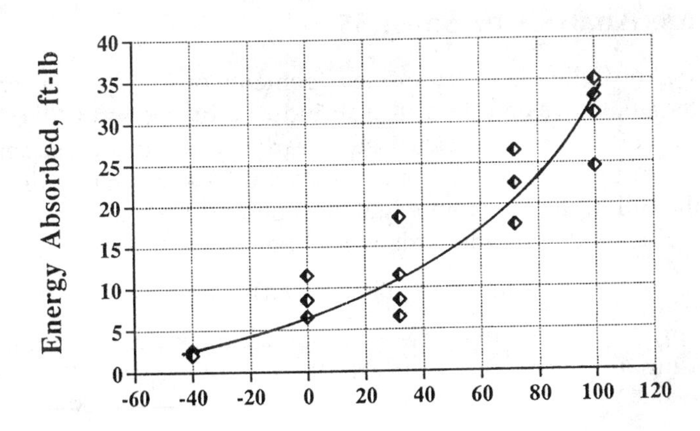

FIG. 18.6 Charpy V-notch fracture toughness of Sheet Pile 55.

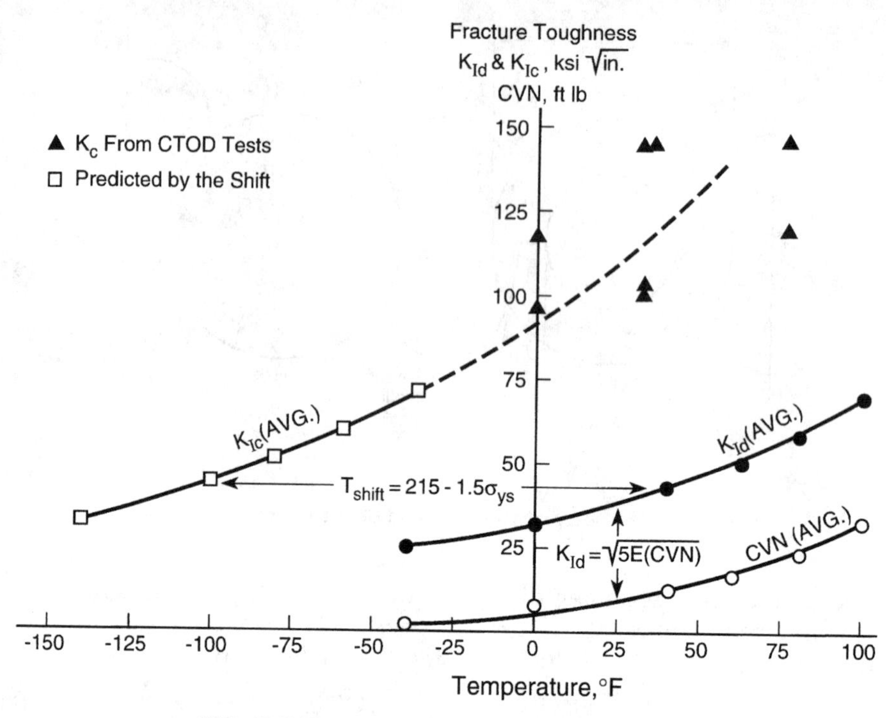

FIG. 18.7 Fracture toughness of Sheet Pile 55.

The calculated K_c values are presented in Figure 18.7 and appear to be a good extension of the K_{Ic} values derived from the CVN data.

18.4 Failure Analysis of Sheet 55

Stress analysis of Cell 68 and strain measurements on similar cells showed that the circumferential stress on the cell wall induced by the sand fill is essentially zero at the top of the cell and increased linearly by about 200 psi per foot from top to bottom of the cell. Consequently, the nominal stress acting on the circumference of the cell at the tip of an edge crack of length a that initiated at the top of the cell and extended downwards would be

$$\sigma_{\text{nominal}} = (200, \text{psi/ft})(a, \text{ft}) \tag{18.4}$$

Assuming that the nominal stress at a given crack depth is uniform, rather than decreasing, up to the top of the cell, the K expression for that crack would be

$$K_I = 1.12\sigma\sqrt{\pi a} \tag{18.5}$$

Because a cofferdam is a statically loaded structure, the critical fracture toughness, Figure 18.7, at 32°F was about 100 ksi$\sqrt{\text{in.}}$ Thus, Equation 5 shows that

$$100{,}000, \text{psi}\sqrt{\text{in.}} = 1.12 \left[\frac{200, \text{psi/in.} \ (a, \text{in.})}{12, \text{in.}} \right] \sqrt{\pi a, \text{in.}} \tag{18.6}$$

or

$$a \simeq 210, \text{in.}$$

This large critical crack length, which is based on the conservative assumption that the applied stress at the tip of the crack was uniform up to the top of the cofferdam, suggests that either the critical stress intensity factor for the material was lower than 100 ksi$\sqrt{\text{in.}}$ or that the driving force to propagate the crack was larger than the applied circumferential stress from the fill, or both.

The removal of a sheet pile by flame cutting during trimming to elevation requires cutting into the interlocking region of the adjacent sheet pile. The crack in Sheet 55 originated at the tip of a 3.5-in.-long sharp notch that was formed when the adjacent sheet was flame cut and removed and the trimming operation was interrupted. Thus, let us consider the assumptions that the surface of the notch where the crack initiated was severely damaged by the flame-cutting process and, therefore, possessed the least possible fracture toughness for steels of 25 ksi$\sqrt{\text{in.}}$ and that this value determined the size of the critical crack that propagated under the influence of the applied stress from the fill. Repeating the preceding calculations with K_{Ic} = 25 ksi$\sqrt{\text{in.}}$ results in a critical crack size of 84 in. These calculations demonstrate that if a small crack initiated in the very brittle flame-cut surfaces, it would have arrested because the circumferential stress from the fill alone was not sufficient to extend it further. Therefore, assuming the least possible fracture toughness for the steel, an external force was required to initiate and propagate a large crack to a critical depth, at which point the circumferential stress from the fill alone would have been sufficient to extend the crack to the bottom of the sheet.

An inspection of the cofferdam revealed several vertical cracks starting at the top of sheet piles and extending about 3 ft in length, Figure 18.8. The most probable cause for these cracks was the front-end loader pushing on the top of the cell wall. Such a force would decrease rapidly with depth. Thus, a crack initiated by the front-end loader at the top of the cell would arrest unless the applied force was large enough to propagate it to a large depth at which point the circumferential stress from the fill would propagate it further. The following presents an analysis of the combined effects of an external force and the circumferential stress from the fill on the fracture behavior of Sheet 55 and determines the magnitude of the force necessary to produce a critical crack that would propagate unstably.

The stress intensity factor expression for an edge crack at the top of the sheet pile subjected to a uniform circumferential stress and to a force at the top of the cell wall, Figure 18.9, can be represented by the equation

FIG. 18.8 Nonpropagating crack in sheet pile.

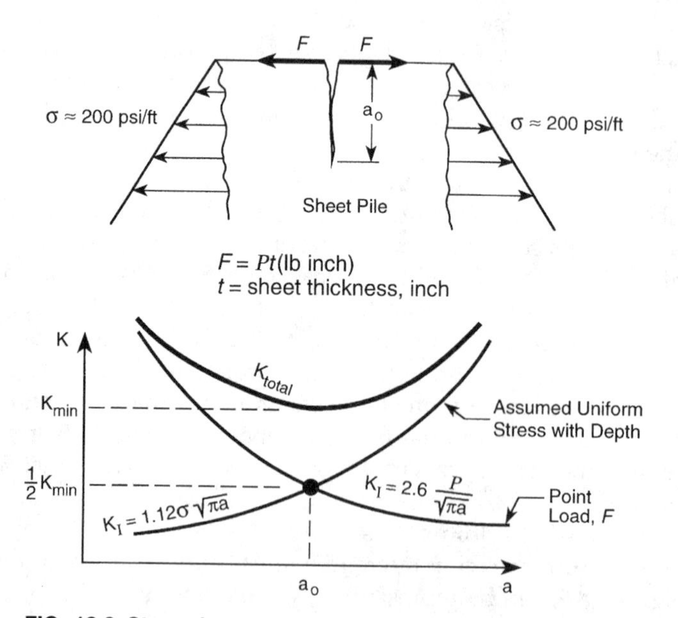

FIG. 18.9 Stress intensity factor analysis for Sheet Pile 55.

$$K_I = 1.12\,\sigma\sqrt{\pi a} + 2.6\,\frac{P}{\sqrt{\pi a}} \tag{18.7}$$

where $P = \frac{F}{t}$, lb/in.

F = force, lb

t = sheet-pile thickness = 1/2 in.

The first term in Equation 18.7 is for an edge crack subjected to a uniform stress, while the second term is for an edge crack subjected to an opening force at its mouth [1]. When the crack is short, the value of K_I is governed primarily by the second term, whereas, when the crack is long, it is governed primarily by the first term. Thus, a plot of stress intensity factor, K_I, versus crack length, a (Figure 18.9), exhibits a minimum value for K_I, K_{min}. This minimum value corresponds to a minimum crack length, a_0, that must exist beyond which the circumferential fill stress becomes the dominant driving force. K_{min} is located at a point on the K_I-a curve where the slope, dK_I/da, is zero. Thus,

$$\frac{dK_I}{da} = \left(\frac{1}{2}\right) 1.12\,\sigma\sqrt{\pi}a^{-1/2} + \frac{2.6P}{\sqrt{\pi}}\left[-\frac{1}{2}\frac{1}{a^{3/2}}\right] \tag{18.8}$$

$$\frac{dK_I}{da} = \frac{\sigma}{\sqrt{a}} - 0.74\,\frac{P}{a^{3/2}} \tag{18.9}$$

Because $\frac{dK_I}{da} = 0$ at a_0, then

$$\frac{\sigma}{\sqrt{a_0}} - 0.74\,\frac{P}{a_0^{3/2}} = 0 \tag{18.10}$$

or

$$a_0 = \frac{0.74\,P}{\sigma} \tag{18.11}$$

Substituting this expression for a_0 in Equation 18.7, we find K_{min} to be

$$K_{min} = 1.12\sigma\sqrt{\pi}\sqrt{\frac{0.74P}{\sigma}} + \frac{2.6}{\sqrt{\pi}}\frac{P}{\sqrt{\frac{0.74P}{\sigma}}} \tag{18.12}$$

$$K_{min} = 3.41\sqrt{\sigma P} \tag{18.13}$$

We already demonstrated that when $K_{Ic} = K_{Id} = 25$ ksi$\sqrt{\text{in.}}$—the lowest fracture toughness for steels—a_0 was 84 in. The minimum K_I value in Figure 18.9, which is assumed to be equal to 25 ksi $\sqrt{\text{in.}}$—the lowest fracture toughness for steels, is the sum of two equal parts from each of the terms in Equation 8. One half the 25 ksi$\sqrt{\text{in.}}$ minimum value is contributed by the circumferential fill stress and the other by the applied force at the top of the cell wall. By using the 12.5

ksi$\sqrt{\text{in.}}$ contribution by the fill stress and the K_I expression for an edge crack, the value of a_0 is determined to be about 48 in. Substituting this value for a_0 in Equation 15, we find that

$$P \simeq 64\sigma \tag{18.14}$$

Substituting this expression for P in Equation 15 we obtain

$$K_{min} = 27\sigma \tag{18.15}$$

For

$$K_{min} = 25 \text{ ksi}\sqrt{\text{in.}}$$

$$\sigma \geq 926 \text{ psi for } a_0 = 48 \text{ in.} \tag{18.16}$$

substituting Equation 18.16 in Equation 18.14 we obtain

$$P \geq 59{,}300 \text{ psi} \tag{18.17}$$

because

$$P = \frac{F}{t} \quad \text{and} \quad t = \frac{1}{2} \text{ in.}$$

Then

$$F \geq 29{,}650 \text{ lbs} \tag{18.18}$$

which is the minimum circumferential force required to initiate and propagate a crack from the top of a cell to a depth a_0 even when the fracture toughness of the steel is assumed to be 25 ksi $\sqrt{\text{in.}}$

The magnitude of the radial force, F_{FL}, applied by the front-end loader against a sheet pile to produce a circumferential force equal to or larger than 29,650 lb can be determined by considering the static equilibrium among these forces. The width of a sheet pile along the circumference of the cell was 15 in., and the radius of a cofferdam cell was 35.5 ft. A radial force applied to the sheet would produce forces at the side of the sheet that must all be in static equilibrium. These considerations, Figure 18.10, show that F_{FL} had to be equal to or larger than about 1200 lb.

18.5 Summary

The preceding analyses demonstrated that even when we assumed the least possible fracture toughness for steels, the circumferential stress from the fill alone would have tolerated a split sheet pile having an 84-in-long crack without causing failure of the cell. Consequently, an externally applied force was necessary to initiate and propagate a long crack from the top of the cell. The only possible source for such a force was the front-end loader pushing sand overboard at the time of the fracture. When the front-end loader applied a radial force less than about 1200 lb, a crack would initiate, propagate, and arrest. Several such cracks were observed in various cells, Figure 18.8. A radial force larger than about 1200 lb caused the crack in Sheet Pile 55 to initiate and propagate to a critical dimension at which length the fill stresses extended it to the bottom of the sheet.

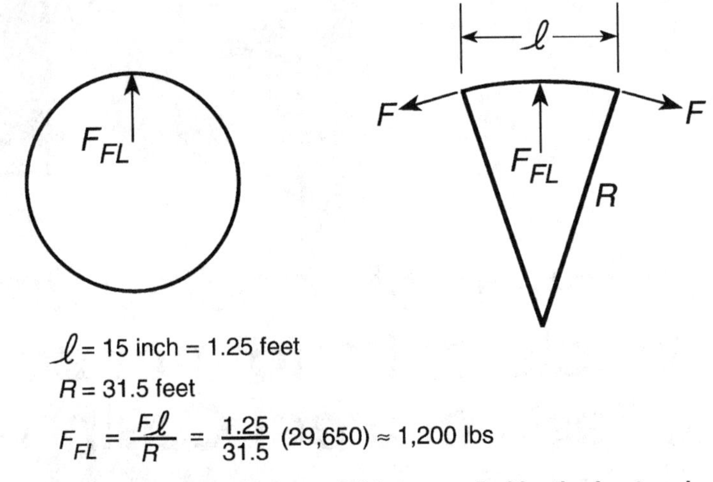

FIG. 18.10 Analysis of the radial force applied by the front-end loader on Sheet Pile 55.

This cofferdam failure analysis presents an example where the design is conservative and where fracture toughness is not a significant parameter for the performance of the structure. However, the fabrication and the construction practices caused a fracture that could have been easily avoided.

18.6 References

[1] Tada, H., Paris, P., and Irwin, G., *The Stress Analysis of Cracks Handbook*, 2nd ed., 1985, p. 2.25.

19

Importance of Loading Rate on Structural Performance—Burst Tests of Steel Casings

19.1 Introduction

THIS CHAPTER PRESENTS chemical and mechanical properties of two J-55 casings selected to have significantly different yield strengths. The chemical composition and tensile properties of the tested casings met the specification requirements for both J-55 and K-55 casing grades. Consequently, the results of the investigation are equally applicable to both grades. Also presented are the results of burst tests that were conducted at Battelle [*1*] on short sections of these casings containing milled notches and the results of a failure analysis of the tested sections conducted at the U.S. Steel Technical Center. Finally, the significance of the results to API Charpy V-notch requirements for J-55 and K-55 casings are discussed.

19.2 Material and Experimental Procedures

Material. Burst tests were conducted on two J-55 casings having 10-in. nominal diameter and 0.450-in. wall thickness. The chemical composition for each casing is presented in Table 19.1.

One casing had a 77.3-ksi yield strength, which is well above the minimum value of 55 ksi required by API 5CT Specification, while the other had a yield strength of 55.5 ksi, which is essentially equal to the required minimum value. The tensile properties for these casings are presented in Table 19.2.

The chemical composition and tensile properties of the tested casings met the specification requirements for both J-55 and K-55 casing grades. Consequently, the results of the present investigation are equally applicable to both grades.

Transverse 2/3-size Charpy V-notch specimens were machined from each casing and tested at temperatures between −50° and +300°F. Figures 19.1 and

TABLE 19.1 Chemical Composition.

C	*Mn*	*P*	*S*	*Si*	*Cu*	*Ni*	*Cr*	*Mo*	*Al*
				Low Strength					
0.38	1.34	0.021	0.007	0.25	0.010	0.031	0.065	0.105	0.022
				High Strength					
0.49	1.14	0.010	0.016	0.065	0.008	0.010	0.024	0.021	0.018

19.2 present impact Charpy V-notch absorbed energy and shear (fibrous fracture) curves for the low-strength and high-strength casings, respectively.

19.3 Experimental Procedure

Specimen Preparation. One section from each strength casing was V-notched longitudinally by using an end mill. Each section contained three notches that were 12.5% of the wall thickness in depth and had a length-to-depth ratio of 10. The geometry of the notches is shown in Figure 19.3.

Each section was instrumented with a clip gage over each notch, direct current electric potential (DCEP) probes at each notch, thermocouples at two or more locations to monitor surface temperature, and pressure transducer to monitor internal pressure. The DCEP probes were tack welded to the casing without precautions, which is not consistent with recommended procedures.

The Drilling Manual [2] of the International Association of Drilling Contractors and the API Recommended Practice for Care and Use of Casing and Tubing [3] caution that "unless precautions were taken, welding may have adverse effects on many of the steels used in all grades of casing, especially J-55 and higher [underline added]. . . . Cracks and brittle areas are likely to develop in the heat-affected zone. Hard areas of cracks can cause failure, especially when the casing is subjected to tool-joint battering. For these reasons, welding on high-strength casing should be avoided if possible" [underline added] [2,3].

Burst Tests. Each casing was tested at 0°F where the impact Charpy V-notch absorbed energy was 2 and 7 ft-lb for the low-strength and high-strength casings, respectively. This test temperature is well below the minimum operating temperature for casings and was selected to ensure conservative test results. The casings were cooled to 0°F by using circulating methanol cooled with dry ice in a supply drum. The temperature of the casing was stabilized for 30 min prior to testing.

A. *Low Yield Strength Casing.* The casing was pressurized to 4250 psig at 0°F, which was calculated by using thin-wall Barlow formula to result in circumfer-

TABLE 19.2 Tensile Properties.

YIELD STRENGTH, ksi	TENSILE STRENGTH, ksi	ELONGATION, %	REDUCTION OF AREA, %
55.5	101.3	24	53
77.3	109.1	27	56

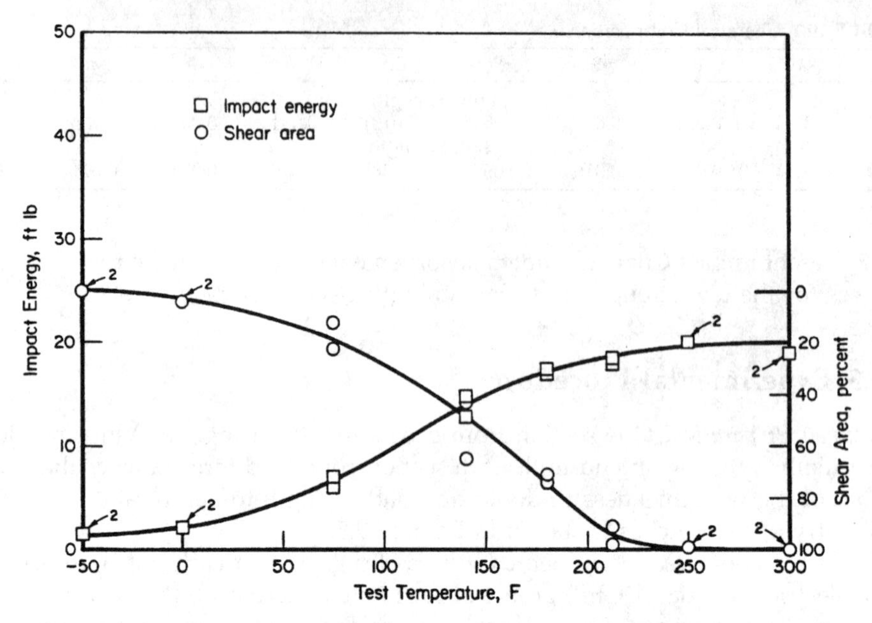

FIG. 19.1 Charpy V-notch impact test results for a low-strength J-55 casing.

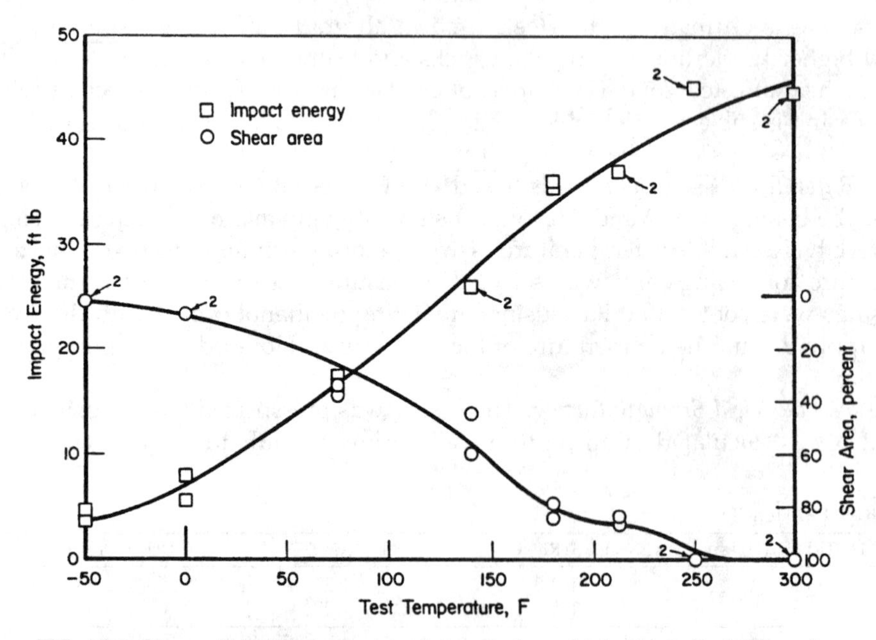

FIG. 19.2 Charpy V-notch impact test results for a high-strength J-55 casing.

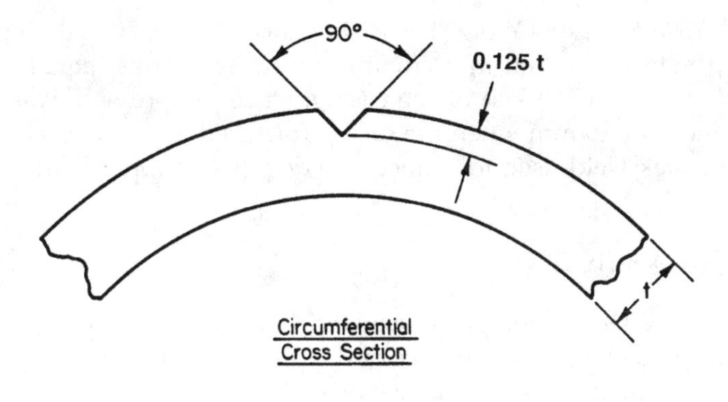

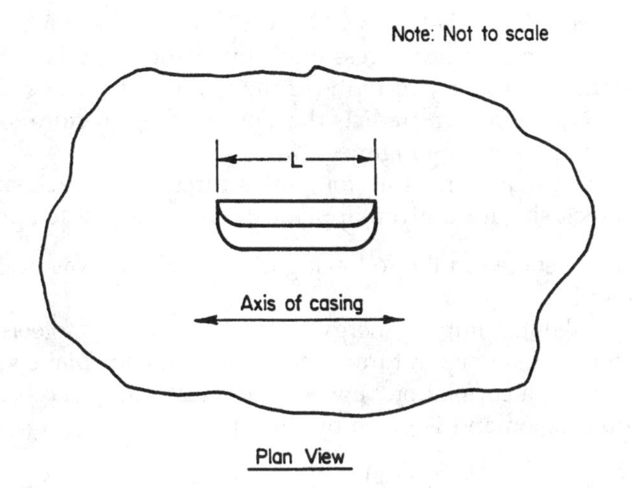

FIG. 19.3 Flaw geometry.

ential wall stress equal to 100% the actual yield strength of the tested section. Once the pressure reached 4250 psig, it was maintained for 1 h without any indication from the instrumentation that the notches extended in depth or length.

The casing was depressurized, one of the notches was lengthened to a length-to-depth ratio of 30, the casing was cooled to 0°F, then repressurized to 4250 psig and held at pressure and temperature for 1 h without failure or any indication of notch extension. Finally, the casing was depressurized, the DCEP probes were removed, the tack welds were ground off, and the surface was polished smooth "to remove any surface anomalies" [*1*]; then the casing was cooled to 0°F and pressurized to failure. The pressure at failure was 5900 psig. The fracture did not originate from the notches. It originated in the tack weld region.

B. *High Yield Strength Casing.* The casing was cooled to 0°F, then pressurized to 6500 psig, which resulted in a circumferential wall stress equal to 100% the actual yield strength (77.3 ksi) of the casing. Once the pressure was reached, it was maintained for 15 min when the casing fractured. The fracture originated at the toe of the tack weld used to connect the dc power supply to the casing wall.

19.4 Failure Analysis

Predicted Behavior. The tensile properties and Charpy V-notch energy absorption were used to predict the performance and behavior of the pressurized casing at 0°F. This prediction is based on the use of:

1. An empirical correlation between Charpy V-notch impact energy absorption and impact (dynamic) critical stress intensity factor, K_{Id}, where K_{Id} is the fracture toughness property in terms of the applied stress and the flaw size.
2. An empirical equation that predicts the slow loading fracture toughness K_{Ic} from the dynamic fracture toughness, K_{Id}.
3. An equation that related fracture toughness, stress, and crack size to predict the critical crack size for a given applied stress at the test temperature.

This procedure is detailed in the following discussion and was presented earlier in Chapters 4 and 5.

The Charpy V-notch impact energy absorption curve for steels undergoes a transition in the same temperature zone as the impact plane-strain fracture toughness, K_{Id}. Thus, a correlation between these test results has been developed for the transition region and is given by the equation:

$$\frac{(K_{Id})^2}{E} = 5\ (\text{CVN}) \tag{19.1}$$

where K_{Id} is in psi$\sqrt{\text{in.}}$, E is in psi, and CVN is in ft-lb. The validity of this correlation is demonstrated by the data presented in Figure 19.4 [4,5] for various grades of steel ranging in yield strength from about 36 to about 140 ksi and in Figure 19.5 [4,5] for eight heats of SA 533B, Class 1 steel. Consequently, a given value of CVN impact energy absorption corresponds to a given K_{Id} value (Equation 19.1). A change in the rate of loading causes the fracture toughness transition curve to shift its location along the temperature axis, Figure 19.6 [4]. The higher the rate of loading, the higher the transition temperature. In other words, the fracture toughness transition for slow loading occurs at lower temperatures than for impact loading. Furthermore, the slow loading fracture toughness behavior can be predicted from the impact (dynamic) fracture toughness behavior by shifting the impact curve to lower temperature using the date presented in Figure 19.7 [4]. These data show that the shift between static and impact plane-strain fracture toughness curves is given by the relationship [4]:

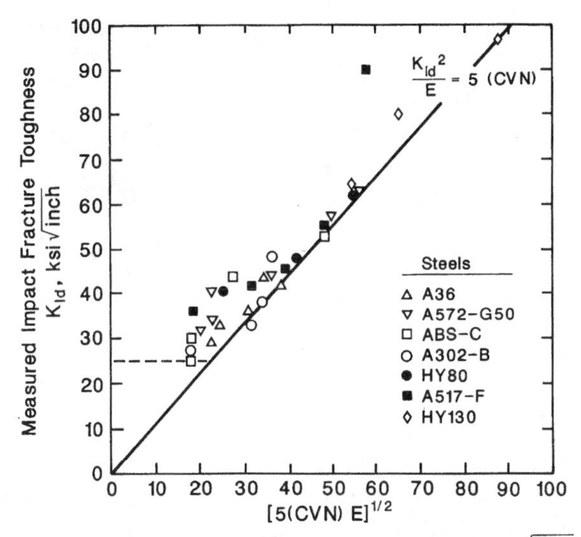

FIG. 19.4 Correlation of plane-strain impact fracture toughness and impact Charpy V-notch energy absorption for various steel grades.

$$T_{shift} = 215 - 1.5\sigma_{ys} \tag{19.2}$$

for

$$36 \text{ ksi} \leq \sigma_{ys} \leq 140 \text{ ksi}$$

and

$$T_{shift} = 0$$

for

$$\sigma_{ys} \geq 140 \text{ ksi}$$

where T is temperature in °F and σ_{ys} is room-temperature yield strength in ksi.

The Charpy V-notch data in Figures 19.1 and 19.2 were used to predict the K_{Id} versus temperature behavior for the tested casings. Because the data in Figures 19.1 and 19.2 were obtained by testing 2/3-size specimens, the data were adjusted to predict the equivalent full-size Charpy V-notch energy absorption values by interpolating the API correction factors in Supplementary Requirements of SR16 Impact Testing (Charpy V-notch) For Pipe In Groups 1, 2, and 3. The corrected data were used with Equation 19.1 to predict the K_{Id} behavior for the casing as a function of temperature, Figures 19.8 and 19.9.

Equation (19.2) in combination with the yield strength data in Table 19.2 results in a temperature shift of 132 and 100°F for the low-yield strength and high-yield strength casings, respectively. These values were used to construct the K_{Ic} curves as a function of temperature in Figures 19.10 and 19.11.

Fracture mechanics technology shows that, for a given crack depth, a very long surface crack is more severe than one having a short surface length. In other

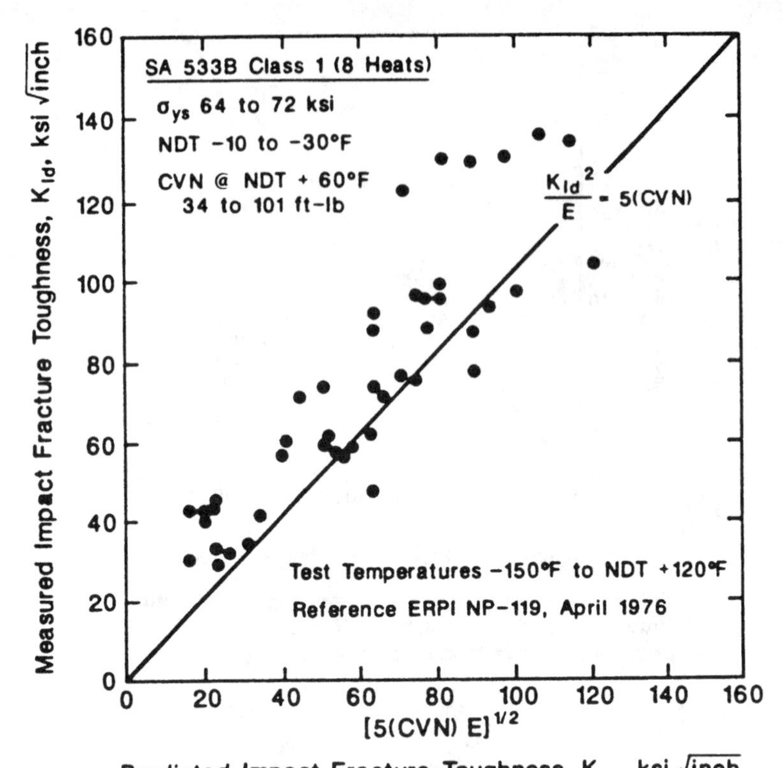

FIG. 19.5 Correlation of plane-strain impact fracture toughness and impact Charpy V-notch energy absorption for SA533B Class 1 steel.

words, if a casing can tolerate a crack 12.5% the wall thickness in depth and having a length-to-depth ratio larger than about 8, it can tolerate a deeper crack whose length-to-depth ratio is less than 8. Also, when the length-to-depth ratio exceeds about 8, the crack behaves as having an infinite surface length. Thus, the notches that were machined in the casing had a length-to-depth ratio of 10 to simulate the most severe notch geometry.

The fracture mechanics equation that relates the critical stress intensity factor, K_{Ic}, to stress, σ, and the depth, a, of a very sharp crack having infinite length is:

$$K_{Ic} = 1.12\sigma\sqrt{\pi a} \tag{19.3}$$

The K_{Ic} values at the test temperature of 0°F for the low-strength and the high-strength casings, Figures 19.10 and 19.11, are 51 and 60 ksi$\sqrt{\text{in.}}$, respectively. Substituting these values in Equation (19.3) and a flaw depth of 12.5%, the casing walls result in a calculated stress at failure that exceeds the actual yield strength of the casings. Thus, in the presence of an infinitely long, very sharp crack having

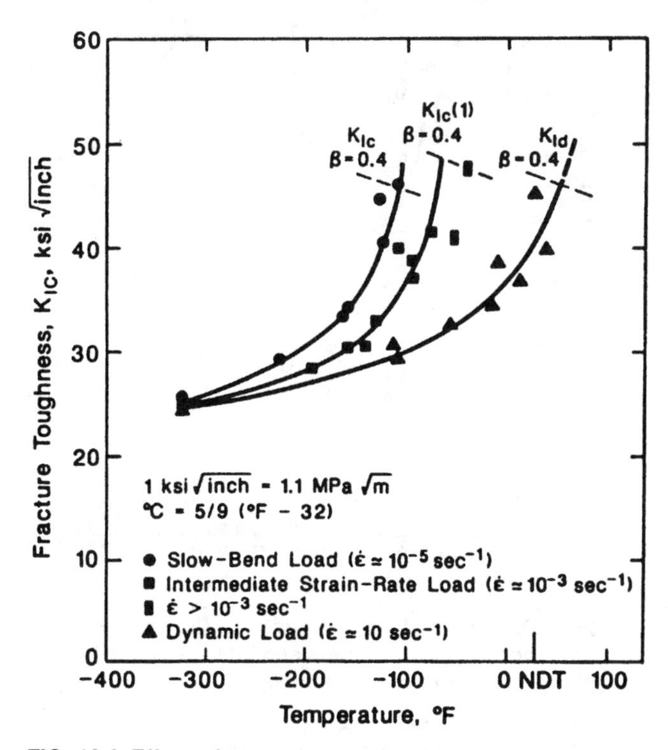

FIG. 19.6 Effect of temperature and strain rate on the plane-strain fracture-toughness behavior of ASTM A36 steel.

a depth equal to 12.5%, the wall thickness, the tested low strength, and high-strength J-55 casings would sustain a circumferential stress equal to their *actual* yield stress at 0°F without fracturing. Also, the tested casings would be expected to plastically deform and bulge prior to fracture. This behavior is predicted even when the impact Charpy V-notch energy absorption for the tested casings were 2 and 7 ft-lb at the test temperature of 0°F.

Equation (19.3) is based on linear elastic analysis and, therefore, is applicable when the applied stress, σ, is less than about 80% the actual yield strength of the material. For stresses higher than about 80% the actual yield strength of the material, Equation (19.3) gives conservative answers. Furthermore, the correlation between impact Charpy V-notch energy absorption and K_{Id} and the predicted K_{Ic} values are obtained by using the temperature shift are based on K_{Id} and K_{Ic} data for very deep cracks. Available data show that shallow cracks exhibit higher fracture toughness than deep cracks, Figure 19.12. Consequently, the casings should behave better than predicted in the preceding analysis. These predications and observations are fully supported by the metallographic analysis of the burst tests presented in the following section.

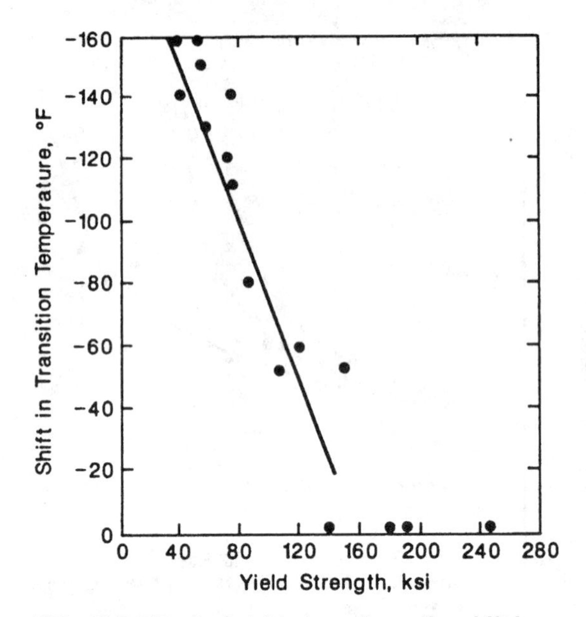

FIG. 19.7 Effect of yield strength on the shift in transition temperature between impact and static plane-strain fracture toughness.

19.5 Metallographic Analysis

General Discussion. The features of fracture surfaces for steels can be understood by examining the fracture-toughness transition curves under static and impact loading, Figure 19.13 [4]. These curves are separated from each other along the temperature axis by the temperature shift calculated by using Equation (19.2). The static fracture-toughness transition curve depicts the mode of crack initiation and the features of the fracture surface at the crack tip. The dynamic fracture-toughness transition curve depicts the mode of crack initiation under impact loading and the features of the crack propagation region under static or impact loading.

In Region I_s, for the static curve, Figure 19.13, the crack initiates in a cleavage mode from the tip of the fatigue crack. In Region II_s, the static fracture toughness to initiate unstable crack propagation increases with increasing temperature. This increase in crack-initiation toughness corresponds to an increase in the size of the plastic zone and in the zone of ductile tearing (shear) at the crack tip prior to unstable crack extension. In this region, the ductile-tearing zone is usually very small and difficult to delineate visually. In Region III_s, the static fracture toughness is quite high and somewhat difficult to define, but the fracture initiates by ductile tearing (shear).

Once a crack initiates under static load, the features (cleavage or shear) of the fracture surface for the propagating crack are determined by the dynamic

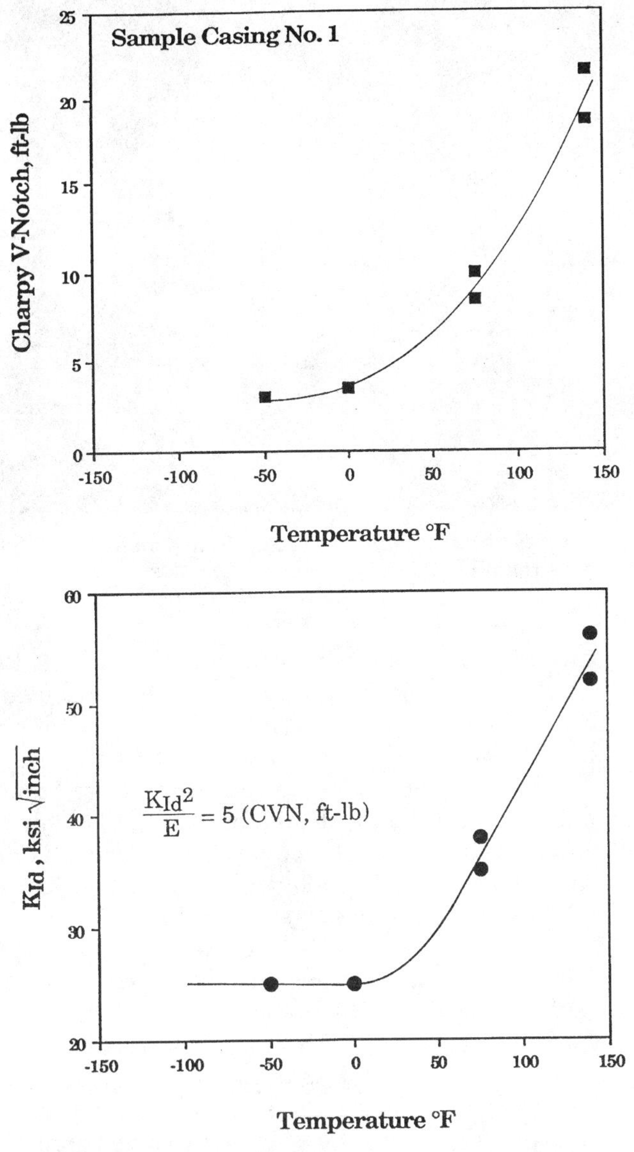

FIG. 19.8 Predicted dynamic fracture toughness from Charpy V-notch test results for a low-strength J-55 casing.

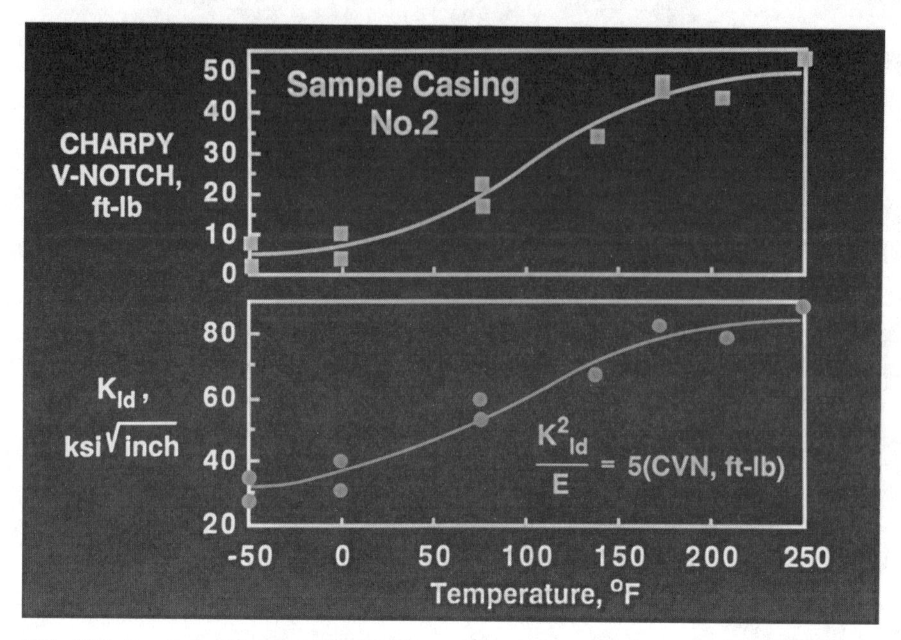

FIG. 19.9 Predicted dynamic fracture toughness from Charpy V-notch test. Test results for a high-strength J-55 casing.

behavior and degree of plane strain at the fracture temperature. Regions I_d, II_d, and III_d in Figure 19.13 correspond to cleavage, increasing ductile tearing (shear), and full-shear crack propagation, respectively. Thus, at Temperature A, the crack initiates and propagates in cleavage. At Temperature B and C, the crack exhibits ductile initiation but propagates in cleavage at Temperature B and in a mixed mode (cleavage plus ductile dimples) at Temperature C. The only difference between the crack initiation behaviors at Temperatures B and C is the size of the ductile-tearing zone, which is larger at Temperature C than at Temperature B. At Temperature D, cracks initiate and propagate in full shear.

Figure 19.14 [4] shows the fracture surfaces and fracture-toughness values (CVN and K_c) for an A572 Grade 50 steel having 50-ksi yield strength. The specimen tested at −42°F exhibited a small amount of shear initiation at a temperature slightly below B in Figure 19.13. The specimen tested at +38°F exhibited increasing shear initiation (between B and C). The specimen tested at +72°F exhibited full shear initiation (Temperature C) and despite the high fracture toughness (49 ft-lb and 445 ksi $\sqrt{\text{in.}}$) still exhibited a large region of mixed mode propagation. Thus, ductile crack propagation should only be expected at Temperature D, which is essentially dynamic upper-shelf Charpy V-notch impact behavior or in components having material properties, geometry, and loading conducive to plastic slip along shear planes.

Burst Tests. Figure 19.15 is a photograph of the burst tests showing location of the fracture origin and the general fracture path. It is clear from the photo-

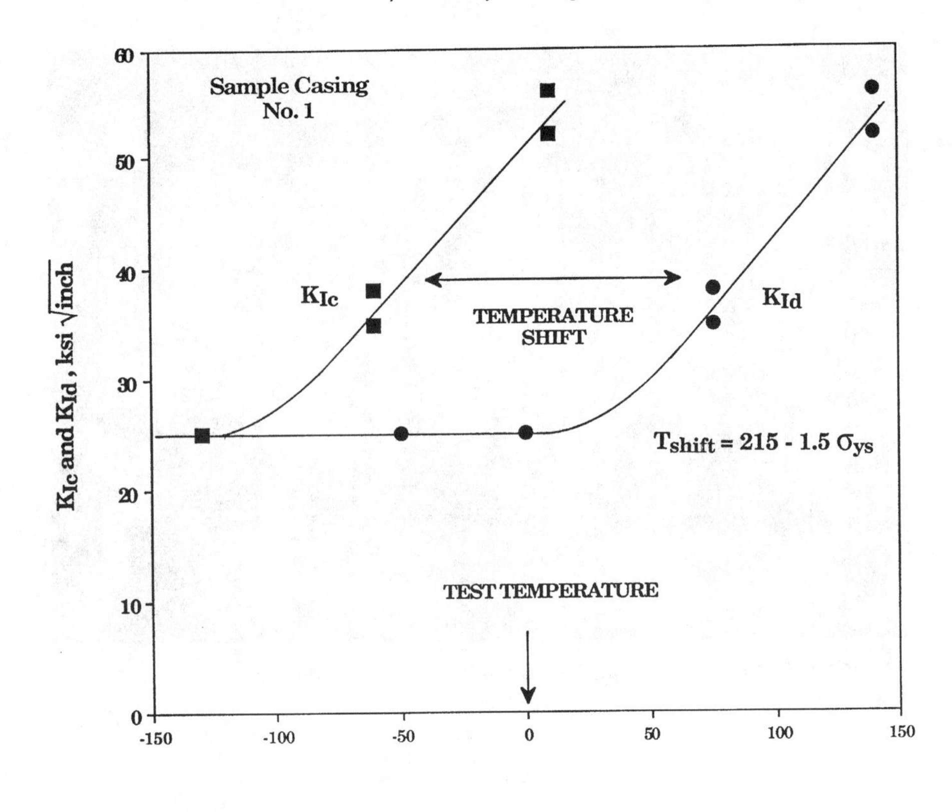

FIG. 19.10 Static and dynamic fracture toughness for a low-strength J-55 casing.

graph that both casings sustained plastic deformation and bulging prior to rupture. A closer look at the low-strength casing, Figure 19.16A, shows the fracture initiation to be in the vicinity of the tack weld that was removed by grinding rather than from one of the notches machined into the casing wall. Figures 19.16B, 19.16C, and 19.16D are increasing magnifications of the pre-existing crack at the initiation site. The crack depth was 16.5% of the wall thickness. Figure 19.16D shows the fracture extending from a pre-existing crack in a brittle manner. The scanning-electron fractograph, Figure 19.17, shows a pre-existing intergranular crack that initiated in a ductile manner as indicated by the ductile dimples at the crack front and extended by cleavage. A light micrograph of a transverse section through the pre-existing crack, Figure 19.18, shows the pre-existing crack residing in a very hard heat-affected zone of the tack weld. The heat-affected zone had a martensitic microstructure with a hardness of 49.3 HRC, whereas the deformed base metal exhibited about 97 HRB hardness.

Figures 19.19 presents a closer look at the fracture origin of the high-yield-strength casing. Figures 19.19A and 19.19B show the tack-welded attachment in place. Figures 19.19B, 19.19C, and 19.19D show the crack initiating in the tack

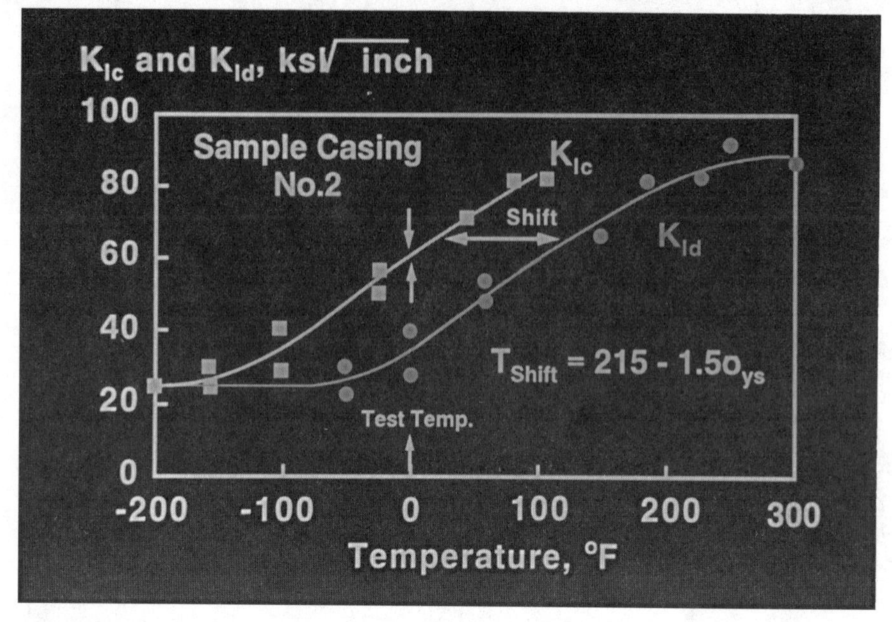

FIG. 19.11 Static and dynamic fracture toughness for a high-strength J-55 casing.

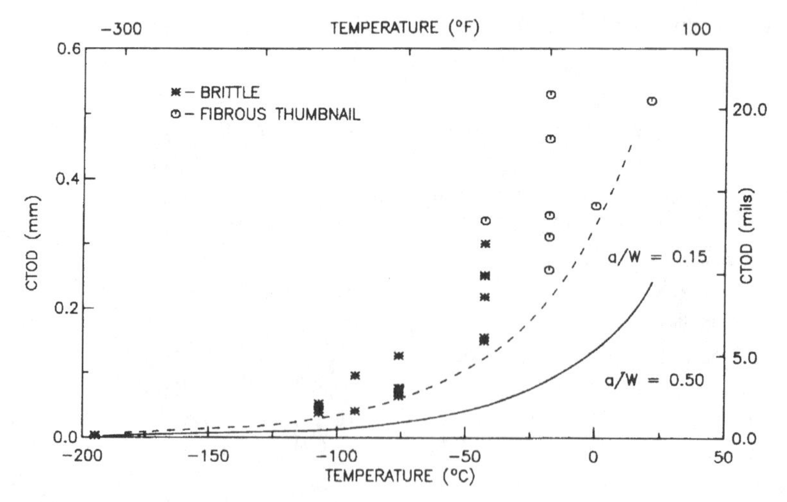

FIG. 19.12 Effect of crack size on fracture toughness of an A36 steel plate.

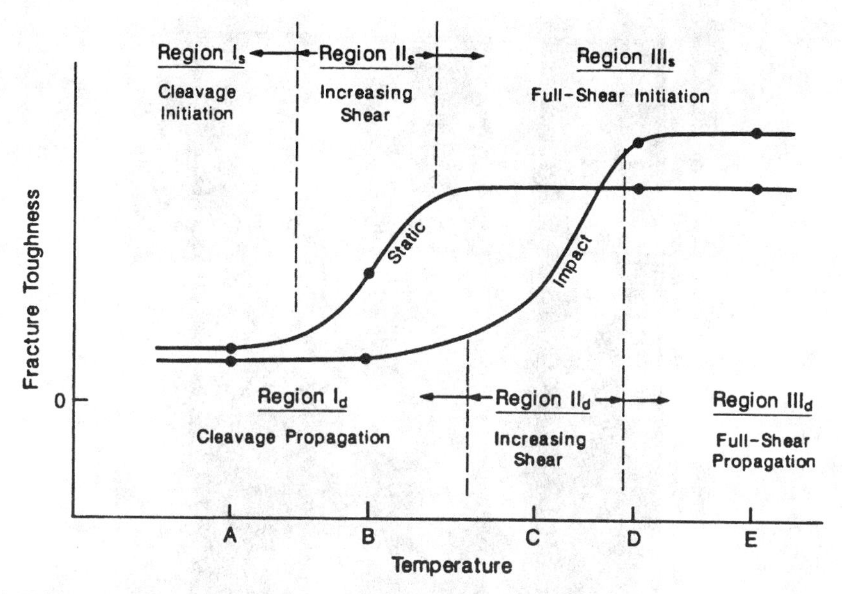

FIG. 19.13 Fracture-toughness transition behavior of steel under static and impact loading.

weld from a pre-existing crack and extended into the casing wall. Figure 19.19D shows the terminal fracture extending from multiple local initiating sites along the front of the pre-existing crack. At high magnification, Figure 19.20, the scanning electron fractographs show an intergranular pre-existing crack from which the final fracture initiated in a ductile manner and propagated in a mixed mode with the cleavage mode predominant. Light micrographs of a transverse cross section, Figure 19.21, show the presence of multiple pre-existing cracks associated with the tack weld. The cracks are very sharp and reside in a hard martensitic heat-affected zone with a maximum hardness of 54.4 HRC, whereas the deformed base metal had a hardness of 97 HRB.

The crack depth as measured from the outside surface of the casing was about 24% of the wall thickness. The effective crack depth, considering the tack weld above the casing surface, is significantly deeper and the severity of the crack was increased by the stress concentration caused by the geometry and proximity of the welded attachment, Figure 19.19.

The preceding metallographic analysis supports the analytical predictions. Also, the burst tests at 0°F where the fracture toughness was 2 and 7 ft-lb for the low-strength and the high-strength casings, respectively, show that the casings sustained stresses in excess of their actual yield stress and burst after permanent deformation and bulging in the presence of very sharp cracks deeper than the maximum 12.5% of the wall thickness permitted by the API specifications.

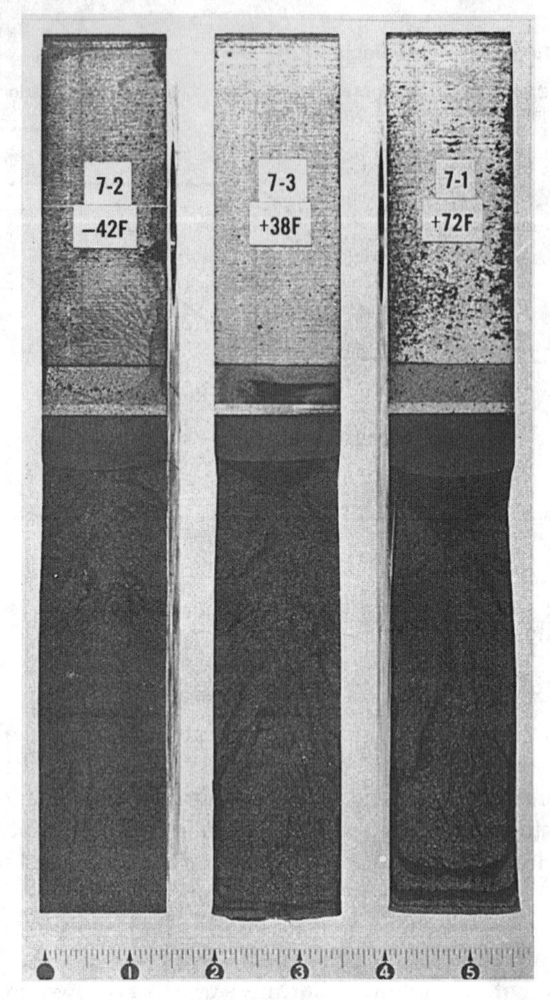

FIG. 19.14 Fracture surfaces of 1.5-in.-thick 4-T compact-tension specimens of an A572 Grade 50 steel plate.

19.6 Examination of API Specifications for J-55 and K-55 Casing

API specifications permit the presence of surface imperfections equal to 12.5% of the wall thickness for J-55 and K-55 casing. Also, at the time the burst tests were being considered, API was considering an additional requirement of 15 ft-lb Charpy V-notch impact energy absorption (full-size specimen) at 70°F. The burst-tested low-strength and high-strength casings with their respective Charpy V-notch values indicate that this Charpy V-notch energy requirement is very conservative. The following discusses the implications of the API fracture toughness requirements in terms of structural performance for J-55 and K-55 casings.

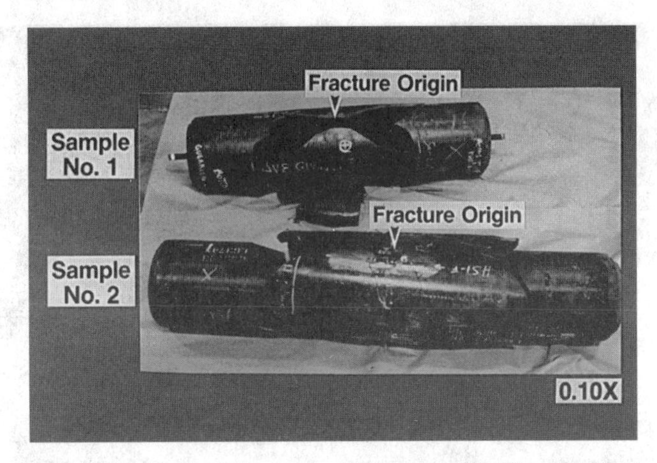

FIG. 19.15 Hydrostatic tests of low-strength (Sample 1) and high-strength (Sample 2) J-55 casings.

Equation (19.1) shows that a 15 ft-lb Charpy V-notch impact energy absorption at 70°F corresponds to an impact (dynamic) critical stress-intensity-factor, K_{Id}, value of 46.6 ksi $\sqrt{\text{in.}}$ at 70°F. For a 70-ksi yield strength material, this K_{Id} value corresponds to a static (slow) critical stress-intensity-factor, K_{Ic}, value of

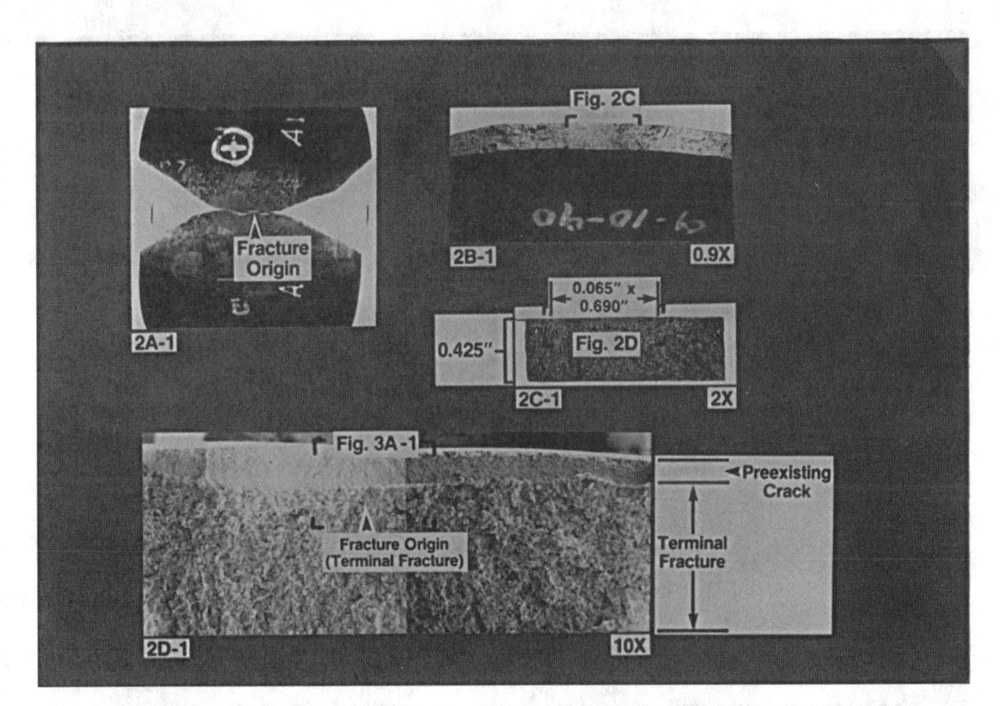

FIG. 19.16 Photographs of the fracture origin for low-strength J-55 casing.

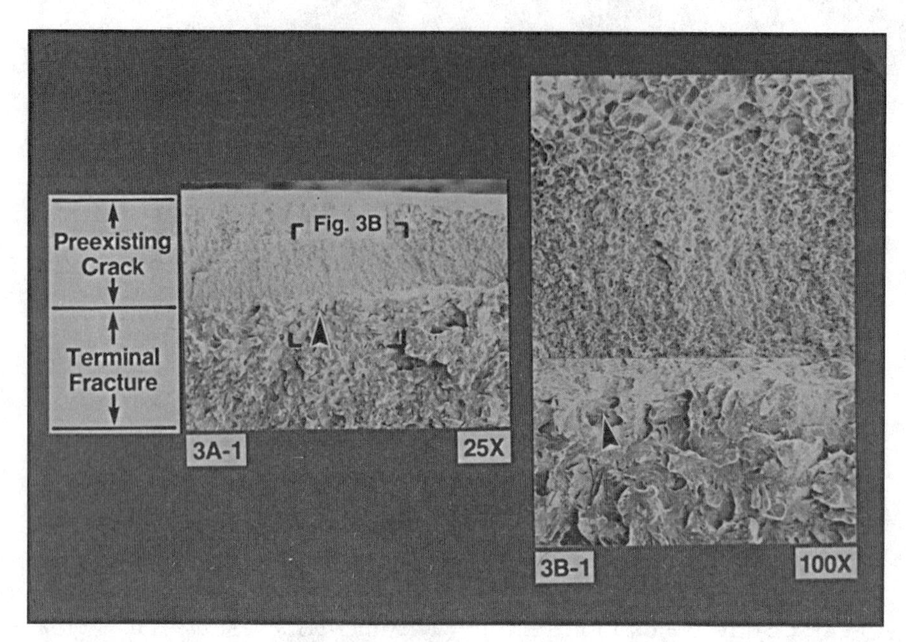

FIG. 19.17 Scanning electron fractograph showing an intergranular pre-existing crack, dimple fracture at the crack front, and cleavage unstable crack propagation.

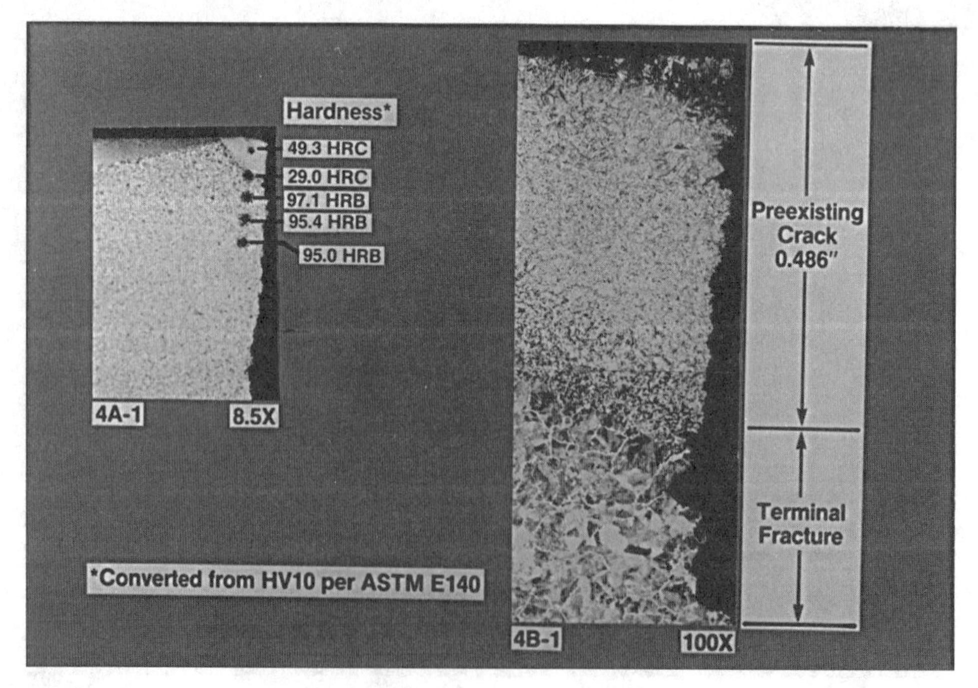

FIG. 19.18 Light micrographs of a transverse cross section through the pre-existing crack in the heat-affected zone of the low-strength casing.

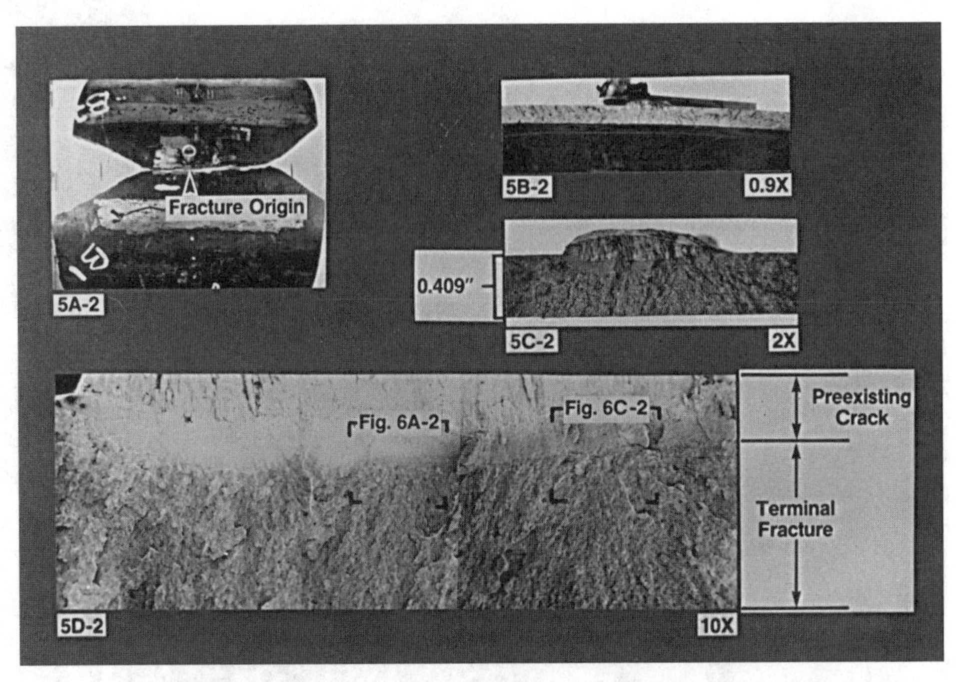

FIG. 19.19 Photographs of the fracture origin for the high-strength J-55 casing.

FIG. 19.20 Scanning electron fractographs for the pre-existing crack front in the high-strength casing showing an intergranular pre-existing crack, ductile fracture initiation, and cleavage unstable crack propagation.

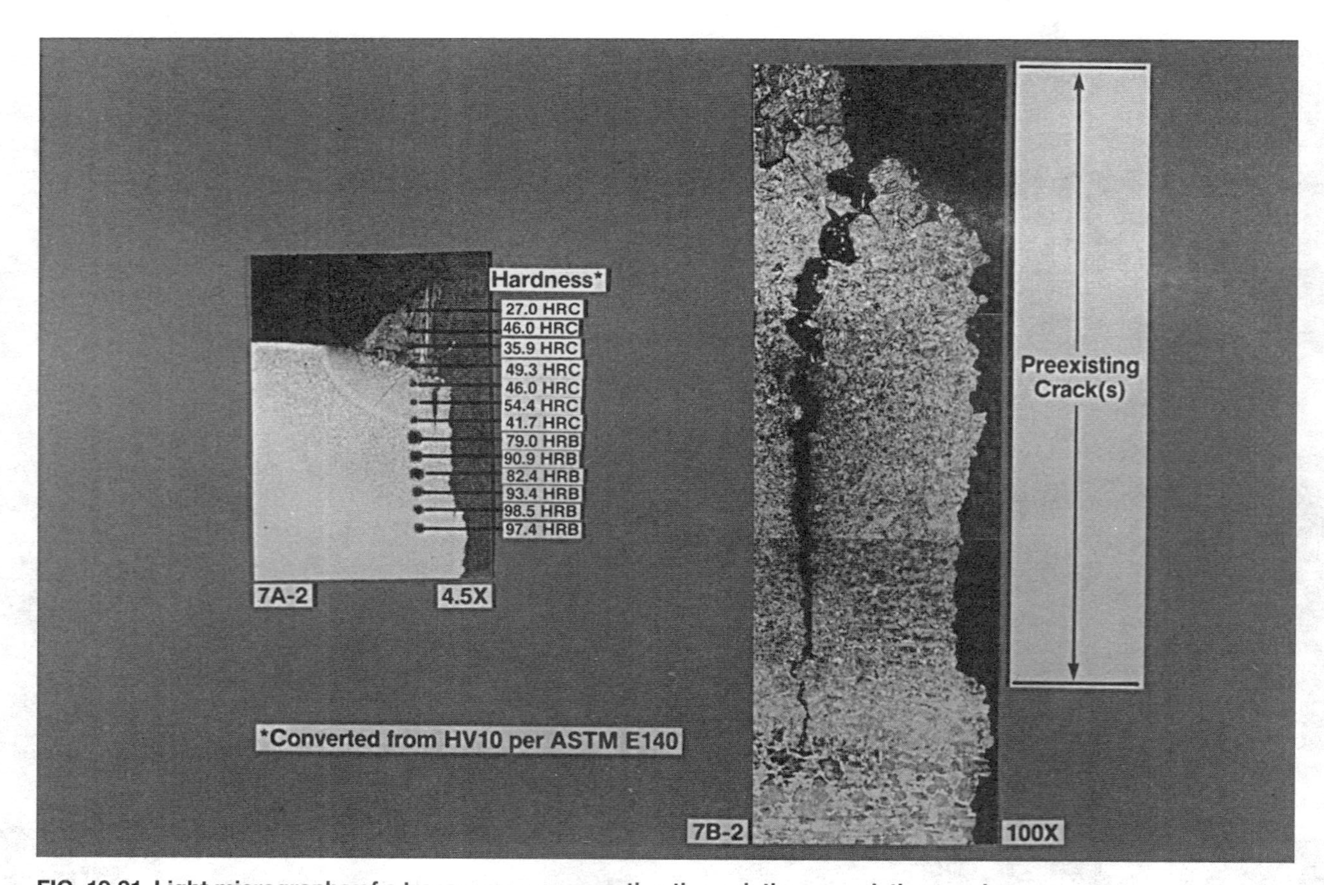

FIG. 19.21 Light micrographs of a transverse cross section through the pre-existing crack in the heat-affected zone of the high-strength casing.

46.6 ksi $\sqrt{\text{in.}}$ at $-40°$ (Equation 19.2). Assuming that the casing is subjected to an applied stress equal to its actual yield strength of 70 ksi at $-40°$, Equation (19.3) indicates that such a casing would tolerate an infinitely long, sharp crack having a critical depth, *a*, of 0.11 in. However, design stresses are limited by the minimum specified yield strength of the casing grade. Thus, the stresses for J-55 and K-55 casing are restricted to values below the minimum specified yield strength of 55 ksi rather than the actual yield strength of a particular casing section. Consequently, substituting a K_{Ic} of 46.6 ksi $\sqrt{\text{in.}}$ and a stress, σ, value of 55 ksi in Equation (19.3) results in a critical crack depth, *a*, of 0.18 in.

The maximum wall thickness for J-55 and K-55 casing is 0.656 in. for a 16-in. outside diameter (OD) casing [6]. Thus, a J-55 or K-55 casing stressed to the minimum specified yield strength of 55 ksi and having 15 ft-lb in a full-size Charpy specimen would tolerate a critical crack depth 27% of the wall thickness of the thickest wall permitted by API. This value of 27% for the 16-in. OD casing would be tolerated at about $-40°$F for a J-55 or K-55 casing having 70-ksi actual yield strength and at about $-60°$F for one having 55-ksi actual yield strength.

The preceding analysis combined with burst test results and metallographic failure analysis indicated that the API Charpy V-notch requirements are overly conservative under normal operating conditions. Also, this failure analysis demonstrates the significant effect of loading rate on the performance of structural and equipment components made of steels and the usefulness of the empirical correlations between static and impact fracture toughness in failure analysis.

19.7 References

[1] Guerrieri, D. A., Jones, D. J., and Kiefner, J. F., "Phase II—The Effect of Toughness on Tubular Good Service Performance," Final Report to American Petroleum Institute, Battelle, Columbus, OH, November 1990.

[2] *Drilling Manual,* Section C1, Tenth Edition, International Association of Drilling Contractors, Dallas, 1982.

[3] *Recommended Practice for Care & Use of Casing and Tubing,* API Recommended Practice 5C1 (RP 5C1), Fifteenth Edition, American Petroleum Institute, Washington, DC, May 31, 1987.

[4] Barsom, J. M., "Fracture Mechanics—Fatigue and Fracture," *Metals Handbook—Desk Edition,* H. E. Boyer and T. L. Gall, Eds., American Society of Metals, Metals Park, OH, pp. 32.2–32.8, 1984.

[5] Barsom, J. M., "Material Considerations in Structural Steel Design," *Engineering Journal,* Vol. 24, No. 3, American Institute of Steel Construction, Chicago, IL, 1987.

[6] *API Specification for Casing and Tubing,* API Spec 5 CT (SPEC 5 CT), Third Edition, American Petroleum Institute, Washington, DC, December 1, 1990.

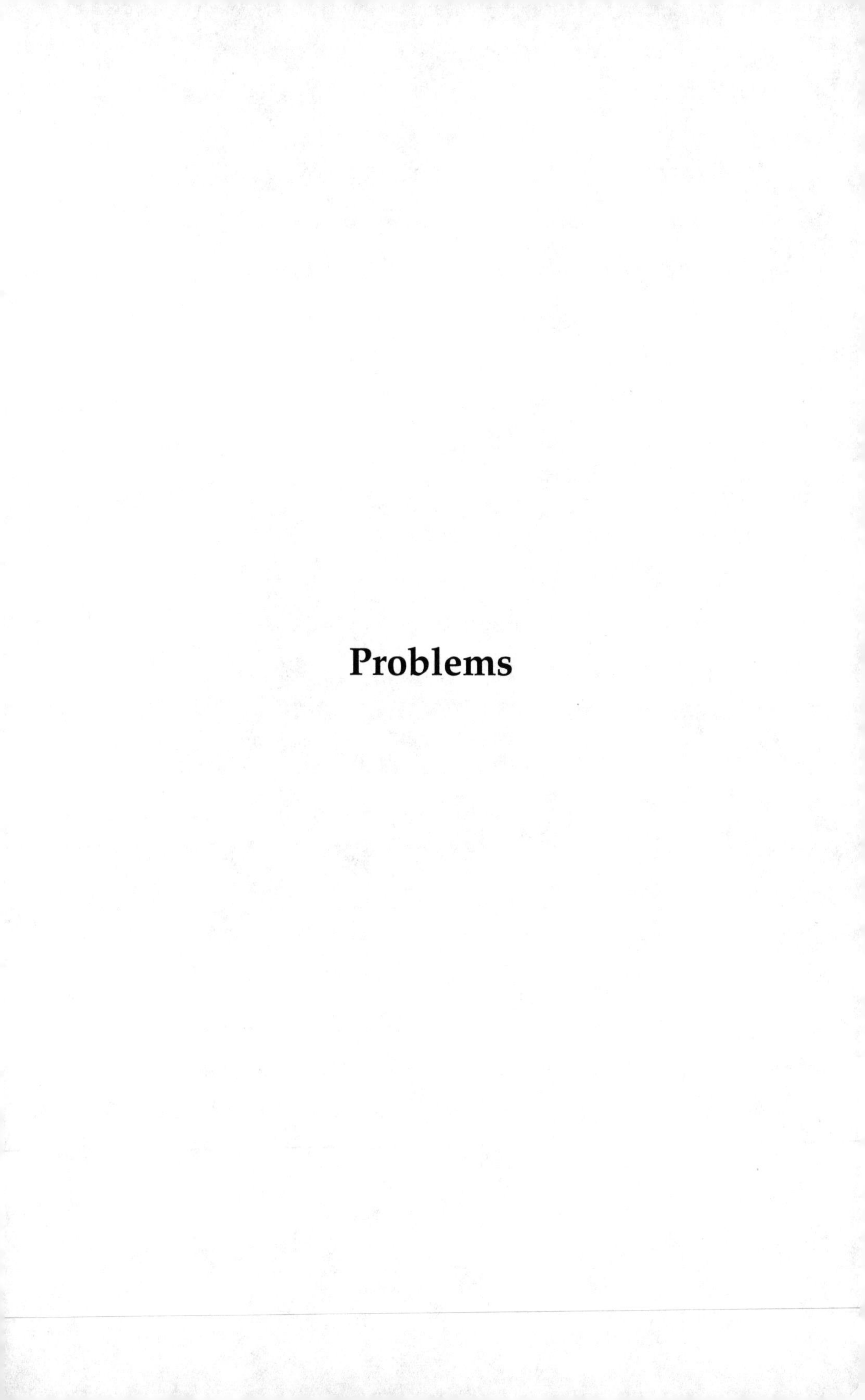

Problems

20

Problems

Part I

Problem 2.1

Given a structural material with a yield strength of 80 ksi and an ultimate strength of 100 ksi loaded to a K_I value of 60 ksi$\sqrt{\text{in.}}$, plot the stress distribution, σ_y, directly ahead of a crack ($\theta = 0$) between $r = 0$ and $r = 1.0$ in.

Problem 2.2

If the crack in Problem 2.1 is an edge crack 1.0 in. long in a 2-in.-thick infinite plate, what is the nominal stress at failure if $K_{IC} = 60$ ksi$\sqrt{\text{in.}}$?

Problem 2.3

Calculate a_{cr} for $\sigma = \sigma_{ys}$, $\sigma_{ys}/2$, and $\sigma_{ys}/4$ for an edge crack in a 4-in.-thick infinite plate in a material that has $\sigma_{ys} = 100$ ksi and $K_{IC} = 120$ ksi$\sqrt{\text{in.}}$

Problem 2.4

For an edge crack in a 30-in.-wide plate with $\sigma_{ys} = 40$ ksi and $K_{IC} = 100$ ksi$\sqrt{\text{in.}}$, calculate a_{cr} for $\sigma = \sigma_{ys}$, $\sigma_{ys}/2$, and $\sigma_{ys}/4$.

Problem 2.5

A steel bridge is being considered for the design of a pedestrian bridge over a busy street. The steel has a K_{IC} value of 80 ksi$\sqrt{\text{in.}}$ at 0°F and a yield strength of 100 ksi. Prepare a curve of allowable design stress versus crack depth for an edge notch in a 40-in.-wide plate loaded in tension.

Problem 2.6

Plot a stress—flaw-size—K_{IC} curve for an edge crack in a plate of infinite width for the following three steels (see table below). Plot all graphs on the same sheet of graph paper.

MATERIAL	K_{IC}, ksi$\sqrt{\text{in.}}$	σ_{ys}, ksi
A	80	260
B	110	220
C	140	180

Problem 2.7

To study the effect of crack geometry on a_{cr}, calculate and draw a_{cr} for the following conditions:

(a) Edge Crack—$K_I = 1.12\sigma\sqrt{\pi a}$

(b) Through-thickness crack—$K_I = \sigma\sqrt{\pi a}$

(c) Surface Crack—$K_I = 1.12\sigma\sqrt{\pi a/Q} \cdot M_K$ for three conditions $a/2c = 0.1$, 0.25, and 0.5.

Given that $K_{IC} = 55$ ksi$\sqrt{\text{in.}}$, $\sigma_{ys} = 100$ ksi, and $\sigma_{des} = \sigma_{ys}/4$, calculate a_{cr} and compare the results using half-scale drawings of the actual crack geometries. Assume a plate thickness of 2 in. and an infinite width.

Problem 2.8

A long 1-in.-thick steel plate loaded in tension is 8 in. wide and has an edge crack 2 in. deep. If the steel has a yield strength of 60 ksi and a K_C value of 200 ksi$\sqrt{\text{in.}}$, what load can the plate withstand before failure? What is the mode of failure? Explain your answer.

Problem 2.9

Using an edge-crack analysis for infinite-width plates, compare four steels that could be used in bridges in Alaska where service temperatures as low as −60°F will occur. The properties of the steels are as shown in the table below:

STEEL	MINIMUM YIELD STRENGTH, ksi	K_{IC} AT −60°F, (ksi $\sqrt{\text{in.}}$)
A36	36	60
A441	50	53
A572	50	100
A514	100	60

Prepare graphs of stress versus flaw size for each of these steels. If the design stress of each steel is 0.6 × the minimum yield strength, compare the critical crack sizes of each steel at the design stress level.

Problem 2.10

Given a longitudinal surface crack in a pressure vessel with the following dimensions and properties, what is the factor of safety against (1) yielding and (2) fracture, if the internal pressure is 3000 psi?

Length = 16 ft $\quad$ $2c = 4.0$ in.

Diameter = 4 ft $\quad$ $\sigma_{ys} = 100$ ksi

Thickness = 2 in. $E = 30 \times 10^6$ psi
$a_i = 1.0$ in. $K_{IC} = 120 \text{ ksi}\sqrt{\text{in.}}$

Problem 2.11

An infinite plate of A36 steel with the properties shown has a 0.4-in.-long crack propagating from a 2-in.-diameter hole. What stress level will cause failure?

$K_{IC} = 50 \text{ ksi}\sqrt{\text{in.}}$
$\sigma_{ys} = 40$ ksi
$E = 30 \times 10^6$ psi

Problem 2.12

An infinite plate of A36 steel with the following properties has cracks propagating from both sides of a 2.0-in.-diameter hole:

$K_{IC} = 50 \text{ ksi}\sqrt{\text{in.}}$
$\sigma_{ys} = 40$ ksi
$E = 30 \times 10^6$ psi

Determine the total crack length that will cause failure at a stress level of 20 ksi and show this crack in a sketch.

Problem 2.13

Given a steel with the following properties.

$E = 30 \times 10^6$ psi
RA = 60%
$\sigma_{ys} = 100$ ksi
$\sigma_T = 120$ ksi
$\sigma_{des} = 40$ ksi
$K_{IC} = 120 \text{ ksi}\sqrt{\text{in.}}$

Assume that this steel is used in an infinite plate that has a 2-in.-diameter hole in the center. Furthermore, assume that cracks are growing uniformly from both sides of the hole. What is the total defect length (crack plus hole) at failure?

Part II

Problem 3.1

Prepare neat curves of σ versus a for the following situations:

(a) A572 steel at −100°F tested dynamically, Figure 4.5.
(b) A572 steel at −100°F tested at an intermediate loading rate, Figure 4.5.
(c) A572 steel at −100°F tested slowly, Figure 4.5.

Plot three curves on the same sheet of paper for each of two crack geometries specified as follows:

1. An edge crack in an infinite plate.
2. A surface crack in an infinite plate with $a/2c = 0.25$.

Finally, compare the actual crack sizes (to scale) for each condition at a stress level of 20 ksi.

Problem 3.2

The following Charpy V-notch impact specimen results shown in the table below have been obtained for an ABS-B steel with a yield strength of 40 ksi:

TEMPERATURE, °F	ABSORBED ENERGY, ft-lb	LATERAL EXPANSION, mils	PERCENT SHEAR, %
−90	1.5	1	0
−90	1.5	2	0
−90	2	1	0
0	6.5	10	5
0	7.5	6	5
32	8.5	13	10
32	12.5	11	15
32	13	10	20
32	15.5	19	10
70	24	23	30
70	28	26	30
70	52	55	50
70	60	65	60
70	68	66	50
120	67	76	80
120	73	67	85
120	87	80	90

Plot the resultant curves on separate pages and sketch the levels of performance on each of the three curves:

(a) Absorbed energy versus temperature.
(b) Lateral expansion versus temperature.
(c) Percent shear versus temperature.

Problem 3.3

Given the following Charpy V-notch impact specimen results (see table below) previously presented in Problem 3.2 for an ABS-B steel, assume the steel has a static yield strength of 40 ksi and a dynamic yield strength of 65 ksi:

TEMPERATURE, °F	ABSORBED ENERGY, ft-lb
−90	1.5
−90	1.5
−90	2
0	6.5
0	7.5
32	8.5
32	12.5
32	13
32	15.5
70	24
70	28
70	52
70	60
70	68
120	67
120	73
120	87

Plot an appropriate curve of K_{ID} and K_{IC} versus temperature.

Problem 3.4

Determine the K_{ID}-K_{IC} temperature shift for:

(a) ABS-B steel $\sigma_{ys} = 40$ ksi
(b) A514 steel $\sigma_{ys} = 120$ ksi
(c) Grade 200 maraging steel $\sigma_{ys} = 220$ ksi
(d) 7076 Aluminum $\sigma_{ys} = 40$ ksi

Problem 3.5

A 1.0-in.-thick deck plate in a ship hull is fabricated from a steel with the following properties:

$E = 30 \times 10^6$ psi
$\sigma_{ys} = 50$ ksi
$K_{IC} = 60 \text{ ksi}\sqrt{\text{in.}}$
$K_{ID} = 40 \text{ ksi}\sqrt{\text{in.}}$
RA = 60%
$\sigma = 65$ ksi

(a) For a design stress of 30 ksi for both static and dynamic loading, what is the maximum crack size, $2a$, that the structure can tolerate before fracture?
(b) If a crack with a total length of 4 in. is discovered at sea, and the maximum stress at that time is 10 ksi, what is the factor of safety against fracture?

Problem 3.6

Determine the critical crack depth for design stress levels of 80, 60, 40, and 20 ksi for a semi-infinite plate with an edge crack. The steel plate has a K_{IC} of 90 ksi$\sqrt{\text{in.}}$ and a yield strength of 120 ksi. Tension test results indicate a percent reduction in area at fracture of 60%. Compare your answers graphically with the values for a steel with K_{IC} = 180 ksi$\sqrt{\text{in.}}$

Problem 3.7

An aluminum used in the aerospace industry has a K_{IC} of 27 ksi$\sqrt{\text{in.}}$ and a σ_{ys} of 70 ksi. Assume that σ_{ys} in tension and compression are equal. This material is to be used in two design situations. For each of these situations, determine the factor of safety against the possible modes of failure for each condition.

Condition I:

Long internally pressurized cylindrical vessel with hemispherical ends.

Pressure = 400 psi
Diameter of vessel = 6 ft
Thickness of vessel = 0.5 in.
Length of vessel = 30 ft
Surface flaw with depth of 0.25 in. and length of 10 in.
Assume crack is longitudinal in direction.

Condition II:

Solid column loaded in compression as shown in Fig. 20.1 (Problem 3.7).

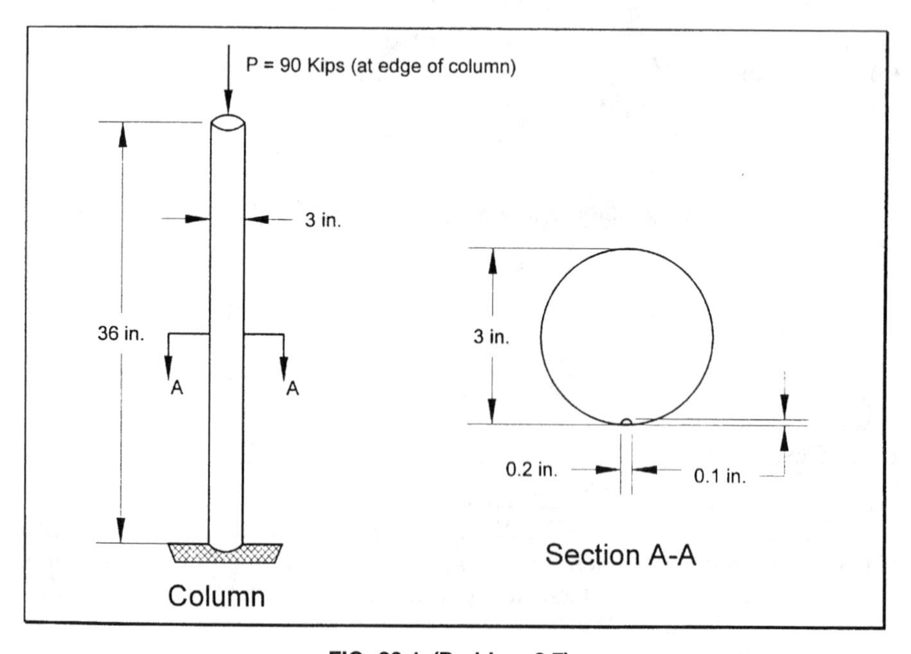

FIG. 20.1 (Problem 3.7).

Problem 3.8

Two steels, A and B, are being considered for use in a 0.4-in.-thick 30-in.-diameter pressure vessel that may have a surface flaw such that $K_I = 1.12\sigma\sqrt{\pi a/Q} \cdot M_K$ (assume $Q = 1.0$). We would like the vessel to be able to be pressurized at a reasonably high pressure while maintaining a factor of safety against yielding of 2. Assume that the fabrication quality and cost are the same for both steels.

Select a steel for this application and a recommended design stress level for that steel (see table below). Justify your answer.

STEEL	K_{IC}, ksi$\sqrt{\text{in.}}$	YIELD STRENGTH, ksi
A	200	150
B	300	200

Problem 3.9

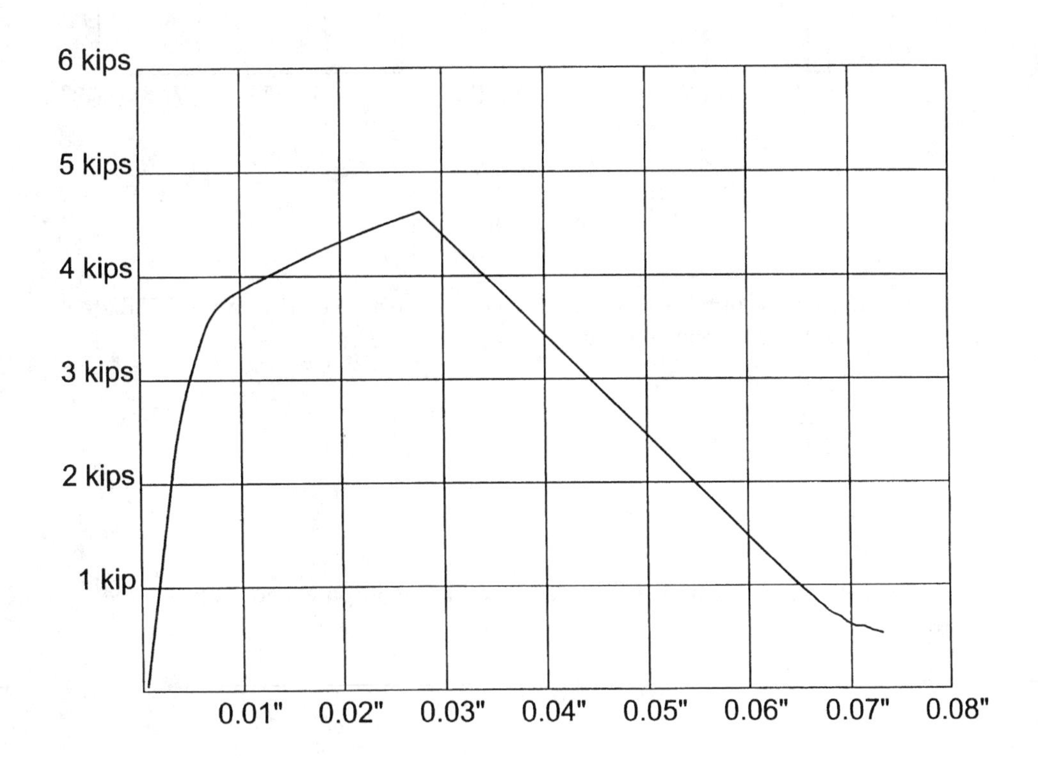

FIG. 20.2 (Problem 3.9)

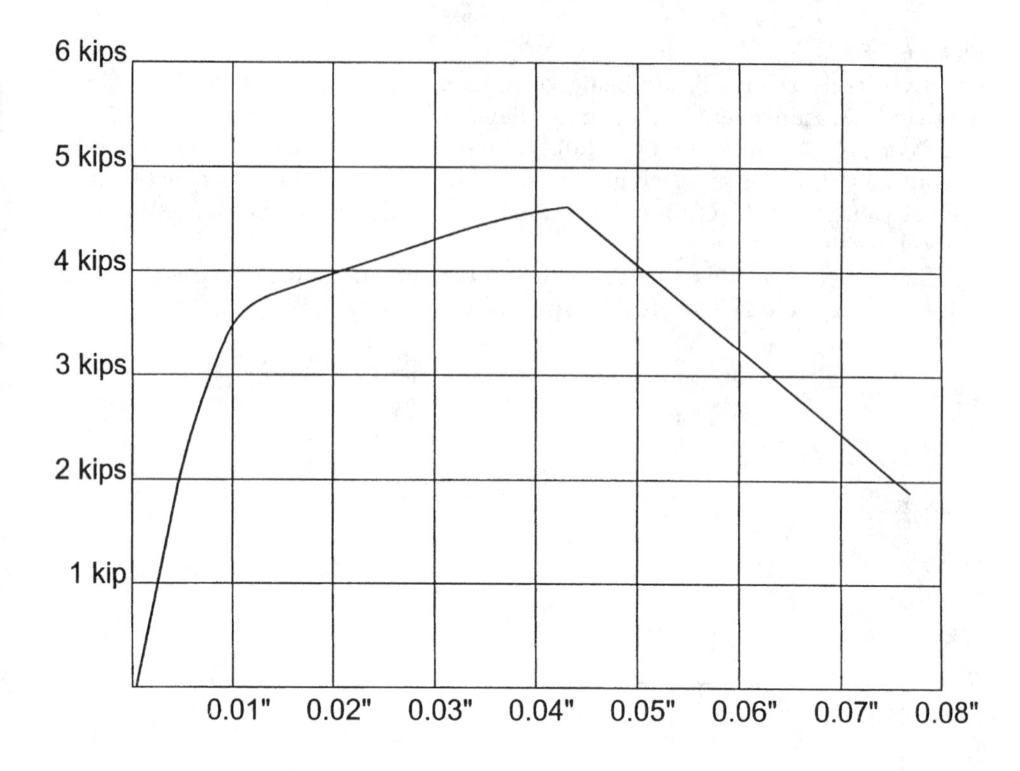

FIG. 20.3 (Problem 3.9)

Two single-edge notch (SEN) bend specimen test records of an A-533B steel are attached. Known properties are:

E = 30,000,000 psi
σ_{ys} = 66 ksi at −40°F
σ_{ult} = 99 ksi at −40°F
W = 0.8 in.
B = 0.8 in.
a = 0.4 in.

Determine K_C from both CTOD and from J. Compare and average the K_C values.

Part III

Problem 4.1

Plot a curve of crack depth versus number of cycles of fatigue loading to failure for the following structural case:

(a) A36 structural steel (ferrite-pearlite).
(b) $K_{IC} = 50$ ksi$\sqrt{\text{in.}}$ at service temperature.
(c) Minimum yield strength is 36 ksi.
(d) $\sigma_{des} = 23$ ksi (live load + dead load).
(e) Live-load stress = 10 ksi.
(f) Dead-load stress = 13 ksi.
(g) Surface flaw with $a_0 = 0.1$ in.
(h) $a/2c = 0.25$ and remains that way throughout the life of the structure.
(i) Plate thickness is 4 in. and is semi-infinite.
(j) Use $\Delta a = 0.1$ in.

Tabulate your results and summarize the initial and final conditions including sketches. Also, determine N, neglecting the R-ratio term $(1 - R)^{1/2}$ to establish the effect of neglecting R-ratio.

Problem 4.2

Prepare graphs of flaw size versus number of cycles for each of the following 2 martensitic steels:

Steel A: $K_{IC} = 90$ ksi$\sqrt{\text{in.}}$
Steel B: $K_{IC} = 180$ ksi$\sqrt{\text{in.}}$

with the following structural conditions:

(a) Edge crack in an "infinite" plate.
(b) $a_o = 0.3$ in.
(c) $R = 0$.
(d) $\sigma_{des} = 45$ ksi.
(e) $\Delta\sigma = 15$ ksi.

Problem 4.3

Prepare a graph of flaw size versus number of cycles on the same sheet of paper for Steel B in Problem 4.2 for the following 2 conditions:

(1) $\Delta\sigma = 15$ ksi.
(2) $\Delta\sigma = 30$ ksi.

for the following structural conditions:

(f) Edge crack in an "infinite" plate.
(g) $a_o = 0.3$ in.
(h) $R = 0$.
(i) $\sigma_{des} = 45$ ksi.

Problem 4.4

Assume that you are responsible for the design of a wide-plate tension member built from a structural aluminum. The member has a center-crack defect ($2a$) and can be subjected to various service conditions. Information on the aluminum is as follows:

$E = 10 \times 10^6$ psi.
$\sigma_{ys} = 40$ ksi.
$$\frac{da}{dN} = 0.66 \times 10^{-8}(\Delta K)^{2.5}$$

Determine the propagation fatigue life for the following design conditions:

(a) Stress range = 20 ksi (0 to tension)
$2a_0 = 0.5$ in.
$K_{IC} = 50$ ksi$\sqrt{\text{in.}}$
(b) Stress range = 20 ksi (0 to tension)
$2a_0 = 2.0$ in.
$K_{IC} = 50$ ksi$\sqrt{\text{in.}}$
(c) Stress range = 20 ksi (0 to tension)
$2a_0 = 0.1$ in.
$K_{IC} = 50$ ksi$\sqrt{\text{in.}}$

Problem 4.5

Discuss briefly the relative importance of the fatigue-crack-propagation life for problems 4.2, 4.3, and 4.4 of:

(1) Lowering the stress range.
(2) Decreasing the initial flaw size.
(3) Using a tougher material.

Problem 4.6

"Arrrhh, Captain!!! We have a problem in the hold!" wails first mate Wiley.

"Well, don't lose your wits, lad! What is it!" you reply.

"Err appears to be a crack in the hull! And eees growing!"

"Of course it's growing you dolt! We're in the North Atlantic and every bloomin' wave that hits the hull causes a stress range of 10 ksi that fatigues the crack. How big is it?"

"Eeet looks like 4 inches sir. . ."

"Well then, quit yer lollygaggin' and do something about it, you lazy barnacle!"

"Shall I call the coast guard for assistance, Captain?"

"Nay, you idiot!!! With our cargo of ill-repute, they'd throw us all in the brig!!! Drill some holes at the ends of the crack tips where it is propagating. Then there will no longer be a sharp tip from which the crack can emanate."

FIG. 20.4 (Problem 4.6).

"Aye Captain!" yells Wiley as he hustles down to the hold. "But wait Captain. . . Errr, if the hull will yield at a stress of 36 ksi, what diameter hole shall I drill, sir, since the likelihood of the crack reinitiating from the hole depends on the yield strength of the material, the stress level, the size of the crack, and the diameter of the hole?"

"Arrr. . ." you scratch your head in bewilderment, "Have you been getting into my fracture and fatigue literature again, Wiley?"

"Uhhh, yes, sir."

"Wiley, my boy," you guilefully reply while patting him on the back, "Then YOU figure it out!!!" you add as you toss him into the hold.

Part IV

(A) Design Problem Regarding Fracture and Yielding

Two-inch-thick plates of A517 martensitic steel with $\sigma_{ys} = 100$ ksi and having the CVN impact test results shown in the following table will be used to fabricate cylindrical pressure vessels having a nominal diameter of 6 ft and an overall length of about 30 ft. This steel has a 0.2% offset yield strength of 100 ksi.

Surface cracks with $a/2c = 0.3$ and a depth of 0.4 in. may go undetected. Note that the plates may be oriented in either the longitudinal or transverse direction and the surface flaws may be oriented in either direction. For a factor of safety of 2.0 against both yielding and fracture, determine the maximum allowable pressure to which this vessel can be subjected. Also, determine the best orientation of the plates. Service temperature will be +75°F.

Assume the ends of the vessels and the connections to these ends are to be analyzed by another division and your concern is only for the longitudinal section of the vessels.

Explain your answer clearly and include clear sketches of the vessel, plate orientation, and crack orientation that controls the design.

TEMPERATURE, °F	$CVN_{transverse}$, ft-lb	$CVN_{longitudinal}$, ft-lb
−100	3	6
−75	5	15
−50	7	28
−25	10	37
0	12	43
25	13	47
50	14	49
75	14	50
100	15	50

(B) Design Problem Regarding Fatigue Initiation and Behavior of a Structure

The structure shown is built from an A517 quenched-and-tempered martensitic steel with the following properties:

$\sigma_{ys} = 120$ ksi
$\sigma_{ult} = 140$ ksi

TEMPERATURE, °F	CVN IMPACT, ft-lb
−150	5
−100	10
−50	25
0	35
50	40
100	40

CVN Impact Properties

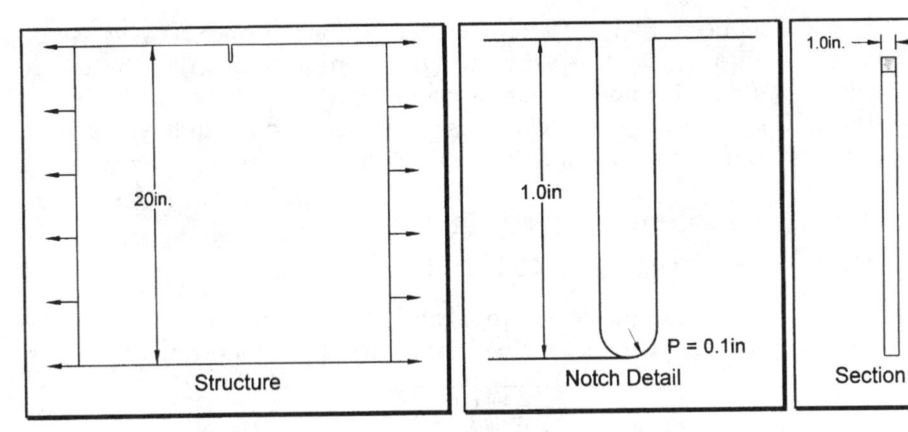

FIG. 20.5 (Design Problem B).

The structure is loaded in fatigue from a minimum stress of 40 ksi to a maximum stress of 60 ksi. To perform its design function, the U-shaped notch shown below was carefully machined into one edge. The structure must operate at 60°F.

(1) How many cycles of loading can it take before total failure? Define failure very specifically.
(2) Estimate the fatigue life if the maximum stress were decreased to 50 ksi.

(C) Design Problem Regarding Fracture and Fatigue

A 48-in.-outside diameter, 1-in.-thick pressure vessel is to be fabricated from Steels A, B, or C. The vessel is 200 in. long with hemispherical ends and will be subjected to an internal pressure of 4000 psi. Assume "perfect" welding with all reinforcement ground smooth. Steels A, B and C have the following properties:

CONDITION	YIELD STRENGTH, ksi	K_{IC}, ksi$\sqrt{\text{in.}}$
A	60	120
B	70	100
C	80	80
D	90	60

(1) Carefully sketch the worst possible location of an external surface flaw that could exist on this vessel.
(2) If a surface crack is 1.0 in. long and 0.3 in. deep, determine the factor of safety against fracture for each steel with the flaw located as sketched in Item (1).
(3) Assuming that Steel A is used, how *deep* can a surface flaw grow by fatigue or stress corrosion before failure occurs, assuming that the $a/2c$ ratio of the crack is 0.3.

(4) If Steel C is used, and a 0.3-in. deep, 1.0-in.-long surface flaw grows by fatigue with a constant aspect ratio ($a/2c$ constant), describe the failure condition. (Hint: Do not forget to account for M_k.)

(5) If Steel C is a martensitic steel and is pressurized from 0 to the maximum pressure with $a_i = 0.3$ in. and $2c = 1.0$, as described in (4), determine N_p.

(D) Design Problem Regarding Fatigue-Crack-Propagation Design Curves

Develop a series of "fatigue-crack-propagation design curves" for a ferrite-pearlite structural-grade steel that can be heat treated to the following conditions:

CONDITION	YIELD STRENGTH, ksi	K_{IC}, ksi$\sqrt{\text{in.}}$
A	60	120
B	70	100
C	80	80
D	90	60

The steel is to be used in a structure that will be subjected to a dead-load stress of $0.2\sigma_{ys}$ and a live-load stress of $0.3\sigma_{ys}$.

Assuming that the design curves for each condition are for an infinitely wide plate with an edge crack, carefully plot a design curve of initial crack size (vertical axis) versus propagation life (horizontal axis). A semilog plot may be desirable. Note that this is not a typical crack-growth curve as plotted before. Rather, it is a series of four design curves showing the relation between initial crack size and propagation life.

(E) Design Problem Regarding Fracture and Fatigue Comparison of Two Steels

Select a steel for use in high-pressure cylindrical containment vessels for the next generation of nuclear submarines.

Two steels are being considered for this application, HY-130 and HY-180, which are both martensitic steels. The material properties for these two steels are shown in the table below.

PROPERTIES	HY-130	HY-180
Yield strength, ksi.	130	180
Tensile strength, ksi	150	190
K_{Ic}, ksi$\sqrt{\text{in.}}$	280	300
K_{Iscc} in seawater, ksi$\sqrt{\text{in.}}$	260	180
Hypothetical cost $/lb	0.50	1.00

Design parameters for the containment vessels are as follows:

(a) Internal pressure = 5000 psi
(b) Internal diameter of cylindrical portion = 30 in.
(c) Overall length of each vessel is 20 ft. Ends are to be hemispherical.
(d) Welded fabrication will be used. Assume that weld metal properties are the same as base metal properties.
(e) The vessel must have a factor of safety of at least 2.0 against both yielding and fracture of a 0.5-in-deep surface flaw. Assume that $a/2c$ is 0.4.
(f) The vessels will be cycled from 0 to full design pressure (5000 psi).
(g) Inspection is such that all flaws greater than 0.05 in. can be found during fabrication. In service, the vessels will be in the forward-flooding zone of the submarine and cannot be inspected, although they can be protected by painting.

On the basis of performance, weight, and cost, recommend which steel you would use. Justify your answer.

Subject Index

D

E

F

V

W

Y